BIOGEOGRAPHY AND ECOLOGY IN AUSTRALIA

MONOGRAPHIAE BIOLOGICAE

EDITORES

F. S. BODENHEIMER
Jerusalem

W. W. WEISBACH
Den Haag

VOL. VIII

Springer-Science+Business Media, B.V. 1959

BIOGEOGRAPHY AND ECOLOGY IN AUSTRALIA

EDITED BY

| A. KEAST | R. L. CROCKER | C. S. CHRISTIAN |
| Sydney | Sydney | Canberra |

Springer-Science+Business Media, B.V. 1959

ISBN 978-94-017-5837-6   ISBN 978-94-017-6295-3 (eBook)
DOI 10.1007/978-94-017-6295-3

Copyright 1959 by Springer Science+Business Media Dordrecht
Originally published by Uitgeverij Dr. W. Junk, Den Haag in 1959.
Softcover reprint of the hardcover 1st edition 1959

CONTENTS

I. The uniqueness of Australia in Biology, by F. S. BODENHEIMER   9

II. The Australian environment, by ALLEN KEAST   15
- Physiography of the Continent   15
- The development of the Australian environment   26
  - The relationship of Australia to other land-masses   26
  - The palaeoclimatology of Australia   29
- The present Australian flora and fauna   32
- Changes in the Australian continent associated with European settlement   33
- References   34

III. Ecology of primitive aboriginal man in Australia, by NORMAN B. TINDALE   36
- References   51

IV. Human ecology in Australia, by GRIFFITH TAYLOR   52
- Environmental controls   52
- The human response   61
  - 1788 to 1820: First steps in exploration   62
  - 1820 to 1840: Expansion from other centres   63
  - 1840 to 1870: Great pastoral expanses. Gold rushes   64
  - 1870 to 1900: End of exploration and of new pastoral occupation   66
  - 1900 to 1956: Federation and industrial development   67
- Future settlement in Australia   68

V. The marsupial fauna: its origin and radiation, by E. LE G. TROUGHTON   69
- The migratory route, and Australian environment   69
- The Papuan region of North Queensland   71
- Origin of the monotremes and marsupials   72
- Classification of the Marsupialia   73
- Adaptive radiation   75
- Insectivorous and carnivorous Dasyuridae   77
- Marsupial anteater and pouched-mole   79
- The bandicoots-Peramelidae   80
- Phalangers-or Australian possums   82
- The koala- and the terrestrial wombats   83

The kangaroo family – Macropodidae	85
Conclusion	87
References	88

VI. Australian birds: their zoogeography and adaptations to an arid continent, by ALLEN KEAST — 89

Introduction	89
Zoogeography	90
The Australian avifauna, origin and relationships	90
Factors governing distribution within Australia	94
Ecological adaptations	96
Adaptive modifications to the seasonal cycle	96
Adaptations ensuring the recovery of population numbers following droughts and heat waves	108
References	113

VII. The reptiles of Australia, by ALLEN KEAST — 115

The Australian fauna	115
The distribution of reptiles within Australia	118
Distribution of families	119
Distribution of species	123
Reptiles and the zoogeographic sub-region concept	129
The various faunal subdivisions that have been advanced	129
The ecological significance of the postulated sub-regions	132
References	134

VIII. The freshwater fishes of Australia, by GILBERT P. WHITLEY — 136

Drought adapted, artesian bore, and hot water fishes	138
Fluvifaunulae	140
Adaptive radiation	144
Spawning	148

IX. The zoogeography of some Australian Insects, by J. W. EVANS — 150

Summary	150
Introduction	150
Frog-hoppers (Cercopoidea)	152
Cicadas	152
Leafhoppers (Cicadelloidea)	153
The Leafhopper fauna of Australia	154
The Australian-Indian association	155
Australian-South African-New Zealand-South American association	155
Endemic Australian groups	155
The late Indo-Malayan invasion	156
Insects with aquatic, or, marsh-inhibiting larvae and with a southern distribution	158
References	162

X. Zoogeographical aspects of the Australian dipterofauna, by S. J. PARAMONOV — 164

The main points of the zoogeography of Australian Diptera	164

The peculiarities of Australian families of Diptera (Distribution and systematic position)	169
Faunal division of the Australian fauna	179
The relationship of the Australian dipterofauna to that of South America (Chile, Patagonia)	187
The relationship of the Australian dipterofauna to that of Tasmania and of New Zealand	188
References	190

XI. The ecology and biogeography of Australian grasshoppers and locusts, by K. H. L. KEY   192

The Australian grasshopper fauna	192
Ecological characteristics of grasshoppers	193
The Australian environment in relation to grasshoppers	196
Grasshoppers of the major habitat categories	203
Grasshoppers and the biotic community	206
Biogeographic or faunal regions	206
References	209

XII. Aspects of the distribution and ecology of Australian termites, by J. H. CALABY and F. J. GAY   211

The Australian fauna and its relationships	211
Ecological factors in termite distribution	213
Aspects of ecology of some Australian genera	215
Remarks on general ecology	221
References	222

XIII. The land and freshwater mollusca of Australia, by DONALD F. MCMICHAEL and TOM IREDALE   224

Nature, number and distribution	225
Freshwater faunal regions	233
Terrestrial faunal regions	235
Ecology and adaptations	238
Origin and relationships	241
Melanesian	241
Newer Asian	242
New Zealand	242
South American and African	243
Older Asian	243
References	244

XIV. The Australian freshwater crustacea, by E. F. RIEK   246

Adaptations	248
Economics	250
Distribution within Australia	250
Zoogeographical regions	252
Origin of the Australian freshwater crustacean fauna	253
Crayfish	255
References	258

XV. La place de l'Australie méditerranéenne dans l'ensemble des pays méditerranéens du Vieux Monde (Remarques sur le climat méditerranéen de l'Australie), par Louis Emberger 259

    Remarques générales 259
    Le bioclimat méditerranéen et la végétation méditerranéenne en Australie 263
    Conclusions 272

XVI. The vegetation of Western Australia, by C. A. Gardner 274

XVII. Past climatic fluctuations and their influence upon Australian vegetation, by R. L. Crocker 283

    Introduction 283
    Tertiary and Quaternary floras – The paleobotanic evidence 283
    Quaternary environmental changes 286
    Quaternary climatic change and the Australian vegetation 287
    Summary 289
    References 289

XVIII. The phytogeography of Australia (in relation to radiation of Eucalyptus, Acacia, etc.), by J. G. Wood 291

    Introduction 291
    The sclerophyllous shrub genera 293
    The genus Eucalyptus 295
    The genus Acacia 297
    Distribution during late Tertiary – Recent 298
    References 301

XIX. Some aspects of soil ecology, by G. A. Stewart 303

    References 313

XX. Recent studies on marsupial ecology, by A. R. Main, J. W. Shield and H. Waring 315

    General Introduction 315
    Circumscribed self contained projects with limited objectives 316
        Economic 316
        Conservation 317
        Trapping and marking studies and analysis of commercial shooting returns 317
    Investigations based on speculations about past distribution 319
    References 330

XXI. Marsupial reproduction, by G. B. Sharman 332

    Introduction 332
    The reproductive system 334
        Anatomy of the female reproductive system 334
        Anatomy of the male reproductive system 337
    Seasonal reproductive periodicity 337

Reproductive periodicity in the female	337
Reproductive periodicity in the male	339
Reproductive cycles	339
The oestrous cycle	339
Pregnancy	345
Lactation controlled delayed implantation	354
Development of young in the pouch	356
Anatomy of the new-born young	356
Composition of marsupial milk	357
Growth rates of pouch young	358
Experimental studies on hormone control	359
Effects of hormones on adult and foetal gonads	359
Effects of hormones on accessory reproductive organs	360
References	363

XXII. The contribution of banding to Australian bird ecology, by ROBERT CARRICK   369

Introduction	369
Ecological significance of territory	371
Trans-equatorial migration	374
Migration and breeding of subantarctic sea-birds	374
Trans-Tasman migration	377
Movements within Australia	377
References	381

XXIII. Ecology of wild ducks in inland Australia, by H. J. FRITH   383

Introduction	383
The environment	383
Waterfowl habitat	384
Habitat utilization	386
Movements	388
Food habits	389
Breeding	391
Summary	394
References	395

XXIV. Ecology of Australian frogs, by A. R. MAIN, M. J. LITTLEJOHN and A. K. LEE   396

Introduction	396
Desert adaptations	398
Biology	398
Breeding biology	399
Adult behaviour	404
Physiology	405
Habitat selection	407
Feeding	407
Adult size	408
Discussion	409
Summary	410
References	410

XXV. Ornithosis research in Australia, by J. A. R. MILES   412

    Early work in Australia   413
    The 1938-39 ornithosis epizootic   415
    Enzootic ornithosis in South Australia   416
    Mutton bird (*Puffinus tenuirostris*) ornithosis   422
    Cross-infection between species   422
    Discussion   423
    References   425

XXVI. Vegetation of high mountains in Australia in relation to land use, by A. B. COSTIN   427

    Introduction   427
    The high mountain environment   427
    Flora and fauna   430
    Description of the vegetation   434
    Land use and economic importance   442
    Hydrological research   448
    Management requirements   450
    References   451

XXVII. Some aspects of ecological research in semi-arid Australia, by N. C. W. BEADLE   452

    General features of the arid environment   453
        The plant communities   454
        The effect of man and grazing animals   454
        Soil erosion and its consequences   455
        Factors regarding the colonisation of bare areas   456
        Levels of soil nitrogen and soil organic matter   457
        Sources of nitrogen   458
    Conclusion   459
    References   460

XXVIII. Species distribution and association in Eucalyptus, by L. D. PRYOR   461

    Distribution of Eucalyptus species   464
    Variability in Eucalyptus populations   465
    Hybridization   466
    The associations in Eucalyptus   471
    References   471

XXIX. The ecology and prevention of soil erosion, by R. G. DOWNES   472

    Introduction   472
    New people in a new environment   472
    The effect of settlement   474
    The problem of soil conservation   475
    Some environments and their problems   477
    A difficult but interesting environment   478
    Soil and ecological surveys as a basis for the determination of land use   482

Agricultural ecology and farm planning for conservation 484
References 486

XXX. The Merino sheep in Australia, by IAN W. MCDONALD 487

Wool production 487
Growth of the sheep population 489
The climate 493
Nutrition 495
References 499

XXXI. Ecological observations on plant communities grazed by sheep in Australia, by R. M. MOORE 500

Introduction 500
The major plant communities grazed by sheep 500
General ecological effects of sheep grazing on Australian vegetation 509
References 512

XXXII. The ecology of sheep blowflies in Australia, by K. R. NORRIS 514

The practical problem of blowfly strike 514
Species involved in sheep myiasis in Australia 516
Local and seasonal variations in the abundance of sheep blowflies 518
Biology and ecology of individual species 520
Inter-relations of blowflies, and other organisms 536
    Sources of blowfly populations 536
    Competition in the limitation of blowfly numbers 537
    Effects of predation and parasitism in the limitation of blowfly populations 540
Climatic influences on distribution and abundance of blowflies 541
References 542

XXXIII. The rabbit in Australia, by F. N. RATCLIFFE 545

Early history – the colonizing spread 545
Consolidation – ecological consequences 547
Biological control – Myxomatosis 549
General ecology and behaviour 556
References 563

XXXIV. The biological control of prickly pear in Australia, by ALAN P. DODD 565

Survey of the problem 565
The Commonwealth Prickly Pear Board and its work 566
    Work in America 567
    The Australian set-up 568
    The host-restriction question 568
    General results 569
The history of *Cactoblastis cactorum* 571
    Life cycle and habits 571
    Introduction, establishment and distribution 572
    Destruction of the dense infestations 574

    The present position   575
    References   577

XXXV. An ecogenetic research programme with introduced plants, by F. H. W. Morley and O. H. Frankel   578

    Aim and purpose   578
    The species   579
    The experimental approach   580
    Definition and assessment of adaptive characteristics   583
    Some results   584
    The dialectics of adaptation   585
    References   586

XXXVI. The eco-complex in its importance for agricultural assessment, by C. S. Christian   587

    Factors influencing the development of methods   588
    The concept of land, land units and land systems   591
    Personnel and survey procedures   593
    Assessment of potentialities   596
    Examples of the eco-complex in potentiality assessment   598
    Economic aspects   603
    References   604

XXXVII. Nature conservation in Australia, by Robert Carrick and Alec B. Costin   606

    Introduction   606
    The case for conservation   607
    The Australian flora and fauna   608
    Legislation   612
    Special problems   618
      Species   618
      Reserves   622
    Research   625
    Conclusion   626
    References   627

Index   628

I
THE UNIQUENESS OF AUSTRALIA IN BIOLOGY

by

F. S. BODENHEIMER

(Hebrew University, Jerusalem, Israel)

To the biologist a first visit to Australia is a great event. This continent with its happy, smiling, generous and virile population has immense problems. To give one illustration: Huge tracts of land offered adequate pasture for large-scale sheep-farming during the earlier years of white settlement, usually settled during a favourable period in the secular trend of the rain cycle. The oncoming drier phases and years or real drought caused overpopulation of the sheep which led to radical reduction of the area of settlement, and on the pastures not only of the numbers of sheep, but also of the good grasses and herbs, leaving in dominance hard herbs, unfit for sheep pasture. This dominance persisted for long after the dry years had passed. The large-scale improvement of these depredated pasture lands aims at the determination of the stable number of sheep in order to prevent future overpopulation and overgrazing. To this aim climatologists, geologists, soil scientists, botanists, hydrologists, agronomists, and plant introduction and animal production workers have co-operated in new ventures to establish the potentialities of under-developed areas (CHRISTIAN). This, one of the big problems of Australia's agriculture, and many others are courageously attacked by the C.S.I.R.O. (Commonwealth Scientific and Industrial Research Organization) from all angles, with a vigour and in teamwork such as are in spirit and in fact rarely seen on this earth. These few words explain the central position of ecology in Australian science. The bigness of the task surpassing by far the forces of any sum of individuals or of specialists did not deter the authorities in question. Experience has shown that even the best individual forces of all the universities available with their many tutorial and other duties are inadequate to undertake such task. The C.S.I.R.O. took over the daring challenge and organised specialists and field equipes for an adequate fieldwork. The integration of work of the various scientists has been developed to a high degree and it is now recognised that each in his own individual field gains so much from collaboration with other scientists of other disciplines that this team work has come to stay. Sir IAN CLUNIES ROSS, the present chairman of the organisation, said correctly that it has spent A. £. 30 million over the years on research, but that the agriculture of the continent benefits already at the rate of over A.

£. 100 million per annum. The real problem is the gigantic one of transforming a dry and relatively sterile continent into a productive land where large scale erosion is no longer present, pastures are not over-exploited, where crops are produced wherever water is available and where the natural resources of the continent are preserved for future generations of Australians.

The varied and fundamental nature of Australian problems explained why it was not possible to select a few papers on animal ecology if we wanted to show Australian ecology in its appropriate setting. This continent is not only striving for knowledge, but it is also attempting to achieve practical results while doing so, at the same time approaching both scientifically in a sound and basic way. Therefore it was necessary to include in this book chapters on soil science with its problems of erosion and soil fertility, on botany with its vegetation maps, agronomy, genetics and plant introduction and others on economic land use itself.

This stimulating frame explains at once the fertility of animal ecology in the fields which are selected here as illustrations.

The zoogeography and the origin and development of the Australian fauna and flora offers still many fascinating problems. The early origin of the faunal components, if from the south or from the north, if certain elements are remainders of earlier elements of worldwide distribution, and of others which are of an old, independent marine penetration into the freshwaters of the continent, is still under debate. All these theories are partly true. They all have found presentation in this book. This still leaves the problem of what an autochthonous Australian fauna and flora really means, to be defined.

Isolation in giant scales has made conspicuous the problem of adaptive radiation in speciation, filling all the eco-niches available and creating in time an harmonious fauna and flora. Relatively old intruders like the placentalian bats and rodents exterminated any vicarious marsupial groups, but else remained within the frame of the existing mammal fauna. With the intrusion of the wild dog, the dingo, about 12.000 years ago, we see the same process in action, as TINDALE shows, in the disappearance of the big carnivorous marsupials in those districts which the dingo has already invaded.

The origin of the Australian fauna has been long a central problem in biogeography. In earlier days the similarities with faunal elements in South America, S. Africa, the hypothetic old Antarctic fauna and Australia were stressed and a common origin from Antarctica or Gondwanaland at an early date was assumed. More recently the conclusions of DARLINGTON (1955) are accepted by many, especially mammalogists and ornithologists, who look more for homologies in the recent homoeotherms in a northern, eastern origin via Asia. We see immediately that recent groups have a different origin than much older ones, and that it was always a mistake to treat all groups coming to Australia in different geological periods according to one theory. THROUGHTON stresses the northern origin for the marsupials. But many students of insects and plants follow the lead of JEANNEL and stress a southern origin of almost all forms, except those modest elements which spread in the Northern Hemisphere from Angara and Lawrentia. Many of the insects have been on the continent for ages. EVANS (l.c.) assumes for the Homoptera that a large group is of distinctly southern origin. Other old stocks have developed from a few remnants of groups which in old times were widely distributed over the earth and are now genuinely autochthonous. PARAMONOV also believes that the Diptera belong to an archaic fauna, which was not split in the mild climate of the continent into many species as in the north (by catastrophic glaciations). He also supports an old antarctic origin of the main fauna of the Diptera. KEY (l.c.) concludes that most Acridoidea are old and autochthonous with a few newer immigrants from New Guinea.

Certain old freshwater groups like the fish (WHITLEY) and crustaceans (RIEK) are obviously old, independent marine elements protruding into the freshwater. The molluscs show diverse stocks arriving from New Guinea, a newer and an older Asian stock, the latter eventually being autochthonous. But other stocks common with New Zealand or with the ancient Gondwanaland are definitely present.

No general conclusion is as yet possible, except that more recent stocks are generally of northern origin. For the rest two alternatives remain: Old stocks developing from forms once widely spread or a southern origin of Antarctica or Gondwana origin. Many freshwater forms are definitely adapted marine forms.

Thus, the problems of biogeography are of an outstanding and peculiar feature: An old fauna has had the time to specialize into all eco-niches of the continent, in soil, desert and tropical forest, and to develop an entirely harmonious fauna, unknown from the marsupials of other continents in past and present. Yet there is one significant exception to this statement: We find no marsupials taking the place of the bats or rodents, of which orders 50 and 70 species respectively succeeded to reach the continent long ago. They have suppressed all earlier marsupials which had developed in their present niches, exterminated them, but apart from that, retained the harmonious structure of the fauna as a whole (BODENHEIMER 1957). The very abundant 'marsupial mice' are all small insectivores, equivalents of the shrews with a dentation not admitting any comparison with the rodents of the rest of the world. TINDALE adds to this picture in studying the influence of the recently introduced dingo, who is — even without the aid of man — on the best way to extinguish the big local carnivores. These have subsisted in the mountain forests of Tasmania and in certain other areas where the dingo has not yet penetrated.

TROUGHTON describes in detail the "unprecedented degree of adaptive radiation during the phylogenetic development" of the Australian marsupials. In birds BODENHEIMER (1957) remarks that according to the revised bird system of MAYR (1951) of the 356 species of Passeres 172 belong to one family-group, the Muscicapidae. If this view is accepted, the large number of flycatchers actually shows radiation to a similar extent to that of the marsupials. This is easily revealed by the wealth of popular names (thrushes, wagtails, shrikes, chats, pipits, etc.) which immediately hints of different biotopes and behaviour (but the names have no relation taxonomically to all these European groups). All, however, remain typical insect-eaters. In the over fifty parrots *(Psittacidae)* are feeders of figs, roots, grass-seeds, nuts, nectar and grubs. A similar radiation we find among the doves: from fruit- to seed-eating species. According to KEAST (i.l.) the parrots are the nearest approach to the marsupial speciation. Yet, the competition among many old established bird families living in various niches has limited species radiation in this order considerably. The old group of reptiles (KEAST), here fully presented, shows little radiative development. The same is true for the freshwater fishes, all of marine origin, in which perhaps the 27 Galaxiidae show some speciation (WHITLEY).

A similar analysis of phytogeographical problems, past and present, mainly for *Eucalyptus* and *Acacia*, is given by GARDNER, CROCKER, WOOD and PRYOR. They all stress the prevalence of humid conditions over all Australia until the recent intrusion of prevalent conditions of aridity. GARDNER stresses the impact of bushfires on the character of the vegetation.

The soils of the continent show a low fertility mainly in consequence of the geochemical nature of its rocks, the low amount of weathering minerals and the continued soil leaching (STEWART). The overwhelming importance of soil erosion is stressed by DOWNES. He omits the great importance of wind erosion in the Central territories and discusses the ecological control and reclamation in the State of Victoria.

Turning to Human Ecology, TINDALE reveals before us the 12000 years and the various strata of palaeolithic man in Australia. He discusses how far man has succeeded in small steps to change in a minor but definite way the landscape, even before the arrival of white man. GRIFFITH TAYLOR sums up the final conclusions of an outstanding geographer. He fought all his life for a moderate but definite environmentalism. He defined in Australia as well as in Canada lines of equal population density in their dependence upon environmental factors. New factors, like the finds of uranium, large irrigation projects, etc., may change these lines which after all are no natural laws. They will change whenever the relative index of productivity is changed.

The many contributions to ecology need no introduction. We mention only the ecophysiological papers on marsupials (WARING, SHARMAN); those on bird populations and on bird breeding with their analysis of vagration of steppe birds and the mechanism of rebuilding of populations after catastrophic droughts (KEAST, CARRICK, FRITH); the ethological studies of CARRICK on the magpie stressing their influence on the maintenance of stable populations; the confirmation of SPENCER'S fascinating reports concerning the adaptation of frogs to life in arid areas (MAIN); the fight of desert areas by nitrogen accumulation in the soil by establishment or improvement of the legume-*Rhizobium* vegetation, the first utilisers of favourable soil moisture (BEADLE), etc.

We have repeatedly made mention of the specific Australian approach to the reclamation of the continent. Here A. COSTIN illustrates well the problems in his study on high mountain vegetation in connection with hydrological research and management. The preservation and its eventual restoration in 30/50 years, its protection before grazing and fires will help to increase Australia's most useful element, the water.

The Merino sheep is the backbone of Australia's animal husbandry. Sheep breeding is mainly limited by the 20/30 inch rainfall, but is improved and 'buffered' against drought and other disasters by modern agronomy. This also keeps the land in a condition better fit to withstand erosion and poor pasturage. The absence of endemic diseases, predators and parasites favoured sheep breeding until the rabbit began to compete. The sheep seems to approach its stable maximum population, in contrast to the great fluctuations in numbers in earlier times by each drought wave. The Merino is a good illustration for TAYLOR'S environmentalism. The studies on sheep pastures and on sheep blowflies show the extent of research (MCDONALD, MOORE, NORRIS). Medical research is not neglected, as the ecology of ornithosis illustrated (MILES). An important program of eco-genetic research on introduced plants has been established, for illustration, on the subterranean clovers, which embraces actually all equi-ecous regions of the world (MORLEY & FRANKEL). The inspiring contribution of RATCLIFFE to the population dynamics of the rabbit in Australia, especially of the past, present and future effect of myxomatosis, is of immediate actual interest. Last, not least, we must mention again the team spirit in which CHRISTIAN describes the mapping of land potential of Australia which has so far covered over 500,000 square miles in the last ten years. The final contribution of CARRICK & COSTIN gives a welcome survey of the present state of Nature Conservation in Australia.

Australian biology is well presented and well described in the present volume. We sincerely hope that these splendid efforts will lead to the success which they so well deserve.

To these two aspects, ecology and biogeography, this book is mainly dedicated. Interesting and fascinating as these two problems are, it would have been impossible to bring them to a satisfactory end without the generous response which our invitation met with in all circles of the frontfighters of Australian science, who did their best to make it representative of the true spirit and the achievements of Australia's science in its present wonderful mood. We sincerely hope that it will encourage and stimulate continued efforts along those lines of the same vitality as the men who created the vigorous atmosphere so promising for Australia and for the whole earth. To all these colleagues our boundless admiration and gratitude is given.

It would be ungrateful to end these general acknowledgements to all, contributors and non-contributors, without a special word to ALLEN KEAST who has helped me to plan and to arrange things from the beginning, who took over most of the local correspondence and shared with me many worries, but never lost his good spirit. I doubt if the outcome would have been as good without this help on the spot. We must also mention the real help which leaders like J. R. NICHOLSON & Sir F. M. BURNET gave by helping to find adequate substitutes in their and other fields, when pressure of work prevented them from participating.

The contributions are all written in a rather factual, modest, scientific style. But we should not forget the long series of glorious Sagas of modern science which are hidden behind this modesty: The Saga of the biological control of the Prickly Pear, which made immense square miles of Queensland again from useless land into good pastures (DODD); that of the discovery of the wild stage of the ornithosis, a virus disease of world wide importance (MILES); the partial history of the Sheep-Blowflies of Australia – which have contributed so much to the Sheep Industry and to general ecology (NORRIS); the Saga of the flooding of Australia by the Rabbit, its destruction by Myxomatosis and its future outlook, (RATCLIFFE); the ecology of the Merino-sheep and of its pasture problems (MCDONALD, MOORE, and others); or in botany the keen approach to the agricultural application hidden in various studies (BEADLE, COSTIN, CHRISTIAN). These Sagas will ever be shining paradigmas how man should tackle and has successfully tackled his fateful present and future.

II
THE AUSTRALIAN ENVIRONMENT

by

ALLEN KEAST

(Curator of Birds and Reptiles, Australian Museum, Sydney)

The Australian continent has a land area of 2,984,000 square miles, some three-quarters of that of Europe (including European Russia) and is thus the smallest of the continents. It is about the same size as the United States and is some twenty-five times the land area of the British Isles.

Australia has been a highly stable land-mass for a considerable portion of its geological history and ancient Archaean rocks outcrop over an extensive area. It is decidedly flat, so that physiographic barriers to distribution and dispersion are virtually non-existent. The geographic position of Australia relative to the Equator and Poles (it lies between 10 deg. and 40 deg. south latitude) ensures a sub-tropical to cold temperate climate. There are no permanent glaciers or snow-fields. Vulcanism ceased during the Pleistocene or Early Recent.

Physiography of the Continent

Geologists recognise three physical sub-divisions in the Australian continent (Fig. 1): the Great Plateau of Archaean Shield (western), the Central Basin, and the Eastern Highlands (HILLS, 1949; DAVID & BROWN, 1950).

The Great Plateau extends over the western half of the continent and has an elevation of 1000—2000 feet. It has a nucleus of a basement complex of Archaean rocks and may have been dry land since before the Cambrian. Consolidated sand dunes cover much of the surface today.

At a number of places on the shield residual masses (composed mainly of schists and gneisses) rise an additional thousand feet above the eroded plain. These low ranges, which tend to be east-west in direction, now constitute floral and faunal refuges.

The Macdonnell Ranges, for example, contain a grove of palms *(Livistona)* whose nearest living relatives are in the coastal forests of Queensland, a thousand miles to the east. They also support a generally richer fauna and flora than the surrounding plain.

The Central Basin, which largely corresponds to the present Great Artesian Basin, like the Great Plateau, appears never to have been elevated to any extent. Throughout the Mesozoic much of it was under water, being covered, from time to time, by a series of huge lakes. There is evidence of two invasions of the sea in the Cretaceous. Most of the Australian dinosaur fossils (Jurassic and Cretaceous), though too few and fragmented to give more than a glimpse of these faunas, occur in this section.

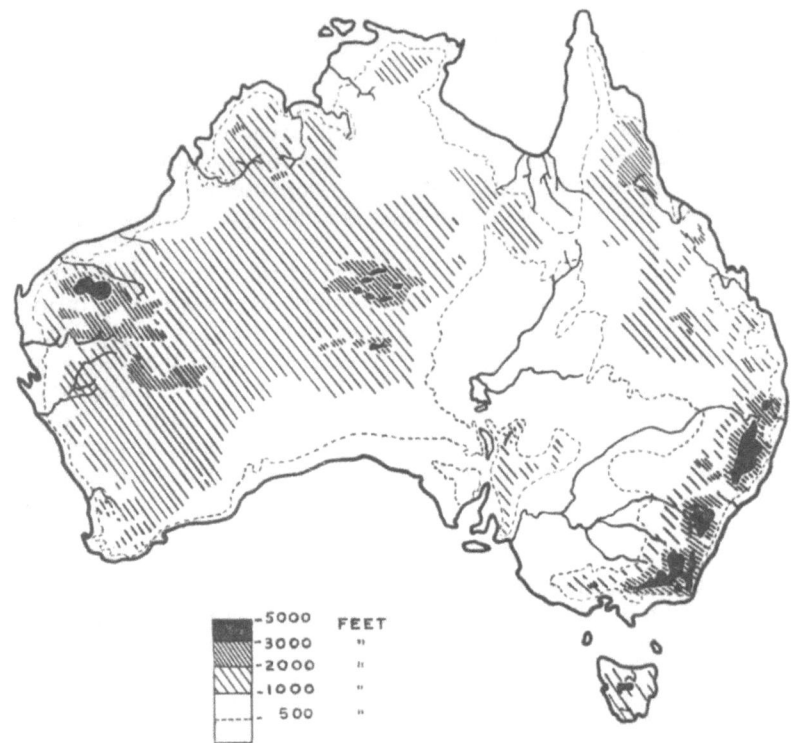

Fig. 1. OROGRAPHIC AND DRAINAGE MAP OF AUSTRALIA (Redrawn from that featured in the article by HILLS (1949) in The Australian Environment, C.S.I.R.O. Melbourne.

The Eastern Highlands (that include the so-called Great Divide) form an elevated strip extending from north to south parallel to the eastern seaboard, including Tasmania, and are composed of a complicated mixture of Palaeozic and more recent rocks deposited in a huge geosyncline. The section has been subject to successive periods of folding, elevation, warping, and erosion, with the present elevation (2000—7000 feet) dating from the late Tertiary. Australia's highest peak, Mount Koskiusko, 7,316 feet in height, lies in the southeastern section of the highlands known as the Australian Alps.

The Great Divide, in being a vast rainfall trap and providing a wide range of different conditions for life is responsible for much of the biological diversity of the Australian continent. All of the rain forests and most of the sclerophyll forests occur between the Great Divide and the sea. The southern half is a refuge area in which a number of the more ancient elements have been able to persist (e.g. *Nothofagus*). In the north it creates conditions that are enabling a wide range of Indo-Malayan rain forest species to colonize the continent.

About a third of the Australian continent is drained by rivers running directly into the sea. The Main Divide, the most important watershed, gives rise to short streams that are mostly swift-flowing and eastwards, and a series of westward-flowing streams that drain either into the Murray-Darling system (mostly), or towards Lake Eyre. These rivers, crossing country of relatively slight gradient, are for the most part, slow-flowing and sluggish. In dry years all but the largest may be reduced to a series of pools. Flooding, by contrast, leads to the inundation of low land, anastomosing and alteration of channels, anabranches, the formation of billabongs (oxbow lakes) and, in places, development of extensive marshes in this section.

Individually, and as a group, the Australian rivers carry relatively little water compared with those of other continents. This can be seen from a comparison of average annual flow:- Murray (largest Australian river) 10.0 millions of acre feet; Nile, 72.0; Columbia (U.S.A.) and Ganges, each 146.0; Danube, 228.0 (figures from AIRD, 1949).

Most of the Great Plateau is of uncoordinated drainage. Many ephemeral salt lakes and playas are scattered over its surface. Part of its eastern section, and much of the Central Basin, are characterised by shallow sandy river beds, whose course is typically marked by a line of eucalyptus (frequently *Eucalyptus camaldulensis*) or acacia. These streams represent the final stage in river degradation. They flow only in exceptional years. Then heavy rain at the headwaters may initiate a marked flow for one or two hundred miles by the end of which distance, through the combination of an evaporation rate of 100 inches per year, and frequently a porous, sandy substratum the water has all disappeared. These rivers, whose beds contain sediments of considerable depth, were apparently deep permanent streams up to, and during, the Pleistocene.

Temperature

The mean annual temperatures of the Australian continent range from over 80 in the tropical north down to 45 degs. F. in the highlands of south-eastern Australia and Tasmania. Normal daily maximum temperatures for January (summer) exceed 100 degs. F. in parts of the centre, range from 85—90 along the north coast, and 60—65 degs. in the far south. Equivalent ones for July (winter) are 70—75 in the centre, 85 in the far north, and 40—55 degs. F. in Tasmania. The daily minimum temperatures for January are 70—75 deg. F. in the central regions, 75—80 in the far north, and 45—50 in elevated regions of the south-east and Tasmania. Respective figures for July are 40—45, 60—70, and 30—35 deg. F.

Approximately one-third of the Australian continent lies within the tropics and the rest is either temperate or sub-temperate. The interior, like that of other continents, has a hot summer and a cold winter.

There are no permanent snowfields in Australia and only a few snowbanks persist throughout the summer in the loftiest of the highlands. Virtually no exodus of animals from the colder areas occurs prior to winter. Likewise, special adaptations to cold are little developed in Australian animals.

Rainfall

No feature dominates the Australian scene to the extent that does rainfall. The continent is essentially an arid one, one-third of the land mass having an annual precipitation of 10 inches or less. Another third comes into the semi-arid category.

A striking comparison of the fertility of the Australian continent with that of Europe and North America could be made by a journey across each. There is no section from, say, The Hague to Stalingrad where agriculture is impossible or that is incapable of carrying forest. The same applies to most of the United States, from the Atlantic to the Pacific seaboard. In the case of Australia, by contrast, over 2.000 miles of the 2.600 mile journey from west to east would be through country of under 10 inches of rain per annum. WADHAM & WOOD (1950) point out that, though Australia and the United States are of approximately the same size, an average rainfall of over 40 inches (400 points) per annum is enjoyed by 826.000 square miles of America but by only 194.000 square miles of Australia.

The over-all pattern of rainfall distribution in Australia is one of concentric zones of steadily increasing rainfall, outwards from a vast arid central region. The 10 inch (100 points) line which approximately outlines the true desert region, cuts the coast both in the central-west of the continent (Shark Bay to south of Point Cloates) and at the head of the Great Australian Bight (Nullarbor Plain area) in the central south. In the north the desert extends to the 15-inch line. The only regions of the continent that could be said to be well-watered are the south-west corner, the south-eastern and eastern coastal sections, and small areas of the northern coastal regions. Over a small section of the north-east (the mountainous Cairns-Atherton area) and in mountainous western Tasmania the rainfall exceeds 80 inches (800 points) per annum and over a few square miles near Innisfail (north-east) 160 inches (1600 points) per annum.

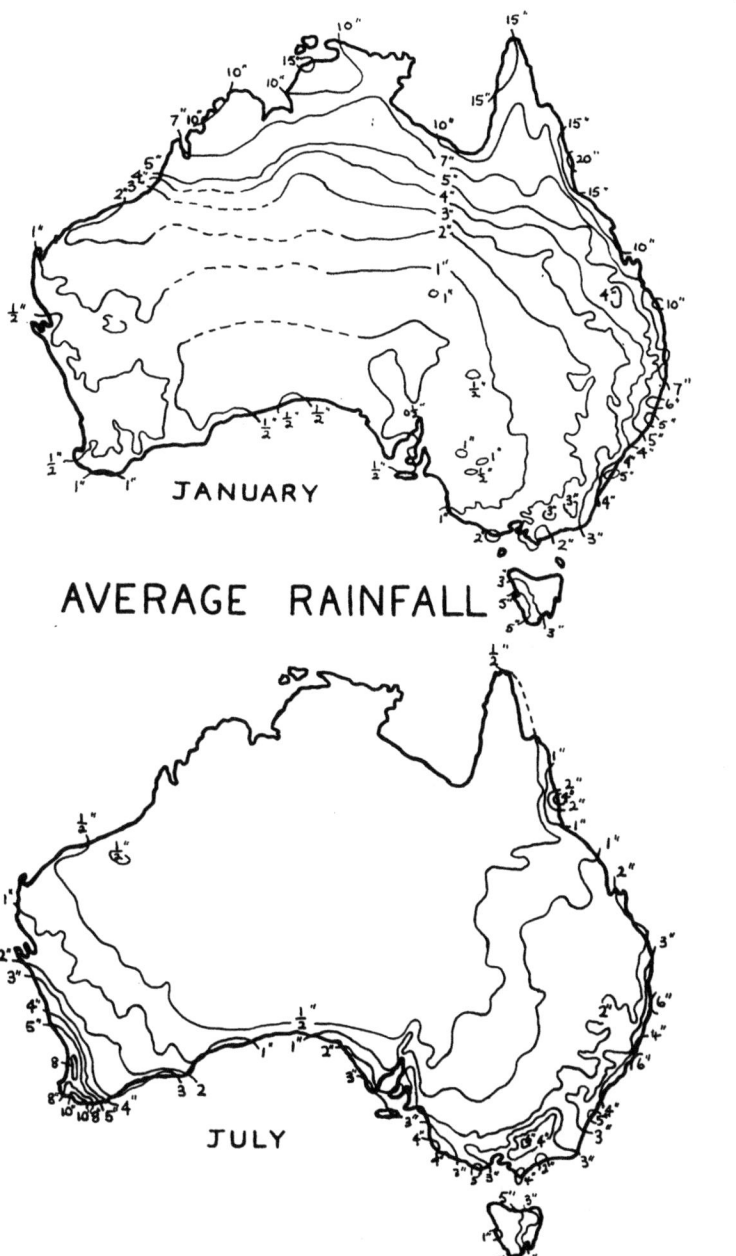

Fig. 2. AVERAGE RAINFALL FOR THE MONTHS OF JANUARY AND JULY (Simplified from the map in Atlas of Australian Resources – Rainfall; Department of National Development, 1952).

Two main systems operate to produce the Australian rainfall, a winter one that moves up over the southern part of the continent (the "Antarctic system") and a summer one (the "Tropical system") that brings monsoonal rains to the north of the continent. (Fig. 2). Hence, speaking generally, the north of the continent has a wet summer and a dry winter, the southern regions a wet winter and a dry summer. A consequence of the wet season being in the summer in the hot north is that the rain there is much less effective than is the winter rain in the cooler south.

The simple south-north and north-south movements of rainfall account, for the aridity of the central region, for this lies beyond the regular tract of the winter system and at the limits of penetration of the northern summer system. Corresponding latitudes of other continents are also dry. The fertility of the eastern coastal section is explained by elevation. Not only does it benefit from both rainfall systems but intermittent rain is likely to fall at virtually any other season. In fact this intermittent rain is enough to give a large area of New South Wales the same average rainfall figures for every month of the year (WADHAM & WOOD, 1950).

A further feature of the Australian rainfall pattern is its unreliability. "Nowhere in the world is there such a huge area of pastoral land of such erratic rainfall as in the pastoral country of Australia" (WADHAM & WOOD). Only the winter rainfall belt from the south-west corner in Western Australia to Victoria has a rainfall more reliable than the world normal. Over half the continent is more than 10% worse than places of the same average rainfall elsewhere in the world. This will be seen by comparing the average rainfall and percentage reliability maps for Australia with the expected variability figures for the world as a whole (WADHAM & WOOD, 1950).

Average Annual Rainfall (inches)	5	10	15	20	30	60
Expected Percentage Variability (world)	31	24	20	18	17	15

Reliability in Australia decreases inland from the east coast as successive isohyets are crossed: Sydney (average annual rainfall 40 inches per annum) variability 20%; Bourke (15 inches), 35%; Birdsville (9 inches), 45%. Thus, the rainfall tends to be least reliable in the central section and towards the limit of penetration of the two rainfall systems, either or both of which may fail.

Reliability is a somewhat theoretical way of considering rainfall. From the agricultural viewpoint "more practically useful maps have been prepared showing how often a given seasonal or annual total will be reached" (LEEPER, 1949).

Evaporation is high over the greater part of the Australian continent and in the central desert areas exceeds 100 inches per annum from a free water surface. Fig. 3, based on a formula derived by DAVIDSON, PRESCOTT, & TRUMBLE (DAVIDSON, 1936; TRUMBLE,

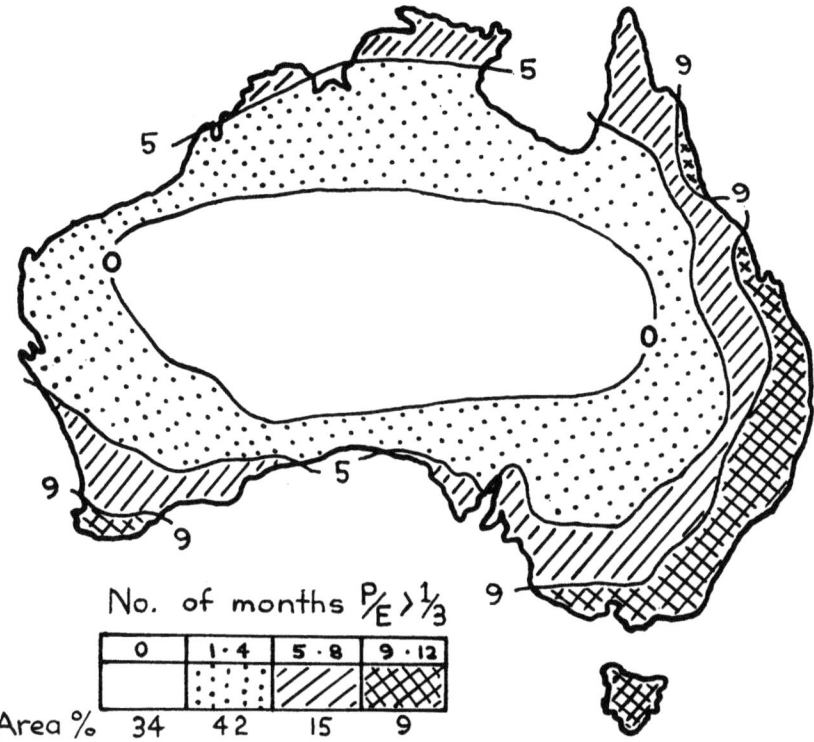

Fig. 3. Areas in which the ratio of precipitation to evaporation exceeds one-third. Five months is taken as the minimum growing period for crops and the nine month line indicates those areas not subject to regular dry seasons.
In the central desert areas of Australia the annual evaporation from a free water surface exceeds 100 inches per annum and evaporation exceeds rainfall everywhere except in the Eastern Highlands a narrow strip along the east coast, and the Tasmanian Highlands (LEEPER 1949). Map taken from LEEPER (1949) and derived by the formulae of DAVIDSON, PRESCOTT, and TRUMBLE.

1937) in South Australia, represents a practical way of expressing the relationship between rainfall and evaporation, that is the effectiveness of rainfall.

The formula was drawn up as part of a program being carried out at the Waite Institute to determine areas of potential agricultural use. It is based on the determination that, in the case of soils of the

Adelaide area, moisture is only available for plant growth when the ratio of precipitation to evaporation exceeds one-third. The number of months in which this minimum is exceeded represents the potential growth period for plants.

Five months is taken as the minimum period for growing crops and the 9 month line marks off those areas not subject to regular dry seasons. Use of this formula is subject to certain qualifications (LEEPER, 1949). For example, in the warmer areas the growing period is shorter, drainage has to be taken into consideration and no distinction is drawn between soaking rain and that which falls in heavy but irregular bursts.

Climate

A number of specialized climatic maps are now available for Australia, including those of DAVIDSON (1936), ANDREWARTHA & BIRCH (1954), derived from DAVIDSON's unpublished data, and the recent comprehensive one in the Atlas of Australian Resources (1954). This last contains, in addition, a series of picturegrams embodying monthly fluctuations in rainfall, temperature, humidity, and frost period.

The climatic indices that are in wide use throughout the world are those of KOPPEN and THORNTHWAITE. When extended to Australia each correctly shows the arrangement of the various intergrading climatic zones but does not properly delimit them (see remarks of LEEPER, 1949). For this the previously-mentioned works have to be consulted. The KOPPEN and THORNTHWAITE indices have the advantage, however, of being directly comparable with maps available for other continents.

The climatic zones of the Australian continent lie in a roughly parallel series from north to south and grade into each other. The northern tropical section is deficient in winter rainfall (Fig. 4). Then follow central desert zones varying slightly but agreeing in year-round deficiency in rainfall. The southernmost zone is subhumid (rainy), mesothermal, and deficient in summer rain. From the centre of the continent to the east coast there is a similar transition from aridity to a humid mesothermal region of uniformly abundant rainfall. The wet, tropical zone in the Atherton area of north-eastern Queensland shows up clearly on the THORNTHWAITE map.

Soils (Fig. 1, Chapter XIX)

A simplified version of the soil map of PRESCOTT (1931), has been reproduced by TAYLOR (1949). As will be noted (STEWART) Australian soils, in general, are weathered and low in essential nutrients.

The characteristics of the various major soil types occurring in Australia are set out in Table I. For map see Chapter XIX, page 308.

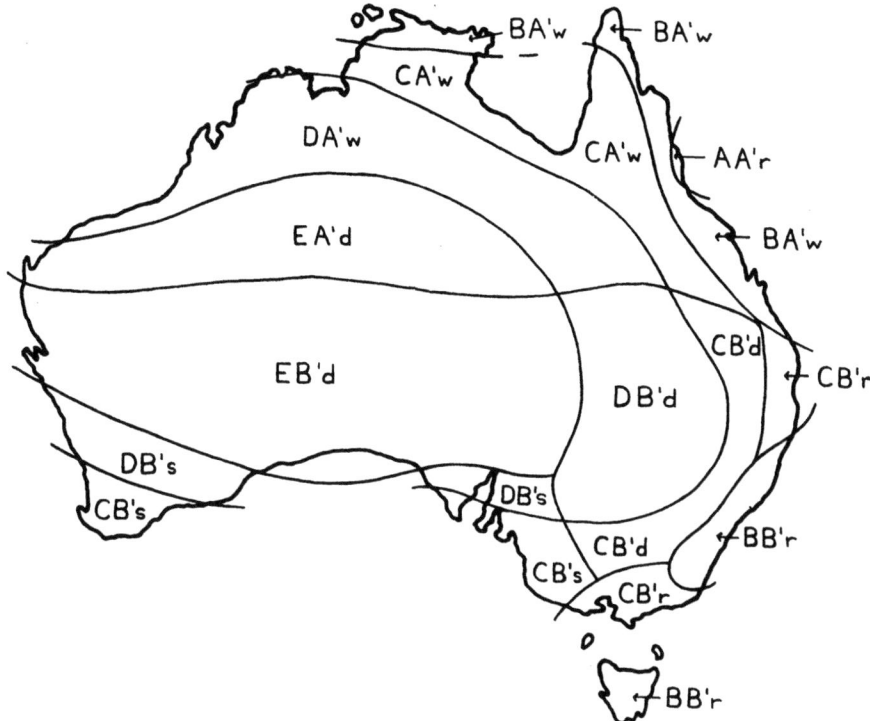

Fig. 4. THE 1931 THORNTHWAITE CLASSIFICATION OF CLIMATES, extracted from the world map in TREWARTHA (1954).

A A'r – Humidity: wet. Temperature: tropical. Vegetation: rainforest. Rainfall abundant at all seasons.
B A'w – Humidity: humid. Temperature: tropical. Vegetation: Forest. Rainfall deficient in winter.
B B'r – Humidity: humid. Temperature: mesothermal. Vegetation: Forest. Rainfall abundant at all seasons.
C A'w – Humidity: subhumid. Temperature: tropical. Vegetation: Grassland. Rainfall deficient in winter.
C B'r – Humidity: subhumid. Temperature: mesothermal. Veg: Grassland. Rainfall abundant at all seasons.
C B's – Humidity: subhumid. Temp: mesothermal. Veg: Grassland. Rainfall deficient in summer.
C B'd – Humidity: subhumid. Temp: mesothermal. Veg: Grassland. Rainfall deficient in all seasons.
D A'w – Humidity: semiarid. Temp: tropical. Veg: Steppe. Rainfall deficient in winter.
D B's – Humidity: semiarid. Temp: mesothermal. Veg: Steppe. Rainfall deficient in summer.
D B'd – Humidity: semiarid. Temp: mesothermal. Veg: Steppe. Rainfall deficient in all seasons.
E A'd – Humidity: arid. Temp: tropical. Veg: Desert. Rainfall deficient in all seasons.
E B'd – Humidity: arid. Temp: mesothermal. Veg: Desert. Rainfall deficient in all seasons.

Table I

AUSTRALIAN SOILS AND THEIR CHARACTERISTICS

(Compiled by Prof. R. Crocker)

(See fig. 1, Chapter XIX)

Soil Type	Characteristics
Stony deserts	A stone pavement overlies a profile which has only a few scattered stones throughout. Usually a loamy A horizon, sharply differentiated from a red-brown structured clay in the B horizon. Lime and gypsum often present in the B horizon.
Tablelands and ranges	Variable, unclassified soils; many skeletal, with little profile differentiation.
Desert sandhills and sandplains.	Wind-piled sands with little or no profile differentiation. Some of the sandplain soils are polygenetic, and the wind spread surface zones overlies an old lateritic Red Earth type profile. May be alkaline or slightly acid.
Low country subject to periodic flooding; tidal marshes and deltaic formations.	Variable; undescribed.
Desert loams	Shallow loam – clayloam A horizon overlies a coarsely structured clay B horizon, which normally contains some lime and gypsum. Nitrogen level low; general fertility low.
Brown soils of light texture	A sandy or sandy loam A horizons changing gradually to sandy clay loams or clay loams in the reddish B horizons. The alluvial zones sometimes contain free-lime.
Grey and Brown soils of heavy texture	Heavy textured soils, with coarsely structured A horizons which are low in organic matter; Alkaline; free lime in sub-soil and throughout profile. No marked contrast in texture between A and B horizons. Some high in NaCl. Heavier members have poor surface structure.
Solonized Brown soils (mallee soils), mallee sandhills, solonetz soils.	Mallee soils; Profile alkaline throughout; much free lime in subsoil. Surface brown or red-brown and sub-soil light reddish brown. Subsoil the heavier. Limy pebbles at about two feet depth common. Fairly saline at times but not usually enough to affect wheat.

Soil Type	Characteristics
Podsols, residual podsols and lateritic sandplain, red loams.	Podsols: chemically poor in nutrient; poor natural drainage. No free lime; acid; grey surface of light texture. Subsoil relatively heavy or heavy and frequently of low permeability, commonly yellow or predominantly reddish. Red. loams: Friable, deep and permeable, drainage good. Rather deficient chemically, not necessarily high in iron.
Red-brown earths and terra rossas	Surface usually slightly acid but may be neutral. Subsoil neutral to alkaline, normally more alkaline with depth. Commonly upper subsoil free of lime. Reddish-brown; upper subsoil red; lower subsoil brown or grey. Sharp transition to heavy subsoil at six inches depth.
Rendzinas and black earths	Surface black; Alkaline, pH 7-8, 8-10% organic matter and excellent granular structure. Chemically rich. Organic matter gradually decreased from surface.
High moor peat	Highly acid, predominantly organic; chemically very poor; resistant to microbial decomposition after draining. May be several feet thick or quite thin (Tasmania and Victoria).

Vegetation Zones (Fig. 1, Chapter XVI)

The characteristics of the major vegetation associations occurring on the Australian continent are set out in Table II. The dominant species comprising them are discussed by STEWART (Chapter XIX).

It will be seen that Australia has a wide range of vegetation associations. The close relationship between them and the rainfall and soil zones will be noted. Secondarily, the vegetation associations govern the distribution of a high proportion of plant and animal species.

Table II

PLANT FORMATIONS AND THEIR CHARACTERISTICS
(modified and enlarged from that in WOOD (1950)).
(See fig. 1, Chapter XIX)

Formation	Characteristics
Gibber Desert (desert/steppe)	Plains with gravel and stone pave, vegetation minimal.
Desert Grassland (desert/steppe)	Sparse tussocks of sclerophyllous grass, especially spinifex *(Triodia)* Chenopodaceous plants. Marked seasonal herbage after rains.

Formation	Characteristics
Mulga Scrub (desert steppe)	Small trees with dense or scattered shrubs, few herbs, and with vast tracts dominated by the Mulga *(Acacia aneura)*.
Mallee Scrub	Associations of dwarf Eucalypts, the trees, characterised by multiple stems arising from a common base; growing in semi-arid regions frequently on soil with characteristic qualities (mallee soil). Growth is generally rather open. Scattered shrubs and tracts of *Triodia* are present.
Sclerophyllous grass steppe and grassy scrub	Fairly dense scrub and shrubs interspersed with spinifex *(Triodia)*.
Savannah Grassland	Grasslands with herbs and a few subfruticose shrubs, interspersed with a few trees, or small clumps of trees. Sometimes intermixed with tracts of savannah woodland.
Savannah Woodland	Rather open communities of Trees, with scattered shrubs and a few herbs. Grassy underfoot and with open areas.
Sclerophyll Forest	Trees of forest form, in closed community; dense undergrowth of hard-leaved shrubs; grass rare. Map includes both wet and dry sclerophyll under this heading.
Rain Forest	Dense assemblage of trees, canopy continuous, wood climbers present.
High Moor and Alpine woodland.	Tussocks of grasses or sedges at high altitudes. Tracts of snow gum.

The development of the Australian environment

The Relationship of Australia to other Land-masses

Australia is the most isolated of the continents and there is geological evidence that it has not been in direct land contact with Asia since the beginning of the Tertiary, a period of over 50,000,000 years. Floral and faunal differences between the two continents are emphasized by the occurrence of an abrupt changeover along the island archipelago between them. Hence, an understanding of the isolation and composition of the Australian flora and fauna depends, to a degree, on an appreciation of the characteristics and geological history of the Indonesian island chain.

The Asian continental shelf extends as far east as Borneo, Java, and Bali, the Australian shelf westwards almost to Timor. Shelf limits, taken as the 100 fathom (200 metre or 600 foot) line do not indicate the extent of land area at the height of Pleistocene glaciation for the sea is believed only to have fallen 300 feet (100 metres) at this time. They do, however, serve to indicate the general nature of former connections and to emphasize the existence of an intermediate deep-water gap.

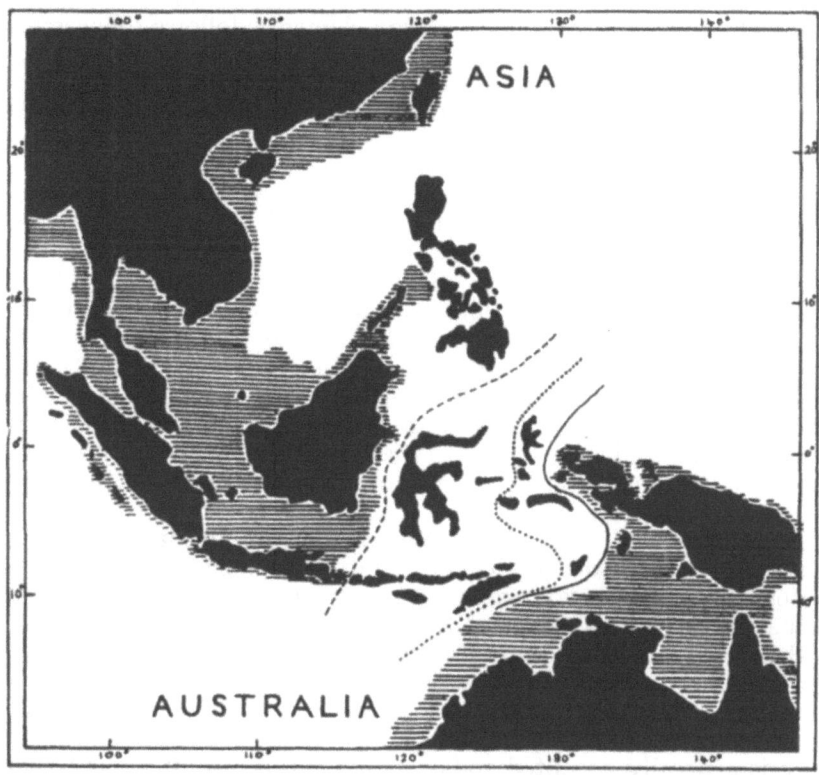

Fig. 5. Asia and Australia to show limits of continental shelves (100 fathom or 300 metre line), and general pattern of connection at the height of Pleistocene glaciation. WALLACE's line (original) is shown (dashes), also WEBER's line (dots), and the western limits of the Australo-Papuan continental fauna (continuous line). (Taken from MAYR, 1945).

The second factor that has obviously influenced the distribution of life is that the Celebes, Moluccas, and Lesser Sunda Islands, within the deep-water section, are in an area of great geological instability, which was subject to violent tectonic activity during the Tertiary. The geological structure here is complex and there is a bewildering array of grabens, geosynclines, geanticlines, and deep sea basins. A feature is that islands only a few miles apart are separated by deep chasms. Thus, the sea between Bali and Lombok, barely fifteen miles distant, extends down to over 300 metres (900 feet). The trough between Borneo and the Celebes, 70 miles apart, exceeds 2000 metres (6000 feet). To the south-east of the Celebes is an extensive basin more than 6000 metres (18,000 feet) in depth.

WALLACE's biogeographic line, (Fig. 5) long considered as one of the world's most marked distributional barriers, marking the eastern limits of many of the dominant Asian elements, follows the western limits of the area of deep water and geological instability. WALLACE's line, though it does correspond to the distributional limits of a number of dominant Asian groups is certainly not the absolute barrier it was once considered to be (MAYR, 1945). To the east of it lies what DARLINGTON (1957) calls a "subtraction-transition" area in which the proportion of forms of western and eastern origin gradually changes. Even so the Celebes, Bali, and Timor, are populated by animals predominantly Asian in origin. The transition area is, as MAYR (1945) describes it "biotically impoverished", i.e. much poorer in terms of number of species than either Borneo to the west or New Guinea—Australia to the east. These islands apparently have not the diversity of life zones and "niches" that would support a rich fauna. It might be noted that, at the eastern end of this archipelago there is an equally sudden transition from the depauperised insular fauna to the rich continental one of Australia—New Guinea (WEBER's Line and line of limits of Australo-Papuan continental forms).

Postulated land connections (Mesozoic or pre-Mesozoic) between Australia and the other southern continents, suggested by various biologists to account for present-day resemblances in the floras and faunas of these areas remain, like the WEGENER hypothesis itself, without present supporting geological evidence. Many workers (reviews in this book) have stressed that the Australian flora and fauna are, basically, a composite of several elements. The persistence of many plant and animal groups in the southern continents only is generally accepted today as indicating nothing more than that they are relicts of formerly cosmopolitan groups. A number of workers in the Southern Hemisphere are, however, strong advocates of a common southern origin for them. Interest in Antarctic radiation is now being revived by geomagnetic findings that the position of the South Pole has changed at various times during the Earth's history.

Of past relationships with smaller land-masses adjacent to Australia none is more uncertain than that of New Zealand, a mere 1,200 miles away and, without a doubt, a continental remnant. These land areas are, however, isolated by a submarine trough extending down to over 2,000 fathoms (4000 metres). DAVID (1950), supporting the palaeobotanist WALKOM, suggests that the two were connected by an intermediate land mass throughout the Mesozoic and possibly even as late as the late Pliocene, or early Pleistocene. In contrast with this the botanist GOOD (1957) points out that the plants of New Zealand are somewhat more closely related to those of New Caledonia and New Guinea to the north, than to those of Australia. The presence of a low submarine ridge connecting these islands is noteworthy. New Zealand, like other isolated islands, has not a rich fauna. At the vertebrate level it has only two frog species, perhaps twenty lizards, and two mammals, both bats. There has been much colonization of New Zealand by Australian birds, however.

Lord Howe Island, some 600 miles to the east of Australia, presents similar historical problems to New Zealand for it supports many "old continental" species. Timor, 350 miles to the north-west was in much closer proximity to Australia at the height of Pleistocene glaciation but, nevertheless, remained separated by a deep sea trough. The flora and fauna of Timor are strongly Asian but there has been, nevertheless, quite a degree of interchange with Australia.

New Guinea and Tasmania, lying on the Australian continental shelf, would be brought into contact by a fall in sea-level of about 30 feet (10 metres) and 150 feet (50 metres), respectively. They are known to have been joined to Australia during part of the Tertiary and apparently twice during the Pleistocene, when the sea-level fell 250—300 feet. This being so, it would seem surprising that the flora of New Guinea bears little resemblance to that of Australia (BURKILL, 1942; GOOD, 1957) and much of its invertebrate fauna is likewise Indo-Malayan, e.g. oligochaetes (MICHAELSON, 1922) and insects (KARNY, 1929; CHEESMAN, 1951; GRESSIT, 1956). In explanation of this it has been suggested that the geologically ancient northern part of New Guinea was once, presumably in the Mesozoic, in direct land contact with Asia (CHEESMAN, GRESSIT). Another proposal is that Australia did not always lie in close proximity to New Guinea but "drifted" into it (GOOD, 1957). In contrast with the "older" groups the reptile, mammal, and bird faunas of New Guinea have strong affinities with those of Australia.

The Tasmanian flora and fauna is Australian in origin but poor in number of species. It is rich in "Antarctic" plant elements such as *Nothofagus*. Several interesting forms of life, moreover, have been able to persist in the insular isolation of Tasmania, e.g. the syncarid shrimp *Anaspides*, the marsupials *Thylacinus* and *Sarcophilus*. The Tasmanian negrito race of man, an ethnic type distinct from the Australian aborigine, survived there until European settlement.

The Palaeoclimatology of Australia

If isolation set the stage for the development and radiation of the Australian flora and fauna it is the geological, climatic, and biological changes within the continent itself that have moulded the environment and life forms we know today.

Tertiary

Physiographic events of importance to the distribution of life were mostly somewhat local. Two, however, must have had wider significance, the long period of stillstand that marked the middle of the period and vertical movements in the east and south. The elevation of the Great Divide took place towards the end of the Miocene and reached its present status towards the close of the Pliocene. Tertiary block faulting, probably starting at the close of the Pliocene (but still going on), initiated the formation of Spencer and St. Vincent Gulfs in the south, new barriers to west-east distribution. At the same time the adjacent Mount Lofty Ranges were elevated. Of fundamental importance to drainage patterns was the formation of the west-east Olary Ridge immediately to the south of Lake Eyre that thereafter made it impossible for the central rivers to reach the southern seabord. Vulcanism commenced and persisted throughout the Tertiary in eastern Australia. Full details of the Australian Tertiary, so far as they are known, can be obtained from DAVID (1950).

The evidence indicates that the climate was warm for much of the Australian Tertiary. A pan-Australian mesic flora (especially *Nothofagus* and *Cinnamomum*) occurred in the first part of the Tertiary, whilst in the late Tertiary *Eucalyptus* and *Acacia* came into prominence (CROCKER, WOOD).

Whilst the Tertiary flora is known in a general fashion virtually nothing is known of the finer details of succession. The origin and Tertiary radiation of the marsupials and monotremes in Australia is, unfortunately, as yet a closed book, for no fossil beds of the right type have been found. On the other hand many Tertiary deposits contain the remains of "giant" animals (see later).

Pleistocene

Features of the Pleistocene in Australia are:

(a) Negligible glaciation. Even during the earliest (and most severe) glacial period the ice sheet covered no more than the western half of Tasmania and some 400 square miles in the Australian Alps.

(b) Minor vulcanism for the first half of the period in the southeast and east of the continent. There was no mountain building. Steady erosion and silting occurred.

(c) Fairly well-defined sea-level (strand line) changes that make it possible to date marine deposits as part of a succession. On at least two occasions the continent was materially larger than at present.

(d) Marked climatic oscillations that expressed themselves, not only in temperature changes, but as alternating periods of "wettness" and "dryness".

Australian Pleistocene chronology can only, with difficulty, be linked with that of the Northern Hemisphere. The order and magnitude of terrestrial changes within Australia have, as yet, not been worked out. Cultural succession in the aborigines is as yet too poorly understood to be of real chronological value and pollen analysis has so far proved unsatisfactory, for many of the prominent myrtaceous plants cannot be identified by their pollens.

There is no doubt that much of the Pleistocene was wet for river and lake sediments are hundreds of feet thick in places. The dry lakes of the interior held great quantities of water and the "dead" rivers actively flowed for much of the period (DAVID, 1950). Crocodile and lungfish fossils (apparently Pleistocene in age) are known from South Australia, thousands of miles to the south of the present range of these animals. On the other hand it is equally certain notwithstanding the inferences of various geologists (BROWNE, 1945; DAVID, 1950) and the remarks of the geographer GENTILLI (1949), deduced from patterns of wind circulation, that desert was never eliminated from the continent. A whole series of today's true desert elements, the unique desert oak, *Casuarina decaisneana* (L. A. S. JOHNSON, personal communication) and to a lesser extent the grass genus *Triodia* (BURBIDGE, 1953) amongst the plants, the marsupial mole, *Notoryctes*, and various lizards, obviously have a history extending well back into the Tertiary.

A more reasonable assessment of climatic succession during the Pleistocene than that the continent as a whole was alternatively "wet" and "dry" is that the arid belt oscillated between the south and the north (KEBLE, 1947). The geologist FAIRBRIDGE (1953) has suggested that, associated with the total of four glacial maxima and three interglacial periods, plus the post-glacial warm period, there was a total of eight "pluvial periods" in the south, with each oscillation producing an arid sweep before and after it. Though the actual occurrence and extent of these changes is unconfirmed, support for climatic successions of this kind during the Pleistocene comes from soil studies (BUTLER, 1956).

The Pleistocene and Early Recent flora is poorly known (CROCKER). Prominent in the fauna, however, were "giant" animals, fossil remains of which occur virtually throughout the continent and as far the Bulolo valley in New Guinea. These remains, many of which are presumable Tertiary, include the ratites *Genyornis* and *Dromornis*, the lizard *Varanus prisca*, a kangaroo with skull 16 inches long and considered to be 10 feet high *(Palorchestes azeal)*, a "wombat" with the bulk of a rhinoceros *(Diprotodon)*, and the marsupial "lion", *Thylacoleo*. Final extinction of the large marsupials appears to have resulted from a combination of aridity and the arrival of aboriginal man with his wild dog, the dingo. There is a radiocarbon date for a lacustrine bed containing giant marsupial bones (western Victoria) as 12,000—13,000 years ago (GILL, 1955), and a provisional date for a horizon containing *Diprotodon* remains at Menindee, N.S.W as 6,570 ± 100 years (R. TEDFORD, personal communication). This supports other evidence that extinction is quite a recent event.

Recent

The Recent in Australia was marked by the onset of widespread aridity. When it reached its peak is not certain, opinions ranging from late Pleistocene to as recently as 4000—6000 years ago (CROCKER & WOOD 1947). At any event the onset was apparently sudden (CROCKER), for soils became freely exposed to wind erosion and were sufficiently unstable for the building up of dune systems. In association with the wholesale drying up of interior rivers and lakes the continent came to assume its present condition of an essentially arid land-mass. The pre-eminence of the xeric flora is, accordingly, a recent phenomenon.

The present Australian flora and fauna

The Australian land flora is composed of some 12,000 species of angiosperms (WILLIS, 1958), 61 gymnosperms, 25 of which are cycads (L. A. S. JOHNSON, personal communication), 36 conifers (National Herbarium estimate), 250 true ferns (TINDALE, 1958), 600 mosses (BURGES, 1958), 1,075 Hepatici (BIBBY, 1958), about 1,000 larger fungi (WILLIS, 1958a), and innumerable small fungi and algae. Australia lacks horsetails (Sphenopsida).

The flora is xeric over the bulk of the continent but the greatest number and diversity of species occur in the wetter coastal areas. Thus, the south-western corner of Western Australia has no fewer than 6,000 species of plants, there being "no richer flora anywhere in the world outside of the tropics" (WENT, 1956) Victoria has approximately 2,200 different native plants (EWART, 1930), whilst coastal Queensland is particularly rich in recent Indo-Malayan rain forest species.

Prominent aspects of the Australian countryside are the absence of extensive tracts of deciduous trees, and the relatively minor development of softwoods and conifers. Woody shrubs and even trees occur, however, in arid areas, an unusual feature (BEADLE). True forests are at a premium in Australia. Thus, for example, the area of New South Wales, (a "moderately fertile" state), that is covered with true forest is only 2.02% of the total, compared to 32,8% for Canada, 59.9 Japan, and 20.2 for New Zealand (ANDERSON, 1947).

The unique aspect of the Australian vegetation lies not in the number of endemic families of which there are only three (with a total of 30 species), but in the large number of endemic genera and species (GARDNER, 1944). Included in this latter are the dominant tree forms of the continent, the endemic (i.e. Australo-Papuan) *Eucalyptus*, that has radiated into some 600 species, and *Acacia*, whose 400 endemic species amount to approximately four-fifths of the world total.

The Australian flora was divided into three elements by DIELS (GARDNER, 1944), an old Antarctic one, a Palaeotropic element that entered the continent by way of India, Malaya, and Indonesia, and an endemic Australian element. GARDNER, however, regards the last as a secondary development only. Within the continent plant distribution is largely in terms of "dispersive opportunity", rainfall, and soils. GARDNER notes that three "phytogeographic zones" or provinces occur, a desert one, a tropical one (rain forest and littoral) of Indo-Malayan origin, and an "old" peripheral southern one. The south-west corner, however, is quite distinct from the forested areas of the south-east, some 75% of total species being endemic to the region (GARDNER). Insular Tasmania, it might be noted, is rich in old "southern" elements, which extend some distance northward along the Eastern Highlands. There is, however, a relationship between certain Tasmanian species and those of the high mountains of New Guinea (GIBBS, 1921).

The land and fresh-water fauna of continental Australia includes a total of 108 placental mammal species, 119 marsupials, 2 monotremes (E. TROUGHTON), 520 birds (A. KEAST), about 380 reptiles (A. KEAST), 112 frogs (A. R. MAIN), but no urodeles, 180 fresh-water fish (G. P. WHITLEY), a total vertebrate fauna of about 1,440 species. Amongst the invertebrates are 750 molluscs (D. F. McMICHAEL), and perhaps well over 50,000 insects (TILLYARD, 1926, gives 37,000).

Some animal groups are characterized by the persistence of primitive forms, e.g. the fresh water shrimp *Anaspides*, the lungfish *Neoceratodus*, and monotremes, others by marked radiation, e.g. marsupials, certain grasshoppers (Morabinae and Cyrtacanthracridinae — KEY).

Australian members of older animals groups would appear to have had a multiple origin. The bird fauna, however, could be adequately explained as a series of colonizations from the north, each superimposed upon the other, or upon an older Australian element.

Distribution of animals within Australia falls into certain basic patterns. Certain "faunal" provinces are recognisable, whilst climato-vegetation zones are of the greatest importance.

Changes in the Australian Continent associated with European settlement

European settlement of Australia in 1788 can be said to have initiated changes almost as drastic as the earlier climatic oscillations, though more selective and less generalized. They included the clearing of forests, extensive burning, introduction of grazing ungalates, and a wholesale colonization by exotic plants and animals. Overstocking, the destructive work of rabbits, and misuse of soil initiated widespread erosion. The unique marsupial fauna, relatively slow to reproduce and unable to compete with the feral fox and cat has, over extensive sections, been materially reduced and various species have been exterminated. Other species, like the larger kangaroos, the phalangers, and some bandicoots, are still reasonably common. The sheep, not the kangaroo, is now the dominant herbivore of the plains. Aboriginal man, long protected by the isolation of the continent has been reduced in numbers from an estimated 275,000 in 1788 to a total of perhaps 30,000 "full bloods" and 50,000 people of mixed parentage.

Acknowledgements

I should like to express my thanks to the following for reading over the present contribution and for the suggestions that have been embodied herein: Dr. L. C. BIRCH, Dr. W. R. BROWNE, Prof. R. CROCKER, Dr. J. W. EVANS, and Prof. F. GRIFFITH TAYLOR and Mr. L. A. S. JOHNSON.

Miss Wendy SLADE drew the maps as well as those depicting the vegetation and soil zones elsewhere in the book.

REFERENCES

AIRD, J. A., 1949. Water conservation and irrigation. *The Australian Environment*: **67**. C.S.I.R.O., Melbourne.

ANDERSON, R. H., 1947. *The Trees of New South Wales*, *5*. Government Printer, Sydney.

ANDREWARTHA, H. G. & BIRCH, L. C., *The Distribution and Abundance of Animals*, *495*. Univ. of Chicago Press.

BIBBY, P., 1958. Liverworts. *The Australian Encyclopaedia* **5**, *343*.

BROWNE, W. R., 1945. An attempted post-Tertiary chronology for Australia. *Proc. Linn. Soc., N.S.W.* **70**, *5*.

BURBIDGE, N. T., 1953. The genus *Triodia* R. Br. (Gramineae). *Aust. J. Bot.* **1**, *121*.

BURGES, N. A., 1958. Mosses. *The Australian Encyclopaedia* **6**, *163*.

BURKILL, I. H., 1942. The biogeographic division of the Indo-Australian Archipelago. 2. A history of the divisions which have been proposed. *Proc. Linn. Soc., Lond.*, 154 Session: *127*.

BUTLER, B. E., 1956. Parna — an aeolian clay. *Aust. J. Sci.* **18**, *145*.

CHEESMAN, L. E., 1951. Old Mountains of New Guinea. *Nature, Lond.* **168**, *597*.

CROCKER, R. L. & WOOD, J. G., 1947. Some historical influences on the development of the South Australian vegetation communities and their bearing on concepts and classification in ecology. *Trans. Roy. Soc. S. Aust.* **71**, *91*.

DAVID, T. W. EDGEWORTH, 1950. *The Geology of the Commonwealth of Australia*. Edit. by W. R. Browne, Edward Arnold & Co., London, Vol. 1, p. 686.

DAVIDSON, J., 1936. Climate in relation to insect ecology in Australia. 3. Bioclimatic zones in Australia. *Trans. Roy. Soc. S. Aust.* **88**, *60*.

EWART, A. J., 1930. *Flora of Victoria*, page *13*, Government Printer, Melb.

FAIRBRIDGE, R. W., 1953. *Australian Stratigraphy*. Univ. of Western Austr. Text Book Board, 2nd Ed.

GARDNER, C. A., 1944. 1. The Vegetation of Western Australia. *J. Roy. Soc. W. Aust.* **28,** *xiii.*
GENTILLI, J., 1949. Foundations of Australian bird geography. *Emu* **49,** *116.*
GIBBS, L. S., 1921. Notes on the phytogeography and flora of the mountain summit plateaux of Tasmania. *J. Ecol.* **8,** *1.*
GILL, E. D., 1955. Radiocarbon dates for Australian archaeological and geological samples. *Aust. J. Sci.* **18,** *51.*
GOOD, R., 1957. Some problems of southern floras with species reference to Australasia. *Aust. J. Sci.* **20,** *41.*
GRESSIT, J. L., 1956. Some distribution patterns of Pacific Island faunae. *System. Zool.* **5,** *11.*
HILLS, E. S., 1949. Physical Geography of Australia. *The Australian Environment, 13.* C.S.I.R.O., Melbourne.
KARNY, H. H., 1929. On the geographical distribution of the Pacific Gryllacrides, *Proc. Fourth Pacific Sci. Cong.,* **3,** *157.*
KEBLE, R. A., 1947. Notes on Australian Quarternary climates and migration. *Mem. Nat. Mus. Vict.* **15,** *28.*
LEEPER, G. W., 1948. *Introduction to Soil Science*: *1* etc. Melbourne Univ. Press.
LEEPER, G. W., 1949. Climates of Australia. *The Australian Environment* p. *23.* C.S.I.R.O., Melbourne.
MAYR, E., 1945. Wallace's Line in the light or Recent Zoogeographic Studies. *Science and Scientists in the Netherlands Indies Surinam, and Curacao.* New York 1945. *243.* Also: *Quart. Rev. Biol.* **19,** *1—14,* 1944.
MICHAELSEN, W., 1922. Die Verbreitung der Oligochaeten im Lichte der Wegenerschen Theorie der Kontinentalverschiebungen. *Verh. Nat. Ver. Hamburg,* **3,** *45.*
PRESCOTT, J. A., 1931. *The Soils of Australia in relation to vegetation and Climate*: *1—82.* C.S.I.R. Bull. Aust. No. 52.
PRESCOTT, J. A., 1952. Second Ed. of the above.
TAYLOR, J. K., 1949. Soils of Australia. *The Australian Environment*: *35.* C.S.I.R.O., Melbourne.
TILLYARD, R. J., 1926. *The Insects of Australia and New Guinea*: page *8.* Angus and Robertson, Sydney.
TINDALE, M., 1958. Ferns. *The Australian Encyclopaedia* **4,** *39.*
TREWARTHA, G. T., 1954. *An Introduction to Climate*: *225* etc. McGraw-Hill Book Co., Inc. New York.
TRUMBLE, H. C., 1937. The climatic control of agriculture in South Australia. *Trans. Roy. Soc. S. Aust.* **61,** *41.*
WADHAM, S. M. & WOOD, G. L., 1950. *Land Utilization in Australia*: *43.* Melbourne Univ. Press. Second Ed..
WENT, F. W., 1956. Report to the Executive of C.S.I.R.O. Melbourne.
WILLIS, J. H., 1958. Plants. *The Australian Encyclopaedia* **7,** *140.*
WILLIS, J. A., 1958a. Fungi. *The Australian Encyclopaedia* **4,** *231.*
WOOD, J. G., 1949. Vegetation of Australia. *The Australian Environment* *77.* C.S.I.R.O., Melbourne.

III
ECOLOGY OF PRIMITIVE ABORIGINAL MAN IN AUSTRALIA

by

Norman B. Tindale

(Curator of Anthropology, South Australian Museum, Adelaide)

This chapter is concerned with the nature and distribution of the aboriginal peoples of Australia and sets out some of the peculiar problems faced in understanding their dispersion and status.

Unlike the generality of mammals and plants discussed elsewhere in this book, man and his attendant dog are not part of the indigenous fauna of Australia but came here from Asia in very late geologic time. Both of them for a time may have assumed dominant roles in the faunal assemblages of the Australian Continent.

Whence, when and how they came are primary questions to which partial answers are possible, and these answers are necessary for an understanding of the late faunal and floral histories of Australia.

The work of Harrisson (1957) in Niah Cave in Sarawak recently has supplied evidence that a type of man, using palaeolithic chopping tools, was living in Borneo, on the margin of Continental Asia 32,000 years before the present (Carbon 14 date 32630 ± 700 B.P.). Although not specifically identical with the tools of the Hoabinhian Culture of South East Asia the Sarawak implements, together with those of the Kartan industry of Late Ice Age date in Australia, fall together into a distinct category among tools of Old Stone Age industries, and from them we may perhaps draw an inference that a type or types of modern man was already strategically placed near the south eastern periphery of Ice Age Asia near the height of the last cold phase of the Pleistocene. During the final phases of the Last Ice Age he was sufficiently well equipped with watercraft to be able to make his way across the permanent but narrow water-gaps of Indonesia and move deeply into the Australian Region. At what date he actually did so is still unknown, but indications of the probable presence of aborigines in Pejark Marsh, about 13,000 years ago (Gill, 1955), and more conclusively the late Pleistocene Kartan Culture suite of Hallett Cove (Tindale, 1937, 1956, 1957) in South Australia (a site documented by several hundred implements of more than one definite type including the "sumatra", the Kartan horsehoof core and hammer-stones) shows us that he had already arrived in parts of the Australian continent remote from Asia well before the beginning of Recent Time. The outline of the subsequent history of culture changes in Australia is now reasonably well known and the implements typical of the several significant stages have been described. TINDALE (1957) gives a recent summary of culture sequence in South Eastern Australia, with some Carbon 14 dates. The accompanying table (Fig. 1) summarises the principal results of work up to the present time and will act as a framework for the discussions in this chapter. It is similar to one published in 1957 but has an additional Carbon 14 date of 1777 ± 175 B.P. recorded

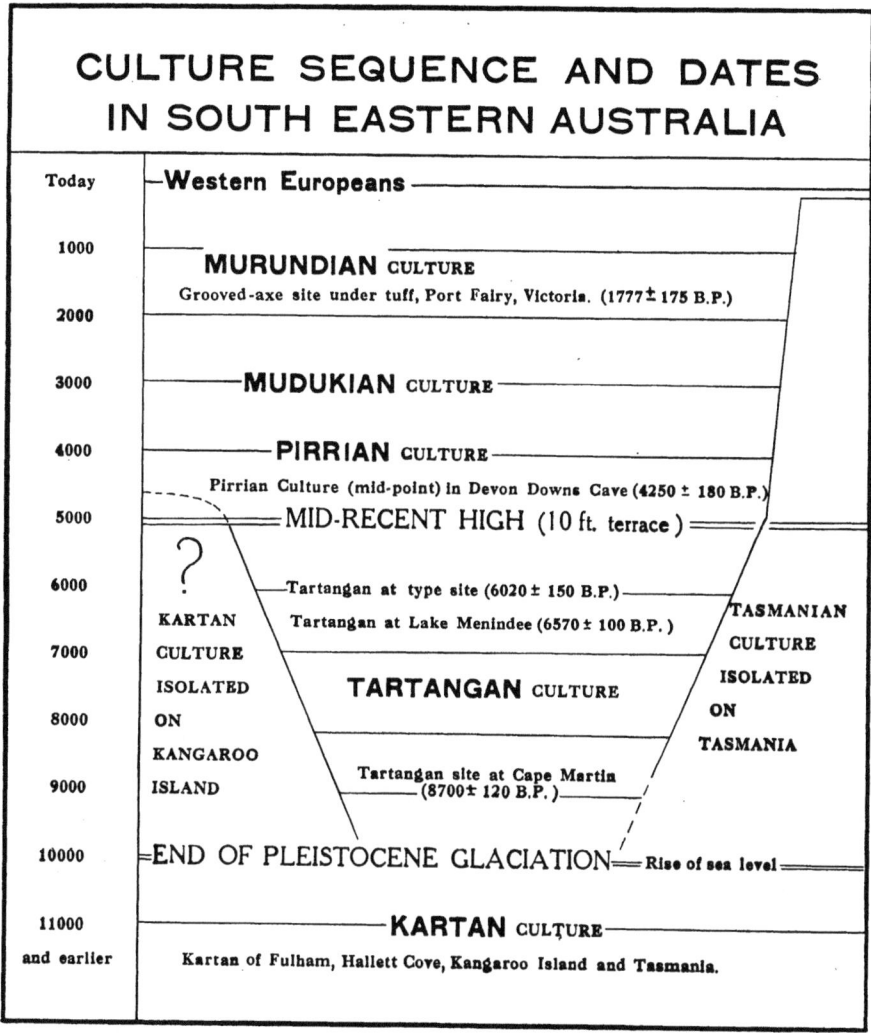

Fig. 1.

by GILL (1955, p. 50) for charcoal from a midden with marine shells under tuff and including a basalt edge-ground axe with hafting groove associated with bone implements.

Collections made in South Australia have produced much of the evidence for the detailed culture succession in Australia. For this some of the credit must be given to the destructive habits of Western man, especially when living on the margins of desert areas. Chronic overstocking with sheep and cattle and over confident cultivation of marginal lands in the production of wheat (so prevalently done in parts of South Australia) has caused soil erosion on a considerable scale. Wherever this occurs archaeological relics of man are likely to become exposed on the surface of the ground. Partly for this reason Victoria with its wetter climate and closely grassed surfaces has yielded particularly little evidence for ancient man except where mining operations or other casual earth-moving activities have brought relics to light. The same is true for Tasmania and other places of higher rainfall. On several visits to Flinders Island in Bass Straits, for example, this author was able to find only two small areas of erosion on the island. Both of these yielded Kartan Culture relics, thus hinting at the possible extent of evidence of ancient occupation, not yet revealed at the surface.

In space available here it is not possible to discuss, even in outline, all the factors governing the use of land and its products by the aborigines, nor the influences which the vegetation, biota, geology, geomorphology, and climate had on his spread and distribution over the Australian continent, once he had access to it.

Some broad generalisations seem safe. Presuming that man came from the general direction of Asia, initial trend of migration was southerly during the Pleistocene and during the Early Recent Period after the Gulf of Carpentaria became established, had also a westerly component, so that Tasmania at first and later the Western Australian coasts came to be the places most remote from Asia. The very size of the Australian continental mass and the smallness of Australian tribal units suggests that exploitation of coast-wise and inland routes of access could not simultaneously have been followed by one connected group of people so that there was rapidly introduced a diversity in modes of exploitation and spread. It may be correct to assume a tendency to utilise the best lands first and that exploration of more difficult environments such as deserts and rain-forests might have proceeded more slowly than occupation of well-watered areas of easy access.

Allowing 25 years per generation for 13,000 years suggests that over 500 generations of men have exploited the continent thus newly gained, using only the tools and equipment of the Old Stone Age (ranging from middle palaeolithic to epi-palaeolithic types). Their movements and shifts have more than dulled the effects of the initial bias and had brought the people on the continent to a quasi-balance with the whole Australian environment, a balance disturbed, particularly in the north, by infiltration of cultural innovations from areas of contact with Indonesia. The Australians were by no means so isolated a people as has been imagined by some, and it seems proper to consider that the impacts of new peoples and ideas from abroad combined with some local inventions, led to the changes which are summarised in the table of culture sequence.

There has been a tendency for one school of thought to view the history of the present day Australian aborigines and their predecessors as if they were so isolated that perhaps they had passed through major stages of their evolution from a primitive type of proto-man while isolated in Australia, and hence they have tended to ascribe to them a semi-independent role in the evolution of physical form as well as of culture. This view is not tenable. Not only are the basic types of Australians represented in South Eastern Asia, but most, if not all of the material cultural possessions of Tasmanians and Australians and perhaps all of the basic forms of archaeological stone implements of their ancestors, are of types once in use outside the Australian region. There must be an increasing tendency to see the Australian continental area as only an enclave into which from time to time were projected small samples of the physical types and successive cultural novelties produced by the main bodies of men on the great adjoining Eur-Asiatic continental land mass. Australian man reflects Asiatic cultures in much the same way as the life activities of the men of the Orkneys tend to reflect and to preserve some of the older forms of the successive culture phases of continental Europe.

If the three types of Australians (BIRDSELL, 1949) and the various culture phases suggested to be found on this continent represent fairly pure samples, drawn off from the whole seething cauldron of Asia at various intervals of time in the Late Pleistocene and Early Recent, they may tend to give us some indications of the kinds of people composing South East Asian populations and of the culture phases prevailing there at the several times when these culture bearers parted company with Asia.

An appreciation of the ecology of the aborigines of Australia is really an assessment of the ways of life, activities and effectiveness of peoples of a series of palaeolithic hunting tribal communities spread over a whole continent. It is a review of the factors and problems that they face in making a living in the varied environments which they encounter between the hot rain-forests of the Australian sub-tropics and the cold and inhospitable lands in Tasmania and among the forests abutting on the once glaciated Kosciusko highlands of South Eastern Australia, from the wet lands of Babinda, with 100 or more inches (260 cm) of annual rain to the dry lands of uncertain rainfall, some with less than three inches (8 cm) per annum, such as are around the shores of Lake Eyre in the northern part of South Australia.

Granted the will to perpetuate their species and allowed such relative freedom from epidemic disease, as often comes from living in small isolated communities, the dominant factors enabling survival in the Australian environment revolved around the successful maintenance of four chief items — food, water, shelter, and territorial integrity.

From the myths and traditions of these hunting people we learn the importance to them of their "Ancestral Beings", often god-like, equally often animal-like Beings, who, for all that, were very human ancestors. They "created" (e.i. discovered) the country and caused its waters, foods, and resources to spring into being, where nothing was before. They demonstrated how, by rational seasonal exploitation of foods and waters, it was possible to have and to continue to enjoy these resources. These "Beings" did not fail to arrange protection from over-exploitation of resources by developing systems of taboo, enforcing restrictions of diet on certain social groups, and on certain age grades, thus enabling a continuous living to be wrested from a given territory. Such "creation myths" hint at a vast experience, in trial and error, leading to an empirical knowledge substantial enough to ensure survival. Thus each tribal territory was such an organic whole that it could be known, in the proper season of the year could be successfully traversed in one or many directions, and could be preserved from usurpation by others, either by diplomatic arrangement, or by power of weapons and fighting skill of the owners. By their very existence these inhabitants showed that they were the territory's owners and demonstrated, by their everyday mastery of the food resources of the locality that it offered a living, and thus that they had both "birth right" and "right over it by reason of knowledge". Little wonder that one of the tests of young manhood was the ability to hunt and in the desert another was to recite in unbroken succession the names of all the water supplies of which he was owner.

The hunting and living territories of primitive man at the level, even of the hunter, are remarkably fixed. In Australia well over 700 such tribal territories can be recognised. Many of these are shown on a map by TINDALE (1940) which is at present being revised and amplified in detail for a second edition. The bonds which tie each group of tribesmen to their own area of soil are many, and varied. Principally they fall into three broad categories; (a) Geographical limiting factors, (b) changes in plant assemblages, hence of food and its availability, and particularly, in desert environments, (c) availability of water, both the temporary waters such as allow the seasonal exploitation of remote bounds, and of the chains of more permanent pools, springs, soaks, and waterholes whose presence and patterns of distribution determine the possibility of survival both of a man's family group, and of the animals on whose security ultimately he depends for livelihood.

Within the fixed areas of tribal territory there are great differences between the completely and in a very real sense compulsorily nomadic hunters of the grass-covered plains and sandhills of the dry Interior of Australia, and the more sedentary, and often indeed only seasonally migratory family groups of more favorably situated tribespeople, blessed by rivers and by well-watered and vegetated coasts and bays, where often whim or mood, or chance social circumstances, such as the death of some member of a family, may well be the chief factor determining whether or not the family or some larger group should for the nonce inhabit this or that place within their territory.

Thus the factors governing the lives of these people are many and varied and the limiting one may be different in one place as compared with another and in one tribe as against another. The Narangga of Yorke Peninsula were compressed into that peninsula and maintained their territorial integrity against pressure from more fully nomadic and often ruthlessly marauding tribes-people from the north such as the Ngadjuri, and the Nukunu. Water was not a primary factor for survival even though their land had few springs except along the coasts. Determined on three sides by the sea their fourth boundary line was one preserved by force of prestige and recourse to arms. The Wailpi territory was determined by their ability to exploit the deep gorges and upland plateaux of the Northern Flinders Range and they were at home only in the Range. Their fears were of starving invaders from the west. While seemingly desolate the flat land running up to the western foot of the Ranges was exploited in normal seasons by the Kujani, using a way of living very foreign to that of the hillsdwellers. The Ngarkat north of the Tatiara country on the borders of South Australia and Victoria were at home only in territory covered by low-growing mallee (*Eucalyptus* spp) scrub, on whose waterbearing roots they were utterly dependent for liquid on an otherwise almost waterless karst plateau. They were denied the Murray River, whose ever flowing waters supplied others who never ventured further from its banks than they could carry skin bags full of water.

Long after the taller and stouter Southern Australoids (Murrayians, (BIRDSELL, 1949)) had spread over vast areas of Australia and apparently had absorbed or eliminated others of their kind from less inaccessible niches in the Australian environment, the people of the dozen or so small Barrinean negrito tribes of the Atherton Tableland in North Queensland maintained their tribal integrity by exploitation of the dark and wet rain jungles shunned by others. In the same way negritic Tasmanians, shut off by the rise of sea level at the end of the Pleistocene Ice Age, remained apart until the events of the 19th Century, after the advent of Western man, threw them into oblivion.

The Australia to which the first Pleistocene visitors were introduced must have been very different from that of today. The climate was far colder, even in the north. The dry and wet latitudinal belts between the tropics and the cold temperate zone must have been very differently placed, and the lands and forests were in an age-old state of balance with a Pleistocene fauna of large and small marsupials and giant flightless birds. It can be assessed that in process of developing a state of equilibrium with this Australian flora and fauna the newcomers must very quickly have eliminated many of the more vulnerable animals of the Pleistocene period and by firestick and digging stick must materially have altered the flora. In using the firestick, man played the same role as that long-time firemaker of nature, lightning. However, lightning often is but an accompaniment to torrential rains which frequently halt the harm it has initiated, before much damage is done. Man, setting fire to large areas of his territory at all times of the year convenient for his hunting, often causes destruction far beyond that done by nature. When Dr. CECIL HACKETT and I accompanied some Pitjandjara aborigines on part of their several months long normal trek around their country, (TINDALE, 1933), we more than once saw *Triodia* grassfires started for the hunting of lizards. On occasion such fires were soon out of control and raced away down-wind in a widening fan towards the horizon, involving many square miles of grass in devastation. Perhaps a half mile's walk over the newly burned sand dunes would yield enough half-cooked lizards, and other small game to satisfy immediate requirements. The rest would be wasted, save as food for the many birds which appeared from nowhere to feast on the food provided for them. Similar destructive activities were common to all of the grassland tribes. I have seen them in Arnhem Land, on Cape York, North Western Australia, and north of the Nullarbor Plain. From aerial photographs taken of the Canning Desert, it can be assessed that such fires often may have drastic effects in breaking down the symmetrical patterns of dune and swale when the protection afforded by vegetation is removed. A good example of fire effected dunes in the Canning Basin is figured by TRAVES and others (1956, p. 10). Thus man probably has had a significant hand in the moulding of the present configuration of parts of Australia. Indeed much of the grassland of Australia could have been brought into being as a result of his exploitation. Some of the post climax rain forests may have been destroyed in favour of invading sclerophyll, as the effects of his firestick were added to the effects of changing climate in Early Recent time.

HERBERT (1938) describes how the *Araucaria* pine-seed gathering aborigines altered the rain forests of parts of the Bunya Mountains into upland savannahs by their continual use of firesticks. Perhaps in the same way American Indian hunters created the so called "balds" of grassland I have seen on the otherwise densely wooded summits of the Great Smoky Mountains in Tennessee.

In the rain forests of the Atherton Plateau there are often to be met such enclaves of grassland as well as curious patches of wet sclerophyll forest. According to the views of local negrito aborigines, as expressed to me in 1938, such areas arise from their occasionally successful practice of setting fire to rain-forest patches during the dry spells which periodically occur and cause the usually wet forest floor to become a giant tinder box.

Since the burning of the rain forest is regarded as a useful hunting expedient, fires are likely to have been lit by many past generations of men, and the cumulative effects of the practice on the forest cover may have been very great. Perhaps it is correct to assume that man has had such a profound effect on the distributions of forest and grassland that true primaeval forest may be far less common in Australia than is generally realised, as indeed it is relatively rare in all lands where man has intruded for lengthy periods of time. Next to the firestick the womans digging stick was probably the most effective instrument in altering the patterns of plant growth, removing a considerable portion of the more edible forms of vegetable life.

The edge-ground stone axe, which was Australian man's most effective chopping implement was probably a relatively late cultural acquisition but was preceded in many places by the kodja chipped edge axe. With the stone axe the aborigines of the Northern Territory today have a marked effect on the patterns of distribution of forest trees since they are constantly felling hollow ones in search of honeycomb and native bees.

The picture of man's effect on the Australian environment, lightly sketched in the above paragraphs, is not seen by all. Some earlier writers on Australian ecology saw the Australian continent as existing in a zoological and botanical equilibrium in which climate, and not man was sole and final arbiter. It was possible thus even as late as 30 years ago to dismiss the Australian aboriginal as an ecological agent. As a "newcomer" in ineffectually small numbers (under 300,000) he had merely scratched a few holes, had destroyed a few plots of forest land with affects which could be dismissed as insignificant. CLELAND (1957) today still considers that, despite over 10,000 years of occupation, aboriginal man had not caused any profound alterations and that it was only the arrival of Europeans with sheep, cattle and the rabbit which caused major changes.

Some zoologists (for example GILL, 1955) think that the elimination of the *Diprotodon* and the other giant animals of Australia may have been an all out effect of a Mid-Recent period of stress or vast aridity, such as was enunciated by CROCKER & WOOD, (1947). These writers having described their "Great Arid" admit to the presence of niches where the flora of Australia survived, although by inference these areas are reduced to such relatively small dimensions that it was possible to suggest that many faunal elements of the Pleistocene were unable to exploit them and had to disappear. This does not explain why all of the animals did not suffer extinction and why for exemple the *Diprotodon* died out in New Guinea. Kangaroo, emu, wombat, the wallabies and even the two forms of blind marsupial mole of the remotest parts of sand deserts in the Canning Basin and around Horseshoe Bend and the Mann Ranges must have kept alive, throughout the supposed times of dire distress.

With one or other of the intrusions of man there came to Australia a second ecological agent, the dog. This animal may have played a role in the modification of the fauna of the Australian environment equal to that of man.

It is not yet proven whether the dingo, evidently a feral escape from man's domination, came to Australia with the first Pleistocene visitors or whether he was brought by later arrivals.

Reports of dingo teeth in the Wellington Caves, which however cannot yet be dated, had led earlier workers to believe that the dog might have had a "considerable fossil history" in Australia, whatever that might mean. Dingo remains may have been contemporary with a horizon at Lake Colongulac dated to the end of the Pleistocene (GILL, 1953). More recently TEDFORD (1955) in reporting on the Early Recent fauna found at Lake Menindee in association with Tartangan relics, one horizon of which is Carbon 14 dated to 6570 B.P., noted what he considered to be the surprising absence of any traces of the dingo and its coprolites. The Lake Menindee fauna included several of the species of animals which were once thought to have died out at the end of the Pleistocene. While further observations must be made, pending which the date of the introduction of the dingo will remain an open question, it may be of interest to consider the following facts, and read their possible implications.

Relics of both Kartan and the Tartangan types of implements have been found in Tasmania and because of this and other lines of evidence the Tasmanians are considered to be survivors of the earliest comers to Australia. They were seemingly without dogs, for these animals did not at any time, so far as we are aware, reach the island of Tasmania, where *Thylacinus* and *Sarcophilus* remained the only large predators. Tasmania evidently was part of the Australian mainland in Late Pleistocene times and remained so until cut off by the rise of sealevel at the end of that Period.

The dingo did not exist on Kangaroo Island when it was first visited in modern times, although stone relics of man of the Kartan culture phase are very common there. This island also was one isolated by rise of sealevel at the end of the Pleistocene. Man himself may have disappeared from Kangaroo Island about 5,000 B.P. (TINDALE, 1957a). If the Kartan folk had a dog with them it must have been eliminated or died out along with them. The kangaroo, a form of emu, and the Kangaroo Island wallaby survived on the island so the dingo if once present, did not perish for lack of food. It may be very reasonable to assume the dingo did not accompany the Kartan and Tartangan Culture holders and if, as TEDFORD deduced, he was not present in the period around 6570 B.P. then the dog may have reached Australia only a short time before or at about the time of the Mid-Recent. TINDALE (1957a) has assumed that differences in faunal lists between the Tartangan of Lake Menindee of about 6570 B.P. and of Tartanga Island at about 6020 B.P. imply that the final decline and disappearance of that part of the so-called Pleistocene mammal assemblage which had survived earlier extinction, may have been taking place in the period around 6,500 B.P. This date could fit in rather well with the very beginning of the Mid-Recent period which culminated in the high sea levels (10ft Terrace) of around 5,000 B.P. The successively higher strata at Tartanga Island in the years around 6,000 B.P. probably record some stages of this rise (HALE & TINDALE, 1930, and TINDALE, 1957a). Leaving aside for the moment the discussion of the pros and cons of the validity of a supposed period of stress and aridity which is postulated to have accompanied this MidRecent event, is it not possible that the introduction of the dog by Australoid man about that time might have had some very material effect on the Pleistocene mammal assemblages. Its rapid spread as a feral animal, may even have been well ahead of the Southward migration of Australoid man himself, whose movements may have been far slower, since he had to contend with and eliminate or absorb earlier Tartangan populations before he could himself spread to the south. The reasonably well documented arrival of Pirrian culture users in the Murray Valley (perhaps around 5,000 B.P. or at some time between 6,020 B.P. and 4,250 B.P.) could have been anticipated in the previous millenium by the appearance in the south of feral representatives of his dog. Dingo remains were not found in the Tartanga Island beds. At Devon Downs Shelter the first indisputable remains of dingo are from Layer IV (Early Murundian). In earlier beds *Sarcophilus* remains were prevalent. Only in the lowest bed of the shelter was a fragment of a two rooted tooth found which was identified as "*Sarcophilus* (or? *Canis*)." No closer identification has yet been possible.

From the trend of the above evidence we may suggest that in Early Recent time Kartan, and after them Tartangan men, in the absence of the dingo, may have come to be in partial equilibrium with the Pleistocene fauna, without eliminating all the giant species.

By their presence, however, they had brought about such an uneasy balance that the arrival of the dingo quickly may have tilted the scales towards destruction of the whole of the remainder of the Pleistocene assemblage. It may be that negrito man also may have suffered a severe population crash at this time as a result of the changes in his available food supply brought about by the new conditions, so that the rapid spread of the Australoids and their Pirrian Culture in the following two millenia was assured.

It will be recalled that in the past century it was the feral cat and the rabbit, reaching out into the Desert far ahead of Western man himself which may have to be blamed for the disappearance of part of the small Australian mammal fauna (native cat, banded-anteater etc.) and even of some of the vegetational cover of quite large areas of Australia (the mulga, *Erythrina* tree etc.) and certainly foreshadowed for Australoid man himself, the end of a regime. Western man indeed seems to have brought about the destruction of the whole economy of the Australoid peoples of the Deserts without the raising of a single rifle against them, and this sometimes even before men of the two races had met.

It is true that some peoples such as the Pitjandjara, Pintubi, Ngadadjara and others of the Desert tribes have for a time held out against this disruption, living on the meat of cat and rabbit for half a century, as a substitute for their older mammals, until the coming of sheep, cattle, horses and Western man himself finally has broken their hold on their old hunting economy.

As has been indicated in the above discussion evidence for the association of man with the earlier fauna of Australia and with the events of the past is largely archaeological but there are some fragmentary recollections in the form of myths and legends which possibly suggest a surviving acquaintance of present day aborigines with the giant animals and birds of the past and of possible remembrance of events of up to 5,000 years ago. Rock carvings also have been noted which seem to register an acquaintance of men with the giant emu (HALL, 1951).

CAMERON (1885, p. 369) describes how in Wiradjuri belief, two bukumari (bookoomuri) who were supposed to be men of an earlier people than the present day aborigines, gave chase to a gigantic kangaroo, which once lived at Hilston on the Lachlan River. Could this be based on folk remembrance of the former existence of such a kangaroo?

In present day aboriginal belief Mt. Gambier in South Australia was the campfire of the Eagle men, who lived on the peak above the present Browne Crater Lake, while the Blue Lake, also a collapsed crater, was the campfire of the Crow Man. Association of these extinct volcanic craters with ancestral camp fires suggests the possibility of traditional knowledge of the former behaviour of these now quiescent craters, while the persistence there of the Eagle and Crow Myth, which is so intimately associated with the Southern Australoid (Murrayian) peoples and with the terminology of their dual social organization in widely sundered parts of Australia, possibly could be taken as a pointer to the type of people who might have been present about the year 4,710 B.P. when the Mt. Gambier eruptions first occurred. Geologists support the idea that Mt. Gambier was formed by a short-lived volcano, so that if these traditions are significant we are likely to be dealing with native beliefs which have been transmitted for at least four millenia.

People possessing the Pirrian Culture must have begun to establish themselves in Australia well before Mid-Recent time although as already indicated they seemingly did not arrive at Devon Downs in the Murray Valley until some time after 6,020 B.P. when Tartangan people were still present. Mid-point of the Pirrian horizon in Devon Downs Rock Shelter is dated to 4,250 B.P.

Pirrian implements probably were the earliest types to be deposited on the ash cone of Mt. Gambier after its eruptions had ceased, and the Pirrian people may well have been there when the eruptions commenced. Impressions of bracken fern *(Pteridium)*, *Banksia*, *Casuarina*, and *Eucalyptus* in the tuff of Mt. Gambier, all of forms still living in the vicinity, suggest that when the eruptions were occurring, near the middle of the Mid-Recent, the climate and vegetational cover around Mt. Gambier can have been but little different from that occurring there today, since bracken requires an annual rainfall of approximately 20—40 inches and does not seem to be a marker either for very hot or for very arid environments. Exponents of the "Great Arid" hypothesis may find these fossils very difficult to explain.

The principal implement types characteristic of the Pirrian Culture are found archaeologically over much of Australia from Arnhem Land to the southernmost point of the South Australian coast and from Goondiwindi in Southern Queensland to Perth and Dampier Peninsula in Western Australia. TINDALE (1957a) has suggested that elements of the Pirrian Culture may survive today in the Upper Murchison area of Western Australia and has expressed an opinion that the pressure-flaked implements of North Western Australia may represent an evolved relict phase of this culture. McCARTHY (1957) has more recently named this the Kimberleyan Culture. Pirrian relics are known from at least 250 sites spread over this vast area of Central and Southern Australia. Even if it is contended that the drift of the culture was from east to west and the camps in the Western Desert may be far later in time than those along the Murray River, it seems likely that the Pirrian folk were not unduly hampered in their spread into Western Australia by any delayed effects of the supposed "Great Arid". In the area south of the Musgrave Ranges the first Pirrian site that happened to be found, in 1933, was on old dunes surrounding an ancient and now dried up lake. At Lake Menindee (TINDALE, 1955) the conditions during Pirrian times must have been no more arid than in the year 1952 for conditions then permitted the filling of Lake Menindee to at least the level achieved in a modern optimum flood year.

Such evidence is of course far from conclusive since man can and does retreat from arid areas and can expand into them again quickly when conditions ameliorate. However, neither Pirrian man nor the Recent mammals are likely to have done so at all readily if some cataclysmic episode had so far removed the whole of the vegetation from vast areas, as envisaged by CROCKER & WOOD, that its restoration was a matter of millenia. They could not have awaited the slow processes by which the supposedly vast wastes of the Interior were subsequently recolonised from the peripheries of Australia, for perhaps even while the Pirrian people themselves were moving west they were in Eastern Australia being modified by accessions of new ideas and culture elements such as are indicated by the implements of the Mudukian phase. The Mudukian culture bearers in their turn were able also to spread across the same Western Desert as far as South Western Australia, at a time which must have been after 4,000 B.P. leaving camp sites distributed in such widely sundered places as also to deny that the possessors were unable to traverse the deserts owing to the effects of any drastic phase of climatic deterioration.

If the presence of the bondi point in the Mudukian culture kit is an indication that they may have sewn skins together for clothing or rug making (as inferred by TINDALE, 1957a) it may be contended that their passage or residence in the Western Desert could have been even easier than it would be today. The present day adze-stone using Murundian culture bearers go naked, seemingly owing to the paramount necessity of reducing their possessions to the absolute minimum for the sake of portability so that spears and a spear-thrower with a few wooden scoops, dishes and digging sticks for their wives, are almost as much as they own. Today the opossums and small rock wallabies which might have supplied skins for Mudukian peoples' rugs are limited to a very few favorable niches in the desert, such as around Lightning Rocks for rock wallabies and Mt. Mann for opossums, so that any further degree of aridity or poverty of vegetation over and above that now apparent would perhaps have denied them any chance at all of life in their arid environment. Certainly the present sparse fauna would have scarcely supported a skin rug using people if they were present in any great numbers.

The presence of Pirrian implements on the sand dunes surrounding the shore line of Lake Eyre, as observed by this writer during a recent expedition there, yields the indication that the present shoreline of this usually dry salt lake may have been the highest to have occurred since at least Mid-Recent times. The very consistent presence of these implements, concentrated on the shore dunes of Lake Eyre North seems to imply that, in a period which could have been as early as 4,500 years ago, climate conditions probably were better and certainly could not have been worse than now. It would seem to us that today the lake shore line would be utterly unusable. We know from 19th century writers that the Tirari aborigines were living in the area when it was first visited. Such occupation as they had was very sparse and if the apparent absence of present day camp-sites is any indication, their living places were not focussed on the lake shore itself. All these lines of evidence seem to point to the absence of special periods of stress in the times from the Mid-Recent onwards and may on the contrary suggest that at times conditions may have been easier than they are at present.

The types of hammer and milling stones present in the various culture levels of the Australians may furnish some indications of the hypothetical changing roles as between open grasslands, induced by mans activities with the firestick, and the climax vegetational communities which in theory may have preceded them.

Kartan man possessed hammer-stones. These usually show the marks of a dual function. One was as a passive nether stone with centrally disposed "bruise" scar arising from crushing blows applied to nuts, shells and other objects broken by another often similar sized and shaped and marginally battered hammer-stone, held in the fist when in use. The type is general and persists in use today so that we have first hand knowledge of the mode of use.

In the Tartangan phase appeared a lighter oblong hammerstone with terminally focussed batterings suggesting gum-hafted hammers (such as appear in the modern Western Australian hafted Kodja axe-hammer, which is a composite tool) as well as larger nether stones like Kartan ones, but often showing minor abrasional surfaces, evidently caused by rubbing action.

From the Pirrian onwards several styles of upper and nether millstones, including ones such as were used in the wet grinding of grass seed food, appear, and culminate in the grass-seed-meal preparing mill sets characteristic of most areas of present day grassland Australia. From this sequence it may be deduced that Late Pleistocene man in Australia tended to be a nut and seed gatherer and bruiser and that he had not yet become dependent upon the grain milling and grass seed bread-making techniques which seem to be characteristic chiefly of peoples who possessed the Pirrian Culture and those having the succeeding modifications of it which we have called the Mudukian and Murundian Culture phases. Among the present day Murundian peoples of the 5—25 inch rainfall areas in Australia where grasslands are most prevalent today, the grass seed milling practices have come to play almost a primary role in aboriginal food economy.

Was the rise of the grassland food economy after Mid-Recent times an expression of the increasing dominant role of man in modifying, with his firestick, the Pleistocene and Early Recent forest cover of Australia, or must we ascribe it to the changing role of climate in destroying the earlier forest vegetation?

Much of the evidence in the above discussion can be held to show that man played some part in the change, although the alternations from the cold of the Pleistocene Ice Age to the warm temperate climate of the Early Recent and the further slight warming up of the earth in the Mid-Recent period around 5,000 B.P. must have played leading roles. The effects of climate in Mid-Recent time probably has been much overrated and it may not be easy in future to maintain that the supposed "Great Arid" of Mid-Recent times ever was more than a hypothesis. This author visualises the northward and southward drift of the arid belts as suggested by GRIFFITH TAYLOR (1918 and later writings) as playing a far more significant role in the climatic history of the past 10,000 years than the "arid versus pluvial" ideas of other writers. That these latitudinal belts are permanent features of the earths meteorology and climate and have always remained interposed between the tropics and cool temperate zones is supported by a vast array of zoological evidence, not the least of which is the distribution of the archaic genera of the moisture requiring Hepialid moths which have been the subject of study by the present writer (TINDALE, 1945 etc.).

Summing up this chapter it is clear that much thinking about the ecology of the aborigines has yet to be done. Effort hitherto has been concentrated on gathering primary data. CLELAND (1940 & 1957) has summarised much of what is known of the basic information. The bulk of it indeed is a result of his own work, in association with that of several colleagues. The interested reader cannot do better than refer for details to his papers and to the bibliographies given therein.

In the present chapter the time factor has been stressed since this is one which has only recently been brought to bear on the subject. Apart from the urgent necessity of recording all possible data available from the aborigines before they become extinct, a primary need is for a better understanding of the role of changing climate, discussions on which at present tend to require undue attention owing to the widely different view points which are still possible in the interpretation of the effects of Late Pleistocene and Recent climates on man, animals and plants.

REFERENCES

BIRDSELL, J. B., 1949. Racial origin of the extinct Tasmanians. *Rec. Queen Victoria Mus.* **2**, *105—122*.
CAMERON, A. L. P., 1885. Notes on some tribes of New South Wales. *J. Anthropol. Inst.* **14**, *344—370*.
CLELAND, J. B., 1940. Some aspects of the ecology of the aboriginal inhabitants of Tasmania and Southern Australia. *Pap. & Proc. Roy. Soc. Tasmania.* **1939**, *1—18*.
CLELAND, J. B., 1957. Our natives and the vegetation of Southern Australia. *Mankind* **5**, *149—162*.
CLELAND, J. B. & TINDALE, N. B., Native ecology of the Haast Bluff area, Central Australia. In preparation, 1958.
CROCKER, R. L. & WOOD, J. G., 1947. Some historical influences on the development of the South Australian vegetation. *Trans. Roy. Soc. S. Aust.* **71**, *91—136*.
GILL, E. D., 1953. Geological evidence in Western Victoria relative to the antiquity of the Australian aborigines. *Mem. Nat. Mus.* **18**, *27—92*.
GILL, E. D., 1955. Radio carbon dates for Australian archaeological and geological samples. *Aust. J. Sci.* **18**, *49—52*.
GILL, E. D., 1955. Problem of extinction with special reference to Australian marsupials. *Evolution* **9**, *87—92*.
HALE, H. M. & TINDALE, N. B., 1930. Notes on some human remains in the Lower Murray Valley, South Australia. *Rec. S. Aust. Mus.* **4**, *145—218*.
HALL, F. J., 1951. Aboriginal rock carvings: a locality near Pimba, S.A. *Rec. S. Aust. Mus.* **9**, *375—382*.
HARRISSON, T., 1957. Giant cave of Niah. *Man* **57**, *161—166*.
HERBERT, D. A., 1938. Upland savannahs of the Bunya Mountains, South Queensland. *Proc. Roy. Soc. Queensland.* **49**, *145—149*.
MCCARTHY, F. D., 1957. Distributional notes on Northern Australian point industries. *Mankind* **5**, *163—168*.
TAYLOR, GRIFFITH, 1918. The Australian environment. Melbourne.
TEDFORD, R. H., 1955. Report on the extinct mammalian remains at Lake Menindee, New South Wales. *Rec. S. Aust. Mus.* **11**, *298—305*.
TINDALE, N. B., 1925—1928. Natives of Groote Eylandt. *Rec. S. Aust. Mus.* **3**, *61—134*.
TINDALE, N. B., 1933. Preliminary report on field work among the aborigines of the N.W. of South Australia. *Oceania* **4**, *101—105*.
TINDALE, N. B., 1937. Relationship of the extinct Kangaroo Island Culture and cultures of Australia, Tasmania and Malaya. *Rec. S. Aust. Mus.* **6**, *39—60*.
TINDALE, N. B., 1940. Distribution of Australian tribes: a field survey. *Trans. Roy. Soc. S. Aust.* **64**, *140—231*.
TINDALE, N. B. & BIRDSELL, J. B., 1941. Tasmanoid tribes in North Queensland. *Rec. S. Aust. Mus.* **7**, *1—9*.
TINDALE, N. B., 1945. *Proc. ent. Soc. Wash.*
TINDALE, N. B., 1955. Archaeological site at Lake Menindee, New South Wales. *Rec. S. Aust. Mus.* **11**, *269—298*.
TINDALE, N. B., 1956. Peopling of South-Eastern Australia. *Aust. Mus. Mag.* **12**, *115—120*.
TINDALE, N. B., 1957a. Culture succession in South Eastern Australia. *Rec. S. Aust. Mus.* **13**, *1—49*.
TINDALE, N. B., 1957b. A dated Tartangan implement site at Cape Martin, South East of South Australia. *Trans. Roy. Soc. S. Aust.* **80**, *109—123*.
TRAVES, D. M., 1956. Geology of the South-Western Canning Basin, Western Australia. *Rept. Commonwealth Dept. of Nat. Development.* **29**, *1—74*.

IV
HUMAN ECOLOGY IN AUSTRALIA

by

Griffith Taylor
(Fellow of the Australian Academy of Science)

Environmental Controls

It is somewhat of a paradox in the field of science that if one studies the effect of environment upon bacteria, beetles or baboons it is accepted as a valuable contribution to science; if however the research is directed towards the effect of the major fields of the environment — such as topography, climate, soil, etc. — upon that fairly important biped called "Man", it is called geography. This latter branch of knowledge is surely an important aspect of Ecology, but I have spent much of my time at such Universities as Sydney and Toronto in the attempt to prove that Geography should be classed with the Sciences rather than with the Humanities.

A very large majority of professional geographers has been trained in the regions of Western Europe or Eastern North America. There, the popular way of looking at geography is based largely on the teaching of leaders (like Vidal de la Blache) who encouraged young geographers to believe that there are many possible ways of making the best of one's surroundings, and it is man's function to choose one of these. This philosophy of "Possibilism" has led to the other point of view (i.e. "environmentalism") being depreciated; so much so that one well known American geographer has declared that it "needs extermination as an obstacle to better understanding".

My own point of view, as is usual, depends mainly upon the surroundings during my academic career. Most of my life has been spent in the rather dry lands of Australia, or in the ice-covered lands of Antarctica or lastly in the cold lands of Canada. In these three large portions of the land surface there is no question that environmental control is paramount as a factor in determining how man shall live and prosper therein. It is this ecological problem, as experienced in Australia, which I propose to discuss in the following chapter.

Few areas in the world can be better suited for the study of man's relation to his environment than Australia. No settlement occurred until 1788, because most of the coastlands discovered before that date were arid or tropical. Yet the "First Fleet" under Captain Phillip in 1788 found settlement at Sydney not at all easy. This was mainly because the latitude was quite different from that of the homeland. Britain is a cool rainy region with fairly fertile soils in its areas of large population. Sydney is almost sub-tropical with a rather erratic rainfall and with a peculiarly sterile sandstone soil. Hence the first attempts to grow crops were quite unsuccessful until the shale soils near Parramatta, and the river silts near Richmond were discovered. Even today the two million folk living near Sydney depend for their main food-supply on wheat, beef and mutton grown on the more fertile plains a hundred or more miles away.

The most characteristic document in connection with the Human Ecology of a land is the map showing the distribution of population. The continent was discovered by Captain Jansz in 1606. His ship sailed along the west coast of Cape York Peninsula, and by the end of the century the coasts from Cape York westward to Cape Leeuwin and along the south coast across the Great Australian Bight had been charted. In all this survey — some two-thirds of the coasts — only in the southwest corner near Cape Leeuwin had wellwatered temperate lands been met with. A glance at Fig. 2 will show that there is no close settlement — even amounting to one-eighth of person per square mile — in all the lands discovered by the Dutch during the 17th century, except in the Perth area. It remained for Captain Cook in 1770 to prove that the unknown south east third of the continent was best suited for close white settlement.

Consequent on the loss of the American colonies, in 1788 the new convict settlements were founded in the coast near Sydney, as recommended by Sir Joseph Banks, who had accompanied Cook. A very instructive map (due to S. H. ROBERTS) shows how the settlement progressed during the 19th century (Fig. 1). For quarter of a century — until 1825 — it was confined to the immediate vicinity of the convict establishment at Sydney. The small area of shales and silts near the city was occupied with farms, but the great belt of sterile sandstones held back expansion most of this period. Then routes to the west reached the Bathurst Plains, while an easier journey opened up the plains to the south near Goulburn and Canberra.

By 1840, as the map shows, independent centres of expansion were settled at Hobart, Adelaide, Perth and Melbourne. In the next 30 years the pastoralists — who were often the actual explorers — had occupied almost all the important pastoral country in the continent. Except for some of the cattle lands in Northern Territory, which were taken up by 1890, all the arid or tropical land within the "1870 line" is still almost devoid of settlement; as can be judged by reference to Fig. 2. It is the purpose of this memoir to explain the natural controls which have led to the definite pattern shown in Fig. 2.

Now that we have the problem stated we may follow the usual plan of the geographer, and discuss the effect of the major environmental controls — such as topography, soils, geology, temperature rainfall etc. — upon the progress of settlement in, Australia.

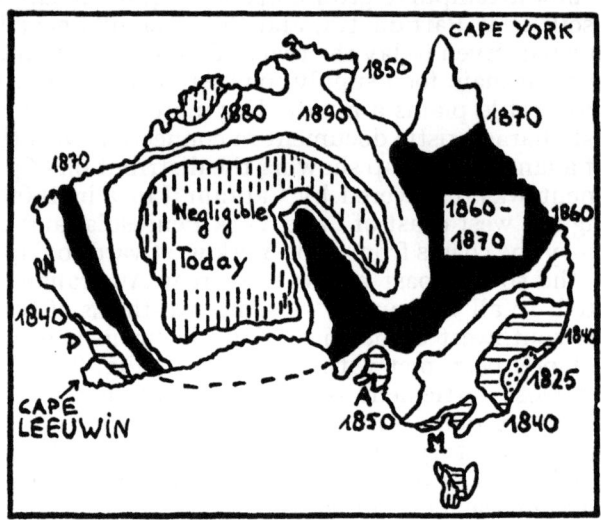

Fig. 1.

Topographic and Geologic Controls

The topography of the continent is probably the simplest of all seven of the major land areas. Like Africa Australia is rather remote from the major mountain-building belts of the world. Unlike Africa the southern continent has experienced few elevations en masse on any such scale as has lifted Southern Africa high above sea level. However Northern Africa in its general environment and latitude is nearer to Australia than is any other large land surface. Only eight areas in Australia are over 2000 feet in elevation (Fig. 3). Four of these plateaux are to be found in the southeast highlands, and being in the temperate zone, their effect is to lower the temperature below what is perhaps the most attractive temperatures, i.e. around 65° F. Three such plateaux occur along the Tropic of Capricorn, but they are in the arid portion of Australia, and so exercise no appreciable effect on settlement. Where a higher elevation would be beneficial, i.e. in the hot wet tropics, we find only one plateau in the northwest corner, known as the Atherton Plateau. Here only is there much improvement as regards close settlement, due primarily to the elevation. The plateaux near Kosciusko and Katoomba have chiefly benefitted tourism. The last elevation of note is in west Tasmania, in so high a latitude that the area is too cold and bleak for settlement. Nature has not been kind to Australia as regards the distribution of highlands.

Australia is one of the continents fringing the Pacific Ocean, and like the other four it conforms to the general plan, and exhibits moderately elevated regions on the east bordering that ocean. But the highest point is far lower than the peaks of the Rockies or Andes, for Kosciusko in the southeast corner is only 7316 feet above sea level. To the west of this moderate belt of highlands is a wide downfold or geo-syncline which includes the enormous Artesian Basin of Queensland; and the chief river basin — that of the Murray — to the south of the Artesian Basin. In agreement with the general continental plan we find a stable portion of the earth's crust — known as a Shield — which builds up much of the Northern Territory and Western Australia. But this latter element of the build of the continent has an average height of little over 1000 feet; and is of small importance in general, since most of it is arid land.

Before turning to the climatic controls, which are the chief determinants of our population, a few words may be given to one aspect linked with topography — that of mining. There is a common phrase that "gold is where you find it"; but as usual it is possible to show that the distribution of gold and other metals is not haphazard in Australia. Of the three structural divisions shown on Fig. 3 metals are absent in the geosyncline, for here the formations

 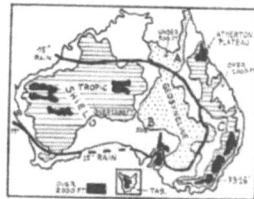

Fig. 2.   Fig. 3.

are all later than Paleozoic, and so have not had time for ore-bodies to develop. The rocks of the Shield are very ancient, and ores are found to be scattered throughout, though the goldfields of Kalgoorlie are outstanding. The continent has been relatively free from earth-crumpling and from late igneous action; but in the core-rocks of the Eastern Highlands many of the older rocks have been exposed at the surface. Here many metal mines such as lead and zinc at Broken Hill and at Mt. Isa (Q.), and in the Victorian Goldfields widespread ores are being worked, and are determining the site of large towns.

Coal is however usually the major factor among minerals as regards population-distribution. Since it is derived from land-plants, and since these did not develop over the earth's surface before Carboniferous times, it is useless to expect coal in the ancient rocks of the Shield. Indeed only in later basins in geological formations were the conditions favorable for the production of coal. These occur mainly along the east coast: — as in the Permian basin around Sydney, the smaller basins of Dawson and Clermont in Queensland, the Triassic basin near Brisbane, and the remarkable seams of Middle Tertiary coal at Morwell in Victoria. Petroleum, which has so altered the environment in many parts of the world, has not so far been discovered in workable amounts in Australia, though many millions have been spent in widespread surveys.

The most valuable "mineral" of all is of course Water. It is rather surprising that the driest of all the continents should have the largest supplies of Artesian Water in the world. The main basin occupies the northern three-quarters of the Geosyncline. It is a triangular area (ABC in Fig. 3) about 1200 miles long and 900 miles wide; and within this area water can be found at various levels as deep as 7000 feet in the centre of the basin. But although it is stated that the carrying capacity for sheep, etc. is doubled as a result of the bores, it is clear from Fig. 2 that this useful supply has not added much to the total human population of Australia. There are a number of other artesian basins besides the one just described, but they are small and are only of local importance. Nowhere have they led to close settlement. About 56,000 folk are directly engaged in the mining industry, which is not a large proportion of the 3,000,000 Australian bread-winners.

Climatic Controls

With its low elevation and smooth bean-shaped outline Australia offers a very simple area in which to study climatic controls. To students in the northern hemisphere it is very helpful to point out its close resemblance in many ways to the conditions in North Africa. Here the belts of rainfall and natural vegetation are more familiar than are those of Australia. The parallels are shown in Fig. 4. Here Australia is supposed to swung into the northern Hemisphere, using the Equator as a hinge. The position of the arid centre of the southern continent agrees with that of the Sahara in the same latitudes; and their origins are much the same. So also the comparison of Nigeria with Northern Territory in Australia is striking. Of course owing to the smaller width of Australia the satisfactory east coast is to be compared with the east coast of the Old World — in far away China; and not with the arid wastes of Libya and the Sudan. The attractive southern coasts of Australia are therefore Homoclimes (allied climates) of the Mediterranean coasts, while the Sydney region is to be compared with southeast China or the Carolinas in the United States.

As regards temperatures we must remember that the heat Equator passes through northern Australia, and Wyndham on the north-west coast is one of the few places in the world with an average annual temperature of about 85° Fahrenheit. The northeast coast owing to the steady southeast Trade winds, off the ocean, is much

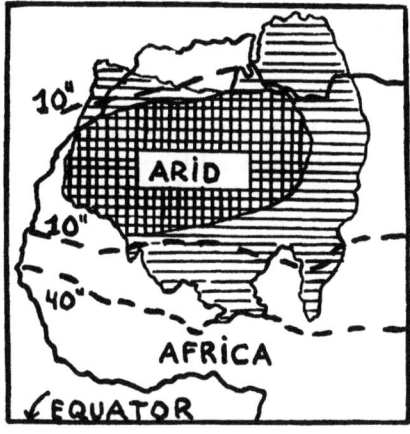

Fig. 4.

cooler than the northwest coast, where the similar east winds are to a large extent heated by the hot continent from which they are blowing. (Fig. 5 at C).

The Tropic of Capricorn runs across almost the middle of the continent, so that some two-fifths of Australia is tropical. Hence the first settlers in much of Australia were meeting environments quite different from anything they had experienced in Britain. Only in Tasmania was the climate cool and damp something like that of the British homelands.

However it is important to realise that there are two kinds of tropical climate. Near the equator the conditions are hot throughout the year and with a fairly uniform rainfall. Hence the humidity is very great; making conditions very favourable for crops, but producing a lack of energy and a tendency to various diseases which are unknown in temperate lands. There is however very little of this type of climate in Australia; and it occurs only in the wet coastlands, roughly between Townsville and Darwin. The other type of tropical climate (Fig. 5 at B) has a very marked seasonal type of rainfall; and in most of the year the hot weather is dry except for a few months only. This is characteristic of most of tropical Australia, and though it limits the possibilities of agriculture very greatly, it is in general healthy and not uncomfortable for the pastoralist, who is the usual dweller in such lands.

Australia's greatest handicap is due to its latitude, in that it lies mainly in the High Pressure belt, while the Trade Winds blow over the northern half of the continent most of the year. Only on the south are the chief weather-bringers the belt of Antarctic Lows (alternating with strong westerly winds) which give a good deal of rain in winter months.

I have heard it stated that if our chief mountains were in the centre of the continent instead of in the southeast corner, it would vastly improve our natural endowment. But in the Chili-Argentine region — in the same latitude — we find very arid lands in spite of the presence of the high Andes. If we could only push the continent ten degrees farther south, it would be in a similar latitude to the best endowed country of all i.e. the United States. Unfortunately in the Southern Hemisphere there is no large land area in this latitude.

The rainfall factor is of course the critical one in our environment. But it is not simply the question of total amounts of rainfall, for the largest areas receiving over 30 inches of rain in a year are found along the north coast (Fig. 5 at A); and in general these, as shown in Fig. 2, are almost devoid of important settlement. We can appreciate how the character of the rainfall varies if we consider the conditions at four stations, all receiving the same rainfall of 15 inches a year. Roeburne is on the west coast, but is marked by the most unreliable rainfall in the continent (50% variation from normal); in 1891 it only received 0.13 inches, while in 1900 there fell 42 inches. With its low average rainfall it is not much use even for stock. Tennant Creek in Northern Territory is chiefly characterised by a totally dry period of seven months, which extends from April to October; however there are a few cattle ranches in the district. Cobar in the centre of New South Wales has its rainfall spread out uniformly, and is just on the arid border of wheat farming. Northam not far from Perth (W.A.) has a very regular rainfall, which falls in the wheat-growing period; and so it is in the middle of the wheat belt of the western state.

Reference to the small maps given at the top of Fig. 5 will make these differences in the character of the rainfall belt (between the 10 and 20 inch isohyets) quite clear. Indeed this chart showing the season of rain (Fig. 5 at B) is one of the most interesting of all, since it will be seen that the closely dotted areas representing uniform rainfall agree more closely with the population map (Fig. 2) than do any others in the climatic field. We may indeed state from this similarity in isopleths, that it is the season of the rain which is the chief determinant of the ecology of man in Australia.

We learn therefore from the history of settlement (Fig. 1) that 15 inches of rain in the southwest are of much greater value than 50 inches of rain in the Darwin region. I have in many publications shown a reliability map of Australia, but space-limits forbid its inclusion here. In summary one may say that a belt of very unreliable rain runs from the south of the Gulf of Carpentaria through

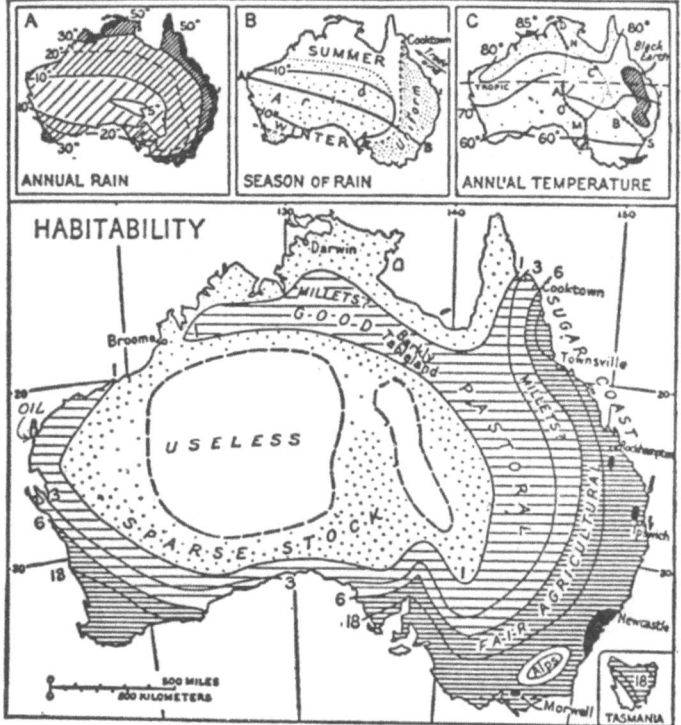

Fig. 5.

Alice Springs and westward to Northwest Cape. The south coast from Perth to Sydney has the most reliable rains, varying from 10% to 20% variation from the normal. It is a good example of the statement "from them that have little, much shall be taken"; for it is precisely the most arid areas which are the most unreliable.

Evaporation is a very potent factor, especially when we come to consider the possibility of Irrigation in arid Australia. Its isopleths depend mainly on two factors: high temperatures and distance from the sea. Hence the resulting map is much like that of "Reliability of Rainfall", with a similar unfavorable major axis from near Cloncurry (Q.) through Alice Springs to North West Cape. In central Australia the evaporation in a year is over 100 inches, while the rainfall is only 10 inches; so that the amount available for plant growth (or irrigation) is very small.

Soil and Vegetation Conditions

Since it now understood that the major determinant in producing a soil-type is the rainfall — with the sort of surface-rock coming second — we are prepared to find that half of the continent's area contains desert-steppe soils, or dunes. There is very little of the moving erg soils of the Sahara, though "fixed dunes" are widespread in the main western desert and in the Arunta desert north of Lake Eyre. The serir type (a layer of desert-pebbles) covers large areas also, as in the arid region south and east of Lake Eyre. Hamada or patches of bare rock etc. are found in a west-east belt south of the Tropic. None of these three types of landscape is of any real economic importance, though cattle in small numbers may graze in the swampy hollows.

PRESCOTT shows that a crescent of podsols results from the rainfall over 30 inches or so, all round the coastlands from the Kimberleys (W.A.) to Victoria. Within this wetter belt is another crescent, where the rainfall lies between 15 and 30 inches per annum, which includes the best soils in the continent. Between latitudes 20° and 32° in the east of this belt is an area of chernozem soils; while to the north the belt is mainly grey-earths, and to the south chestnut soils. Unfortunately the wide geo-syncline (Fig. 3) — where one would expect the best soils to have collected, — is mostly so dry that it is not rich in plant food. Throughout Australia it is stated that the soils are unusually deficient in phosphorus. This must be remedied by large imports of guano and superphosphate. Of late years the value of trace-elements — such as copper, zinc, and molybdenum — in improving the native pasture has been widely demonstrated, especially in the saltbush area east of the Flinders Range (S.A.).

To the student of Human Ecology in a new country, the Natural vegetation is the best clue to occupation by man and beast, and this for two reasons. It gives a pretty clear indication of where crops and fodder may be grown, and secondly its major features are easily detected — without instruments — by natives and settlers alike. The map given in Fig. 6 shows the main features of the Natural

Fig. 6.

Vegetation, though it is somewhat simplified. Roughly speaking, there are three main belts, which of course agree pretty closely with the rainfall. The outer belt is labelled "Forest"; and it consists of rather open eucalypt forest with small trees in the northern region of winter drought, of small patches of true Malayan Rain Forest where the rainfall is over some 50 inches, and of much larger eucalypts in the region of good uniform rains (Fig. 5 at B). Perhaps the finest forests of all are those of Jarrah and Karri (both eucalypts) in the corner of West Australia which is sometimes called "Swanland". Along the south coast the eucalypt belt is also present, but in the drier areas it consists of a dwarf close-set forest of "Mallee".

Within this eucalypt belt is another of drier lands which are covered with grasses and acacias. There is also a very valuable fodder plant called "Saltbush" *(Atriplex)* with dense grey leaves which are much relished by stock. Unfortunately when it is eaten off during droughts it takes a long time to recover. In the northeast there is the valuable Mitchell Grass *(Astrebla)*, probably the best of the native grasses. In the dry west the grass is less abundant and the acacias are more abundant, usually of the species *Acacia aneura* or Mulga. This acacia-grassland on its drier side is of course of no value for crops, but is excellent for cattle and sheep, provided it is not overstocked. I have given it the name of Sparselands. (Fig. 2). By foreign geographers it is often called Desert, and it lies on the dry side of the ten inch isohyet in the south, and south of the twenty inch isohyet in the north. It is a surprise to most Australians to be told that this area of Sparselands (and Desert) which comprises half of Australia's area only feeds 3% of her cattle and sheep. The human population is negligible, being about one-third of one per cent of the nine million Australians.

The Desert area is approximately charted in Fig. 6. It will be seen that some authorities link the smaller eastern desert ("Arunta") to the large Western Desert by an arid region near Tennant Creek (N.T.). There are a few ranches hereabouts, and I have usually added this "link" to the Sparselands area.

In two quite small areas the climate — owing to its elevation — is cold, wet and bleak. One of these surrounds Mount Kosciusko, and the other is larger and includes the rugged mountain plateau in the west of Tasmania. They offer a little summer grazing, and are attractive to tourists, but are becoming of more importance as hydro-electric power is developed in both areas.

The Human Response

The spread of settlement throughout Australia may be considered in five stages, each characterised by rather special influences. Thus from 1788 to 1820 we have the infant colony confined to the immediate vicinity of Sydney for the most part (Fig. 1). Exploration had not opened up the hinterland to any important degree. In the next twenty years from 1820 to 1840 other centres of expansion besides Sydney became of importance; such as Hobart, Adelaide and Perth. In the third stage from 1840 to 1870 there was a wide spread of pastoral interests, and the gold rush started about half way through this period. From 1870 to 1900 we find the state boundaries soon finally settled; notable exploration ends with the end of the century; and the gold of West Australia greatly affects the southwest of the continent. In the fifth stage Federation has been accepted in 1901; and thereafter enough research took place so that soon an accurate appreciation of the potentialities (and disabilities) of the Australian environment was now possible. These five stages will be briefly considered in the concluding pages of this memoir.

1788 to 1820: First steps in exploration

Brief mention has already been made of the unfavorable environment of Sydney as a site for farm lands. Except for a circular patch of shale lying just west of Sydney, and about 20 miles across, the new settlement was hedged in by a belt of sterile and rugged sandstones. This was about 100 miles from north to south, and some 40 miles wide. This sandstone belt was first traversed by Blaxland's party in 1813 due west of Sydney, and Evans crossed the Divide and reached the fertile Bathurst Plains soon after. To the south Hume found a much less rugged route to the Goulburn Plains near Canberra in 1817. Thereafter the western slopes of the highlands were not difficult to traverse, and pastoral occupation soon spread far from Sydney. (Fig. 1).

Phillip had great difficulty at first with the crops. There were no competent farmers in the First Fleet; and they tried to grow wheat and rear sheep on the poor soils near Sydney. Today both the sheep and wheat belts have migrated to the western slopes some 200 miles to the west. Vegetables, some orchards of peaches and oranges, and a few thousand acres of maize now characterise the better soils in the Sydney district.

There seemed a strong probability that French occupation of outer parts of Australia might develop. Hence Collins in 1803 was sent to find a suitable site for settlement near Port Phillip. But he was not favorably impressed, and carried his company to Tasmania where Hobart was founded in 1804. Here better timber, a cooler climate for grain, and a profitable seal fishery were all attractions wanting at Sydney. Coal had been discovered in the cliffs both north and south of the capital, and small mines were opened at Newcastle about 1804.

Surveyor Oxley was the first to explore the further western slopes in his effort to find the outlets of the west-flowing rivers. He found that the Lachlan River in that year ended in swamps, while the country beyond was dry and unattractive. The variable character of the rains was not understood in those days, and wheat grows over much of the area which had little appeal in 1817. However the Murrumbidgee — a much stronger stream — was discovered in 1821, near where Canberra now stands. At the end of our first period Van Diemens Land (as Tasmania was known) possessed more sheep than New South Wales, and was actually exporting wheat to the mother colony.

1820 to 1840: Expansion from other centres

Free settlers became relatively numerous early in the new century, and in 1824 two important Land Companies were granted large areas of land in Tasmania in the northern coastal lands, and in New South Wales north of Newcastle. In both cases it was found that the pioneers had chosen districts with very poor soils, and later the grants were exchanged for better sites. In the northern colony the Company obtained valuable coal deposits, and also took up good pastoral properties north-west of the main divide in the vicinity of Tamworth. We see here that it was being recognised that sheep and cattle were more profitable than crops grown on the small fertile areas on the seaboard.

Meanwhile Hume in 1824 had been sent to explore the region between Sydney and the coasts near Port Phillip. He crossed the Divide near Goulburn, at the broad Lake George Gate, with no difficulty; and keeping along the western side of the mountains reached Geelong. They traversed good pastoral country throughout, and in spite of the Governor's attempt to keep settlers within reach of Sydney, the pastoralists by 1840 had driven their sheep to the Queensland border in the north, and southward to what is now Victoria. (Fig. 1).

Cunningham made a journey to the north keeping to the western slopes of the highlands, as did Hume, and reached Brisbane. The eastern slopes have been deeply cut by rivers eating back into the highlands; so that roads (and later railways) were linked up in the west far sooner than in the better-watered but rugged coastlands. In 1830 Sturt in a boat followed the Murray to the sea. Mitchell penetrated far into the more arid western plains, and explored much of the Darling River in 1835. Next year his most famous journey led him down the lower Lachlan River and south across the Murray near Balranald. Here he soon crossed the 15 inch isohyet and approached the Victorian coastlands. These were cooler and much less rugged than the coastlands near Sydney, and were blessed with an adequate rainfall. He named this south western part of Victoria "Australia Felix", and today it carries the densest sheep and some of the best wheat crops in the whole continent. At Portland he was amazed to find settlers with houses and crops and stock. They antedated the settlement of Melbourne by a couple of years.

Military conditions led to the first settlements on the north coast, first at Melville Island and then at Port Essington in 1828. This latter outpost was maintained with intervals till 1848. No real attempt to make this garrison self-supporting took place; and indeed the humid summers, long winter drought and sterile soils made it a very difficult enterprise. Today however there seems some hope that rice may be grown on the flats to the south of the former military outpost.

Stirling explored some of the region near Perth, and in 1829 he was made Governor of a new Colony in the west of the continent. Here a "Land Company" under Peel took up thousands of acres south of Perth. But the soil was in general sterile or very densely timbered, so that about 1830 — though over a million acres had been alienated — only 160 acres were cultivated (S. H. ROBERTS).

In South Australia a new colony was proclaimed in 1834, and Adelaide was founded in 1836. Owing largely to the theories of GIBBON WAKEFIELD land was not alienated in the wasteful fashion of earlier settlements. Sheep and wheat soon spread through the open savanah woodlands near Adelaide for the winter rains just suited wheat-growing; and good deposits of copper ore helped the opening up of this colony about 1842.

Towards the end of this period convicts were no longer sent to New South Wales. Perhaps in all 100,000 convicts were sent to Australia; but by 1830 there were far more free settlers than convicts in the continent.

1840 to 1870: Great pastoral expansion. Gold Rushes

By 1840 it was realised that arid country lay on the inland side of the settled areas in the south of Australia. Eyre had penetrated the desert country near Lake Eyre in 1840, and crossed the dry Nullarbor Plains near the Bight in 1841. Sturt barely survived the dry season in the far northwest of New South Wales in 1884, and later explored "Sturt's Stony Desert" near that region. However, the area north of Brisbane receives good rains fairly uniformly distributed through the year, and accordingly we see a remarkable extension of the pastoral industry throughout Queensland between 1860 and 1870.

Gold was discovered in the ancient rocks in the Eastern Highlands as early as 1823 near Bathurst, but the authorities kept the discovery secret. However Hargraves in 1851 found rich gold deposits in the same district, while even more spectacular ores were discovered in similar rocks — or in the alluvial above them — in the Ballarat region north of Melbourne. Both of these fields were in good farming lands, and after the easy alluvial gold was exhausted the miners settled in these well-watered areas. The vast immigration to Australia soon benefitted the pastoral industry, leading to a great demand for meat and wool.

It is to be noted that the pastoral holdings, which follow closely on the first exploration, by the year 1860 (Fig. 1) had occupied almost all the area which 100 years later was to contain the vast proportion of Australia's population. Notice how closely the 1860 isopleth on Fig. 1 agrees with the "one person per square mile" isopleth in Fig. 2. This again should be compared with the limits of "Uniform rainfall" as shown in Fig. 5 at B. It was clear to the writer some fifty years ago that the vast area labelled "Empty Australia" in Fig. 2 was never destined to be a region of close settlement, though this was a very unpopular thesis for many years.

It was during this period that all of Australia except the most arid regions were explored. Leichhardt traversed the well-watered region between Brisbane and Darwin in 1844, leading to the vast pastoral expansion of the next 20 years. Major Mitchell found that the lands of southern Queensland were in general much more attractive than the area along the Darling which he had discovered ten years earlier. The isohyets bend to the northwest from central New South Wales (Fig. 5 at A), so that central Queensland is better favoured than the lands along the Darling.

From 1859 to 1862 McDouall Stuart made repeated efforts to cross Australia from Adelaide through the centre of the continent. He followed the line of mound springs (the overflow from the artesian basin) for a hundred miles near Lake Eyre, and so reached the Macdonnell Ranges in the centre. Though the almost-desert country near Tennant Creek foiled him for a time, he managed to reach the coast near Darwin in July 1862. He was fortunate in missing the deserts which lie both east and west of his track, (Fig. 6) and his account of the springs in the south and of the dune-free country near Alice Springs gave a somewhat too favourable impression of the arid region.

By 1859 when Queensland was made a separate state, the political boundaries were almost completed. Northern Territory was accredited to South Australia in 1863, having been part of the "Mother State" until that year. (In 1911 it became Federal Territory.) Although the five larger states differ so much in area, it is interesting to note that three of them possess about the same area of attractive lands (i.e. 90,000 sq. miles) as Victoria. The remainders in the large states of West Australia, South Australia, and New South Wales are so dry that they add little to the economic value of the states. Queensland — in this first approximation — is better endowed with rainfall than the other states. Tasmania is too small and rugged, and Northern Territory too dry, to enter into competition.

1870 to 1900: End of exploration and of new pastoral occupation

By 1870 all of "Economic Australia" had been explored, and apart from some widely scattered cattle ranches in the north west, which were developed from 1870 to 1890, all the pastoral country was known and in part occupied. True, some of the most difficult exploring took place in the seventies; for Warburton crossed the western desert by the hottest and driest route in 1873, while Forrest and Giles made several crossing in the less inclement desert areas further south. But practically no part of this region has been utilised since. The margin of the Arunta Desert, some 700 miles long to the east of the Overland Telegraph through Alice Springs (Fig. 6), was explored by Barclay (1878) and Lindsay (1885), and Madigan flew across it in 1929. It seems to have no prospects of pastoral occupation.

Tropical agriculture may be said to have started about 1862 when some sugarcane was planted by Louis Hope near Brisbane. Cotton was also introduced a few years later, but this latter crop was abandoned when American cotton recovered after the war. Since 1911 however cotton is grown profitably near Rockhampton. About 46,000 Kanakas from the Pacific islands were employed on these plantations until their importation was prohibited in 1890. Sugar grows all down the east coast from Cairns to Grafton, where the average rainfall is over 40 inches. Near Rockhampton however the rainfall is not sufficient even on the coast.

In Tasmania with its restricted lowlands the colony progressed slowly until the seventies and eighties, when rich mines for copper tin and silver-lead were found in the mountainous west. The same stagnation took place in Western Australia, and ceased on the discovery of the rich gold fields between Perth and Kalgoorlie from 1887 to 1893. Mining has many times helped Australian settlement, but it is a robber industry; and if the mines are situated in arid country as is often the case, the population migrates when the ores are exhausted.

After about 1865 most of the better pastoral lands and almost all the croplands were known, so that the "closer settlement" of exploited areas became the rule. By 1890 the numbers of cattle and sheep had risen to figures which were not exceeded for many a decade. Indeed the terrible drought of 1902 produced a loss of nearly half of the sheep, and not till 1930 did the numbers reach 100 millions again. In 1915 and 1946 there were further large losses, due to drought, but not to the same extent; for pastoralists are now better prepared for the onset of dry seasons in the chief pastoral areas. It is not generally realised in Australia that the sheep and cattle are grazed mainly in the smaller states. Tiny Victoria is as important as the three giant areas of West Australia, South Australia and Northern Territory added together. This fact merely illustrates the statement that it is rainfall not area which counts in Australia.

1900 to 1956. Federation and Industrial development

Federation of the various states came about in 1901, but its implications are political rather than ecological. The pattern of population-distribution has not altered much in the last half century, save that centres of industrial activity have become very much more important. In 1947 out of 3 million occupied persons less than half a million were engaged in rural industry. This shift of the population to the towns has become more marked with each decade. In Australia as in other lands the presence of coal or hydro-electric power has become of paramount importance. Thus in New South Wales the three leading cities are Sydney, Newcastle and Wollongong, and they are the chief industrial towns. The fourth is Broken Hill where no coal is present, but where what is perhaps the richest lead-zinc mine in the world has been exploited in nearly desert lands. In Victoria Mildura in similar dry country has become fifth town, owing to extensive irrigation from the Murray River. In Western Australia Kalgoorlie is the second town, owing to the very rich goldfields in the vicinity, though the rainfall is only 10 inches. It is watered by the lengthy pipe from dams 330 miles away on the coastline.

Irrigation has been mentioned as an important feature in certain arid areas; but unfortunately it can never have much effect on the million square miles of arid Australia. At present only 600,000 acres of croplands are irrigated, i.e. about 1000 square miles. Since one cannot see large new water resources which can be conserved, even if we double or treble our irrigated areas in the distant future, it is clear that Australia will always be a "pastoral continent".

As mentioned earlier, great improvements in pasture — and to a lesser degree in crops — have resulted from the use of "trace elements". Very small quantities of copper and zinc (added to superphosphate) have turned useless heath country into valuable grazing for sheep, especially in rather dry lands to the east of Adelaide. A new crop in recent years has been rice, grown on the alluvial valleys east of Darwin. The rich chernozyems of east central Queensland have hardly been used yet, and would seem to be suited to a maize-swine economy as in so much of eastern U.S.A. The most spectacular development in recent years is of course the Snowy Mountains Power and Irrigation Plan in the well-watered plateau north of Mount Kosciusko. This will cost 500 million sterling, and it is expected that it will produce power up to 772,000 kilowatts.

Future Settlement in Australia

For the last fifty years I have been trying to forecast where the Australian population will dwell after a few generations. By comparison with other lands with similar environments — homoclimes as I named such — I produced the main map shown in Fig. 5. The method is explained in my advanced text on Australia (7th edition, Methuen, London, 1955). Consideration of the presence of abundant coal in other lands led me to place the densest future populations where the black patches occur in the map. The wetter agricultural lands (where not too rugged) may acquire a density of 18 per square mile, in place of the "one per square mile" of today. Marginal croplands may rise to 6 per square mile; pastoral lands of value should rise to "one per square mile". Sparselands and deserts, though forming nearly half of the area, will never in my opinion support an important proportion of Australia's future population.

Today there are some nine million Australians; and the authorities of late years are wisely encouraging the immigration of British and other nationals in numbers almost equal each year to the natural increase. It seems well however to bear in mind that our best resources are being rather fully utilised, so that I still hold to my forecast (in 1920) of thirty or forty millions as the likely future population, if we hold to standards of living such as obtain at present.

V
THE MARSUPIAL FAUNA: ITS ORIGIN AND RADIATION

by

E. LE G. TROUGHTON

(Past Curator of Mammals, Australian Museum, Sydney)

Scientific opinion differs on the zoogeographical dispersal of marsupial life, but of far greater significance is the fact that Australia, in prolonged prehistoric isolation except for New Guinea, fostered the greatest phylogenetic deployment of a single mammalian Order that the World can ever know. Because of the complexity of premise and overwhelming literature thereon, this contribution will not be concerned with arbitrary conclusions on prehistoric routes of migration, or a detailed survey of authoritative opinion on the phylogeny of the Marsupialia.

The origin and genetic relationships of the marsupial Order have been expounded by BENSLEY (1903), GREGORY (1910) and SIMPSON (1945). Unfortunately, SIMPSON's work on the evolution and classification of mammals overlooked the comprehensive and stimulating review by ABBIE (1941) of the "Marsupials and the Evolution of Mammals". Covering the origin, basic classification, and adaptive radiation of the Australian marsupials, ABBIE's paper, and the book by DARLINGTON (1957) on zoogeographic distribution, are herein treated as essential works, the extensive bibliographies of which eliminate the unnecessary repetition of references.

The Migratory Route, and Australian Environment

Regarding the zoogeography of mammals, DARLINGTON states that although mammals originated in the Jurassic they remained until the end of the Mesozoic insignificant animals, overshadowed by the reptiles, and with so limited a fossil record over that period as to scarcely warrant zoogeographical discussion. It was only about the beginning of the Tertiary that mammals achieved a dominance, and began leaving significant traces of their general deployment.

Recognizing the American ancestry of Australian marsupials, ABBIE (1941) points out that most authors favour their northern approach along the land chain once connected with Asia, though some postulate a southern route across a great Antarctic continent, formerly uniting the southern ends of America, Australia, and Africa. Wingless birds, ABBIE says, inhabit South America and Australia, but also occur in New Zealand and South Africa which lack marsupials or their remains. If the connexion with New Zealand and South Africa became lost in the interim (as between birds and marsupials) why did it persist between South America and Australia? And if it existed in the time of birds why did not some primitive premammals of South Africa reach Australia? Such objections seem to offer insuperable obstacles in accepting the southern route hypothesis. According to ABBIE, no such objections have been raised against the northern approach, and until they are the northern route must be accepted as that followed in the passage of marsupials to Australasia.

Regarding a possible South American point of marsupial deployment DARLINGTON (p. 324) writes: "It is unexpected and noteworthy that the oldest South American mammals do not reach the southern tip of the continent. Marsupials are unknown below about 47° S. (300 or 400 miles from the Straits), and edentates (armadillos) apparently do not extend even that far south."

However, the basic flaw in the theory of migration via Antarctica appears to lie with the assumption that the southern deployment of miniature marsupials could be contemporaneous with that of infinitely more primitive terrestrial invertebrates, under ecological conditions mainly antecedent to the age of mammals. According to GILL (1957), with the dawning of the Cainozoic "modern" era of plants, some 70 million years ago "by the uranium clock", there was an epochal change in Australian vegetation from the cycad and conifer forests of dinosaur days.

The distinctive continental flora thus became "recognizable" in contrast with the earlier floras appearing as of another planet. During this vegetational change, in the mid-Cainozoic, grasses covered vast areas of the earth — the prairies of North America, the African veldt and Eurasian steppes — and clothed the inland plains of Australia. Because of the long isolation from other land masses there were no grazing ungulates but only the ancestral marsupials to utilize this new "ecological setting".

It was the ancestral macropods that availed themselves of the grassy ecological field, and it was not before the Pleistocene, or recent Ice Age, that wattles *(Acacia)* and eucalypts came into greater prominence. They provided the ancestral phalangeroids with the arboreal background for their physiological transition from the basically insectivorous to a nectar, pollen, and leaf-eating diet. And so it happened that in this land of great antiquity was preserved the most intriguing assortment of archaic, ancient, and early-recent representatives of the plant and animal kingdoms.

Over vast areas only the strongly xerophytic adaptation of flora has preserved any kind of vegetation, whereas by contrast the northern tropical forests can compete with those of New Guinea in luxuriance. Such contrasts profoundly influenced the adaptive radiation of ancestral marsupials, and, indeed, influenced WALLACE (1880) in postulating two separate and isolated zones of evolution only latterly joined in a single continent. Of lesser and more recent significance, geologists have postulated a "drying-up" of the greater part of the interior, in a period of intense aridity about 8,000 years ago — a "searing period" following the Pleistocene Ice Age in which sedentary mammals and birds perished. Only "pockets" of vegetation were left, such as Palm Valley in central Australia, whereas the eastern and south-western coastal zones were sustained by rains from seaward.

However, in stressing the distinctive "peculiarity" of Australian flora GOOD (1957) maintains that the supposition that Australia and New Guinea have long been parts of the same larger landmass has apparently "gradually become established without critical consideration of the real evidence..." In possible explanation of the present close proximity, as not due to positional permanence, GOOD postulates a variant of WEGENER's hypothesis by which Australia had "drifted into contact with New Guinea, which as a result has been pushed out of alignment with its Malaysian neighbours."

On the contrary, surely no biogeographical theory of drifting or "colliding" land-masses could explain away the Torresian shallows and the rich radiation and inter-distribution of marsupials.

The Papuan Region of North Queensland

Despite the fact that HEDLEY (1894) proposed the biological region of "Papuan" to cover the obvious relationship of the "very distinct fauna and flora" developed in coastal Queensland and northern New South Wales, such biological unity of the Atherton Tableland and mountains of south-eastern New Guinea has been either overlooked or discounted in the hypothesis of a southern intrusion of marsupials. But as HEDLEY stated, the types of vegetation, reptiles, birds, and mammals of the Atherton Tableland are such that any traveller in the heart of the rain-forest "could scarcely answer, from the surroundings, whether he were in New Guinea or Australia."

In the dense "jungles" of Atherton Tableland, north to Cooktown, occur two species of tree-kangaroo *(Dendrolagus)*, also two species of Phalanger-possum or cuscus, and the highly specialized striped-possum *(Dactylopsila)*, all of which attain their maximum generic radiation within New Guinea. The *Phalanger* genus is represented by two species in Celebes. There is, likewise, an ecological affinity between the lowlands of the north-eastern mainland and the coastal region of the Western and Central Divisions of Papua. The appearance of the soil, eucalypts, and vegetation is remarkably similar to parts of the mainland south to Cloncurry and the mining region of

Mt. Isa, so that a traveller unaware of the Torresian "barrier" would be equally uncertain as to habitat.

In New Guinea, the commonest northern Australian macropod, *Wallabia agilis*, is represented by a geographical race in the Western Division of Papua, from the Oriomo River opposite Daru to Merauke across the Dutch New Guinea border, and inland to Lake Daviumbo on the middle Fly River; this wallaby also occurs in the Central Division along the narrow coastal strip, from about Hall Sound to the Rigo district east of Port Moresby. In the Western Division, from the Oriomo and Wassi Kussa rivers, the Archbold Expedition also collected a small "scrub" wallaby *(Thylogale)* which TATE (1948) could distinguish only as a race of the eastern mainland species *(stigmatica)* ranging from Cape York to northern New South Wales. Fossil remains occur in New Guinea of mainland *Nototherium*, rhinoceros-sized contemporary of *Diprotodon;* also of a giant species of kangaroo akin to *Palorchestes*.

However, apart from such biological evidence of the permanence of the land connexion of Australia and New Guinea, proponents of marsupial migration via Antarctica seemingly overlook the biogeographical and geological evidences of intermittent land connections from the northward, advanced by MERRILL (1946). Expressing the view that the Australian mammals as a group are essentially American and Eurasian in character, he points out that the distribution of plants, and infiltration of mammals, was greatly complicated by the existence of two great continental shelves: the "sundaland" or Malaysian extending from south-east Asia for about 1,500 miles — and the Sahul or "Papualand" shelf extending from north Australia. The submergence of these "shelves" is remarkably even at an average depth of 100 ft. The probability of earlier connexions of the land-masses must therefore be realized — since a sea-level drop of 150 ft. could reunite the Malaysian zone with the Philippines and Asia, while a drop of about 65 ft. would reunite Australia with New Guinea, the levels being well within the eustatic range.

Origin of the Monotremes and Marsupials

Confined to Australia and New Guinea, the egg-laying mammals exhibit a remarkable mixture of reptilian and mammalian features. As ABBIE states, little is known of extinct monotremes, but their living representatives are too highly specialized to be in the direct line of ascent to the eutherian mammals. From evidence collected by ANDREWS in the region of Outer Mongolia, ABBIE postulates the evolution of a group of small insectivore-like animals. They retained the egg-laying habit and structure, but developed a brain rather better organized than a reptiles. Regarding the brain and reproductive system, the Monotremata may therefore be considered as the basis of mammalian evolution.

The original group spread throughout the world, and some members became extremely specialized to produce the platypus and spiny anteaters of today, while retaining the primitive egg-laying habit. The main stem, however, progressed as the advantages of viviparous reproduction were imposed on other survivors of the original group. The functional change entailed adaptations that characterize marsupials and which, despite their intermediate limitations, proved eminently satisfactory in evolving the unique radiation of marsupial life within the prehistoric isolation of Australia and New Guinea.

Some of the primitive polyprotodonts, avoiding any too exclusively a marsupial specialization, evolved the mechanism for full intrauterine maturation; and with the acquisition of the corpus callosum they emerged as the ancestors of the Eutheria. As ABBIE concludes:

"Of the two innovations the one offered better prospect of survival to the young, but the other carried the almost incredible endowment of cerebral possibilities which led to the rise and dominance of the Eutheria.

"It is true that the scheme presented here does not depart to any great extent from that most widely accepted... What does appear to be new is the demonstration that the structure of the reproductive system and the brain can be fitted into such a scheme, fixing definitely the intermediate status of marsupials between monotremes and eutherians. The Marsupialia, then, are not merely an aberrant evolutionary sideline: they are living representatives of the pioneers which carry the impetus of evolution through to the Eutheria and to man."

Classification of the Marsupialia

According to GREGORY (1910) OWEN confirmed DE BLAINVILLE's view by separating the marsupials into two groups — the Polyprotodontia with eight to ten upper incisors and never less than six lower ones — and the Diprotodontia with never more than one pair of lower incisors. These divisions are now generally accepted. Some workers, however, emphasize that the polyprotodont bandicoots (Peramelidae) share with diprotodonts the character of syndactylism of the 2nd and 3rd digits of the pes. It was suggested that the bandicoots should be grouped with the diprotodont phalangers and kangaroos, but a survey of the evidence had convinced OLDFIELD THOMAS (1888) that such alliance would be phylogenetically unsound. Subsequently WOOD JONES (1923) revived the question, premising that because no didactylous (simple-toed) diprotodonts have been found syndactylism must have preceded diprotodontism, and provides a more basic distinction.

He therefore placed the Peramelidae with the Diprotodontia in a group called Syndactyla and the remaining marsupials, all polyprotodonts, in a group Didactyla. WOOD JONES was contending that the syndactyle paired nails of the bandicoot pes functioned as a primitive fur-comb, along with the many nibbling incisors. It is demonstrable, however, that just as the syndactyle foot developed independently in many birds, and lemur-monkeys, primarily as an arboreal adaptation, so syndactylism was independently evolved by groups of ancestral marsupials such as the phalangers, burrowing wombat-cousin of the koala, hopping kangaroos, and the many-incisored bandicoots.

However, ABBIE has shown that the bandicoots do not possess the necessary cerebral qualification for diprotodont status, and that their syndactylism is merely a further example of parallel evolution "reflecting the play of environment and circumstances upon similar organizations". In confirming the view of ELLIOT SMITH that a cerebral character provides an absolute distinction between the polyprotodonts and diprotodonts, ABBIE points out that the Diprotodontia achieved a distinct advance by developing an additional bundle of cerebral fibres — the fasciculus aberrans which "overrides all other taxonomic criteria." And so, to obviate the confusion arising from classification upon dental and pedal characters, ABBIE divided marsupials into the Simplicicommissurala and Duplicicommissurala — those without and those with the fasciculus.

In considering the Australian marsupials, however, the dental subdivisions of Polyprotodontia and Diprotodontia correspond adequately to the above in basic classification. As ABBIE states, America was probably the original home of the diprotodonts, but since no living representatives have been discovered in that continent, the "experiment" apparently did not flourish outside Australia. The progressive development disclosed by the fossil record of the South American marsupial carnivores, *Borhyaena*, indicates the dangers to which such newly-emerging creatures were exposed. The prototypal diprotodont evidently enjoyed a measure of success, in spreading from the New World at least to Australasia.

But about this time the eutherian advance had probably commenced so that the extra-Australian expansion of the diprotodonts was relatively short-lived. It seems likely, therefore, that together with the weaker of the polyprotodonts they were compelled to migrate to avoid destruction while in a comparatively primitive state. Thus Australasia was doubtless invaded by relatively weak and defenceless members of both groups, destined to attain their fullest radiation in the sub-terminal sanctuary of New Guinea and ultimate refuge of the Australian continent.

Adaptive Radiation

It is now a generally accepted fact that in the course of their evolutionary deployment marsupials have contrived to exploit practically every economic resource of the sub-desert, sub-tropical, and tropical environments of Australia and New Guinea. The magnitude of this marsupial radiation is evidenced by the fact that, prior to white settlement in 1788, only rats and bats of the sixteen Orders of non-marsupial or eutherian mammals listed by SIMPSON (1945) inhabited Australia; excluding the wild dog or dingo, introduced by native man some 12,000 years ago.

By contrast the marsupial Order, represented on the American continent by only the two living didelphid and caenolestid families, has evolved no less than nine families *(sensu lato)* within the varied ecological conditions of Australia and New Guinea. In his Classification, SIMPSON takes the modern view of including the marsupial-wolf *(Thylacinus)* and remarkably aberrant marsupial anteater *(Myrmecobius)* as subfamilies of the Dasyuridae. Correspondingly, as expounded by WOOD JONES (1924), the extraordinarily specialized koala is bracketed, along with the ringtail possums and greater-glider possum, as a subfamily of the Phalangeridae.

Such groupings have an obvious advantage in phylogenetic exposition, but they also tend to obscure the view of ecological radiation and the workings of parallel or convergent evolution. For example, the mole-like *Notoryctes* is accorded its lone family status as either an extremely aberrant peramelid or a highly modified dasyurid. It would seem consistent, therefore, to accord *Myrmecobius* and *Thylacinus* family status respectively as remarkably aberrant and specialized marsupials. The koala likewise surely deserves consideration of family status in view of the number of anatomical features homologous with those of the terrestrial wombat, isolated in the Family Vombatidae? But the phylogenetic issue is clouded by placing the koala in a subfamily of the Phalangeridae because of the crescentic (selenodont) cusping of the molars, possibly another example of convergence. However, the fact remains that the six families of living Australian marsupials listed by SIMPSON include no less than twelve extremely distinctive subfamilies.

In the absence of placental competition ancestral marsupials adapted themselves to various terrestrial, arboreal, and even aerial ways of life which in most instances have paralleled those of the eutherian mammals. Producing convergences of the greatest evolutionary interest, such adaptive radiation developed in marsupials parallel resemblances to the placental insectivores, edentates, rodents, carnivores and, in some respects, even the ungulata, and the lemuriform Aye-Aye *(Daubentonia)* of Madagascar.

The remarkable physical and dental resemblances, evolved under the general "law" of convergent or parallel evolution, not only made a profound impression on the earliest explorers and colonists of the "great southern continent", but inevitably misled scientists of the 18th Century, who could not evaluate the effects of convergent evolution and the basic unity of the Marsupialia. Such resemblances also exercised a profound influence on the vernacular and taxonomic naming of the more strikingly specialized marsupials, resulting in such combinations as *Phascolarctos* or pouched-bear for the sloth-like koala, and *Thylacinus cynocephalus* meaning a "pouched dog with a wolf's head" for the marsupial wolf or tiger of Tasmania, once known to colonists as the "Zebra Opossum" and "Zebra Wolf".

In a historical sense, it is notable that the first description of one of the kangaroo family, made in 1629 by the Dutch navigator PELSART when wrecked on Houtmans Abrolhos, off Western Australia, referred to the insular form of Tammar Wallaby *(Thylogale eugenii binoe)* as being: "a species of cats... about the size of a hare... their head resembling the head of a civet-cat; the forepaws are very short... resembling those of a monkey's forepaw. Its two hindlegs, on the contrary, are upwards of half an ell in length, and it walks on these only, on the flat of the heavy part of the leg, so that it does not run fast. Its tail is very long, like that of a long-tailed monkey; if it eats, it sits on its hind legs, and clutches its food with its forepaws, just like a squirrel or monkey."

Another Dutchman, SAMUEL VOLCKERSEN, referred in 1658 to the Short-tailed Wallaby *(Setonix)* on Rottnest Island, off Fremantle, as "a wild cat, resembling a civet-cat, but with browner hair." The British explorer WILLIAM DAMPIER while at Sharks Bay in 1699, on his second voyage to the western coast, stated in his Journal: "The Land-Animals that we saw here were only a Sort of Raccoons, different from those of the West Indies, chiefly as to their Legs for these have very short Fore-Legs, but go jumping upon them as the others do (and, like them, are very good meat)." Sometimes claimed to be the first account of a kangaroo, the description referred to the Banded Hare-Wallaby *(Lagostrophus fasciatus)*, which reminded DAMPIER of the raccoons seen by him about the Spanish Main.

It was not until about three-quarters of a century later, in June 1770, on the eastern coast at Cooktown 1,500 miles north of Botany Bay, that Captain COOK's party observed the marsupials that the aborigenes called "kangooroo", and the naturalists BANKS and SOLANDER likened to a "giant jerboa". Despite such striking convergences arising from similarity of environments and habits, Australian marsupials are indubitably the diversified descendents of prototypal forms such as the mouse-like insectivorous dasyures *(Antechinus* and *Phascogale)* and the pigmy phalangers *(Cercartetus* and *Eudromicia)*, so fundamentally paralleled by the caenolestid "rat-opossum" of South America.

Common ancestral heritage is shown in the reproductive system resulting in the "premature" embryonic birth, possession of epipubic or "marsupium" bones, and location of mammae within the pouch area. Also in cranial features lacking in eutherian mammals, such as the inwardly inflected angles of the mandible and other pecularities described by ABBIE, and the common tendency to develop vacuities in the bony palate.

DARLINGTON (1957) provides the following "Summary of the Native Land Mammals of Australia and New Guinea", stating that the figures are mainly from various papers listed as by TATE. The species of rodents, he says, may be rather more finely split than those of marsupials, while the species of bats "are admittedly lumped" by TATE; genera in general are more finely split than in SIMPSON's Classification.

Table I.

	Australia (and Tasmania etc.)			New Guinea (and close islands)		
	Families	Genera	Species	Families	Genera	Species
Monotremes	2	2	2	1	2	3
Marsupials	6	52	119	4	24	47
Rodents (murids)	1	13	67	1	20	56
Bats	7	21	41	6	21	45

The ecological factors and adaptive radiation of the Australian families of marsupials, in their widest sense, may be briefly summarized in the following way:

Insectivorous and carnivorous Dasyuridae

The members of this group *(sensu lato)* include about thirty species of mouse- and rat-like marsupials, the so-called "native cats", and the largest carnivorous marsupial in existence, the thylacine or pouched "wolf" now confined to Tasmania. Main characteristics of the family are the full (polyprotodont) set of incisors typical of insectivores and mixed-feeders, and the more primitive type of marsupial pes, with toes remaining separate and independent of each other. It is in this combination of characters that the dasyures approach more nearly the primitive rather unspecialized American opossums (Didelphidae).

Consideration of form, dentition, and foot-structure suggest that the least specialized genus of marsupial-mice *(Antechinus)* closely resembles the prototypal form of marsupial life. These broad-footed pouched-mice parallel the habits and general structure of the rat-Opossum *(Caenolestes)* of the Andean region of South America.

Regarding origin, however, ABBIE and SIMPSON have shown that available evidence is positively more consistent with the caenolestid derivation from didelphoids than in common with any Australian group.

The dasyurid dentition is of particular interest in illustrating an evolution from the needle-cusped insectivore-teeth, through the omnivorous dentition of the native "cats" to the canine type of premolars and molars of the marsupial "wolf". Variation in diet has resulted in a remarkable variation in physical structure, the marsupial-mice blending the outward appearance of foreign shrews and small rodents, while the native cats, Tasmanian devil, and thylacine resemble generally the foreign weasels, cats, and canines.

Some of the broad-footed marsupial-mice are active climbers with more or less serrated sole-pads, nesting in tree-hollows or on cave-ledges, while others are terrestrial burrowers in sub-desert regions; mainly insectivorous, they have a decided carnivorous tendency. The smallest genus *(Planigale)* has a lizard-like flattening of the skull for sliding into crevices in sun-cracked earth or between rocks, and amongst the spinifex-spikes and coarse inland grasses. The smallest species *(subtilissima)* with a total length of about $2\frac{3}{4}$ in., is probably the smallest existing marsupial; a somewhat larger species is described from the Astrolabe Range in Papua. The largest of known broad-footed pouched-mice *(Neophascogale)* inhabits New Guinea, along with a number of other species.

The narrow-footed species of the genus *Sminthopsis* are almost exclusively insectivorous and terrestrial, favouring a hopping action as indicated by the apical sole-pads. An allied genus *(Antechinomys)* has evolved an extreme modification of the hind-limbs for saltatory movement, with extremely slender kangaroo-like feet with apical pads, and a very long brushy tail. It parallels the appearance of the foreign rodent jerboas, and it is of ecological significance that it shares the sandy habitat, and possibly the burrows, of the Australian true hopping-rats *(Notomys)*. Thus the savannah country and sub-deserts of Australia have fostered saltatory adaptation in a polyprotodont dasyurid, the diprotodont kangaroos, and an indigenous genus of rodents.

The dasyurid "cats", apart from their uniquely spotted fur, more nearly resemble the foreign weasels in appearance, dentition, and carnivorous habits. Weasel-like, once sampling the blood of poultry they persist in the nightly raids with insensate killing of many birds. The so-called "tiger cat" *(Dasyurops)* is an expert climber, as shown by the strongly-serrated sole-pads and retention of the hallux or great-toe. Attaining a total length of 4 ft., it has been observed launching itself from tree-tops and catching birds on the way to the ground. These native-cats are, however, useful mixed-feeders on insects such as grasshoppers, reptiles, mice, and rabbits. The much smaller north Australian native-cat *(hallucatus)* agrees with the widely-ranging New Guinea species in retaining the hallux, and in having eight teats, instead of the six common to other Australian species.

The Tasmanian "devil", excepting for its short dasyurid tail, looks more like a miniature bear with its stout body, broad head, and plantigrade feet. It also has a rather bear-like coloration of the coarse blackish-brown hair, with several whitish patches. Lone member of its genus *(Sarcophilus)*, with widely-gaping jaws and extremely massive dentition for the crushing of bones, most authorities have avoided the phylogenetic issue by including this highly specialized dasyurid in the Subfamily Phascogalinae with the native cats.

But it is the thylacine or marsupial-wolf that provides the most startling example of convergent evolution in its dog-like form and digitigrade stance, and the canine appearance of the skull and dentition. The carnivorous transformation of the premolars and molars is such that, but for the possession of eight instead of six upper incisors, palatal vacuities, and inflected mandible-angles, all but the trained observer could be deceived by the "canine look" of the skull. However, TATE (1947) in attempting to distinguish between ancient homologies and more recent convergences actually postulated derivation of the thylacine from a New Guinea genus *(Murexia)* of marsupial-mice. Such extreme specialization anyway surely warrants at least the Subfamily status (Thylacininae) accorded by most authorities.

Regarding past theories that the predacious South American marsupials, the borhyaenids or "sparassodonts" were ancestral to the Australian dasyurid marsupials, SIMPSON (1941) stated: "Even a hypothetical borhyaenid combining all known *Thylacinus*-like characters of the group would not make an acceptable ancestor... because all known borhyaenids do have specializations that are absent in *Thylacinus*." With reference to the abundant seemingly conclusive evidence that the thylacine is merely a specialized dasyurid, SIMPSON endorses CABRERA's idea of independent derivation of the dasyurids and borhyaenids from a common didelphid ancestry. Where that common ancestor lived, says SIMPSON, is a different question but from a summary of evidence he concludes that it "removes any particular reason for postulating a connection directly between Australia and South America, although in itself not disproving such postulate."

Marsupial Anteater and Pouched-Mole

Though possibly of a separate family heritage these marsupials are bracketed together as providing most remarkable examples of the specializations evolved in the adaptive radiation of ancestral marsupial stock within Australia. Regarding the banded anteater or "numbat" of the aborigines, the view of BENSLEY (1903) is now generally accepted that *Myrmecobius* is more probably an aberrant dasyurid than a descendent of the Jurassic trituberculates, and that the high number of molars and their irregularly tricuspidate character resulted from degenerative processes accompanying adoption of the ant-eating habit.

More aptly described as a termite or "white-ant" eater, *Myrmecobius* has no less than five upper and five to six lower molars, the entire dental series ranging from fifty to fifty-two. Supposedly, the degenerate teeth were scarcely used in mastication but FLEAY (1942) observed that while smaller termites were swallowed whole, his anteater captive rapidly and audibly chewed the soldier caste which have formidable brittle jaws. It was also noted that the anteater's very extensible cylindrical tongue is perfectly adapted for extracting the termites "cleanly" from their timber "galleries". Thus in the marsupial anteater we find evident degeneration allied with effective specialized function.

The marsupial-mole *(Notoryctes)* presents an extreme example of adaptation in that every feature, excepting the genitalia and brain, has evolved an extraordinary resemblance to the true mole *(Talpa)*, the Cape Golden Mole *(Chrysochloris)*, and the rodent "mole" *(Sphalax)*. Its muzzle is protected by an oblong horny shield for thrusting through sandy soil, scooped aside by the spade-like 3rd and 4th fingernails, and thrown back by the flattened toenails. Sightless, all trace of the embryonic eye has degenerated to a mere lens-like disc; tiny ear-openings beneath the fur have no lobes, and the nostrils are mere slits. There are two teats in a quite well-developed pouch which, necessarily, opens rearwards.

Except for the short breeding and resting burrow, *Notoryctes* does not construct a tunnel-system but literally dredges its way beneath the sandy soil, when its movements may be traced by aborigines, who also follow the surface-tracks after rain. Its genetic affinity was first thought to lie with the bandicoots (Peramelidae) because the pes retains traces of a former syndactyle condition, now regarded as due to parallelism. However, with *Myrmecobius*, the marsupial-mole is now considered to be a highly-modified form of dasyure but so complete is its fossorial adaptation that it is listed as sole representative of the Family Notoryctidae.

The Bandicoots – Peramelidae

These small omnivorous marsupials form one of the lesser yet most interestingly varied groups of Australian pouched-mammals. They occupy a peculiarly intermediate genetic position in possessing the syndactyle pes, restricted otherwise to the diprotodont phalangers and macropods which have only paired lower incisors, whereas the peramelid incisors conform with the truly polyprotodont condition of the dasyures. The more typical species resemble miniature kangaroos with less remarkably elongated hindquarters, so that the hopping action is less pronounced and an "all-fours" gallop is commonly used. The tail consequently is relatively very short and rat-like instead of being developed as a rudder in leaping, except in the rabbit-bandicoot which has a long and strongly crested tail.

However, all bandicoots are recognizable by their long and sharply tapered snouts, associated with the primarily insectivorous habit of extracting beetle-larvae and worms from conical pits dug with the sharply-elongated central fingernails; the molars are also of the needle-cusped insectivore type, excepting in the rabbit-bandicoot which has the flattened molar crowns indicative of its more omnivorous diet. About eighteen recognizable species occupy eight genera, of which four inhabit the mainland and two Tasmania, and four occur in New Guinea and adjacent islands. The general diet is truly mixed including insects, snails, lizards, mice, and a varying amount of vegetable matter.

The size ranges from that of a large rat to a rabbit or hare and it was doubtless the rodent "look", noted by colonists about Sydney, which suggested the name "bandicoot", actually a corruption of the "native" name for the large "pigrat" inhabiting southern India and Ceylon. The less specialized and more exclusively insectivorous genus *Perameles* includes some species with a quaint grid-marking of dark brown on the brightly-coloured rump, providing a sort of camouflage pattern. The stouter and somewhat shorter-nosed *Isoodon* species, according to WOOD JONES (1924) have a more pugnacious and definitely carnivorous disposition, a tame captive having killed a larger *Perameles*, almost "plucking" its body in the process overnight.

The unique Pig-footed Bandicoot of the sub-desert inland, apparently a specialized offshoot of the small "bar-backed" peramelids of the open plains, exhibits an almost ungulate parallelism in the quadrupedal development of its limbs. In its readaptation to a cantering or loping rather than a hopping action, the manus or "forefoot" has mimicked the cloven, and hindfoot the solid, structure of the hooves in certain groups of ungulates, as indicated by the genus name *Chaeropus* or "hog-foot".

There are two species of the remarkable rabbit-bandicoot or bilby, distinguished by their enormous rabbit-like ears *(Macrotis)*, and soft chinchilla-grey fur, which resulted in trade-exploitation and its disappearance from populous areas years ago. Allied with the more active hopping habit, after small mammals, lizards, and insects, the tail is longer and stouter than in other bandicoots, and even more distinctive in having a black-haired base and white-crested outer half. A small naked "spur" at the tail-tip may result from the gathering-up of nesting materials or "propping" action while digging, the rabbit-bandicoots being the only ones excavating deep burrows.

Phalangers – or Australian Possums

The diprotodont family of Australian possums (Phalangeridae) presents if possible an even more diverse radiation than the polyprotodont Dasyuridae. The family is of great evolutionary interest because some "pigmy" phalangers combine vestiges of the primitive insectivorous dentition of mouse-like terrestrial ancestors with arboreal adaptations of limbs and tail, while the tail of the koala has atrophied during a prolonged period of terrestrial existence. Otherwise, the tail is invariably long and thoroughly adapted for arboreal existence, either prehensile in a monkey-like way, or well-brushed and even feather-like as a rudder for aerial leaps and glides.

An important climbing aid, apart from the strong sabre-like nails, is the nailless great-toe (hallux), opposable to the other toes in grasping branches in a thumb-like way. The term phalanger, from the Greek for a finger or toe, actually refers to the syndactyle condition of the 2nd and 3rd toes, webbed together but with separate nails. Though primarily an arboreal adaptation it is this character, plus dentition, which distinguishes all phalangerids from the separate-toed American opossums. Entirely arboreal, with rare exception in rocky places, the diet of the phalangers is basically insectivorous with the addition of nectar and pollen from blossoms, the larger kinds feeding on foliage and fruits, and with a decided carnivorous tendency in the brushtail and the phalanger-possums (cuscus).

The pigmy "honey possum" *(Tarsipes)*, with a 3 in. head and body, and an equally long prehensile tail, is so completely specialized for obtaining insects, nectar, and pollen from blossoms in the manner of honey-eating birds that it occupies the Subfamily Tarsipedinae by itself. The snout forms a proboscis with lip-flanges which overlap, forming a sucking tube, while the whip-like tongue is bristled at its tip and is protrusible for at least an inch. The insectivorous needle-cusped molars have degenerated by change to a pulpy diet, much as in ant-eating mammals.

Three distinctive groups of phalangerids provide excellent examples of adaptive radiation in the evolution of gliding membranes, thus paralleling the gliding equipment independently evolved in some foreign non-marsupial insectivores and rodents. The smallest glider is the Feather-tail *(Acrobates pygmaeus)*, derived from the group of pigmy or "dormouse" phalangers *(Cercartetus* and *Eudromicia)*. It makes spring-like leaps of more than a yard with side-flaps stretched between the limbs, and the tail forming a feather-like rudder, fringed with long hairs on each side. The larger "squirrel" gliders *(Petaurus)* evolved from another group of the Subfamily Phalangerinae, possibly *Gymnobelideus* which has a similar dentition and bushy tail, but no gliding membranes.

Largest of the "flying-possums" is the Greater-Glider *(Schoinobates volans)* with the body averaging 17 ins. and the slender fluffy tail about 20 ins. It glides for up to 120 yards from a tree-top to the base of another, the landing curve being so low that the glider may be impaled on barbed-wire fencing, or caught by a fox. Entirely a leaf and blossom eater, similarity of dentition shows the greater-glider actually to be a ringtailed phalanger *(Pseudocheirus)* with gliding-membranes, and the tail functioning as a furry rudder instead of the much more prehensile appendage. The side-flaps reach to the elbow, instead of the hand as in *Petaurus*, thus avoiding interference with the ringtail agility of the greater-glider. These gliding adaptations of the phalangers have obviously been influenced by the character of the eucalypts with trees and branches so widely spaced.

The ringtailed phalangers and greater-glider, along with the extremely specialized and tailless koala, are grouped in the Subfamily Phascolarctinae because of the crescentic-cusped or selenodont molars, irrespective of whether this ill-assorted union provides a mere "marriage of convergence" or not! However, the less specialized brushtail-possums *(Trichosurus)* and scaly-tailed cuscuses *(Phalanger)* are included in the Subfamily Phalangerinae, with rounded bunodont molar-cusps, along with the pigmy-possums, lesser gliders, and the Tasmanian fossil *Wynyardia*.

The latter Subfamily provides a truly remarkable example of parallelism in the "striped possums" *(Dactylopsila* and *Dactylonax)* of the rain-forests of north-east Queensland and New Guinea. These marsupials mimic strikingly the primitive lemur-monkey of Madagascar, the Aye-Aye *Daubentonia*, in having the fourth finger remarkably elongated and with a small hooked nail for extracting larvae from timber. The front upper incisors of the striped phalanger are enlarged to meet the strong diprotodont lower ones, paralleling the front incisors of *Daubentonia*, for the rodent action of gnawing out the insects. In both marsupial and lemur the insect quarry is located and disturbed by a rapid vibrating of the slender fingers along the branches. In keeping with the skunk-like marking, the striped possum exudes a strong glandular odour which, though not ejected, must afford the slightly-clawed marsupial protection by rendering it and its nesting place offensive to predatores.

The Koala – and the Terrestrial Wombats

Although consensus of opinion has placed the koala with the ringtail possums and greater-glider in a Subfamily (Phascolarctinae) of phalangers, while according the wombats a Family (Vombatidae) to themselves, these remarkably specialized marsupials are here bracketed together in order to emphasize their common phylogenetic heritage. The basic reason for placing the koala with the ringtail phalangers is the possession of similar selenodont molar-cusps, whereas in *Phalanger* and *Trichosurus* the molars have bunodont cusps. However, since such authorities as BENSLEY and TOMES, while admitting certain dental resemblances between the koala and wombat, regard them as mere evidence of convergence, there appears no reason to doubt that the selenodont molar cusps in the koala and ringtail phalangers may also be independently adaptive in character.

SONNTAG (1922) in reviewing the Myology and Classification of the Wombat, Koala, and Phalangers provides a relative summation of anatomical and dental characteristics, leading to the following conclusions: The koala cannot be included with the Phalangeridae because, although dentition and hand-structure suggest relationship with *Pseudocheirus* there are many important differences, and allying them by superficial characters would obscure the series of fundamental characters uniting the koala and wombat. These characters enumerated by SONNTAG are regarded as far outweighing the resemblances between *Phascolarctos* and the phalangerids, while not sufficient to place the koala in a separate family from the wombats. He therefore expresses agreement with WEBER and WINGE that there should be only two families, one for the phalangers, and the Phascolarctidae with the Subfamilies Phascolarctinae for the koala, and Vombatinae (ex Phascolomyinae) for the wombats.

Classification apart, the wombat and koala provide extraordinary examples of specialization from arboreal ancestors for an age of terrestrial existence, and subsequent divergence owing to readaptation of the koala for arboreal existence. As evidence of common arboreal ancestry, the earthbound wombat retains the syndactyle condition of the foot, together with vestiges of the musculature for control of the prehensile tail, subjected to similar atrophy in the koala during its prolonged period of terrestrial life. Other features in common are the granulation of the palm- and sole-pads, striated in the ringtails and prototypal phalangerids, the possession of vestigial cheek-pouches, structure of the marsupium, and single pair of mammae.

In the course of its fossorial adaptation as a powerful digger, the wombat has evolved the extreme in diprotodont dentition, with rodent-like paired incisors adapted for gnawing through root-obstructions, and feeding on the more tender roots and coarse herbage. In general form, skull, and dentition the wombat may be regarded as paralleling the eutherian beaver, minus an aquatic tail. It is notable also that the marsupial was known to the earliest settlers as "the badger" and a certain kind of tussocky grass is still called "badger grass" in wombat haunts south-west of Sydney.

By contrast, the koala may be described as a kind of plantigrade bear turned tree-climber, in the course of which readaptation it has evolved the maximum degree of opposability of the pollex or index finger to the other digits. This modification, combined with the grasping ability of the nailless great-toe, resulting from the syndactyle condition of the foot, plus the sabre-like finger- and toe-nails, provides a powerful climbing equipment in compensation for loss of the prehensile tail. In effect, nature has imposed upon the bear-like koala a somewhat sloth-like habit and movements, although living and descending trees "right side up". Finally, in its restriction to a diet of eucalypt leaves the koala acquired the extreme in selenodont cusping, and an enormous extra caecum or pseudo-appendix to aid ingestion of the bulky food. A maximum glucuronic acid excretion, as an aid to toleration of the exclusively eucalypt diet, has been recorded by HINKS & BOLLIGER (1957), and there is also a weaning period when the juvenile feeds on the emulsified droppings of the parent.

The Kangaroo Family – Macropodidae

On the palaeontological evidence, there can be no doubt as to the pre-Australian evolution of the Diprotodontia and that, according to the dentition and syndactyle pes, the unique saltatory radiation of the kangaroos (Macropodidae) was evolved within Australia from a prototypal phalangerid, in adaptation to the ecological conditions of sub-desert plains, more or less rugged savannahs, and sub-tropical forests.

It is therefore of great genetical significance that the most primitive living macropod, the Musky Rat-Kangaroo *(Hypsiprymnodon)*, inhabits the rain-forests of northern Queensland where, as in New Guinea, the phalangers have achieved their maximum radiation. This miniature rat-kangaroo with a total length of about 18 ins., almost subequal limbs and naked scaly tail is usually regarded as the lone survivor of a macropod Subfamily. However, in the view of some authorities *Hypsiprymnodon* should be accorded distinctive family status as annectent between the phalangers and kangaroos, because it is the only "kangaroo" that retains the functional hallux or great-toe, while its outwardly inflected secatorial premolars are also indicative of its insectivorous diet and prototypal phalangeroid heritage. Feeding on insects and larvae scratched from the jungle humus, the term "musky" refers to its individual odour due to habitat and diet.

It is notable that the so-called "Marsupial-Lion" *(Thylacoleo carnifex)* showed an extreme development in its huge secatorial premolars, from which RICHARD OWEN concluded that it had been the carnivorous exterminator of its much larger fossil contemporaries. But subsequent research on the fossil remains has suggested that the marsupial-lion was actually a herbivorous diprotodont, probably with carnivorous tendencies, like some living phalangers, and occupying an annectent position between the phalangers and macropods — such as assigned to *Hypsiprymnodon*. However, as ABBIE says, whether or not one may legitimately surmise that *Thylacoleo* had its actual affinities with the rat-kangaroos (Potoroinae), all surviving members share the specialized development of the hypsiprymnodont premolars.

The rat-kangaroo muzzle, in comparison with that of small wallabies, is much more tapered, especially in the more primitive genus *Potorous*, in which the short ears are more broadly rounded, and the tail relatively shorter and thicker. The nails of the much longer middle fingers are decidedly larger than the outer ones, for scratching-out the larvae, fungi, and succulent grass-roots forming the main diet. Some rat-kangaroos make burrows, but most lie-up in surface hollows, to which they transport nesting materials held in bundles by the supple tail. There are three species of *Potorous*, and four shorter-nosed *Bettongia* in the sub family Potoroinae; represented also in Tasmania, while several mainland species may now be extinct. The larger Rufous Rat-Kangaroo *(Aepyprymnus)* has a hairy instead of naked muzzle-tip, and the unusual marsupial condition of a complete instead of perforated bony palate.

The more "typical" kangaroos of the Subfamily Macropodinae include the somewhat intermediate "hare-wallabies" *(Lagorchestes)*, and the "nail-tailed" wallabies *(Onychogalea)* with a horny tail-spur, possibly due to their rapid darting movements, or gathering-in of nesting materials. There are about ten mainland species of rock-wallabies *(Petrogale)* with the shorter feet having granulated sole-pads for landing on rock-ledges, and the relatively pliable tail brush-haired as a leaping rudder. Plants and succulent roots are eaten as well as grasses.

The unique short-tailed wallaby *(Setonix)*, the aboriginal's Quokka of south Western Australia, is about the size of a hare; the pes-length is barely 4 ins, and the tail-length scarcely twice that of the head, and the rounded ears barely show above the long fur. Feeding on the coarse vegetation of its normally swampy habitat, the very long secatorial premolars, and low-crowned molars, show a decided approximation respectively to those of rat-kangaroos, and the foliage-eating dorcopsid wallabies and tree-kangaroos of New Guinea.

The tree-kangaroos *(Dendrolagus)* exhibit a remarkable reversal in evolution by their readaptation to the arboreal habitat of primitive ancestors, even more successfully than if a plantigrade bear had turned tree-dweller. They are now characterized by the almost subequal proportion of the limbs, and stout but pliable feet with roughened anti-skidding soles and strongly curved sabre-like nails. The long cylindrical tail functions as a climbing prop and rudder in leaping amongst branches, and as much as 60 ft. to the ground.

The remaining Macropodinae are divisible broadly into three generic groups: The small pademelon or scrub-wallabie *(Thylogale);* the medium-sized kangaroos or brush-wallabies *(Wallabia*);* and the giant kangaroos of the genus *Macropus (sensu lato)*. The mountain kangaroos or "wallaroos" may browse on foliage, but the great-grey foresters, and red-kangaroo of the plains, are grazing marsupials filling the place of the foreign hoofed herbivores. In profile, a kangaroo's skull is superficially like that of a horse, with three upper incisors opposed to the diprotodont lower ones, and the diastema for transferring cropped grasses to the grinding molars.

As already noted, Captain COOK's naturalists applied the "gigantic jerboa" comparison to the two species of "kangooroo" captured at Cooktown in 1770, without realizing the marsupial affinity implied by their recognition of the ringtailed phalanger as a kind of "possum". However, to this day eminent geneticists are profoundly impressed by the spectacular examples of remarkably varied saltatory adaptation evolved within the kangaroo family, paralleling in physical structure the rodent jerboas of Africa and America, and larger Springhare *(Pedetes)* of Africa. In Australia, furthermore, a maximum degree of saltatory adaptation has been evolved by the dasyurid pouched-jerboa *(Antechinomys)*, and the indigenous hopping rodents *(Notomys)* which share its habitat but naturally exceed their marsupial mimic in abundance.

Conclusion

The foregoing general summary discloses an unprecedented degree of adaptive radiation of the original marsupial stock in response to the variable ecological niches, as between the mainland and New Guinea, and within the Australian continent. In assessing the range and significance of such adaptations, the prehistoric conditions under which they evolved must ever be borne in mind. The original marsupials doubtless suffered little opposition on reaching the continent because miniature eutherians such as bats and rodents made little impression, especially with development of the carnivorous marsupial dasyurids. Fossil evidence proves that many marsupials flourished to an extreme resulting in their extermination, like a number of eutherians, but the survivors entered into a secure occupation of their ecological inheritance.

Thus the many examples cited above illustrate the remarkable

---

* Authors' recent use of OWEN's genus *Protemnodon*, introduced for a giant fossil kangaroo, is absolutely invalid as applied to this group of living wallabies.

degree of phylogenetic resiliance whereby the marsupials achieved a standard of adaptation admirably adjusted to the far from luxuriant ecological conditions of the continent. As ABBIE concludes, it would seem illogical therefore to mourn the marsupials' failure to achieve a higher status than was either necessary or expedient. Actually, both polyprotodonts and diprotodonts surmounted the utmost demands of environment to a degree never surpassed by even the most highly specialized of eutherian or higher mammals.

REFERENCES

ABBIE, A. A., 1941. Marsupials and the Evolution of Mammals. *Aust. J. Sci.* **4**, *77—92*.

BENSLEY, B. A., 1903. On the Evolution of the Australian Marsupialia; with remarks on the relationships of the marsupials in general. *Trans. Linn. Soc. Lond.*, 2nd ser., **9**, *83—217*.

DARLINGTON, P. J., 1957. Zoogeography. John Wiley & Sons Inc., N.Y.

FLEAY, DAVID, 1942. The Numbat in Victoria. *Victorian Naturalist*, **59**, *3—7*.

GILL, E., 1957. Whence Australia's Mantle of Green? *Aust. Mus. Mag.* **12**, *217—220*.

GOOD, R., 1957. Some Problems of Southern Floras with Special Reference to Australia. *Aust. J. Sci.*, **20**, *41—44*.

GREGORY, W. K., 1910. The Orders of Mammals. *Bull. Amer. Mus. Nat. Hist.* **27**, *1—524*.

HEDLEY, CHARLES, 1894. The Faunal Regions of Australia. *Rept. Aust. Ass. Adv. Sci.* **5**, *444*.

HINKS, N., & BOLLIGER, A., 1957. Glucuronuria in Marsupials. *Aust. J. Sci.* **19**, *228*.

JONES, F. WOOD, 1923—25. The Mammals of South Australia. Handbook of South Australian Branch of British Science Guild, Adelaide. 3 pts., 458 pp.

MERRILL, E. D., 1946. Plant Life of the Pacific World. The MacMillan Co., N.Y.

SIMPSON, G. G., 1941. The Affinities of the Borhyaenidae. *Amer. Mus. Novitates*, **1118**, *1—6*.

SIMPSON, G. G., 1945. The Principles of Classification and a Classification of the Mammals. *Bull. Amer. Nat. Hist.* **85**, *1—350*.

SONNTAG, C. F., 1922. On the Myology and Classification of the Wombat, Koala, and Phalangers. *Proc. Zool. Soc. Lond.* **2**, *863—896*.

TATE, G. H. H., 1948. Studies on the Anatomy and Phylogeny of the Macropodidae. *Bull. Amer. Mus. Nat. Hist.* **91**, *237—351*.

THOMAS, OLDFIELD, 1888. Catalogue of the Marsupialia and Monotremata in the British Museum, *1—401*.

TROUGHTON, E. Le G., 1954. Furred Animals of Australia. 5th Edition. Angus & Robertson, Sydney.

WALLACE, A. R., 1880. Island Life: or, the phenomenon and causes Insular Faunas and Floras, including a revision and attempted solution of the problem of geological climates. 1 Vol., London; rev. Ed. 1892.

VI
AUSTRALIAN BIRDS: THEIR ZOOGEOGRAPHY AND ADAPTATIONS TO AN ARID CONTINENT

by

ALLEN KEAST

(Curator of Birds and Reptiles, Australian Museum, Sydney)

Introduction

Zoogeography

The avifauna of Australia is made up of 531 species of breeding land and freshwater birds, and is the smallest numerically of the continents. There is a high degree of endemism. Australia and New Guinea together constitute a distinct zoogeographic region.

Adaptations to an Arid Continent

The seasonal cycle of Australian desert and dry-country birds is compared with the "typical" one seen in inhabitants of the temperate regions of the Northern Hemisphere. Rainfall is of the greatest importance in initiating the various phases. Light, however, would appear to be important in some instances.

(1) Nomadism is the dominant form of seasonal movement, the main need being to get to where rain has provided temporarily favourable conditions. South-north migration, which has as its basis the avoidance of cold (or accompanying food shortages) is, by contrast, little developed on the mild Australian continent.

(2) Breeding is in the spring in those areas where conditions are uniformly good at that time. In the interior, however, suitable conditions are dependent on the irregular precipitation and breeding follows rain, irrespective of time year. Recent research on the relationship of rainfall to breeding in Australian interior birds is discussed.

(3) The moult is also involved in the over-all pattern of adaptation and the "rule" that the major moult never occurs simultaneously with breeding, but follows it, is not necessarily maintained. Recent work has shown that when a drought renders breeding impossible the "post-breeding moult" occurs nevertheless and it does so on schedule (various species). On the other hand a sudden shower of rain may immediately initiate nest-building in apostle-birds *(Struthidea)*, not withstanding that they are at the climax of their moult. It is suggested that, in such cases, either moulting and breeding proceed simultaneously to completion, or else the former must be temporarily suspended to admit a period of breeding.

The Effects of Droughts on Bird Numbers

Droughts: (a) Totally stop breeding (mostly), or reduce the number of eggs laid (and young reared).

(b) Occasionally lead to catastrophic exterminations of populations.

The Build-up in numbers following a drought is accomplished by:

(a) Concentrating for breeding in areas where conditions are most suitable to the rearing of young.

(b) Having a greater number of broods than normal (the breeding season may be protracted or else there may be both a spring and an autumn period of reproduction).

(c) An increase in clutch size (in certain species).

It is pointed out that the various adaptations to aridity seen in Australian birds are not unique. Though they have not been properly studied elsewhere the literature suggests that they are duplicated to at least extent in African and/or Central American birds.

Similar cases in rodent and marsupial mice are on record.

ZOOGEOGRAPHY

The Australian avifauna, origin and relationships

The number of bird species on the Australian list is given by MAYR & SERVENTY (1944) as 651. Of these 83 are visitors that do not breed on the continent and 37 are seabirds that nest on islands off the coast. This leaves a total of 531 species of breeding land and fresh-water birds, somewhat fewer than the fauna of the tropical island of New Guinea to the north (566 species).

In terms of numbers of species the Australian continent compares with the other continents as follows:

Australia	– 531	North America	– 750
Africa	– 1,750	Europe and Asia	– 1,110
South America	– 2,500		

Hence, in accord with its being the smallest land mass Australia has the smallest number of bird species. If however, New Guinea be included (over three-quarters of whose birds do not occur in Australia) the number of species would considerably exceed that of North America. The richness of the New Guinea avifauna is in keeping with that of other densely vegetated tropical regions of the world, the fauna of the Amazon basin accounting, in the main, for the high South American figure.

Endemic Elements in the Australo-Papuan Avifauna

In its composition the Australo-Papuan avifauna is rich in distinctive elements, as follows: (see also DARLINGTON, 1957):

Casuariidae (cassowaries) – New Guinea mainly, one species reaching Australia. Large flightless Ratites.

Dromaeidae (emus) – Australia only, one species. Large flightless Ratites.

Megapodiidae, mound-builders which depend on the heat generated by the sun, rotting vegetation, or volcanos, to hatch their eggs. Extend as far as Nicobars and Philippines. The arid-country Australian species *Leipoa ocellata* probably has the most elaborate heat-regulating system of all.

Pedionominae (collared hemipodes) – Australia, one species only.

The related sub-family Turnicinae is well developed in Australia, where 7 of the 15 known world species occur.

Gouriinae (giant ornamented pigeons) – New Guinea only.

Loriinae (honey lories) – most diverse and numerous in New Guinea but extends as far as Bali, Celebes, and Philippines.

Micropsittinae (pigmy parrots) – New Guinea and Solomons only.

Kakatoeinae (large seed-eating cockatoos) – most diverse in forests and on plains of Australia. Extend to Bali, Celebes and Philippines.

Menuridae (lyrebirds – 2 species) – restricted to coastal forests of south-eastern Australia. These large birds, superficially pheasant-like, are remarkable for the elaborateness of the male's dancing displays, diversity in songs, and degree to which mimicry is developed.

Atrichornithidae (scrub-birds – 2 species, one extinct) – restricted to coastal forests of south-eastern and south-western Australia. Primitive passerines with aberrant sternum and syrinx muscles, and with loud, ringing call-notes.

Malurinae (Australian warbler-like birds). This sub-family was created and placed in the Muscicapidae by MAYR & AMADON (1951) to cover a series of genera that were formerly included in many small families. Thus defined, the subfamily extends from Australia to New Guinea, Polynesia, and New Guinea, and contains 25 genera and 85 species. Australian members (64) are chiefly in the genera *Malurus* (wrens), *Acanthiza* (thornbills), *Gerygone* (warblers), *Epthianura* (chats).

Meliphagidae (brush-tongued honeyeaters) – Australia and New Guinea to New Zealand, Bali, Celebes, and Hawaii. Of the 159 known species 67 occur in Australia proper and about the same number in New Guinea: the two areas between them accounting for almost three-quarters of the 160 members of the family. *Promerops* of South Africa, generally placed in a sub-family of the Meliphagidae may not really be related to them (MAYR & AMADON, 1951).

Cracticidae (Australian magpies and butcher-birds) – mostly Australian. The genus *Cracticus* has the shrike-like habit of wedging the bodies of victims, not needed immediately as food, into forks and thorns. Most are splendid songsters.

Grallinidae (Australian mud-nest builders) Into this family (see MAYR & AMADON, 1951) have been placed three diverse genera (2 monotypic) that build large, cup-shaped, mud nests in trees. Three Australian, one New Guinea species.

Ptilonorhynchidae (bower-birds) – 17 species, about equally divided between Australia and New Guinea. Group is remarkable for elaborate display structures and associated ornamentation. fresh flowers, lichens, coloured feathers, bones being used. Prior to and during the breeding season the male spends much of his time arranging and renewing the play objects. Courtship posturing is elaborate and exaggerated.

Paradisaeidae (birds of paradise) – New Guinea (43 species), only three reaching Australia. The worlds' most elaborate and diverse display plumes are to be found in the males of this group. The display is exaggerated and bizarre. Polygamous.

The avifauna of continental Australia can be looked upon as being a blend of relatively old elements and successsively recent arrivals from the Oriental region to the north-west and New Guinea to the north. There is, of course, no way of dating the successive waves of colonization. Two hundred and thirteen species of land and fresh-water birds, some 41% of the total species, belong to the endemic families and sub-families listed above, showing that isolation of the continent has been relatively complete. This is further borne out by the considerable radiation that has occurred in many of these groups, and the intimate way in which they are bound up ecologically with other organisms in the environment. Thus, the Meliphagidae (67 Australian species) are apparently the main pollinators of many species of *Eucalyptus*, *Banksia*, and *Grevillea*. The group is, however, equally insectivorous and much of its adaptation is to habitat. The 55 different parrots are specialized for a wide range of foods from nectar to seeds, nuts, and roots. Again, some of the black cockatoos fulfil a woodpecker-like role in digging into treetrunks for wood-boring grubs. The insectivorous malurine genera *Acanthiza* and *Malurus* contain 11 and 10 species respectively. They occupy warbler-like niches in different kinds of forest and grasslands.

Apart from the essentially Australo-Papuan families many widely-ranging families must have extended to Australia a considerable time ago, judging from the extent of their development here. There are 17 species of ducks and geese (Anatidae), and they include several aberrant, monotypic genera — *Chenopis* (Black Swan), *Cereopsis* (Cape Barren Goose), *Anseranas* (Pied Goose), *Malacorhynchus* (Pink-eared Duck), *Biziura* (Musk Duck). The typical flycatchers (Muscicapinae) are well represented. There has been much radiation in the pigeons and doves (Columbidae), particularly in the development of interior, ground-dwelling species. There are 17 Australian species of weaver-finches (Estrildinae). An early Australian offshoot of the flower-peckers (Dicaeidae), the genus *Pardalotus*, which subsists largely on tiny insects from amongst the leaves, has broken up into a number of species occupying a range of habitats and climatic zones.

Direct Relationships with other Faunas

Australia has no close ornithological relationships with South America. Likewise, supposed direct relationships with Africa (vide *Promerops* with the Australian Meliphagidae) are now open to question (MAYR & AMADON, 1951). Many families, do, however, extend through from Asia, or through Asia to Africa and Europe. Thus, the Turnicidae, Campephagidae, Zosteropidae, and Ploceidae, are equally well developed both in Africa and Australia. By contrast, a whole range of families well-developed in Asia and Africa have only one or two representatives in Australia: Coraciidae, Meropidae, Pittidae, Timaliinae, Sylviinae, Motacillidae, Nectariniidae, Sturnidae, Oriolidae, Dicruridae, and others.

WALLACE's Line is now no longer regarded as the fundamental division between the Oriental and Australian Zoogeographic Regions, analyses by MAYR (1945) and others having shown that:

(a) There is a gradual change in the percentage of species of western and eastern origin through the island groups from west to east (vide Borneo to New Guinea). There is no "sudden jump". In this regard it might be noted that three Old World families reach New Guinea but not Australia: Laniidae (true shrikes), Hemiprocnidae (tree-swifts), and Bucoteridae (hornbills).

(b) WALLACE's Line actually merely corresponds to the area of sudden drop between a rich continental fauna (to the west), and a depauperised insular one in the Celebes-Molluccas section. There is an equally sudden change from the depauperised fauna to a rich continental one (vide Australia) as the eastern end of the Archipelago.

It is this depauperised zone, not a narrow but mysterious tract of sea that has stopped a lot of important oriental families from reaching Australia, vide the pheasants, trogons, barbets, woodpeckers, broadbills, bulbuls (which reach the Moluccas), fairy bluebirds, and true finches.

Relationships of Australia with surrounding Islands

New Guinea:

This land has been connected with Australia at various times (probably twice during the Pleistocene), and interchange of species continues between the two. At the present time, however, this is mainly in the form of the colonization of the savannah woodland areas of New Guinea by Australian savannah species and of the rain forest areas of Australia by New Guinea jungle forms. Nevertheless, each has a high level of endemism.

Torres Strait, and the arid north of Australia, are the important distributional barriers, not only in preventing faunal interchange, but in initiating isolation and speciation. Together with the fact that most birds inhabiting the two land masses are quite differently adapted, these barriers explain the continuing basic differences between the avifauna of Australia and New Guinea. Of bird species common to Australia and New Guinea (191), 92 would appear to be of Australian origin, 66 of New Guinea origin and the rest are either Asian or could equally well be of Asian as of Australo-Papuan origin.

Timor:

As noted by MAYR (1944) there is some faunistic interchange between Australia and Timor today but the fauna of the island is essentially Oriental is origin.

New Caledonia:

The 68 species composing the land-bird fauna of this island include at least 18 species of Australian origin. The bulk of the species are Polynesian. The unique flightless kagu (monotypic family Phynochetidae) is confined to New Caledonia.

There are no cases of colonization of Australia from New Caledonia, so far as can be determined.

New Zealand:

The avifauna of these islands is characterized by a number of unique families and sub-families: the now extinct Dinornithidae and Anomalopterigidae (moas); Apterygidae (kiwis — three species); Strigopinae (unique, flightless parrots); Nestorinae (heavy-bodied, long-billed parrots, that include the kea, famous for its secondary adaptation to flesh-eating — three species); Callaeidae (New Zealand wattled "crows" — 3 species); Xenicidae (New Zealand wrens, once regarded as being related to the Conopophagidae of South America

(MAYR & AMADON, 1951), four species. There are also a large number of endemic genera. A distinct Australian influence is, however, obvious (e.g. Meliphagidae, flycatchers of the genera *Petroica — Miro* group, and *Rhipidura*. There is a continuing steady colonization of New Zealand by Australian birds (see FALLA, 1953). The reverse is not, however, occurring.

Factors governing distribution within Australia

This subject will be dealt with in brief only as it has recently been covered in another paper (KEAST, 1959a) and the subject of zoogeographic provinces or sub-areas is being discussed elsewhere in this book.

The distribution of an animal species depends upon its specific needs and tolerances to a number of variables.

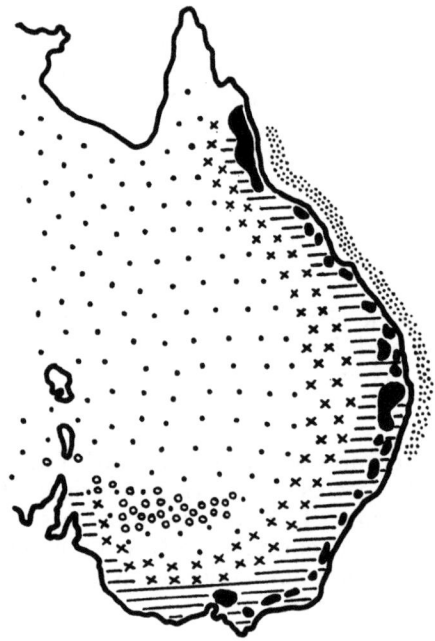

Fig. 1. Habitat specialization in Australian honeyeaters of the genus *Meliphaga*. The distributional arrangement is into parallel zones corresponding with successive vegetation associations. In its large number of species and degree of habitat "specialization" *Meliphaga* is an exceptional genus but the distributional trends seen here occur in other genera. The species are as follows: *M. fasciogularis* (mangroves – finely spotted areas); *M. lewini* (rain forest and wet sclerophyll forests – black areas); *M. chrysops* (sclerophyll forest and wetter savannah woodland – cross-hatched areas); *M. fusca* (savannah woodland – crosses); *M. virescens* (mulga and savannah grassland – large spots); *M. ornata* (mallee – circles).

The map does not show limits, as there is a certain amount of interdigitating of both vegetation type and bird range.

Many bird species are restricted to the north of the continent, and others to the south. Some are restricted to well-watered regions, others to the desert. In a smaller number of cases bird species range over the whole continent.

The study of bird distribution in Australia is considerably simplified by the environmental variables tending to be grouped in characteristic ways. These are typified by the various major vegetation associations. Each of the vegetation associations extends over a considerable area, is associated with fairly constant types of physical environment, and creates a biotic habitat with consistent kinds of shelter, characteristic food plants and insects, and certain types of nesting sites. Accordingly, bird speciation and evolution in Australia has been largely along the lines of adaptation to, and specialization for, one or another kind of habitat. An example of this will be seen on Fig. 1, six of the various species of *Meliphaga* (honeyeaters) occurring on the continent being specialized for life in mangroves, rain forest, sclerophyll forest, savannah woodland, savannah grassland and mulga, and mallee, respectively.

The major vegetation associations occurring in Australia are listed below and their distribution will be seen from the vegetation map in the Chapter XVI of the book (page 275). Rain forest, mangroves, and marshes, cover only negligible areas (certainly less than 1% each of the total land surface of Australia). A provisional estimate made by the author (KEAST, 1959a) of the areas covered by the vegetation associations as they effect bird distribution is as follows: Rain forest, negligible; sclerophyll forest, 7%; savannah woodland, 24%; savannah grassland, 16%; mallee, 7%; mulga, 30%; desert grassland (spinifex), 14%; mangroves, negligible; marshes, negligible. The number of bird species specific to, or reaching their greatest abundance in each of the major vegetation associations, or part thereof, is set out below:

Table I
Specialization of bird species in terms of habitat

Vegetation Association	Number of bird species	Percentage of total avifauna
Rain forest	81	16
Sclerophyll forest	84	16
Savannah woodland	about 146	about 28
Savannah grassland	about 59	about 12
Mallee	9	2
Desert mulga	24	5
Desert spinifex	17	3
Mangroves	16	3
Swamps and marshes	74	14
Miscellaneous	14	3

In view of the restricted land surface covered by them the relative richness of the rain forest and marsh habitats, in terms of number of bird species, is obvious.

ECOLOGICAL ADAPTATIONS

Adaptive modifications to the seasonal cycle

The "typical" seasonal cycle of birds

Birds have a stereotyped annual cycle of behaviour that typically includes a spring period of reproduction followed by a postnuptial moult, outward migration (in some species), frequently a period of gregariousness in winter, then return migration and (in some species), a prenuptial moult. Breeding itself involves a chain of events that include male territorial song and aggressiveness, pair formation, choice of nest site, nest-building, incubation, and feeding of the young. In some species the first brood may be followed by a second, or even a third.

The timing and order of the major events in the annual cycle is no accident but the result of a long evolutionary development. Under the "normal" conditions of Europe and North America the spring is the time when most insects are available for the young. Autumn is the ideal time for the moult as the birds are than free from the stresses of reproduction, the climate is yet warm, and warm body and maximally efficient flight feathers are assured for the ensuing winter months. Migration serves to take the population away from a place of deteriorating climate and food supplies to one where the latter are assured. The prenuptial moult produces, in many of the species in which it occurs, special colouring and plumes for use in courtship.

Of the aspects of the annual cycle that have attracted experimenters none has had the appeal of those governing the timing of the various events, for over vast areas of the Northern Hemisphere the regularity of their occurrence from year to year and individual to individual is remarkable. Foremost amongst the workers in this field was ROWAN (1929), the first to demonstrate that it was the changing daylength that, when it reached certain thresholds, was a fundamental environmental "trigger" and "synchronizer". Both migration and gonad development were initiated in early experiments.

Subsequently, various workers began artificially manipulating the daylength of captive birds such that, by 1938, no fewer than 100 species of birds, as well as various mammals and even a couple of reptiles, had been shown to respond to light changes (ROWAN, 1939). In most cases the winter enlargement of the gonads by artificially creating spring day-lengths was the end sought and achieved. Though, in many of the experiments, the light intensities and increments used were highly artificial, and the work has been criticised on this score, the premise that under normal conditions light is the "environmental trigger" initiating breeding over vast areas of the Northern Hemisphere can be regarded as correct. (It might be noted that the moult can also be experimentally induced by reproducing day-lengths appropriate to the moulting season — see especially MIYAZAKI (1935) for the Japanese white-eye *(Zosterops palpebrosa)*. In the Southern Hemisphere the spring moult of the Australian white-eye, *Zosterops lateralis*, has also been induced by manipulating the day-length (KEAST 1956)).

The initial enthusiasm for regarding changing daylength as the all-embracing environmental regulating factors has had to be modified somewhat in later years. Thus, BAKER (1939), pointed out that (a) in parts of the world only rainfall could fulfil the role of being the "trigger" to breeding and (b) many birds must have an innate cyclical breeding rhythm. He based the latter on the fact that several aviary species transported from Southern to Northern Hemisphere continued to breed at the time of the southern spring. Again, various tropical animals, living where daylength is constant, are remarkably regular in their time of breeding.

BAKER's suggestions with regard to rainfall being the factor initiating breeding in various places have been subsequently borne out — note papers on breeding seasons in inland Australia (especially KEAST & MARSHALL, 1954; MARSHALL & SERVENTY, 1957); parts of Africa (MOREAU, 1950; SMITH, 1955); the Galapagos Islands (LACK, 1950), and central America (SKUTCH, 1950 and WAGNER, 1957).

The present chapter reviews the timing of the major events in the life of the bird in inland Australia, and reviews in detail how behaviour and pattern of adaptation differ from that which has come to be regarded as typical. The relationship of rainfall to seasonal movements, breeding, and the moult is discussed and attention directed to recent and current investigations in this field.

The Australian Environment

Of the characteristics of the Australian continent the following have a profound influence on aspects of the annual behaviour cycle:
(a) Australia is a temperate continent. There are no extensive areas of permanent snow.
(b) It is essentially arid, only one third being suitable for agriculture and close settlement, whilst one third is desert (rainfall of under 10 inches per annum). Inland from the well-watered east coast approximately parallel north-south isohyets indicate the rapid transition from fertile to semi-arid and arid zones. The over-all rainfall pattern of Australia is

one of concentric isohyets of increasing rainfall outwards from a vast central arid zone.

The distribution of the different types of vegetation in Australia, as elsewhere, is closely associated with rainfall.

(c) Corresponding with the falling-away in over all rainfall from coast to centre is increased variability in the rainfall. Thus, at Sydney on the east coast, it is under 20%, at Bourke 450 miles inland 35%, and at Birdsville, 900 miles from the coast 50%.

(d) Australia is a continent at low relief. Physical barriers to dispersion are virtually non existent and the all important ones are climatic.

Within the Australian continent there are sections where conditions are uniformly fertile. Others are uniformly arid. Across vast tracts of the continent, however, conditions are quite unpredictable and amount to a jumble of wet (flood), dry (drought), and intermediate years, following in irregular succession. The varied nature of different parts of the continent and the range of climatic conditions represented enable animal groups to adapt from one extreme to another and permit an understanding of the intermediate steps.

The Seasonal Movements of Australian Birds

Australian land and fresh-water birds fall into three categories, (a) sedentary species, that is those that remain throughout the year in the same area, (b) south-north migrants, whose movements take them to areas where the winter temperatures are warmer and (c) nomads, whose movements are irregular and vary from season to season. (See KEAST 1956a for a semi-popular account of seasonal movements in Australian birds.)

The outstanding single feature of Australian birds is the small number of true migrants, reflecting the absence of severe winter temperatures on the continent, and the high proportion of nomads, birds that follow a form of movement based either directly on rainfall or (in that rainfall influences the flowering of plants) indirectly on it. For a variety of reasons an absolute percentage of the two types cannot be derived (populations within species may differ in their behaviour, some species are sedentary except in drought years, species that undertake "local" movements can only be classified as sedentary, and some seasonal movements are a blend of south-north migration and nomadism). A reasonable estimate, however, is as follows:

Sedentary species:	66%
South-north migrants:	8%
Nomads:	26%

These figures differ profoundly from those for northern Europe and Northern America, where declining temperatures and not irregular rainfall are the hazard. There, perhaps, 60% of the species are true latitudinal migrants and only a handful could be classed as nomadic.

Sedentary species:

The bulk of these are inhabitants of those sections of the continent where conditions are reasonably good and that permit a regular spring breeding (vide the forested areas of the east, south-east, and south-west). Notwithstanding this, these habitats contain many "blossom nomads". Towards the interior of the continent the proportion of sedentary species falls off rapidly. Even the arid central desert, however, has its share of sedentary birds. These are the "true" desert genera and species such as: *Amytornis*, spp., *Oreoica gutturalis*, *Malurus leucopterus*, *M. callainus*, and *Artamus melanops*.

South-north migrants:

This form of movement is seen in a number of tropical species that are obviously recent colonizers of the Australian continent — e.g. *Eurystomus orientalis* (roller), *Tanysiptera sylvia* (kingfisher), *Aplonis metallica* (shining starling), and that winter in the islands to the north of the continent. In addition the populations of a variety of species breeding in the colder parts of the continent (Tasmania and south-eastern Australia) migrate north in winter, e.g. *Zosterops lateralis* (white-eye), *Hirundo neoxena* (swallow). The majority of south-north migrants are inhabitants of the better-watered coastal strip in the east and they keep within this strip during their movements. In contrast with interior nomads, members of the species (though not necessarily particular individuals) return to the same areas to breed from year to year.

Interior nomads:

Within the nomads there is every kind and gradation. Some are nomadic over only part of their range, or they may remain and breed in an area for years until a drought causes them to move away (vide the cuckoo-shrike *Pteropodocys maxima* in the Mudgee district of New South Wales).

There is a well-defined segment of "extreme nomads", whose irregular wanderings take them all over the continent (recent banding work has shown teal ducks to come into this category). Flock Pigeons, *Histriophaps histrionica*, disappear into the uninhabited interior for years on end, until finally a combination of climatic events brings them out into on to the fringes of settlement in one or another part of the periphery. Successive "occurrences" of the species may be thousands of miles apart.

An important group of interior nomads has a distinct south-north bias to its autumn movements, their travels bringing them to that section of the continent watered by the summer monsoon. Included in these is the chat, *Epthianura tricolor*, and the wood-swallows *Artamus superciliosus* and *A. personatus*. The songlarks *(Cinclorhamphus)*, by contrast, only go north in those autumns when food supplies are scarce. On their return to the south these various species do not go to regular areas but concentrate for breeding where-ever conditions happen to be best. Thus, on the average of every 3—4 springs the two wood-swallows extend a hundred miles to the east of their normal range to breed near Sydney.

Not only do environmental conditions initiate widespread movements in Australia but they alter whole patterns of distribution. Thus the dry year 1957 (when inland swamps and billabongs were low) saw a great influx of waterbirds to coastal areas. The number of bird species inhabiting the Ayer's Rock region of central Australia doubled between May and September, 1952, associated with the breaking of a drought (KEAST & MARSHALL, 1954). In June of that year less than ten species were recorded on the drought-stricken Alexandria Downs, Northern Territory, whereas at Camooweal, 120 miles to the east and where there had recently been about 100 points of rain the number of species was in the vicinity of 40. INGRAM (1907) has recorded 91 bird species from Alexandria Downs in a good season.

The Environmental Factor initiating Bird Movements

Both day-length and immediate environmental conditions would appear to serve to initiate seasonal movements in Australian birds. Like latitudinal migrants in the Northern Hemisphere those of southern Australia vacate their breeding grounds well in advance of the cold winter (mostly in February and March). Accordingly, it seems reasonable to suppose that these birds are responsive to a shortening daylength. The movements of the interior and blossom nomads which are dictated by the local environment at the time and are irregular in time of occurrence, cannot however, be initiated by changes in daylength. Many interior nomads, as has been noted, move off only when conditions become untenable. Again, good rains in the arid areas will start such species off in search of improved pastures. The all-important stimulus in these birds must be a combination of rainfall, food supplies, and degree of "uncomfortableness" of the environment.

The Breeding Seasons of Australian Birds (Fig. 2)

Heavily-forested coastal regions of the east, south-east, and south-west.

Breeding season: spring (August-December mostly), but with a few specialized species nesting at other times (e.g. *Menura* and *Elanus* in winter; *Dicaeum* and *Grantiella* in late summer). This is the section of the continent over which conditions tend to be uniformly good for a winter rainfall ensures a fertile spring. The east, in addition, has a rainfall peak in late summer. This initiates, in certain years, a limited amount of autumn nesting (especially in *Grallina* and certain finches). Autumn nesting would appear, however, never to exceed 5—10% of the normal spring nesting.

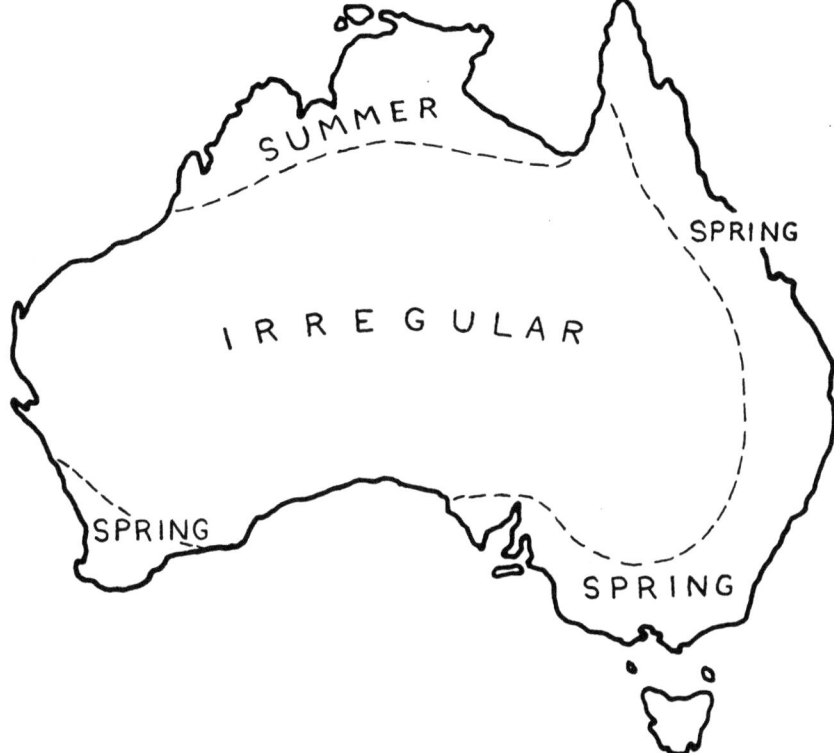

Fig. 2. Breeding seasons of birds in different parts of the australian continent. The map is meant to give only a general impression. Centralwards from the coast are extensive intermediate zones where the birds give every evidence of "wanting" to breed in the spring and in fact do so then when the spring is fertile. The north of the continent has only one wet season, in summer.

Northern coastal regions of the continent

Although much more work will be necessary before breeding in in this section is properly understood there would appear to be no doubt that the peak is in mid-summer in association with the monsoon rains of that season (see SERVENTY & WHITTELL, 1948). Summer is the time of peak insect life in the Northern Territory, the autumn, winter, and spring being fairly dry. In north-eastern

Queensland, despite its having a good summer rain, breeding is largely a spring phenomenon suggesting that, where this is possible, the birds have a natural preference for nesting in the spring.

Desert centre of the continent

Here, where the rainfall is sparse and irregular, the breeding of the birds follows rain, wherever it may fall. This is aptly summarized by McGilp (1924), a resident for fifteen years in the arid Lake Frome area of northern South Australia:

"In bad years few birds attempt to breed, in fact, most of them seek better quarters but, when a break-up of the drought occurs, they return and commence nesting operations at once. The best rains usually fall in the summer months, so that birds which breed in the spring about Adelaide are often found nesting at the end of the summer or in autumn or winter. In good seasons some birds nest practically all the year, and in this way make up for the non-breeding years."

A feature of some desert species is the striking rapidity of their response to rainfall. One such instance is given by McGilp (1919). A Desert Chat *(Ashbyia lovensis)* began nest-building five days after the breaking of a drought and ovulated on the eleventh day. This case, in the light of recent observations by the writer in western New South Wales, is possibly not exceptional.

The "intermediate" zones

These may be defined as those parts of the continent in which the birds have a spring gonadal recrudescence (see discussion later) and breed then if the environment permits. Spring reproduction would appear to occur, to at least some extent, every year. Breeding, however, may be induced at other times, provided the rainfall stimulus be sufficiently strong. A smaller quantity of rain (say, half an inch or 50 points) will induce territorial behaviour and song). Summer and autumn breeding is possibly most widespread when the preceding spring has been dry. On other occasions, however, a wet summer following a normal spring will lead to a continuation of the spring breeding into summer and autumn, a larger number of broods than usual than being reared (e.g. Cunnamulla, Queensland 1945-46 — M. Schrader, personal communication).

The "intermediate zone" represents the area of transition from normal spring to a "desert type" of breeding season.

Recent Research on Rainfall and Breeding in Australian birds

The localities in which recent work has been carried out are shown on Fig. 3 (see "A", "B", "C", and "D").

The drought winter of 1952 in central Australia (Keast & Marshall, 1954).

Until August and September, when good rains fell, 1952 was a year of severe drought in central and northern Australia. As at April, when collections were made at Ayers Rock, there had been no breeding for many months. Gonads were minimal in size. In July,

Fig. 3. Localities mentioned in text in connection with breeding and moulting studies. The main recent investigations have taken place in the areas marked A, B, C and D.

however, an inch of rain fell, followed by a second inch in August. There was an immediate influx of species into the area and widespread breeding occurred in August and September.

On the Barclay Table land in the Northern Territory the death of approximately a million cattle in 1952 testified to the severity of the drought. Over one small section (Camooweal, Queensland), however, an inch (100 points) of rain fell in early July (three weeks prior to collections being made) and produced a temporary local alleviation of the conditions. This made it possible to compare histologically the gonads of a variety of species from where the rain fell (Camooweal) with those from an area slightly to the north and west (Alexandria) where the drought was still in progress and there had been no rain.

The gonads of the Alexandria birds were of minimum size and without development. At Camooweal, however, sperm were being actively proliferated in several species, whilst the testes of all showed some degree of enlargement. Surprisingly enough, however, a wide range of developmental abnormalities were also found: asymetric tubule maturation, premature invasion of the interstitium by capillaries and the dissolution of individual Leidig cells, and premature collapse (i.e. whilst still proliferating sperm) of the whole organ. The inference, with respect to these species, was that the rainfall stimulus had been sufficient to initiate breeding behaviour and testicular development but, in that there had been no "follow-up" rain, insufficient to enable breeding to take place and premature testicular regression was occurring.

No development had taken place in the case of the females.

The influence of the autumn cycles of 1953 and 1955 in south-western Australia (MARSHALL & SERVENTY, 1957).

As noted, the south-west corner of the continent invariably has a wet winter and dry summer and autumn. The spring is accordingly fertile and it is then that breeding takes place. Inland, there is a rapid transition to arid conditions and breeding is less regular.

In March, 1953, and again in 1955, atypical autumn cyclones swept down from the north and abundant rain fell over a large area, creating ideal conditions for breeding. The inhabitants of the more inland areas responded immediately, and a considerable nesting followed. The coastal inhabitants did not respond at all, however, and connecting the two sections was a geographic gradient of lessening response. This study is significant in indicating markedly different adaptation in coastal and inland birds. The possibility must not be overlooked, however, that the responsiveness of the latter may have been accelerated by only limited breeding having been possible in the previous spring.

The paper of MARSHALL & SERVENTY includes a histological analysis of a wide range of testes, both from areas where the stimulus was, and was not, effective in initiating breeding. In addition, a detailed analysis is given of factors influencing breeding in Western Australia and an impressive array of data assembled to indicate that rainfall, not day-length, is the all-important proximal factor.

The drought spring of 1957 in the interior of New South Wales (KEAST, unpub.).

This year, one of severe drought, was marked by widespread inhibition of breeding inland and by an exodus of inland birds into coastal areas. In mid-October three inches of rain fell over one small section (Bingara). Accordingly, field work was initiated at the beginning of November in three areas with widely-differing rainfall histories: Bourke, Coonabarrabran, and Bingara (see "C" on Fig. 3). Monthly rainfall figures for 1957, normal (average) monthly figures (in brackets), together with those for Sydney, are set out in Table II.

Table II
Rainfall figures for the drought year of 1957
(in points)

	Feb.	Mar.	Apr.	May.	Jun.	Jul.	Aug.	Sep.	Oct.
Bourke ...	205	17	2	—	49	23	31	—	—
	(121)	(75)	(91)	(100)	(127)	(88)	(61)	(63)	(76)
Coonabarrabran	343	287	225	7	163	83	177	4	16
	(230)	(225)	(173)	(159)	(246)	(258)	(159)	(167)	(175)
Bingara ...	234	349	207	15	199	37	148	22	291
	(276)	(252)	(180)	(189)	(228)	(232)	(157)	(178)	(241)
Sydney ...	294	447	98	14	141	505	569	68	69
	(315)	(444)	(565)	(498)	(368)	(489)	(241)	(277)	(280)

In each of the three areas a study was made of general behaviour and a range of specimens of various species collected (a) to detect the presence in the population of immature birds (by the skull vacuolation technique) and (b) for histological analysis of gonads.

The state of breeding in the three areas may be summarized as follows:

Bourke area:

Of thirty species sampled, breeding had occurred in only three, these being normally exceptionally early breeders: *Meliphaga penicillata* (honeyeater), *Pardalotus striatus* (diamond-bird), and *Smicrornis brevirostris* (warbler). No species were nesting at the beginning of November except for a single pair of apostle-birds *(Struthidea cinerea)* which are mud-nest builders and had rehabilitated an old nest. Ovulation was, however, imminent in two species, the migratory kingfisher *Halcyon pyrrhopygia* and the seed-eating pigeon *Ocyphaps lophotes*. All other species were in pairs and though dispersed there was neither active territorial behaviour, nor courtship.

An interesting event was an overnight shower of some 30 points which stimulated a wagtail *(Rhipidura leucophrys)* to loud territorial song, and a flock of *Struthidea* into a frenzy of nestbuilding activity. By the second day, however, the small pool from which they were drawing mud had dried up and the partly built nests had been abandoned.

Coonabarrabran area.

The same species as at Bourke, plus the following, had nested in early spring: *Myzantha melanocephala*, *Meliphaga fusca* (honeyeaters), *Microeca fascinans* (flycatcher). The butcher-bird *Cracticus destructor*, had a nest containing three large young. Other birds, including the newly returned migrants, had neither bred nor were showing any interest in breeding, though they were in pairs.

Bingara area.

Widespread nest-building (some 20 species) had resulted from the 3 inches of rain a fortnight before. Territorial aggressiveness and song were most pronounced.

The age-structure of the Bingara population indicated wide breeding at the beginning of the season (early nesters) but, except in *Cracticus* and a couple of other species, a virtual cessation until the October rain.

The drought spring of 1957 in the Sydney area of N.S.W.

Breeding was widespread and heavy during the first part of the season. The drought effects showed in (i) a delay of almost a fortnight in the ovulation date of many species, (ii) an early tapering off in the nesting and failure to rear second broods. The returning migrants (that normally start in late October) were severely affected. Thus, the majority of pairs of the warbler, *Gerygone olivacea*, and whistler, *Pachycephala rufiventris* attempted to nest but were forced either to abandon efforts or else lost young (apparently from heat) in the nest. The cicada-bird *(Edoliisoma tenuirostris)* did not nest at all.

Heavy rain in late summer (early 1958) initiated a limited amount of autumn nesting, as normally occurs under conditions of heavy summer rain.

The Environmental Factor initiating Breeding.

MARSHALL & SERVENTY (1957) have pointed out that all parts of the continent in which spring breeding is the rule have a winter rainfall. Hence, they argue, rain could actually be the initiating stimulus here as well as in the interior, and the fact that the day-length is increasing at this time may be purely coincidental.

Three observations indicate, in the opinion of the writer, either the presence of an internal rhythm or that light cannot, out of hand, be discarded as at least an accessory an "environmental trigger" in Australia: (a) There was spring testicular recruidescence in the Bourke area in 1957, (b) In north-eastern Queensland, despite the bulk of the rain being in the summer, breeding still takes place in the spring, (c) The silvereye, *Zosterops lateralis*, has been experimentally shown to be responsive to changing day-length. It is obvious that breeding can only take place when a range of conditions are fulfilled and, in the long run, the most important of these is the immediate environment. More experimental work, however, is necessary before the question of whether rainfall is essentially dominant as the environmental trigger in Australia, or whether light has a role, can be finally resolved.

The Moult in Australian Birds

It has long been known that the major moult of birds in the Northern Hemisphere occurs at the end of the breeding season and, in those species that have a second moult, this takes place immediately prior to nesting season in spring (see DWIGHT, 1900 et al.). The constancy of this succession, especially the fact that reproduction and moulting never overlap, is so invariable as to amount virtually to a physiological "antagonism" between breeding and moulting.

As yet the moult of only one Australian bird has been studied in detail *(Zosterops lateralis,* (KEAST, 1952)) and this species, an inhabitant of the fertile east coast, is as regular in starting date and time taken to complete the process as are passerine birds in northern Europe and North America. *Zosterops* has two moults, a complete one in late summer — autumn, and a partial one in July—August. Possibly the majority of bird species inhabiting the eastern and southern coastal forests have moults as regular as these in *Zosterops*.

The only worker to pay any attention to moulting in birds inhabiting the dryer parts of Australia is SODERBERG (1918). This worker, after some ten months residence in the Kimberleys and though somewhat limited in the number of specimens of each species he was able to collect, was able to draw attention to the moult being both more individual (in time of starting and finishing) and protracted than is the case in Europe. Again, different species, he noted, varied in moulting time relative to the summer "wet" season, some moulting before it, others during it, and in some cases the moult spread throughout the entire summer. In those cases where there was a second moult it typically occurred at the opposite time of the year, that is in the winter. SODERBERG's work is significant in revealing basic information on the moult of a variety of species. Unfortunately, however, he did not appreciate the significance of the relationship between moulting and breeding and accordingly missed the opportunity of getting fundamental physiological information.

Recent work by the writer in inland and northern Australia has indicated that the moults are intricately bound into the over-all adaptational pattern in various species. Various desert species were found still to be in moult in central Australia in April and May, 1952, that is somewhat after the time of completion of the process in the south of the continent and the observation of SODERBERG that several species moulted in the north of the continent in June and July was confirmed (KEAST, 1959b). An interesting observation, indicating geographic variation in moulting time within a species, was that specimens of the honeyeater, *Myzomela obscura*, collected at Port Keats in the Northern Territory in July were in full moult whilst those taken only a couple of weeks later at Cairns, in the north-east, had obviously been in full feather for some time. The breeding season of this species is not precisely known in the north-west of the continent but it is probably in summer, a month or two later than in the north-east.

Collections of birds made at Bourke, N.S.W. in November, 1957, indicated that a variety of species were then starting to moult. Interestingly enough, this included not only those that had successfully bred early in the season but several that had been prevented by the drought from so doing. That is to say these latter were "moulting to schedule" notwithstanding that the breeding that had preceded the moult had been eliminated from the cycle. A group of apostle birds that started nest-building with the overnight fall in rain were in full moult at the time. Gonadal studies showed that not only were the testes of maximum size and actively proliferating sperm but the ovaries gave every indication of being able to discharge ova at the appropriate time. It was obvious that, in this species at least, it was inevitable that there would be a period when the two processes must proceed together.

With reproduction depending on rainfall, as it does in various parts of Australia, and with it liable to occur at any seasons, and with response to rain rapid, it would appear that, on occasions, either moulting and breeding must, in point of fact, occur simultaneously (note observation of SODERBERG that on one occasion he shot a cuckooshrike, *Pteropodocys maxima*, from its eggs and found it to be in full moult), or that the moult must temporarily be suspended, to be resumed after the period of reproduction. These matters have yet to be investigated in Australia.

Modification to the moult as part of an adaptation to life under arid conditions is not unique to Australia. The recent observations of WAGNER (1957) with respect to twenty species of Central American hummingbirds should be noted. These birds, he found, moult at any time of the year as a result of the irregularity of their breeding (which is dependent on rainfall). Significantly enough, however, in hummingbirds no moulting occurs as long as a bird is in breeding condition so that "moulting is fitted into pauses between breeding periods".

Adaptations ensuring the recovery of population numbers following droughts and heat waves

Numerical Losses during Droughts

During periods of drought population numbers fall due to two causes

(a) Species do not breed, or else only produce a few young, so that the natural wastage due to death from old age and predation, is not made good.

(b) Notwithstanding the various adaptations to prevent this, death due to starvation, thirst, or extreme heat, may occur. Undoubtedly some individuals perish in this way during every drought. Infrequently, however, the deathroll reaches such catastrophic proportions as to be recorded in the literature.

Failure to Breed and Reduced Clutch Size during Droughts

There are various accounts in the literature of birds not breeding in protracted droughts. Thus, BERNEY (1906) refers to "nesting being out of the question" for most species during 1905 (Hughendon, Queensland), it "taking them all their time to keep alive". Again, in the seventeen months up to mid-February 1925 (western-central and western-northern Queensland) no birds nested (BERNEY, 1928). Few birds, if any, bred during the drought years of 1921 and 1922 in the Lake Frome area of northern South Australia (McGILP, 1924). Virtually no water-birds bred in south-eastern Australia during the drought year of 1957, they being concentrated on the coast, far from the dry inland swamps in which they normally breed (H. J. FRITH, M. SCHRADER, and various other correspondents). When droughts do not eliminate the nesting of species clutch-size may be reduced (McGILP, 1924; BERNEY, 1928). Thus, the latter refers to the parrot *Kakatoe roseicapilla* having only two eggs instead of 4—5, the honeyeater *Myzantha flavigula* 2 (normal clutch 3—5), the warbler *Smicrornis brevirostris* (1 instead of 2—3) and the magpie *Gymnorhina tibicen* (2 instead of 3—5) in central Queensland during the drought year of 1926. It might be noted, however, that in the "fertile" Sydney area though the drought spring of 1957 did interfere with breeding it did not influence clutch size to any extent.

An interruption to breeding in dry years is known from Africa (MOREAU, 1950), Central America (SKUTCH, 1950), and other places. Instances of reduced clutch size in dry years in other Continents are hard to find but a few instances are recorded from Africa (MOREAU, 1944; LACK, 1954; BOURNE, 1955).

"Catastrophic" Losses of Bird Life during Droughts

Birds, being mobile animals, commonly respond to a deterioration in conditions by moving out of the area concerned. When, however, a drought is sufficiently severe and widespread this course is not open to them. Thus, BARNARD (1917, 1927), a long-term resident of central Queensland, lists five drought periods when there was a marked extermination of birds: 1902, 1915, 1919, 1922 and 1926. The first of these was of exceptional severity, the whole of central Queensland being "without a vestige of grass" and even the timber died over miles of country. It almost "wiped out" the birds over a wide section (see Fig. 4). Restocking took years, no wrens *(Malurus melanocephalus)* being seen for three years, and even up to 1917 not reaching their former numbers. Kookaburras *(Dacelo novae-*

guineae) were rare for several years and a parrakeet *(Psophotus pulcherrimus)*, formerly a common species, was never seen again in the area (this species is now apparently extinct). Repopulation of the area was, of course, accomplished in two ways, by colonization from without, and by reproduction.

Fig. 4. Instances of the wholesale extermination of bird-life due to heat and drought. The droughts of 1902 (especially), 1905, 1915, 1919, 1922, 1926, caused a heavy death-roll in central Queensland (area "A"). A record heat wave in January-February, 1932, caused "catastrophic" loss of bird-life at various places within area "B".

An equally great loss of bird life is recorded for northern South Australia (area marked on Fig. 4) during the record heat wave of 1932 (see various accounts in *S. Aust. Ornith.* vol. 11, parts 6 and 7). At this time the temperature exceeded 116° F for sixteen consecutive days in one area and did not drop below 100° F for two months in another. Most deaths took place in the vicinity of water-tanks and dams, at which the birds concentrated in an attempt to keep cool. Over twenty species were recorded as dying but the bulk were small, (length, 7½ inches) grass parrakeets *(Melopsittacus undulatus)*, 60,000 of which dropped dead at one dam, and on one verandah a 40 gallon petrol drum was filled with bodies in an afternoon.

The Build-up in Numbers during Good Seasons

The evidence indicates that the mechanism for building up population numbers following droughts is accomplished in the following ways:
- (a) By concentrating in areas most favourable to the rearing of young.
- (b) A more protracted period of breeding than usual or even by having two periods of breeding, that is to say an increased number of clutches is reared.
- (c) Increased clutch size.

The Concentration of Birds in Areas Favourable to the Rearing of Young

Breeding alone underlies much of the nomadism in Australian interior birds for species concentrate for breeding wherever conditions happen to be best. Examples of how widely separated areas may serve as breeding grounds in successive years have recently been given by the writer in the case of wood-swallows *(Artamus)* and chats *(Epthianura)* — (KEAST, 1958a, 1958b).

Increase in the number of Broods

It is well-known that, in a good season following a drought, nesting will be concentrated and protracted. Thus McGILP (1924) has written of the Lake Frome area "In good seasons some birds nest practically all the year..." The year 1946 was such a one in the Cunnamulla district of western Queensland (M. SCHRADER, personal communication). Unusually good summer rains led to an upsurge of breeding towards the end of the normal spring nesting and this continued well into the autumn. About this time the river levels rose and marshes and billabongs were flooded, leading to a striking nesting of water birds that continued throughout the autumn, winter and following spring. There was also a period of summer-autumn nesting following the spring one this same year in the Finley-Deniliquin area of southern New South Wales (J. HOBBS, personal communication).

Whilst it is generally accepted that birds have more broods than usual in bountiful years specific examples of multiple broods by a single pair are rarely properly authenticated. There are, however, a few recorded examples of isolated pairs of wagtails *(Rhipidura leucophrys)* successfully rearing three broods in rapid succession — Collarenebri, N.S.W., 1946 (J. LAWSON, personal communication) south-western Australia, 1934 (SEDWICK, 1947). In the former case the birds laid again in the old nest three days after the second brood left it. Again, GREENHOW (1948) records a pair of thornbills *(Acanthiza chrysorrhoa)* successfully raising four broods in the same nest between November 1946, and February (1947) in south-western Australia.

The ability of various bird species to nest in spring and again in autumn if rain falls is probably fairly widespread in Australia. It has been commented upon by ROBINSON (1954) and CARNABY (1954) in the Hamersley-Gascoyne section of Western Australia and is not abnormal in a few species near Sydney *(Grallina cyanoleuca, Poephila castanota, P. bichenovii, Meliornis novaehollandiae)*. The crimson chat, *Epthianura tricolor*, nests in the south of the continent in spring and, in favourable years, again in autumn in central and northern Australia (KEAST, 1958a).

Increased Clutch Size

It has been known to egg-collectors for a long time that in the areas of marked climatic fluctuations the clutch-size of various species may increase in a good year. Thus WHITE (1931) states that the normal clutch-size of magpies *(Gymnorhina dorsalis)* in south-western Australia is 3—4 but that in the "remarkably favourable spring of 1912" he received details of no less than six clutches of five eggs each. McGILP (1924) lists six species of birds whose clutches were larger than normal in 1923 in the Lake Frome area of South Australia, following the breaking of a drought. In each case the net increase was one egg, 3 to 4, or 4 to 5. The trend, it might be noted, is by no means confined to Australia. For references to papers on food-clutch size relationships in Africa and Europe see LACK (1954).

LACK (1947, 1954) discusses factors controlling and influencing clutch-size in birds. It is pointed out that in some species the number of eggs is fixed by heredity but in others varies with latitude, age, availability of food, and so on. The influence is that in the case of many of the species inhabiting the area under discussion here nutrition is the all-important factor.

Increased clutch size is probably only an incidental mechanism for building up numbers after a drought. The evidence is, moreover, that it only occurs in certain species (McGILP, 1924).

SPENCER (1896) produces evidence that, in a good year following drought, the desert marsupial mouse *(Sminthopsis crassicaudata)* will have a larger litter than normal, and that there may be two breeding periods. MORRISON (1945) states that the mouse *(Mus)* plagues that periodically ravage southern Australian occur in the first good years following a drought and result from the animals having larger families at more frequent intervals. The mechanism for numerical build-up in certain mammals would, accordingly, appear to be similar to that in birds.

REFERENCES

BAKER, J. R., 1938. The evolution of nesting seasons. *In Evolution: Essays on Aspects of Evolutionary Biology:* p. *161.* Oxford Univ. Press.
BARNARD, C., 1917. Bird life as affected by drought. *Emu* **16**, *234.*
BARNARD, H. G., 1927. Effects of droughts on bird-life in central Queensland. *Emu* **27**, *26.*
BERNEY, F. L., 1906. Drought and flood in Queensland. *Emu* **6**, *26.*
BERNEY, F. L., 1928. Birds and drought in central-western Queensland. *Mem. Qld. Mus.* **9**, *194.*
BOURNE, W. R. P., 1955. The birds of the Cape Verde Islands. *Ibis.* **97**, *508.*
CARNABY, I. C., 1954. Nesting seasons of Western Australian birds. *W. Aust. Nat.* **4**, *149.*
DARLINGTON, P. J., 1957. Zoogeography: the Geographical Distribution of Animals: p. 236. John Wiley & Sons, New York.
DWIGHT, J. JR., 1900. The sequence of plumages and moults of the passerine birds of New York. *Ann. N.Y. Acad. Sci.* **13**, *73.*
FALLA, R. A., 1953. The Australian element in the avifauna of New Zealand. *Emu.* **53**, *36.*
GREENHOW, R. R., 1948. Multiple broods of Yellow-tailed Thornbill. *W. Aust. Nat.* **1**, *151.*
INGRAM, C., 1907. On the birds of the Alexandra district, North Territory of South Australia. *Ibis* **(9)** 1, *387.*
KEAST, A., 1952. The Physiology of the avian moult as evidenced by a study of the moulting cycles of the Silvereye *Zosterops lateralis.* M. Sc. thesis, University of Sydney unpub.
KEAST, A. & MARSHALL, A. J., 1954. The influence of drought and rainfall on reproduction in Australian desert birds. *Proc. Zool. Soc. Lond.* **124**, *493.*
KEAST, A., 1956a. Migration in Australian birds. *Aust. Mus. Mag.* **12**, *59.*
KEAST, A., 1956b. The moulting physiology of the Silvereye *(Zosterops lateralis) Proc. 14 Int. Cong. Zool. Copenhagen* 1953, p. *314.*
KEAST, A., 1958a. The relationship between seasonal movements and the development of geographic variation in the Australian Chats. *Aust. J. Zool.* **6**, *53.*
KEAST, A., 1958b. Seasonal movements and geographic variation in the Australian Wood-swallows *(Artamidae) Emu* **58**, *207.*
KEAST, A., 1959a. Bird speciation on the Australian Continent. *Bull. Mus. Comp. Zool.* in press.
KEAST, A., 1959b. Body fattiness, moult, and distribution of birds in central and north-western Australia during the drought winter of 1952. *Emu.* in press.
LACK, D., 1947. The significance of clutch-size. *Ibis.* **89**, *302.*
LACK, D., 1950. Breeding seasons in the Galapagos. *Ibis* **92**, *268.*
LACK, D., 1954. The Natural Regulation of Animal Numbers. Clarendon Press, Oxford: 35.
MCGILP, J. N., 1919. Notes on nest and eggs of the desert bushchat. *(Ashbyia lovensis). Emu.* **19**, *56.*
MCGILP, J. N., 1924. Seasonal influences on the breeding of native birds. *Emu* **24**, *155.*
MARSHALL, A. J. & SERVENTY, E. L., 1957. Breeding periodicity in Western Australian birds: with an account of unseasonal nestings in 1953 and 1955. *Emu* **57**, *99.*
MAYR, E., 1944. The Birds of Timor and Sumba. *Bull. Amer. Mus. Nat. Hist.* **83**, *127.*
MAYR, E., 1945. Wallace's Line in the light of recent zoogeographic studies. Science and Scientists in the Netherlands Indies. p. 241. Published by the Board for the Netherlands Indies, N.Y.
MAYR, E. & AMADON, D., 1951. A classification of recent birds. *Amer. Mus. Novitates* No. 1496.

MAYR, E. & SERVENTY, D. L., 1944. The number of Australian bird species. *Emu* **44**, *39*.
MIYAZAKI, H., 1935. On the relation of the daily period to the sexual maturity and to the moulting of *Zosterops palpebrosa japonica*. *Sci. Rep. Tohoku Imp. Univ.* **9**, *183*.
MOREAU, R. E., 1944. Clutch-size: A comparative study, with special references to African birds. *Ibis.* **86**, *286*.
MOREAU, R. E., 1950. The breeding seasons of African birds. 1. Land Birds. *Ibis.* **92**, *223*.
MORRISON, P. C., 1945. Mouse plagues past and future. *Wild Life* **7**, *11*.
ROBINSON, A. H., 1954. Nesting seasons of Western Australian birds — further contribution. *W. Aust. Nat.* **4**, *187*.
ROWAN, W., 1928. Experiments in bird migration. 1. Manipulation of the reproductive cycle; Seasonal histological changes in the gonads. *Proc. Boston Soc. Nat. Hist.* **39**, *151*.
ROWAN, W., 1938. Light and seasonal reproduction in animals. *Biol. Revs.* **13**, *374*.
SEDGWICK, E. H., 1947. Breeding of the Black and White Fantail. *W. Aust. Nat.* **1**, *14*.
SERVENTY, D. L. & WHITTELL, H. M., 1948. A Handbook of the Birds of Western Australia: p. 6 Patersons Press, Perth.
SKUTCH, A. F., 1950. The nesting seasons of Central American birds in relation to climate and food supply. *Ibis* **92**, *185*.
SMITH, K. D., 1955. The winter breeding season of land-birds in Eastern Eritrea. *Ibis.* **97**, *480*.
SODERBERG, R., 1918. Studies of the birds in north west Australia. Results of Dr. E. Mjobergs Swedish Scientific Expeditions to Aust. 1910—1913. *Svensk. Veta. Al. Handl.* **52**, No. 17 : 3.
SPENCER, B., Report on the work of the Horn Scientific Expedition to Central Aust. Mammalia, II, p. 2 Dulau and Co., London.
WAGNER, H. O., 1957. The moulting periods of Mexican Hummingbirds. *Auk.* **74**, *251*.
WHITE, H. L., 1931. Do variations in seasons effect the sieze of eggs?. *Emu* **13**, *48*.

VII
THE REPTILES OF AUSTRALIA

by

ALLEN KEAST

(Australian Museum, Sydney)

The Australian fauna

The components that make up the Australian reptile fauna are as follows:

Crocodiles:	1 species (plus 1 marine species)
Land and fresh-water tortoises:	about 10 species (plus 4 marine turtles)
Lizards:	230-240 species
Land and fresh-water snakes:	130-140 species (plus 20 sea-snakes)
Total:	360-380 species

The land and fresh-water reptile fauna of Australia amounts to 8% of the living reptiles of the world.

New Guinea to the north has some 240 species and New Zealand to the west about 20. Eliminating duplication, although the three areas have little in common at the specific level, the reptile faunas of Australia, New Guinea, and New Zealand, together total some 11% of the world species.

The families represented in Australia and the number of species relative to those occurring elsewhere in the world are set out in Table I.

Table I

The families of reptiles occurring in Australia

Family	Number of species in Australia	Total world fauna
Chelyidae - chelyid fresh-water tortoises (Aust. & South America)	4 genera, 13 species	10 genera, 30 species (MERTENS & WERMUTH, 1955)
Agamidae - dragon lizards (warmer parts of Old World especially Asia)	8 genera (6 Australo-Papuan), 30 species (WAITE, 1920)	30 genera, 280 species (SMITH, 1935).
Gekkonidae - geckos (cosmopolitan)	about 14 genera, 35 species (WAITE, 1920)	70 genera, 700 species (SMITH 1935; LOVERIDGE, 1945).
Pygopodidae - legless lizards (Australo-Papuan.	8 genera, 13 species (KINGHORN, 1926)	8 genera, 14 species (KINGHORN, 1926).

Family	Number of species in Australia	Total world fauna
Scincidae - smooth-skinned lizards (cosmopolitan)	About 10 genera, perhaps 120 species (WAITE, 1920)	40 genera, over 600 species (SMITH, 1935)
Varanidae - goannas (Africa, south-east Asia, Australia)	1 genus, 15 species	1 genus, 24 species (MERTENS, 1942)
Typhlopidae - worm snakes (cosmopolitan)	1 genus, 19 species (KINGHORN, 1956)	4 genera, 200 species (DARLINGTON, 1957)
Boidae - pythons and boas (Tropicopolitan)	3 genera, 8 species (KINGHORN, 1956)	22 genera, 57 species (DARLINGTON, 1957)
Colubridae - common snakes (cosmopolitan)	10 genera, 14 species (KINGHORN, 1956)	Perhaps 180 genera, 2.000 species
Elapidae - front-fanged snakes (cosmopolitan)	15 genera, about 70 species (KINGHORN, 1956)	30 genera, perhaps 300 species.
Crocodylidae - crocodiles (tropicopolitan)	1 genus, 2 species (one endemic)	3 genera, 13 species (MERTENS & WERMUTH, 1955)

Australia has only one endemic reptile family, the legless "scale-footed" lizards (Pygopodidae), which also occurs in New Guinea. The number of endemic genera is, however, high and almost two-thirds of the species belong to them. There has been much radiation and speciation in several of the endemic genera, especially in small, swift-running dragon lizards of the genus *Amphibolurus* (Agamidae), various geckos and skinks and in the case of some of the elapid snakes (e.g. *Denisonia*). Genera equally developed elsewhere in the world that contain numerous species in Australia include the skink lizards *Lygosoma*, the worm snakes *Typhlops*, and the goannas, *Varanus*.

Amongst the more unusual endemic lizards is the primitive gecko genus *Nephrurus*, which has a grotesque bony head. The monotypic, desert devil lizard *(Moloch)*, 4—6 inches long, is characterized by a curious armour of long, thorn-like spines, and subsists solely on small black ants. It has a superficial resemblance to the American desert horned toad *(Phrynosoma*, Iguanidae). The monotypic Frilled Lizard *(Chlamydosaurus kingi)* is characterized by a neck bearing a distensible skin-collar that, in resting position, lies in neatly folded symmetrical pleats along the side of the body. The frill, elevated with the suddenness of an opening umbrella when the lizard is frightened, is brightly coloured and gives the reptile a huge and formidable appearance.

The snake fauna of Australia is characterized by the great development and radiation of the front-fanged poisonous snakes (Elapidae), in contrast to the rest of the world where the Colubridae (either fangless or with rear fangs) are dominant. This would seem to be somewhat of an anomaly as the Elapidae are generally believed to be a more recent development than the Colubridae and to be derived from them (DARLINGTON, 1957). The majority of Australian elapids are small and their venom insignificant to anything larger than a rodent. Included in the fauna, however, are several species highly dangerous to man: the Taipan *(Oxyuranus scutellatus)*, the Tiger Snake *(Notechis scutatus)*, the Death Adder *(Acanthophis antarcticus)*, the Brown Snake *(Demansia textilis)*, and the Copperhead *(Denisonia superba)*.

Because of the variables involved it is virtually impossible to compare the potency of the venom of the more dangerous Australian snakes with those of other continents. N. H. FAIRLY (1929) who, together with A. H. KELLAWAY has carried out extensive work on venoms in Australia, states, however, that the average venom yield of the Tiger Snake at one milking would suffice to kill 118 sheep; the Death Adder, 84; Indian Cobra, 31; Copperhead, 8; Russel's Viper, 2. The killing power of the Tiger Snake is thus 3.5 and the Death Adder 2.7 times that of the Cobra. TIDSWELL & FERGUSON (FAIRLEY, 1929) give the recorded death-rate for bites from the various species (prior to the introduction of modern treatments), as: Death Adder (10 recorded bites), 50% deaths; Tiger Snake (45), 50%; Brown Snake (70), 8.6%. The Taipan, growing to 11 feet in length and with fangs of up to half an inch, is undoubtedly one of the most dangerous snakes of the world. KELLAWAY (1929) states that the glands have a venom content of up to 400 mg, equal to 200 lethal doses. The human death-rate from bites stands at well over 50%.

There does not appear to be any valid ecological reason why such a high proportion of Australian snakes have venom glands, it is rather that the front-fanged snakes happen to be the dominant group.

WALLACE's Line and the area of geological instability to the east correspond to the general transition zone between the Asian and Australian continental reptile faunas. Several Asian lizards like *Draco* (Agamidae) and *Mabuia*, (Scincidae) and the tortoise *Trionyx* (Trionychidae) extend to Timor (DE ROOIJ, 1915). The vermiform lizard *Dibamus* (Dibamidae), the snake *Cylindrophis* (Aniliidae), *Mabuia*, and other species reach New Guinea (LOVERIDGE, 1946), the arboreal *Goniocephalus* (Agamidae) and various water snakes New Guinea and Australia. Such genera are to be compared with a series that are as well developed in Australia as in Afro-Asia *(Varanus, Physignathus, Phyllodactylus)*, or over the warmer regions of the world as a whole *(Lygosoma, Typhlops)*. To these genera, the place of origin of which is uncertain, WALLACE's Line does not appear to have constituted a distributional barrier at all.

The westward movement of Australian reptile genera has been only on a reduced scale. The continental elapid *Acanthophis antarcticus*, reaches the Halmaheras. The giant skink *Tiliqua gigas*, extends across WALLACE's Line to Java and Sumatra.

The Australian fresh-water tortoises (Chelyidae) are of particular interest in that they differ from those of the Orient (the latter extending to Timor), and agree with those of South America. The 10 genera and 30 species in this family are distributed as follows: South America, 6 endemic genera and 11 species; Australia, 4 genera (2 endemic, 2 shared with New Guinea), 13 species; New Guinea, no endemic genera, 6 endemic species with an additional 4 occurring also in Australia (MERTENS & WERMUTH, 1955).

Australia has relatively little in the way of "living relics" amongst reptiles. Several such occur, however, on the islands adjacent to Australia. Thus the unique *Sphenodon* (Sphenodontidae), the lone living representative of an Order of reptiles widespread as fossils up to the Lower Cretaceous (perhaps 135,000,000 years ago), persists on a few islands off the New Zealand coast. The monotopic Pitted-shelled Turtle *(Carettochelys)* of New Guinea is the only surviving member of its family. Related genera are known from the early Tertiary of Eurasia and North America (DARLINGTON, 1957). The occurrence of a live iguanid species on Fiji, it and those on Madagascar being the only members of the family living outside of the Americas, pose a fascinating problem for zoogeographers. The distribution of fossils of the extinct, giant, horned land tortoises *(Meliolania*, Meliolanidae) in eastern Australia, Lord Howe Island, New Caledonia, and South America is equally puzzling.

The distribution of reptiles within Australia

The following section was derived from distribution maps drawn up for a sample, approximating 250 species, of the reptiles occurring in Australia.

Various field workers have kindly supplied data, especially R. MACKAY, H. COGGER, J. CALABY, and A. R. MAIN. The groups covered by the survey are as follows: Gekkonidae, Varanidae, Agamidae, Pygopodidae, 30 of the 120-odd Scincidae (the genera *Tiliqua, Leiolopisma, Egernia*), Elapidae, Colubridae, and Pythonidae. The bulk of the skinks are in need of a modern taxonomic revision and hence there is no point in attempting to include them. Likewise, the distributional limits of Chelyid tortoises are poorly known. In making interpretations, allowance has had constantly to be made for deficiencies in the older systematic reviews, particularly in not stating which forms are geographic counterparts and should be grouped as belonging to super species. Little habitat data is available for Australian reptiles and distributional knowledge is inadequate for many species. Hence, a number of species may prove, in the future, not to fit as neatly into some of the distributional categories as they now appear to do.

The present review seeks to determine if:
(a) Reptile distribution in Australia is reducible to a number of generalized patterns.

(b) Hypotheses as to the past history of groups on the Australian continent can be drawn from their ranges today.
(c) General conclusions can be drawn as to the ecological basis of distribution of species.

Virtually no physiological or ecological studies have as yet been carried out on Australian reptiles, hence no assistance is forthcoming from that source.

Distribution of families

The number of species of each of six reptile families occurring in a selection of 12 regions of the Australian continent are set out in

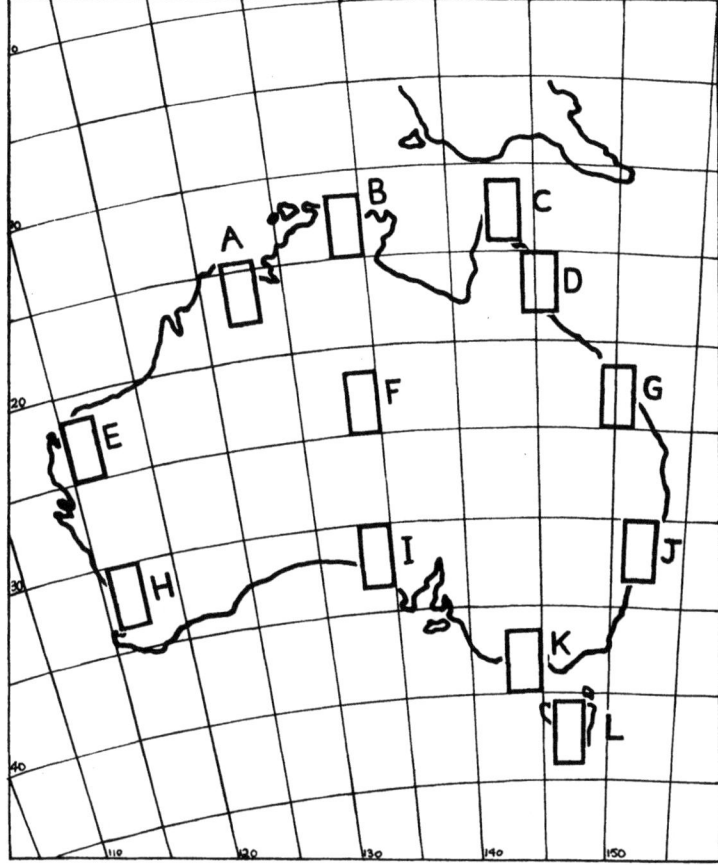

Fig. 1. Distribution of reptile families in Australia. Rectangles represent a series of areas of differing latitude, temperature, and rainfall. The number of species of Elapidae, Colubridae, Pythonidae, Typhlopidae, Gekkonidae, and Varanidae, recorded from these areas are set out in Table II.

Table II. The regions, that are shown on Fig. 1, represent the full extremes of temperature and rainfall conditions:

Table II

The distribution of reptile families on the Australian continent

		Elapidae	Colubridae	Pythonidae	Typhlopidae	Gekkonidae	Varanidae
Kimberleys	(A)	10	5	5	7	6+	4
Arnhemland	(B)	10	9	5	3+	7	4
Cape York	(C)	9	9	5	5	7	7
Cairns - Atherton	(D)	15	9	5	6	6	3
Hamersley area	(E)	10	—	4	4	9	3
Central Aust.	(F)	8	—	3	4	10	7
East-central Queensland	(G)	16	3	3	6	6	3
South-west corner	(H)	16	—	3	5	8	3
South-western South Aust.	(I)	8	—	1	2	6	2
Central-eastern N.S.W.	(J)	19	2	1	4	10	2
Melbourne area	(K)	12	—	1	3	3	2
Tasmania	(L)	3	—	—	—	—	—

The distribution of the various reptile families occurring in Australia may be summarized as follows:

Elapidae

Apart from the degree of radiation achieved the fairly long history of this group in Australia is shown by the extent of habitat specialization and the number of species occurring in the colder south. The greatest number of species occur in the ecologically diverse south-east (where there is an intermingling of northern, southern, and inland forms) and south-west, which has about 3 endemic genera and 6 endemic species. The dryer areas have less species of elapid snakes than the wetter ones. Tasmania has only 3 species.

Colubridae

This family has only recently arrived in Australia and is restricted to the north, except for a couple of species that penetrate down the east coast. All are either water snakes or tree-snakes, that is to say they occupy niches not filled by elapids. None enters the dryer country.

Nine species have been recorded each from Arnhemland and Cape York, but 5 are different in the two areas, indicating separate invasion pathways into the north-west and north-east respectively. Only in a couple of instances, however, do the Australian populations differ morphologically from the parental ones in Indonesia and New Guinea.

Pythonidae

Pythons likewise are a tropical group with the greater concentration of species in the north of the continent. In contrast with the colubrids, however, pythons have spread far out into the desert and colder southern limits of the continent. Thus, three occur in the south-western corner.

Pythons are of two main ecological types, ground pythons that mostly inhabit the more open country and shelter in old mammal burrows and rocky crevices, and tree-pythons. Two of the latter, inhabiting the rain forests of the north-east are recent immigrants, the giant (20 foot long) *Python amethystinus* having reached race status, and the Green Python, *Chondropython viridis* of Cape York, not yet having done so.

Typhlopidae

Approximately the same number of species have been recorded from the various parts of the continent, except Tasmania. These small, secretive reptiles cannot be considered adequately known, however. A couple of the northern "species" are recorded from one area only and others only from widely separated localities.

There is a continuing colonization from New Guinea.

Gekkonidae

With the exception of Tasmania (no species) the number of gecko species occurring in the different parts of the continent is fairly uniform. The group is very well developed-in the central desert. Geckos live under logs and stones on the ground, where there are rocky outcrops, on cliff faces, and under bark on dead and living trees. They shun damp habitats. Thus, rain forests and wet sclerophyll forests do not suit geckos well.

The 10 species shown as occurring in central eastern New South Wales merits comment. H. COGGER informs me that only 4 species occur in the coastal section between Sydney and the Blue Mountains, whilst one occurs near Singleton, two about the dry head of the Hunter valley, and an additional 3 species enter the section along the cold New England Tableland (Tamworth). That is to say, no more than 4 occur in any one district.

The number of distinct endemic gecko genera (8—9) confirms other evidence that the family has had a long history in Australia.

Varanidae

This group of large lizards, recently reviewed by MERTENS (1942) is least developed in the south, richest in number of species in the central desert and on Cape York. Four species are restricted to the desert (two of them "pigmy" goannas). The high figure for Cape York is because that continues to be a point of entry for New Guinea species, e.g. *V. salvator* and *V. prasinus.*

Different species of goannas are specialized for terrestrial life, when they live in burrows (most species), arboreal life (especially *V. varius*), or else are water-goannas *(V. mertensi).*

A Pleistocene goanna *(Megalania (= Varanus) prisca)* is estimated, on skull dimensions to have reached 15 feet in length. This is a size considerably in excess of that of the largest living lizard of today, the Komodo Dragon *(V. komodoensis)* males of which reach 10 feet in length.

Pygopodidae

This endemic family of eight genera (3 monotypic) has a highly aberrant distribution in Australia. Only 3 genera extend throughout the continent and only 5—6 species have a wide range. One species reaches Tasmania. Two monotypic genera *(Pletholax* and *Ophioseps)* are confined to south-western Australia and a third of two species *(Aprasia)* obviously originated there. The monotopic *Ophiocephalus* is only known from the desert region of northern South Australia and the monotypic *Paradelma* from central Queensland *(*KINGHORN, 1926).

The pygopodids are a true autochthonian element as is shown not only by their confinement to the Australo-Papuan region but by the number endemic to south-western Australia.

Agamidae

This family reaches its greatest development under desert conditions, a number of species (especially in *Amphibolurus)* being confined to there. The monotypic *Moloch* is also a desert species. *Physignathus* (water dragons) occur in the north, desert and east. The arboreal dragons, *Goniocephalus*, Indo-Malayan rain forest species, are confined to the eastern coastal strip.

There has been much radiation in the Australian agamids and the family. There are, however, only 3 endemic genera, an additional 2 being shared with New Guinea.

Scincidae

This family has obviously had a long history on the continent and has undergone much radiation. There are some 8 endemic genera, though a couple of these have been reduced to subgeneric status by recent workers.

Chelyidae

There are two main groups in Australia, the long-necked tortoises *(Chelodina)* and the short-necked tortoises *(Emydura)*, representatives of both extending widely over the continent. In addition there is a monotypic genus in the north *(Elseya)* and another in the south-west *(Pseudemydura)*.

The bulk of the species are northern and eastern, or eastern, in distribution, whilst there has been some interchange of species with New Guinea. Precise distributional limits are not known for the bulk of species.

Crocodilidae

The endemic fresh-water crocodile, *Crocodilus johnsoni*, extends through the lagoons and rivers of the northern fringe from the Kimberleys to Atherton, and the large Indo-Malayan saltwater crocodile over a somewhat similar range. A fossil crocodile, presumably of Pleistocene age, is known from Balladonia, South Australia, 1,000 miles to the south of the present range of the family (DAVID & BROWNE 1950). Together with a fossil lungfish, and other evidence, this indicates the former occurrence of warmer temperatures and permanent streams in this section of the continent.

Distribution of species

The common patterns of distribution occurring in Australian reptiles are shown in Fig. 2. These are as follows, the numbers corresponding to those on the map:

Peripheral Distribution Patterns

(1) Tasmania and coastal strip of south-eastern Australia, sometimes as far westward as Adelaide, but typically restricted to the highlands in the northernmost parts of the range (New South Wales). This is shown on Map "A" of Fig. 2.

At least 5 species, in the groups surveyed, have this kind of distribution, e.g. *Denisonia superba*, *Amphibolurus diemenensis*, *Tiliqua nigrolutea*. Of the two lizard species restricted to Tasmania, *Leiolopisma pretiosa* and *L. ocellata*, one is hardly specifically distinct from a counterpart in the Koskiusko region of New South Wales (LOVERIDGE, 1934).

(1a) Like the above but with, in addition, an isolate in the forested south-western corner of Western Australia. There are some 5 of these, e.g. *Notechis scutatus* and *Denisonia coronoides*.

(2) South-western corner of Western Australia. (Map "B" on Fig. 2). At least 15 distinctive species, about 8 belonging to endemic, monotypic, genera, are restricted to this section. Included in these genera are: *Rhinhoplocephalus*, *Elopagnathus* (snakes), *Ophioseps* and *Pletholax* (legless lizards), and *Pseudemydura* (tortoise).

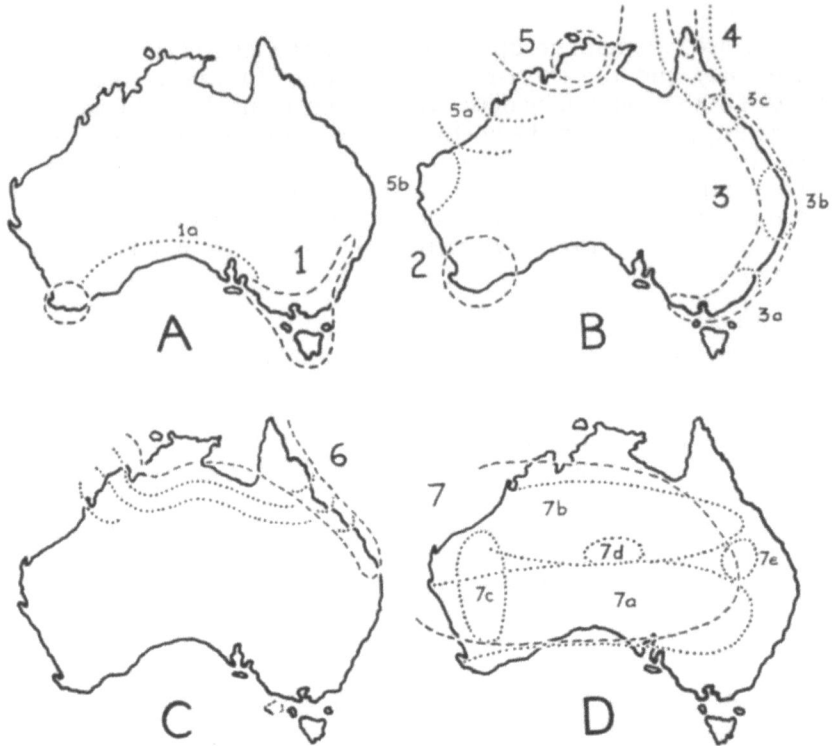

Fig. 2. Common distribution patterns of Australian reptile species (see text).

(3) Eastern coastal strip (seaward side of Dividing Range) of Australia. About 8 species range throughout the section southwards from Cairns, sometimes to south-eastern South Australia, e.g. *Pseudechis porphyriacus, Denisonia signata* and, apparently, *Varanus varius*. (Map "B").

(3a) A couple of species are apparently restricted to the more southern section of this belt: *Typhlops nigrescens*, the race *Morelia spilotes spilotes)*, and *Denisonia flagellum* (Victoria with an isolate in the Mount Lofty Ranges).

(3b) A dozen or more species have ranges of variable size and type in the centre of this eastern strip (i.e. northern New South Wales to central Queensland); *Hoplocephalus bitorquatus, Brachyurophis australis, Tropidechis carinatus* (all elapines). A few species inland from coastal Queensland into the upper parts of the Darling basin, e.g. *Denisonia maculata* and *Leiolopisma vertebralis*, the coastal mountains being low here.

(3c) Northernmost parts of section (especially Cairns-Atherton).

This rain forest section has certainly not been properly surveyed. The monotypic gecko *Carphodactylus laevis*, tree dragon *Gonyocephalus boydii*, and skink *Egernia frerei* are restricted to it, whilst the goanna *Varanus semiremex* has a slightly wider range.

(4) Cape York only, or extending southwards to Cairns and beyond, (Map "B"). These are reptiles that have only recently colonized the continent from New Guinea. At least 15 species are involved and they are typically inhabitants either of rain-forest, the shoreline, or river margins. Many belong to genera not otherwise represented in Australia, e.g. several colubrid snakes, the python *Chondropython viridis*, and gecko *Thecadactylus australis*.

(5) North-western coastal section of continent (Kimberleys and Arnhem-land). About 10 species, several of them colubrid snakes that have recently entered the continent from the north-west, have this distribution pattern. *Varanus mertensi* and *Tympanocryptis uniformis* are endemic examples whilst the python *Liasis olivaceus* extends well to the south down the west coast. Some species that extend well out into the desert proper possibly radiated from this section of the continent, e.g. the goannas *Varanus timorensis* (also on Timor) and *V. acanthurus*.

(6) North coast of continent (from Kimberleys or Arnhem-land) to Cape York, and thence for a variable distance down the east coast. (Map "C" on Fig. 2.). There are some 20 such species, including the blind snake *Typhlops unguirostris*, the python *Liasis fuscus*, the tree-colubrids *Boiga fusca* and *Ahaetulla punctulatus*, the elapids *Denisonia pallidiceps* and *Oxyuranus scutellatus*, skinks *Egernia major* and *Leiolopisma pectoralis*.

Several of these species also occur in New Guinea, most have direct counterparts there. In some cases the range if apparently interrupted by the dry country at the head of the Gulf of Carpentaria. Others, such as the dragon *Chlamydosaurus kingii*, however, extend well out into the dryer country of the interior.

Inland Distribution Patterns

(7) Arid and semi-arid interior of the continent as a whole (Map "D" on Fig. 2.). Up to 20 species come into this category, e.g. the large elapids *Pseudechis australis* and *Demansia nuchalis*. Most desert species, however, are more restricted in their distribution. The dragon-lizards *Amphibolurus reticulatus* and *Moloch horridus* tend to be associated with spinifex-covered sand-desert. Mr. F. J. MITCHELL informs me that the association is particularly close in the case of the skink *Tiliqua branchiale*.

Desert reptiles characteristically extend to the arid western seaboard of the continent, frequently to the arid head of the Great Australian Bight in the south, and sometimes pierce the more fertile south-western corner where physiographic barriers are minimal.

Whilst a great deal less is known about reptile distribution in dry areas than in the more settled regions several sub-categories are recognisable, including:

(7a) Southern section of continent, from near the seaboard in the south-west corner (typically) eastwards for a variable distance towards the east coast. Of the 15-odd species coming into this grouping the gecko *Phyllurus milii* reaches the eastern seaboard, the skink *Tiliqua rugosa* the western slopes of the Dividing Range and *Egernia stokesii* far-western New South Wales, *Demansia affinis* and *Tiliqua luctuosa* no further east than central Australia.

(7b) Northern or northern-and-central interior section of the Continent. A few species have such a distribution, e.g. the goanna *Varanus spenceri*, is confined to the north whilst *V. giganteus* has a somewhat wider range.

(7c) Dry country of south-west and west of continent. About 6 species would appear to come into this category, e.g. the goanna *Varanus caudolineatus* and the gecko *Diplodactylus alboguttatus*.

(7d) Mountainous areas of central Australia. As there are distributional refuges in the case of various plant and animal species and contain a greater diversity in habitats than the surrounding plain it is likely that a number of reptile species may be found to be confined to them. The blind-snake *Typhlops endoterus* would appear to be such an example. Mr. F. J. MITCHELL informs me that there are various isolated populations of agamid lizards, whose taxonomic status is still in doubt, confined to the mountain ranges of central Australia.

Continent wide Distributions

(8) About 20 species, amongst those sampled, are distributed from north to south and from coast to inland, to an extent where they can only be described as having a continent-wide range. Examples are the elapid snakes *Acanthophis antarcticus* and *Aspidomorphus diadema*, the skink *Tiliqua scincoides*, and the dragon-lizard *Amphibolurus barbatus*.

Specialized and Restricted Ranges

(9) A variety of species are restricted, so far as is known, to relatively local areas in different parts of the continent. Included in these are various species of *Leiolopisma* in parts of eastern Queensland (MITCHELL, 1953) and the blind-snake *Typhlops minimus* to Groote Eylandt.

Factors Governing Distribution in Reptiles.

It is significant that over two-thirds (180 out of 250) of the species covered in this distributional survey can be grouped into less than a dozen categories and that, in many parts of the continent, there is a tendency for ranges to be delimited at certain points (e.g. Great

Dividing Range in the east). Whilst it is premature to attempt to discuss factors governing distribution it is obvious that several general effects are operative:

Temperature

The confinement of certain south-eastern species to the highlands in the northern parts of their range indicates that they or their food source, are limited by increasing temperatures. Though the evidence is less direct it is reasonable to suppose that decreasing temperatures are per se a barrier to the southward spread of tropical reptiles.

Rainfall and Humidity

These are of basic importance but the pinpointing of cases and determination of the extent to which individual species are dependent on them must await physiological studies. Arid-country species are frequently ensured against excessive water-loss by being covered with spines (outstanding example *Moloch horridus*) or hard plates *(Tiliqua rugosa)*. There is an increased tendency towards rock-dwelling (note spiny tail of *Varanus acanthurus* that enables it to wedge itself tightly into fissures), or towards a subterranean way of life with accompanying reduction in limbs (various skinks).

In some wet country geckos and skinks, on the other hand, it is apparent that the eggs must assimilate a considerable amount of moisture from the atmosphere during their development, those of *Leiolopisma guidenoti* swelling to almost three times their volume (F. J. MITCHELL, personal communication).

Contrasting adaptations in desert and wet-country reptile species would appear to include different capacities for taking up moisture through the skin, the laying down of fat stores, and for undergoing aestivation.

River Systems and Surface Water

The water-tortoises are distributed according to surface-water and several self-contained river systems have species confined to them, e.g. *Chelodina steindachneri* to the Murchison-De Grey segment, and *Pseudemydura umbrina* to the south-west. The lack of suitable streams may be the main factor limiting the southward movement of various tropical colubrids.

Vegetation

The major vegetation associations represent areas of relative constancy in the physical and biological environment. Notwithstanding this, and in contrast to birds and mammals, the vegetation associations appear to exercise only a moderate influence on reptile distribution in Australia. The tree-dragons *(Gonyocephalus)* are apparently confined to rain forests, as is the shade-living, ground skink *Egernia major*. The tree-snakes *Boiga fusca, Ahaetulla punctulata* and the tree-goanna, *Varanus varius*, only occur in areas of somewhat dense tree growth (i.e. rain forests and sclerophyll forests). Various desert reptiles, especially several dragon lizards of the genus *Amphibolurus, Moloch horridus* and *Tiliqua branchiale* tend to correspond in distribution to that of the xeric porcupine grass or spinifex *(Triodia)*, with which they are associated.

Physiography, Surface Rock and Soils

With few exceptions reptiles are ground-dwellers and their link with the substratum is a direct one. Hence, apart from the physical environment and availability of food, the extent to which the substratum provides cover and shelter is undoubtedly an all-important factor in reptile distribution. Thus species may be linked with soils of a particular texture, areas where there is substratum litter, or rocks of certain types.

Distributional Barriers and Speciation

Superficial inspection reveals the existence of a considerable number of morphologically differentiating isolates in the Australian reptile fauna, and that active speciation is taking place. These isolates occur on either side of the Torres and Bass Strait water barriers (in various species), and of arid areas such as the Nullarbor Plain (e.g. in the snake *Notechis scutatus* — GLAUERT, 1951). The elapid *Brachyurophus*, as at present defined, has distinctive geographical isolates in the south-west of the continent, in Arnhemland, and on Cape York, all sections separated by tracts of arid country. The genera *Leiolopisma, Denisonia, Typhlops*, and others, contain distinctive isolates that are either approaching, or have apparently recently reached species status. The various populations of *Amphibolurus* in the mountains of central Australia have been noted in this regard (F. J. MITCHELL, personal communication). The findings of COGGER (1957), that various geckos of the genus *Oedura* are broken up into discontinuous local populations by tracts of country deficient in the necessary cover, and of the acquirement of habitat-differences contemporaneous with speciation, will be noted.

The Distribution of Reptiles compared to that of Birds

Reptiles differ from birds (KEAST, 1955) in their distributional limits being less closely associated with those of the major vegetation associations occurring on the continent, and in being relatively less dependent on vegetation per se. A markedly smaller number of species are apparently specialized for life in rain forest and sclerophyll forest. There is, by contrast, a relatively high proportion of "true" desert species, possible four times as many as in birds. The number of pairs of populations, races, or species, isolated from each other in the forested regions of the south-east and south-west respectively is considerably smaller in reptiles than in birds (there being perhaps a dozen compared to over fifty). On the other hand the south-west has at least 8 endemic reptile genera compared to 1 in birds. Tasmania has a particularly poor reptile fauna (20 species as against perhaps 80 in birds), and has only 2 endemic species, compared to about 10 in birds.

There appear to be proportionately many more species with small, restricted ranges in reptiles than in birds, a result of the relative immobility of the smaller species.

As with birds a number of south-eastern Bassian species are restricted to higher altitudes in the more northern parts of their range. Again, many genera contain both an eastern Bassian and an interior Eyrean species, as has been noted by SERVENTY (1953) in birds. Examples are the elapids *Pseudechis porphyriacus* and *P. australis*, and the goannas *Varanus varius* and *V. gouldii*. In both reptiles and birds various inland species take advantage of the absence of physiographic and biotic barriers to extend into the forested south-western corner of the continent, whereas in the eastern segment they rarely extend beyond the western slopes of the Dividing Range.

Reptiles and the Zoogeographic sub-region concept

A zoogeographic region may be defined as a geographic subdivision of the earth that is the home of a peculiar fauna. "Such a region is characterized by the presence of many endemic genera and families and by the absence of the characteric genera and families of other zoogeographic regions". (MAYR, 1945). A sub-region or faunal province by contrast can be held to be a somewhat lesser division, characterized by a series of endemic forms of lesser degree.

A number of workers, mostly of the older generation, have postulated various sub-regions or faunal provinces within the Australian continent. Each author had a considerable experience in his special field and hence his findings can be expected to be reliable for that animal group. If the sub-region concept has any real basis in Australian zoology, however, the same or similar patterns of distribution should, theoretically, be discernible throughout.

The various faunal sub-divisions that have been advanced

The main systems that have been advanced and their authors are set out in Fig. 3. These are as follows:

(A) Division of the continent into a south-eastern coastal Bassian, a northern coastal Torresian, and an interior Eyrean Zone.

This scheme was advanced by BALDWIN SPENCER (1896), being partly derived from the earlier works of TATE & HEDLEY. It has since been modified slightly by SERVENTY & WHITTELL (1951), who point out that, so far as birds are concerned, the south-western

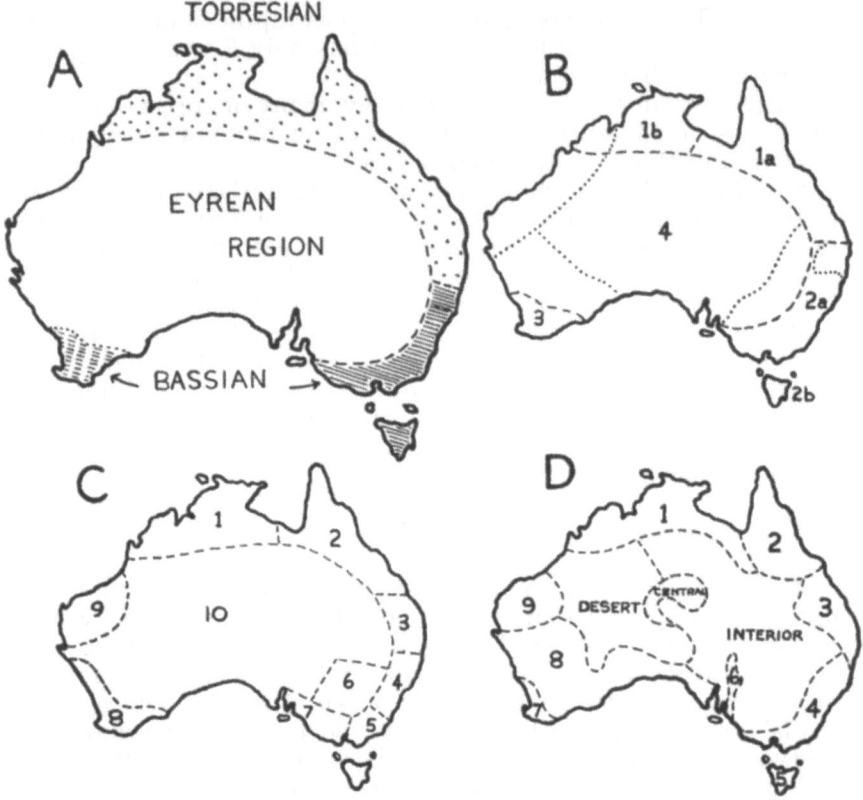

Fig. 3. The zoogeographic sub-region concept in Australian zoology. Maps show the schemes that have been advanced by different workers. A. BALDWIN, SPENCER (1898), as modified by SERVENTY & WHITTELL (1951). B. IREDALE (1929) indicated by dashes, and (1937), dotted lines. C. SLOANE (1915). D. CAMPBELL (1943).

Four major areas of origin are indicated in Australian reptiles plus several lesser ones.

fauna is not really Eyrean but is a blend of Eyrean and Bassian elements.

The SPENCER scheme has a solid basis for included in its proponents are a mammalogist (SPENCER), a conchologist (HEDLEY), and ornithologists (SERVENTY and WHITTELL).

(B) Division as above but demarkation of the south-west as a distinct region and the Kimberleys and Tasmania as sub-divisions of the Torresian and Bassian zones respectively (IREDALE, 1929, working from molluscs and birds).

In a subsequent map devoted solely to molluscs IREDALE (1937) has made various changes to his earlier scheme (see dotted lines in map "B", Fig. 3). These include an additional area (central-south Queensland) and extension of the Kimberleyan zone as far south as Shark Bay. The south-eastern and south-western zones are also made relatively larger.

(C) Division of the continent into 10 zones, eight of them being peripheral (5 in the east and 3 in the west), and separation off of the Murray-Darling Basin. This is the scheme of SLOANE (1915), and was derived from the study of carabeid beetles.

(D) Division of the continent into 12 zones (CAMPBELL, 1943). This scheme was built up from an analysis of distribution patterns in bird species. Actually, this scheme is not satisfactory from the ornithological viewpoint, for several of the postulated areas are supported by no more than races.

It is apparent, at the outset, that a common basis characterizes these various schemes. All authors distinguish peripheral faunas from interior ones, and note that the changeover in the south is a relatively sharp one. Furthermore, a northern coastal fauna is distinguished from an eastern coastal one. The main points of difference are in the number of subdivisions allowed within these basic elements. Three of the four schemes separate off the south-western corner of the continent. Three mark off the north-west from the north-east and three allow a section in south-central Queensland. Two distinguish a Tasmanian area and two an area in the north-western (Hamersley) section of Western Australia.

There is a reasonable level of correspondence between the range patterns of reptile species as listed above, and the "faunal province" categories of one or another of these workers. At the same time a considerable generalization is necessary to get the bulk of reptiles to fit. Category "1" corresponds to SPENCER's Bassian region, category "2" to the south-western section of IREDALE, SLOANE, and CAMPBELL. Category "3" corresponds to part or all of the fertile eastern coastal strip, with a "pocket" of species in south-central coastal Queensland. Category "6" accords with SPENCER's Torresian zone and numbers "4" and "5" to the north-eastern and north-western regions of several of the authors. The semi-arid and area region as a whole "7" approximates to SPENCER's Eyrean section.

It is, to some extent, a matter of individual choice as to which one of the zoogeographic sub-region schemes be held to be the most applicable to reptiles. The SPENCER scheme accords fairly well with the distribution of various major components. The IREDALE system is satisfactory in that it recognises a distinct south-western element but not in recognising Tasmania as distinct, which island has a reptile fauna of only 3 snakes and 15 lizard species, a mere two of which are endemic (Lord & Scott 1924). With some justification a zone could also be recognised in coastal Queensland. The extension well inland of the south-eastern and south-western zones in the later 1937 Iredale map does not accord well with reptiles distribution.

The Sloane & Campbell schemes are of little significance from the reptile viewpoint. Admittedly the north-western (Hamersley) segment of Western Australia has a few endemic species (e.g. tortoise *Chelodina steindachneri* and the goanna *Varanus brevicaudus*). The odd species distributions that might be fitted into the extra segment of these authors are certainly not sufficient to warrant them being given the title of "Faunal regions" from the viewpoint of the herpetologist.

The Ecological Significance of the Postulated Sub-Regions

Relationship to Vegetation and Climatic Zones

As has been noted by Dr. K. H. L. Key (chapter in this book) there is distinct link between the Bassian, Torresian, and Eyrean sub-regions, as postulated, and climatic and vegetation zones. This is particularly the case with the Bassian zone, the sharp delimitation of which to the west corresponds to the Dividing Range with its drop in rainfall and transition from sclerophyll forest to savannah woodland. The Torresian zone in the north corresponds fairly well with the tropical savannah belt there. The Eyrean zone is characterized by dry, open conditions and vegetation associations, ranging from savannah woodland to sand desert.

The smaller sub-divisions of Iredale, Sloane, and Campbell, are not associated with any particular kinds of vegetation, or at least to any extent.

It is obvious that, so far as the Spencer scheme is concerned, that whilst the sub-regions may also be centres of origin of faunal elements they are maintained to at least some extent, as recognisable units, by also having a climatic and vegetation basis.

Sub-regions and Areas of Minor Isolation

The true nature of many of the "provinces" put forward is actually that they are "refuge zones", areas where isolation and speciation are currently occurring. This can be seen from a review of speciation in the bird genus *Climacteris* (Keast, 1957), which genus has distinctive isolates in the areas "1", "2", "3—7", "6", "8", "9", and "10" of Sloane's map, as well as area "6" on the Campbell map. Many of these areas likewise correspond to areas of isolation of populations in other bird genera, e.g. *Acanthiza*

(MAYR & SERVENTY, 1938), *Neositta* (MAYR, 1950), and *Platycercus* (CAIN, 1954). There is currently a great deal of differentiation going on as between forms isolated in the south-west and south-east of the continent (SERVENTY & WHITTELL, 1951) and north-west and north-east (KEAST, 1956, 1957). As yet no studies equivalent to these have, however, been carried out on reptiles.

It is thus noteworthy that most of the IREDALE & SLOANE areas are of significance in the higher vertebrates, though it is of a considerably lesser than those of a zoogeographic sub-region. The zones may well, hoewever, be characterized by forms of a higher level of differentiation in less mobile invertebrates.

Areas of Origin of Major Faunal Components

The following areas have given rise to major faunal components in reptiles: (1) Torres Strait area, being point of entry of New Guinea and Indo-Malayan forms from the north; (2) the arid interior; (3) the forested south-western corner of the continent; (4) the forested south-eastern and eastern seaboard of the continent. These correspond respectively to the Torresian, Eyrean, Bassian regions of SPENCER and the autochthonian region of IREDALE. The matter is not altogether simple, however, for there are various lesser areas characterized by genera (sometimes) and groups of species: the north-west corner, south-central Queensland, the Atherton area, whilst the semi-arid and arid fauna is also a composite.

In that three to four main centres of origin can be recognised reptiles accord with other groups. The Bassian zone is populated, to at least a degree, by old endemic, inhabitants of the cold temperate forests, though there is a continuing intrusion of tropical elements from the north and by at least some elements from the interior. The Torresian region is populated to a large degree by species that have only recently colonised Australia from the north. The semi-arid and arid regions, that make up two-thirds of the total land-area, are occupied by a diversity of forms adapted to the variety of conditions to be found there. They are a combination of somewhat old Australian elements that, like *Moloch*, have obviously had a long history in the desert, or are derived from Bassian or Torresian elements. The south-western corner of the continent is relatively more distinct zoogeographically in reptiles than it is either in birds or mammals. In the case of reptiles there are at least eight endemic genera whilst there is only one in birds *(Purpureocephalus)* and three or four in mammals *(Tarsipes, Setonyx,* probably *Myrmecobius).*

The review of reptile distribution indicates that, whilst it is legitimate to use the zoogeographic sub-region concept within the Australian continent it should be used only in broad context. It is undesirable to try to give precise limits to the zones for not only have the inhabitants of each radiated to a variable degree but there are several areas of origin within each. The distribution of many species, furthermore, bears no resemblance to them. The origin of these cannot be stated. In that the term "sub-region" or "faunal province" indicates a generalized grouping of elements or an area of origin, it has a value. To think of it to any degree in a static or absolute sense is quite misleading.

Acknowledgments

The writer would like to express his thanks to Mr. R. MACKAY with whom he worked out the summaries of family and species distribution within Australia and to Mr. JOHN MITCHELL, Drs. A. R. MAIN & D. RIDE, and Mr. J. CALABY, for suggestions.

REFERENCES

CAMPBELL, A. G., 1943. Australian faunal regions. *Emu* **42**, *242*.

COGGER, H. G., 1957. Investigations in the gekkonid genus *Oedura* Gray. *Proc. Linn. Soc. N.S.W.* **82**, *167*.

DARLINGTON, P. J., 1957. *Zoogeography: The Geographical Distribution of Animals. 177.* John Wiley & Sons, New York.

DAVID, T. W. E. ed. by BROWNE, W. R., 1950. *The Geology of the Commonwealth of Australia 614.* Edward Arnold & Co. London.

FAIRLEY, N. H., 1929. The present position of snake-bite and the snake bitten in Australia. *Med. J. Aust.* **1**, 16th year: *296*.

GLAUERT, L., 1950. *A Handbook of the Snakes of Western Australia. 13.* West. Aust. Nat. Club, Perth.

GLAUERT, L., 1951. A note on the western tiger snake. *West. Aust. Nat.* **3**, *43*.

GLAUERT, L., 1952. Herpetological miscellanea — notes of some forms of *Diplodactylus*. *West. Aust. Nat.* **3**, *166*.

GLAUERT, L., 1954. Herpetological miscellanea: Western Australian geckoes. *West. Aust. Nat.* **4**, *174*.

GUNN, S. B., 1956. Notes on some Australian snakes. *Mem. Qld. Mus.* **13**, *141*.

IREDALE, T., 1929. The avifaunal districts of Australia. *Rep. Hobart meeting of Aust. Assoc. for Adv. Sci.* **1928**, *244*.

IREDALE, T., 1937. A basic list of the land mollusca of Australia. *Aust. Zool.* **8**, *287*.

KEAST, J. A., 1955. Bird speciation on the Australian continent. Ph. D. Thesis, Harvard University.

KEAST, J. A., 1956. Variation in the Australian Oriolidae. *Proc. R. Zool. Soc. N.S.W.* **1956**, *23*.

KEAST, J. A., 1957. Variation and speciation in the genus *Climacteris*. *Aust. J. Zool.* **5**, *474*.

KEAST, J. A., 1958. Variation and speciation in the Australian flycatchers. *Rec. Aust. Mus.* **14**, *73*.

KELLAWAY, C. H. & WILLIAMS, F. E., 1929. The venoms of *Oxyuranus maclennani* and of *Pseudechis scutellatus*. *Aust. J. Exp. Biol. Med. Sci:* *155*.

KINGHORN, J. R., 1926. A brief review of the family Pygopodidae. *Rec. Aust. Mus.* **15**, *40*.

KINGHORN, J. R., 1956. *The Snakes of Australia.* 2nd. ed. *46*. Angus & Robertson, Sydney.

LORD, C. E. & SCOTT, H. H., 1924. *A Synopsis of the Vertebrate Animals of Tasmania 104.* Oldham, Beddome, Meredith, Hobart.

LOVERIDGE, A., 1934. Australian reptiles in the Museum of Comparative Zoology, Cambridge, Massachusetts. *Bull. Mus. Comp. Zool.* **77**, *243*.

LOVERIDGE, A., 1948. New Guinean reptiles & amphibians in the Museum of Comparative Zoology & United States National Museum. *Bull. Mus. Comp. Zool.* **101**, *305*.
MACKAY, R. D., 1955. A revision of the genus *Pseudechis*. *Proc. Roy. Zool. Soc. N.S.W.* **1953—54**, *15*.
MAYR, E. & SERVENTY, D. L., 1938. A review of the genus *Acanthiza*, Vigors and Horsfield. *Emu* **38**, *245*.
MAYR, E., 1945. Wallace's Line in the light of recent zoogeographic studies. *Science & Scientists in the Netherlands 241*. Board for Netherlands Indies, Surinam, and Curacao, New York.
MAYR, R., 1950. Taxonomic notes on the genus *Neositta*. *Emu* **49**, *282*.
MERTENS, R., 1942. Die Familie der Warane (Varanidae). *Abh. Senckenb. naturf. Ges.* 462, 465, 466.
MERTENS, R. & WERMUTH, H., 1955. Die rezenten Schildkröten, Krokodile und Brückenechsen. *Zool. Jb.* **83**, *323*.
MITCHELL, F. J., 1948. A revision of the lacertilian genus *Tympanocryptis*. *Rec. S. Aust. Mus.* **9**, *57*.
MITCHELL, F. J., 1950. The scincid genera *Egernia & Tiliqua* (Lacertilia) *Rec. S. Aust. Mus.* **9**, *275*.
MITCHELL, F. J., 1951. The South Australian reptile fauna. *Rec. S. Aust. Mus.* **9**, *545*.
MITCHELL, F. F., 1953. A brief revision of the four-fingered members of the genus *Leiolopisma* (Lacertilia) *Rec. S. Aust. Mus.* **11**, *75*.
MITCHELL, F. J., 1955. Preliminary account of the Reptilia and Amphibia collected by the National Geographic Society-Commonwealth Government. Smithsonian Institution Expedition to Arnhem Land (April to November, 1948). *Rec. S. Aust. Mus.* **11**, *373*.
DE ROOIJ, N., 1915. *The Reptiles of the Indo-Australian Archipelago 74*. E. J. Brill, Leiden.
SERVENTY, D. L., 1953. Some speciation problems in Australian birds. *Emu* **53**, *131*.
SERVENTY, D. L. & WHITTELL, H. M., 1951. *A Handbook of the Birds of Western Australia*. 2nd Ed. *44*. Patersons Press: Perth.
SLOANE, T. G., 1915. On the faunal subregions of Australia. *Proc. Roy. Soc. Vict.* **28**, *139*.
SMITH, M. A., 1935. Reptilia & Amphibia. In: Fauna of British India including Ceylon & Burma. **2** (Sauria) p. 1. Taylor and Francis, London.
SPENCER, B., 1898. *Report on the work of the Horn scientific Expedition to Central Australia. Narrative. 197*. Dulau and Co., London.
THOMSON, D. F., 1935. Preliminary notes on a collection of snakes from Cape York Peninsula. *Proc. Zoo. Soc. Lond.* **1935**, *723*.
WAITE, E. R., 1918. Review of the Australian Blind Snakes (Fam. Typhlopidae) *Rec. S. Aust. Mus.* **1**, *1*.
WAITE, E. R., 1929. *The Reptiles & Amphibians of South Australia. 32*. Government Printer, Adelaide.
ZIETZ, F. R., 1920. Catalogue of Australian lizards. *Rec. S. Aust. Mus.* **1**, *181*.

VIII
THE FRESHWATER FISHES OF AUSTRALIA

by

GILBERT P. WHITLEY, F.R.Z.S.
(Curator of Fishes, The Australian Museum, Sydney)

Whilst the seas around Australia are inhabited by well over two thousand distinct species of fishes, the freshwater rivers are the preserves of a hardy few, only about 180 kinds in all the continent, if we exclude introduced forms, such as trout, and occasional stragglers into freshwater, like gobies, soles, stingrays, toadoes, anchovies, and other estuarine fishes. So only about 8% of the Australian fish-species are freshwater. By comparison with other continents, Australia has not a large freshwater fauna, due to the absence of many large rivers and the prevalence of droughts, and some curious local water-conditions. We find no trace of certain families of fishes commonly found in the rivers of countries the other side of WALLACE's Line. There are no native members of the carp tribe (Cyprinidae), no killifishes (Cyprinodontidae) and very few catfishes. In Australia, we have no torrent-inhabiting fishes with special suckers on their fins or bodies for adhering to rocks, as some countries have (unless we except the discs of the Lampreys). Also our fishes are not troubled by freezing. *Galaxias findlayi* enjoys life above the snowline but swims to lower warmer levels during the winter.

What we lack in variety, however, we make up for in interest as only the fittest of our freshwater fishes have survived and some of these have persisted for millions of years, long after their relatives had died out in other parts of the world. Such "fossil fishes" as the Queensland Lungfish *(Neoceratodus)* and Burramundi *(Scleropages)* and probably, the freshwater Blackfish or Slippery *(Gadopsis)* are examples. Others like the Murray Cod, certain gudgeons, freshwater herring, perches and grunters have evidently been derived long ago from marine ancestors but are now purely fluviatile. Lampreys and freshwater eels still spend part of their lives in the sea as do one or two species of Native Minnow (Galaxiidae).

Papua has affinities with northern Australia, through the former connection across Torres Strait, and the freshwater fishes of Southern New Guinea and the Aru Islands are just like those of tropical Australia. But as we travel farther north, to northern New Guinea, the Celebes and Philippines, the freshwater fishes differ more and more from ours.

My first desire is to impress upon everyone the unique character of the Australian native freshwater fishes and plead for their conservation. They are as Australian as koalas and kangaroos, waratahs and gum-trees, and I must admit prejudice against the introduction of fishes and other forms of water-life which might cause them harm or destruction. Trout, redfin, perch, gambusia, carp and other aliens which have been acclimatised are therefore excluded from further consideration.

Why are Australian freshwater fishes different from those elsewhere? We all know that Australia is an island-continent and that in comparatively recent ages (as the geologists reckon time!) it was joined to Tasmania across Bass Strait and to Papua across what is now Torres Strait. But our Australian continent has much longer, possibly permanently, been separated from the other continents, the cleavage from Asia being along what zoologists call WALLACE's Line.

WALLACE's line, the boundary between the rich Oriental fauna and that of Australia and New Guinea, was named after ALFRED WALLACE, the great English naturalist and contemporary of DARWIN. This line passes between Bali and Lombok in Indonesia, up through Macassar Strait between Borneo and Celebes and turns into the open Pacific Ocean between Mindanao in the Philippines and the Sangjir Islands. As WALLACE himself wrote, a century ago, "South America and Africa, separated by the Atlantic, do not differ so widely as Asia and Australia."

In addition to WALLACE's Line, modern biologists recognize WEBER's line, named after a Dutch zoologist, which threads its way through island groups somewhat eastward of WALLACE's Line and marks particularly the boundaries beyond which freshwater fishes do not trespass. So it is as if an impassable "wall" of salt water surrounded the Australian continental shelf, embracing the Aru Islands and New Guinea, but excluding Timor, Ceram and Halmahera. This puzzling, very ancient but invisible barrier is responsible for the nonappearance of elephants and tigers and monkeys to the east of WALLACE's line. It is also the reason why we do not find the extraordinary variety of carps, loaches, freshwater catfishes and other kinds which are so prolific in Asia. Isolated for we know not how many million years, Australia has had time, not only to develop her gentle fauna (marsupials, the ancient platypus, the emu and other well-known animals) almost free from beasts of prey, but as well, her rivers have become the unique haunts of some fishes which have survived from great antiquity and of other fishes which may only have made their way from the surrounding seas in the last million years or so, a mere ticking away of the last few minutes as Nature regards Time!

Some of our freshwater fishes have interesting alliances with overseas genera and species. The lamprey *(Geotria)*, grayling *(Prototroctes)* and so-called "mountain trout" *(Austrocobitis, Galax-*

ias and *Brachygalaxias*) have close relatives in South America. A few Galaxiidae are found in South Africa also. Perhaps *Pseudaphritis*, which lives in fresh or salt water in Australia, had South American affinities long ago. *Neoceratodus* has only fossil relatives, widely distributed over the world back to the Permian. *Scleropages* has a close living relative in South-east Asia and more distant cousins in Africa.

Potamalosa used to be considered to be close to the American fossil herring *Diplomystus* but is nowadays regarded as generically distinct.

Retropinna has cognate species in New Zealand.

Drought adapted, artesian bore, and hot water fishes

Accounts of the water-conditions in inland Australia, including floods, artesian water, storms, etc., can be consulted in The Australian Encyclopaedia just published. With so much of Australia arid, the adaptations of animals to drying watercourses must be of vital importance. However, this field has as yet been little studied.

The manner in which fishes survive the drying of watercourses by aestivation, lying buried in mud, etc., was a phenomenon first recorded by THEOPHRASTUS (who lived from about 372 to 287 B.C.) according to BASHFORD DEAN's Bibliography of Fishes iii, 1923, *323*. Many different kinds of fishes have been observed to do this in overseas countries, the most famous being the African Lungfish, *Protopterus*, which lies for a long time in a torpid condition enclosed in a mud cocoon. No fish in Australia is known to aestivate like this, but several different Australian freshwater fishes have been found buried in the ground, mud or moist soil, notably Gudgeons (family Gobiomoridae) and Native Minnows (*Galaxias* spp.) *

Do fishes fall in rain? Can such a method of distribution explain the wide dispersal of some of our freshwater fishes, the Spangled Grunter, for example. The hypothesis is that fishes are in some way caught up in water and carried through the air by strong winds and later deposited far away. An Indian legend has it that waterspouts are elephants' trunks which suck up fish and later let them fall in rain. The first case of this kind was referred to in the 3rd century A.D. in the Greek classic Deipnosophists, and there have been innumerable reports of "rains" of fishes, frogs, and other water-animals since then. The cleverest book on the subject is simply entitled "Lo!" by CHARLES FORT. There have been over a score of Australian instances of showers of fishes, or fishes found stranded after rain,

---

* See T. S. HALL, 1901, *Vict. Nat.* **xviii**, *65–66;* J. FLETCHER, 1906, *Proc. Linn. Soc. N.S.W.*, **xxxi**, *430–431; 497–498;* and G. P. WHITLEY, Australasian Aqua Life **ii**, 4, April, 1957, *10*, and **ii**, 5, May 1957, *6*.

and museums have examples of various freshwater fishes *(Philvpnodon, Carassiops, Craterocephalus, Reporhamphus, Madigania, Mogurnda* and *Nannoperca)* so found. Fish can be dropped from a height without much injury. Some Boy Scouts (no doubt doing a good deed by helping scientific enquiry) dropped some fish from the high memorial obelisk at Washington, U.S.A., and the fish were still alive afterwards. Also trout waters have been restocked by dropping fish from aeroplanes.

During heavy rainstorms, fishes such as *Madigania* disperse widely over flooded paddocks and are left behind in waterholes afterwards. They have also been seen swimming rapidly along the water in cart-ruts *.

Artesian water was first discovered in Australia at Kallara Station, northern New South Wales, in 1879, and later over widespread areas, approximately one-third of Australia having been found to be underlain by artesian water. This is a potential habitat for fishes but the field is far from properly explored.

There has been much argument as to whether or not fishes have been ejected in very hot water from artesian bores from far underground. The late D. G. STEAD (1909, *Proc. Linn. Soc. N.S.W.*, **xxxiv**, *116)* showed specimens of the Spangled Grunter *(Madigania unicolor)* which were said to have come up through an artesian bore from a depth of 943 feet at Corella, north-western New South Wales. It was noticed that in most of the fishes the eyes were damaged and the opinion was expressed that they did not live and breed in the subterranean waters but had individually found their way thither. OGILBY & McCULLOCH (1916, *Mem. Qld. Mus.*,**v**, *112)* refer to New Zealand experiments on young trout which indicated that the protrusion of the eyeball in fishes (so-called "pop-eye" disease) was caused by excessive gas in the water, identical with conditions observed in the Corella bore fishes and they wrote, "Under these circumstances, we see no reason to accept Mr. STEAD's suggestion that these fish came up the bore, but assume that they were merely aestivating somewhere in its vicinity, and upon being again vivified by the bore water, came under the influence of its contained gases and so developed pop-eye." During drought, many freshwater fishes can aestivate in mud in a state of "suspended animation" and Mr. STEAD found that aestivating gudgeons which had become dehydrated revived on being placed in warm water and were soon swimming. When artesian water flows over the soil it may therefore revive and release buried fishes and cause people to think that they came up through the bores.

The Australian Museum has specimens of Grunters *(Madigania*

---

* P. S. WILSON, Bowen, Queensland, "Independent" newspaper, Jan. 29, 1954 and BRUCE SHIPWAY, *West. Aust. Nat.* 1 (2), Sept. 1947, *47*.

unicolor) from the Corella bore, also from Weilmoringle bore drains (11/7/1908) and from artesian water from the Walgett hospital grounds (May 1910) all in New South Wales and from a well, 63 feet deep, east of Carnarvon, Western Australia. In October, 1938, the Australian Museum received a series of another kind of Grunter *(Hephaestus fuliginosus)* from the bore drain, Offham Station, Cunnamulla, central southern Queensland, taken at a temperature of 102° F. The donor, Mr. N. GEARY wrote: "If these fish are put into cold water, they die very quickly." The same species of grunter was found on 5 Oct. 1938 (about a month after the first series) in a drain five miles from the bore and in cold water; it had evidently been able to acclimatize gradually to the reduction in temperature.

Regarding other parts of the world there are records, stated to be authentic, of fishes and also crabs and molluscs, having been brought up in bore waters (G. F. ROLLAND, 1881, *C. R. Acad. Sci. Paris*, **xciii**, *1090—3*). The only blind fish ejected from an artesian well was a Catfish, *Trogloglanis pattersoni* (EIGENMANN, 1919, *Proc. Amer. Phil. Soc.* **lviii**, *397—400*), from underground waters of Texas, where, unlike Australia, there is an abundant cave fauna.

As regards the highest temperatures which freshwater fishes can stand, there are few data concerning Australian species and Mr. N. GEARY's figure of 102°, quoted above, is apparently the maximum. The subject of "Hot Springs and Fish Life", was carefully reviewed by R. R. MILLER (*Aquarium J.* **xx**, 11, Nov. 1949, *286—288*) who discounted some obviously exaggerated maximum temperatures given by earlier authors, but mentioned fishes *(Cyprinodon)* which he had himself collected in water of 104° F, to the east of Death Valley, California. This is regarded as the highest temperature at which fish can survive and he points out that water below the surface is colder than this and that fishes are usually found near the bottom or some distance from the hottest layers of water. In Steinhart Aquarium, San Francisco, specimens of *Cyprinodon* were accidentally subjected to 110° F.: some were cooked but a few survived and recovered as the temperature was lowered to the usual 85° F. or so.

Fluvifaunulae

Australia's purely fluviatile native fishes inhabit fairly definite limits, determined, in the first place, by the extent of river-systems, and by climate and land-barriers. However, the zoogeographical regions inhabited by our freshwater fishes do not fit so exactly into the various river-systems to enable usage of the rivers' names, so an alternative scheme whereby the freshwater animals are classed into fluvifaunulae has been proposed (IREDALE & WHITLEY, 1938, *S. Aust. Nat.* **xviii**, *64—68* and map).

Fig. 1. Zoogeographical borderlines in the Malay Archipelago and the freshwater regions or Fluvifaunulae of Australia. JOHN BEEMAN del.

A fluvifaunula — derived from the Latin *fluvius*, a river, plus the diminutive of fauna — is a consociation of animals found in a river or a series of rivers. Each was named in honour of a naturalist or explorer historically associated with the fluvifaunula concerned. Nine Australian and one New Guinea fluvifaunulae have been defined below.

Subsequent study of more material, not only of fishes but of molluscs, crustacea, river-tortoises, sponges, etc., has confirmed, with slight modifications, the fluvifaunular limits proposed in 1938. Naturally our divisions of Australia into areas were not intended as hard-and-fast outlines, and, since the ana-branches and upper reaches of our rivers interlock like clasped fingers in some regions, or river-captures* may result from intruded barriers, it is impossible to do more than broadly outline these zoogeographical limits on a small scale, but the latest outlines are as shown on the accompanying map.

1. Leichhardtian Fluvifaunula

Rivers from about Broome, Western Australia, around the Northern Territory and Gulf of Carpentaria to the tip of Cape York and also taking in rivers of southern New Guinea. Named after LUDWIG LEICHHARDT, explorer and naturalist.

Animals characteristic of the Leichhardtian Fluvifaunula: Sawfish *(Pristiopsis)*, 2 species of Herring *(Fluvialosa)*, Burramundi *(Scleropages)*, many catfishes (Plotosidae), a few Belonidae, Soleidae and some characteristic Melanotaeniidae *(Quirichthys, Amneris, Aidaprora)*, many grunters (Terapontidae), two eels *(Synbranchus* and *Anguilla bicolor)*, a number of percoids and gudgeons (7 genera of Gobiomoridae) and the curious *Kurtus gulliveri*.

The mollusca are mentioned in MCMICHAEL and IREDALE's contribution to this work.

2. Greyian

Rivers from about Ninety Mile Beach to south of Shark's Bay, Western Australia. Named after Sir GEORGE GREY.

A very little known area. Characteristic fishes, etc.: The northern tortoise, *Chelodina steindachneri* and a mussel *(Lortiella)*. A blind gudgeon *(Milyeringa)* inhabits one well. Other fish species are *Melanotaenia australis, Craterocephalus cuneiceps, Shipwayia aurea*, and a gudgeon and some grunters trespassing from other regions.

3. Vlaminghian

Fresh waters of south-western Australia. From WILLEM VLAMINGH, Dutch pioneer.

Characteristic fishes, etc. Two lampreys *(Geotria* and *Yarra)*, 3 kinds of Galaxiidae, 3 atherines, some unique percoids *(Nannatherina, Bostockia* and *Edelia)* and gobies. The freshwater eel is absent. A sponge *(Ephydatia)*, the tortoise, *Chelodina oblonga* and some crustacea are noteworthy *(Chaeraps* spp., *Palaemonetes australis* and a cladoceran, *Daphnia thomsoni*, otherwise occurring only in New Zealand and South Africa). A *Galaxias* and a frog *(Hyla cyclorhyncha)* of the Albany-Esperance region are closely allied to Tasmanian forms. Viviparine molluscs do not enter this fluvi-

---

* Notably in south-eastern Queensland and northern New South Wales where "Murray River" fishes are found in rivers flowing into the Tasman Sea.

faunula, though Bullinids are present, and the mussel, *Westralunio*. *Coxiella* frequents the salt water lakes.

4. Sturtian

Central Australian waters, westward of the Darling system. From CHARLES STURT, Explorer.

Characteristic fishes, etc. A goby *(Chlamydogobius)*, and central Australian relatives of peripheral species of *Craterocephalus*, catfishes *(Neosilurus)*, grunters, etc. "The molluscs show specialisation, a Viviparine evolution, *Centrapala*, being noticeable, as also the widely spread Bullinid genus, *Isidorella*, and the Mussel genus *Centralhyria... Coxielladda* is found in the salt water lakes."

5. Mitchellian

The whole of the Murray River system with its tributaries from Queensland to South Australia. After Sir THOMAS MITCHELL.

Characteristic fishes, etc.: A lamprey *(Mordacia)*, and various fishes such as *Fluvialosa richardsoni*, at least one species of *Galaxias*, *Tandanus tandanus*, *Melanotaenia fluviatilis*, two *Craterocephalus*, and one species each of the genera, *Macquaria*, *Maccullochella* (the Murray Cod), *Plectroplites*, *Blandowskiella*, *Nannoperca*, *Bidyanus*, *Gadopsis*, *Pseudaphritis* and *Carassiops*. Eels are absent. The Bullinid shell, *Isidorella*, and mussels, *Alathyria* and *Hyridunio* are notable.

6. Lessonian

Rivers of eastern New South Wales, Victoria and northern Tasmania. Named after RENE LESSON, French naturalist.

Characteristic fishes, etc.: Lamprey *(Mordacia)*, Herring *(Potamalosa)*, Grayling *(Prototroctes)* and a considerable variety of Galaxiidae, percoids and gudgeons, a mullet *(Trachystoma)*, the Bullrout *(Notesthes*, a scorpaenid) and two species of eels *(Anguilla reinhardtii* and *australis)*. There are numerous species of Bullinid and Unionid mollusca.

7. Tobinian

Southern Tasmanian rivers and lakes. After GEORGE TOBIN, a sailor-naturalist with BLIGH.

Characteristic fishes: Lampreys *(Mordacia* and *Geotria)*, Smelt *(Reptropinna tasmanica*, distinct from the Lessonian *R. semoni)*, *Nannoperca tasmaniae*, and six Galaxiidae. Mussels are entirely missing from the Great Lakes and south Tasmania. Potamopyrgid molluscs, with New Zealand affinities, flourish, but the most characteristic mollusc is the ,,Limpet", *Legrandia* (syn. *Tasmancylus)*. The Great Lakes show speciation in the Galaxiidae and might well be regarded as a new sub-fluvifaunula to be named the Smithian, after GEOFFREY SMITH, author of "A Naturalist in Tasmania", 1909, and an early student of Tasmanian freshwater crustacea and other animals in the field.

8. Krefftian

For the Mary and Burnett Rivers, Queensland, haunts of the Queensland Lungfish, discovered by GERARD KREFFT.

Characteristic fishes: besides the Lungfish *(Neoceratodus)*, grunters *(Scortum hillii* and *parviceps)* and the perchlet, *Priopidichthys marianus*, seem to have become separate species here. Amongst molluscs are the mussel, *Cucumerunio* and the Viviparine *Larina*.

9. Jardinean

Rivers of eatern and north Queensland, named for the pioneering JARDINE family of Cape York.

Characteristic fishes, etc.: Eels *(Anguilla reinhardtii* and *obscura)*, catfishes and forms of Melanotaeniidae, grunters, percoids and gudgeons slightly different from the Leichhardtian species, the Belonid, *Stenocaulus kreffti*, and an island gudgeon *(Lindemanella)* may be noted. A freshwater mussel, *(Rugoshyria aquilonalis)* and a globose Rissoid mollusc *(Jardinella)* are distinctive.

10. Gaimardian

This is not strictly Australian, applying to the northern parts of New Guinea. Named after JOSEPH GAIMARD, an early French naturalist.

Characteristic fishes: Eel *(Anguilla interioris)*, specialised sunfishes of the family Melanotaeniidae *(Glossolepis, Centratherina, Chilatherina, Anisocentrus,* etc.), a variety of Catfishes, a Soldier Fish *(Apogon abo)*, a Gudgeon *(Mogurnda bloodi)* and other species.

Adaptive Radiation

There is a good deal of research to be done on our fishes to discover which kinds, if any, favour certain biological niches. Because of Australia's unstable freshwater conditions, the fishes have to be hardy and versatile, unlike those of more constant river-systems in other countries. Instead of speaking of fishes specially modified for bottom-living or other haunts, we should think rather of rivers which may dry up, or be opaque with mud and debris, ponds which may be almost full of decaying matter, saltpans and lakes which may dry up for years on end and whirlwinds (willy-willies) which may even blow fishes away. Gudgeons of several genera and species are to be found swimming together in some rivers. No adaptive radiation or fitting into various niches seems to be exhibited by Australian fresh-water fishes, except perhaps to a very limited extent in the Galaxiidae. The other families are represented by one or only a few species. These show a good deal of variation amongst themselves (notably in fin and scale-counts and in bodily proportions, especially

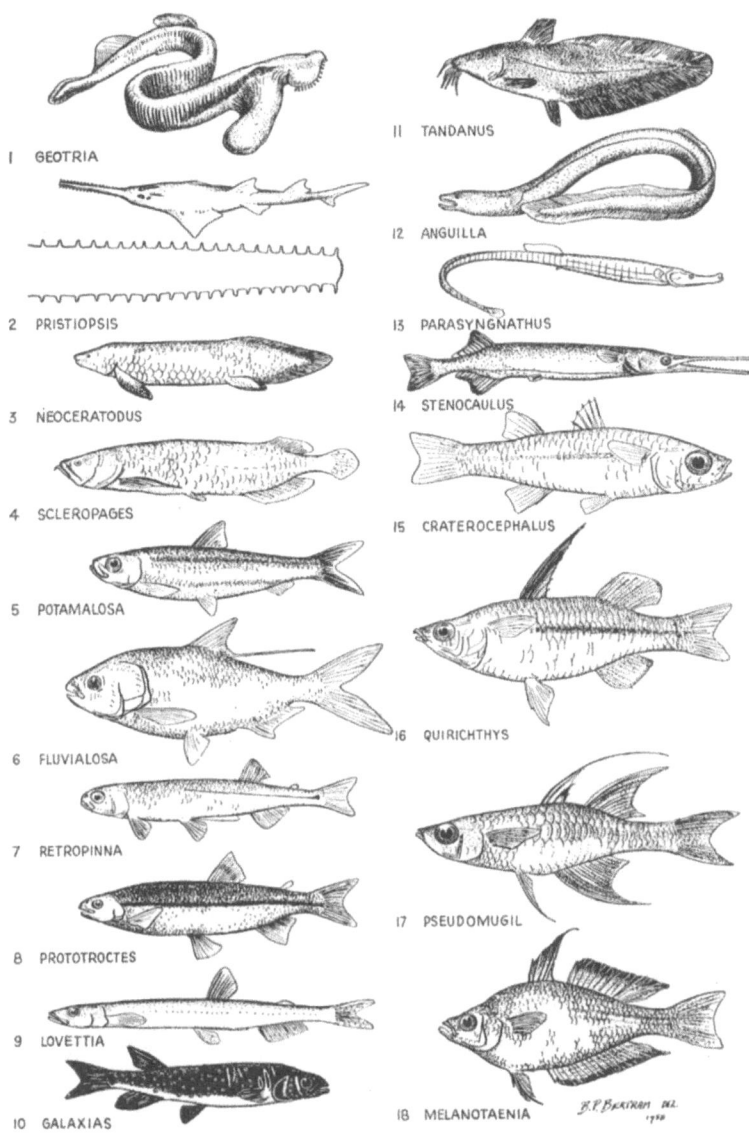

Fig. 2. Principal genera of Australian freshwater fishes.

1. Lamprey — *Geotria*
2. Sawfish — *Pristiopsis*
3. Lungfish — *Neoceratodus*
4. Burramundi — *Scleropages*
5. Herring — *Potamalosa*
6. Hairback Herring — *Fluvialosa*
7. Smelt — *Retropinna*
8. Grayling — *Prototroctes*
9. Derwent Smelt — *Lovettia*
10. Galaxias — *Galaxias*
11. Catfish — *Tandanus*
12. Eel — *Anguilla*
13. Pipefish — *Parasyngnathus*
14. Long Tom — *Stenocaulus*
15. Hardyhead — *Craterocephalus*
16. Blackmast — *Quirichthys*
17. Blue Eye — *Pseudomugil*
18. Sunfish — *Melanotaenia*

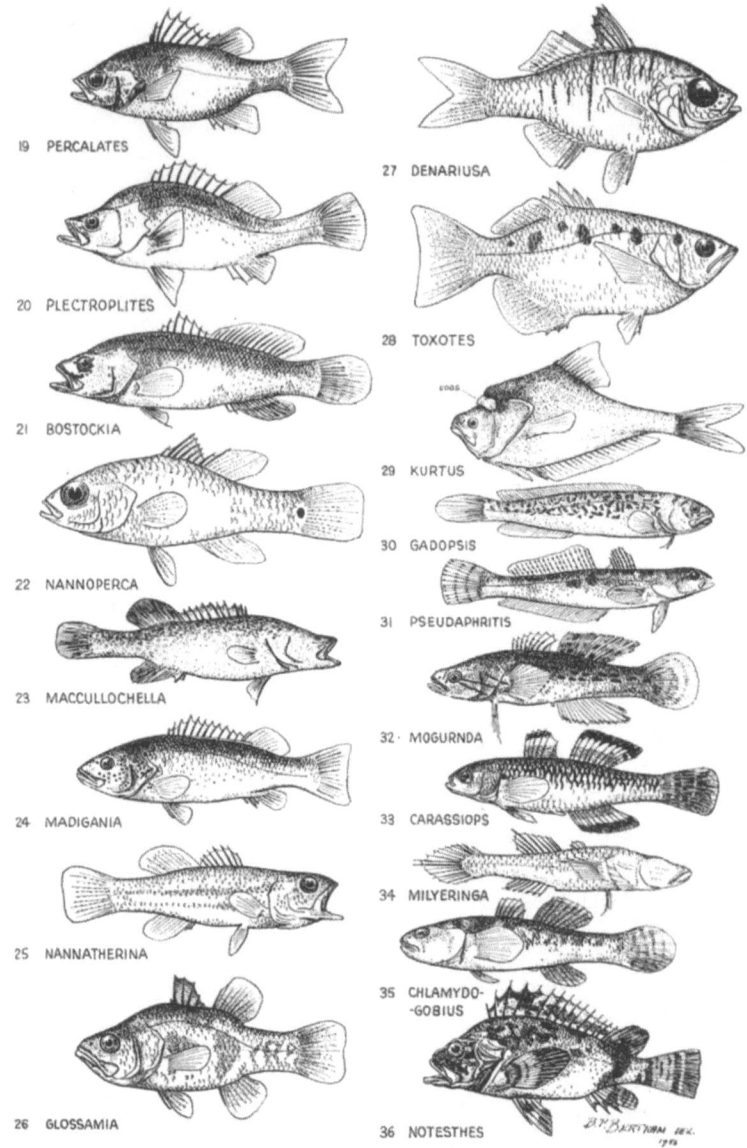

Fig. 3. Principal genera of Australian freshwater fishes - continued.

19.	Bass	*Percalates*	28. Archer Fish	*Toxotes*
20.	Callop	*Plectroplites*	29. Nursery Fish	*Kurtus*
21.	Night Fish	*Bostockia*	30. Blackfish	*Gadopsis*
22.	Pigmy Perch	*Nannoperca*	31. Congolli	*Pseudaphritis*
23.	Murray Cod	*Maccullochella*	32. Gudgeon	*Mogurnda*
24.	Grunter	*Madigania*	33. Gudgeon	*Carassiops*
25.	Perchlet	*Nannatherina*	34. Blind Gudgeon	*Milyeringa*
26.	Soldier Fish	*Glossamia*	35. Goby	*Chlamydogobius*
27.	Penny Fish	*Denariusa*	36. Bullrout	*Notesthes*

in *Melanotaenia*) and wellknown polymorphism is shown by *Percalates* and *Maccullochella*, but practically the whole of the Australian freshwater fish-fauna seems to be of too recent origin to have radiated as, for instance, the marsupials have done on land. However, one eyeless gudgeon *(Milyeringa)* has been found in a well in the North-West Cape area. So far no fishes have been discovered in the underground waters of the Nullabor Plain. An introduced trout *(Salmo irideus)* in New South Wales managed to penetrate 300 feet below the surface to the underground waters of Coolamon Cave. Sharks, sawfish, lampreys, Murray cod and possibly other fishes may swim hundreds of miles up or down a river. Tagged Murray cod have been observed to have travelled up to 150 miles.

Here may be inserted a Table of the genera of Australian native freshwater fishes. Only one species of each of these genera is known to inhabit freshwater in Australia, unless the number of species is stated in brackets after the generic name. An asterisk* after a name indicates that the genus is found in New Guinea as well as Australia.

Table I

Order	Family	Genera
Hyperoartia	Geotriidae	*Geotria, Yarra*
Hyperoartia	Mordaciidae	*Mordacia*
Batoidei	Pristidae	*Pristiopsis* (*)
Dipneusti	Ceratodontidae	*Neoceratodus*
Osteoglossoidei	Osteoglossidae	*Scleropages* (*)
Isospondyli	Clupeidae	*Potamalosa*
Isospondyli	Clupanodontidae	*Fluvialosa* (4 spp.)
Isospondyli	Retropinnidae	*Retropinna* (3)
Isospondyli	Aplochitonidae	*Prototroctes, Lovettia*
Galaxiiformes	Galaxiidae	*Galaxias* (21), *Austrocobitis, Brachygalaxias* (2), *Saxilaga* (2) and *Paragalaxias*.
Nematognathi	Plotosidae	*Tandanus* (3), *Neosilurus* (6), *Lambertichthys* (*), *Anodontiglanis*, and *Porochilus* (*).
Nematognathi	Tachysuridae	*Nemapteryx* (*), *Pararius* and *Neoarius*.
Synbranchia	Synbranchidae	*Synbranchus* (*)
Apodes	Anguillidae	*Anguilla* (*) (4)
Thoracostei	Syngnathidae	*Oxleyana* and *Parasyngnathus*
Synentognathi	Hemiramphidae	*Reporhamphus*
Synentognathi	Belonidae	*Stenocaulus* (*)
Percesoces	Atherinidae	*Atherinosoma* (3) and *Craterocephalus* (6)
Percesoces	Melanotaeniidae	*Quirichthys, Pseudomugil* (*) (3), *Rhadinocentrus* (2), *Amneris* (*), *Aidaprora, Rhombosoma* (*) and *Melanotaenia* (*) (4).
Percesoces	Mugilidae	*Trachystoma* and *Moolgarda*

Order	Family	Genera
Rhegnopteri	Polynemidae	*Polynemus*
Percomorphi	Macquariidae	*Percalates* (2) and *Macquaria*.
Percomorphi	Plectroplitidae	*Plectroplites*
Percomorphi	Bostockiidae	*Bostockia*
Percomorphi	Nannopercidae	*Nannoperca* (4) and *Edelia*
Percomorphi	Maccullochellidae	*Maccullochella*
Percomorphi	Terapontidae	*Madigania, Amniataba, Bidyanus, Hephaestus* (4), *Mesopristes* (*) (2), *Papuservus* (*), *Scortum* (4), *Amphitherapon, Leiopotherapon* (*) and *Pingalla* (*)
Percomorphi	Nannatherinidae	*Nannatherina*
Percomorphi	Duleidae	*Dules* and *Herops*
Percomorphi	Apogonidae	*Glossamia* (2) and *Kurandapogon*
Percomorphi	Chandidae	*Austrochanda* (2), *Blandowskiella* (4), *Acanthoperca* (*), *Priopidichthys* and *Denariusa*
Percomorphi	Latidae	*Lates* (*)
Percomorphi	Lutjanidae	*Lutjanus* (enters freshwater)
Percomorphi	Sciaenidae	*Johnius* (enters freshwater)
Percomorphi	Monodactylidae	*Monodactylus* (*) enters freshwater
Percomorphi	Leptobramidae	*Leptobrama* (*)
Percomorphi	Toxotidae	*Toxotes* (*) (3) and *Protoxotes* (*)
Percomorphi	Kurtidae	*Kurtus* (*)
Percomorphi	Prenidae	*Selenotoca* (*) (2) and *Prenes* (*)
Heterosomata	Soleidae	*Trichobrachirus* (*) (2) & *Liachirus* (*)
Gadopseiformes	Gadopsidae	*Gadopsis*
Jugulares	Pseudaphritidae	*Pseudaphritis*
Gobioidei	Gobiomoridae	*Culius* (*) (2), *Shipwayia, Bunaka* (*), *Oxyeleotris* (*), *Prionobutis, Meuschenula, Gobiomorphus, Mogurnda* (*), (4), *Philypnodon, Carassiops* (4), *Ophieleotris* (*) *Ophiocara* (*) and *Lindemanella*.
Gobioidei	Milyeringidae	*Milyeringa*
Gobioidei	Gobiidae	*Arenigobius, Chlamydogobius, Glossogobius* (*) (2) and *Lizagobius*.
Cataphracti	Scorpaenidae	*Notesthes*
Plectognathi	Tetraodontidae	*Sphaeroides* (enters freshwater).

Spawning

The life-histories of Australia's freshwater fishes have been very incompletely studied and, whilst it is assumed that all are oviparous, we still do not even know whether some breed in fresh, brackish or salt water. Some of those of rivers flowing into the Tasman Sea *(Retropinna, Philypnodon)* probably breed in salt or brackish water. Yet species of *Retropinna, Fluvialosa* etc., in the upper reaches of the Murray-Darling river system must surely breed in fresh water as otherwise we should have to postulate enormous migrations to the mouth of the Murray in South Australia and these are not known to occur.

Anadromous kinds, ascending rivers to breed, are the Tasmanian Troutlet *(Lovettia)* and probably all the lampreys.

Catadromous species are the eels *(Anguilla)*. Those of north-western Australia probably breed near Sumatra, whilst the eastern species probably travel to near New Caledonia to lay their eggs.

Austrocobitis is evidently catadromous and has been observed in New Zealand to spawn at the edge of tidal waters but the closely related *Galaxias* spp. must breed in freshwater *(G. coxii* has been proved to do so); the idea that *Galaxias* breeds in the sea, favoured by some zoogeographers, is probably fallacious.

Brachygalaxias breeds in peaty, brown freshwater in Western Australia.

Fishes which are known to lay their eggs in freshwater are: *Neoceratodus*, the Melanotaeniidae, *Macquaria*, *Percalates*, *Plectroplites*, *Maccullochella*, *Gadopsis* and their polymorphs, at least some of the gudgeons, and the anadromous kinds mentioned previously. Many more must spawn in freshwater but we have no definite data.

Scleropages and *Glossamia* practise buccal incubation, the eggs being carried in the father's mouth until hatched. Some Papuan catfishes have a similar habit but, so far as is known, there is no oral gestation in Australian catfishes. Those of the genus *Tandanus*, on the other hand make "nests" of sand or pebbles for their eggs, and perhaps the lampreys do as well. In *Kurtus* the eggs are carried by the male on a hook on his head, like a bunch of grapes.

There is much to be learnt about the fascinating freshwater fishes of Australia, yet little research is being undertaken, and all the time introduced alien fishes are replacing the native ones.

IX
THE ZOOGEOGRAPHY OF SOME AUSTRALIAN INSECTS

by

J. W. Evans

(Australian Museum, Sydney)

Summary

Reasons are given for regarding insects in general and the Homoptera in particular as useful indicators of past geography. It is suggested, from evidence based on Permian and Triassic Homoptera, that for periods during these eras Australia may have been connected by land with the northern hemisphere.

Most Upper Triassic Homoptera differ considerably in facies from recent forms; hence it is assumed that the characteristics of the present Australian leafhopper fauna have been determined by geological events subsequent to the commencement of the Jurassic.

Australian leafhoppers (Cicadelloidea) are grouped into several components based on their geographical distribution. Four such components are discussed. These comprise respectively: leafhoppers which it is supposed are of pre-Tertiary Indian origin; those with close relatives confined to other southern land areas; endemic forms; and ones which entered Australia at the time of the late Tertiary Indo-Malayan invasion.

As the source of origin of the southern faunal component is the subject of controversy this problem is discussed. In this connection particulars are given of several groups of insects belonging to eight Orders, all of which are confined to the southern hemisphere.

It is deduced from the evidence that there can be only two possible explanations to account for the present distribution of these insects. These are, either that all were formerly of universal occurrence and that the northern hemisphere representatives have become extinct, or else, that as late as towards the end of the Mesozoic, the various southern land areas had direct continuity with each other. Reasons are given for supposing that the second possibility is the more probable. The galaxiid fishes, leptodactylid frogs and marsupials conform to the same pattern and it is considered that they, too, probably entered Australia from the south.

Introduction

Australia abounds with animals and plants of great interest and of all its fascinating fauna and flora, the mammals undoubtedly hold pride of place. For this reason they have probably been more intensively studied than any other group and much thought has been given to the zoogeographical problems they present. Nevertheless, views still differ in regard to the direction from which marsupials entered the continent. Thus SMITH (1909) was of the opinion that they travelled from their place of origin in the northern hemisphere through South America and Antarctica into southern Australia. On the other hand, as ABBIE (1941) has pointed out, most authors favour an approach from the north.

Marsupials and, in fact, all mammals are, like birds, but creatures of yesterday, since although they had their beginnings during the Mesozoic era it was not until Tertiary times that they came to play a dominant part in the life of our planet.

Insects on the other hand have been in existence since the Palaeozoic and the major evolutionary development of many groups was completed well before the dawn of the Mesozoic.

It is customary for students of vertebrates to ignore evidence for paleogeographical conclusions based on invertebrates in general and insects in particular. They are aware that much of the insect fauna of the world is still but little known and that by far the greater part has not yet been subjected to critical study on a world-wide basis. Then many biologists, and entomologists are included in their number, consider that insects are unsuitable as indicators of past geography. This is because they may be transported by wind currents in the upper air and on rafts of floating vegetation. Furthermore, they are particularly favourable organisms for delayed dispersal as many pass through quiescent stages in their life-cycles which sometimes are of long duration. Then, compared with vertebrates, the insect fossil record is of a fragmentary nature. Finally, insect faunas are known to exist on oceanic islands and these must have been carried there by adventitious means.

On the other hand, it can be argued that insects, by reason of their very abundance, diversity and age, are particularly good palaeogeographical indicators. Those which can survive transport over long distances and become successfully established in new environments are exceptional and not typical representatives of the Class. The majority have such specialised physical and nutritional requirements that, even were they able to survive periods of enforced travel, they would be unlikely to discover a favourable ecological niche which would permit subsequent establishment.

The particular insects which have been chosen for discussion in this Chapter belong to the Order Homoptera. This Order comprises aphides, scale insects (Coccidae), cicadas, and an abundance of forms known variously as leafhoppers, frog-hoppers and plant-hoppers (Cicadelloidea, Cercopoidea, Fulgoroidea). Many are injurious to crops and some are of great beauty, but possibly none, apart from cicadas, are well-known to others than entomologists. Although related to insects in the Order Heteroptera, which includes stink bugs and assassin bugs (Pentatomidae and Reduviidae), the relationship between the two groups is not a close one as they have followed separate lines of evolutionary development since Carboniferous times.

The Homoptera are particularly favourable insects to serve as palaeographical indicators for the following reasons: they are all plant feeders and the majority are restricted to particular plants or groups of plants; apart from the egg stage, they lack a period of dormancy in their developmental cycle, hence they are less likely than many other groups to be able to withstand prolonged adventitious transport; they have abundant fossil representation extending from Lower Permian to Jurassic times; from evidence based on comparative morphology it is possible to differentiate between those genera that have changed but little in form since the Mesozoic and others which owe their distinctive characteristics to prolonged periods of Tertiary isolation.

Some features of the distribution of three groups of Homoptera are discussed below.

Frog-hoppers (Cercopoidea)

The immature stages of these insects which are known as "spittle bugs", are more familiar than the adults. Not all frog-hopper nymphs live surrounded by froth and those in one family, the Machaerotidae, which occur in the Oriental region, tropical Africa and Australia, live in calcareous tubes of their own making, immersed in their own liquid excretions.

The earliest known Cercopoid *Belmontocarta perfecta* (EVANS, 1957) has been found in Upper Permian rocks in New South Wales and the wings of several belonging to different genera have been recorded from the Upper Triassic of Queensland (EVANS, 1956).

In the northern hemisphere the first recorded Cercopoid is from the Liassic of Central Asia (BEKKER-MIGDISOVA, 1949, EVANS, 1957, b).

This insect, *(Mesotracis reducta)* is undoubtedly closely related to, and possibly generically identical with, some Australian Triassic species *(Dysmorphoptiloides* spp.). The family to which they belong (Dysmorphoptilidae) has no representatives living at the present day.

The evidence provided by these fossils suggests that during the early part of the Mesozoic there was, at some time, direct land access between Australia and the northern hemisphere and that the Tethys Sea did not at all times prevent an interchange of fauna.

Cicadas

All those who live in countries with warm climates are familiar with the song of cicadas and it is generally known that the males alone are "vocal" while both sexes have auditory tympana.

In eastern Australia and Tasmania respectively, there are two species of cicadas which differ from all others in several features (EVANS, 1940).

These insects *(Tettigarcta* spp.) lack tympana; but both sexes have tymbals and tymbal muscles. This characteristic is an earlier development than the more specialised condition found in other cicadas and *Tettigarcta* is undoubtedly a relict form.

The family Tettigarctidae comprises no other living insects and its known extinct representatives, which are four in number, have been recorded from Lower Liassic beds in Central Asia and the Lower Cretaceous of Belgium (BEKKER-MIGDISOVA, 1949, EVANS, 1957 b).

The occurrence in Australia of living insects related to ones which existed in the northern hemisphere in the Mesozoic, parallels the occurrence of the lungfish *Neoceratodus* in Queensland and it suggests, as has been pointed out by MACFARLANE (1923) in a discussion of the distribution of fishes, that during Triassic-Jurassic times eastern Australia and the continents of the northern hemisphere were at some time inter-connected.

Leafhoppers (Cicadelloidea)

The earliest recorded leafhoppers (Archescytinidae) are known from Lower Permian beds in Kansas and from the Upper Permian of Russia and eastern Australia. Representatives of a family, the Scytinopteridae, which is probably derived from these, occur also in the same beds in Russia and Australia.

A close resemblance exists between a few Scytinopterids from the northern hemisphere and some from the Permian of Australia and those which indicate possible close relationship between the two faunas have a generalised wing-venation pattern. More numerous wings have been found in both hemispheres which have a specialised venation and these indicate separate evolutionary trends and hence faunal isolation.

Apart from a single poorly-preserved Jurassic leafhopper from New South Wales, the most recent Mesozoic Homoptera, so far discovered in Australia, are from Upper Triassic beds in Queensland. All but two of the very numerous leafhopper wings which have been obtained from these beds have an early Mesozoic facies and differ considerably from present-day representatives of the group. The two which have a more modern appearance seem to belong respectively to the families Cicadellidae and Eurymelidae. This suggests the possibility that representatives of these two families may already have been in existence at this early date but that they were not an important element in the fauna.

It can be assumed, accordingly, that such geological events as may have influenced the occurrence and distribution of the present leafhopper fauna of the continent probably took place subsequent to the commencement of the Jurassic era.

The Leafhopper Fauna of Australia

The present-day leafhopper fauna of Australia can be separated into the following components: ancient groups which are represented elsewhere in the tropics of the eastern and western hemisphere and which are particularly abundant in India; ancient groups which are represented elsewhere in South Africa and New Zealand; endemic groups which in some instances have sparse representation also in New Guinea and New Caledonia; groups of probable Australian origin which occur also in New Guinea, the Phillipines, China and Malaya; representatives of possibly more recently evolved groups of tropical origin which occur particularly in north-eastern Australia; cosmopolitan groups.

Leafhoppers in the last mentioned category can be separated into those in which the Australian representatives have endemic genera; grass or generalised feeders belonging to cosmopolitan genera which may have arrived by adventitious means; and those established since the time of arrival of Europeans.

In order to present a broad picture of possible events and to seek to avoid confusion, some leafhoppers in four only of the above

Fig. 1. Supposed Post-Triassic chronology of Australian geography.

categories will be discussed. These are the old groups with Indian associations; old groups with southern associations; endemic groups; and groups of recent northern origin. The supposed chronological sequence of the above faunas is indicated in Fig. 1.

The Australian-Indian Association

Reasons for regarding some groups of leafhoppers as representing earlier stages of evolutionary development than others have been given elsewhere (EVANS, 1946, 1947). These reasons are based on comparative morphology and not on fossil evidence.

There is an ancient group (Ledrinae) which has its richest representation in India and which occurs, as well as elsewhere in the Oriental region, also in Australia, South and tropical Africa, and in Central America, with derivatives in North and South America. It has, as well, a single representative in Europe.

It is assumed that the Ledrinae were established already in Australia prior to its Tertiary isolation because there are in this continent as well as insects which belong to the same tribe (Ledrini) as the Oriental forms, also some apparent derivatives from this tribe which are comprised in two other tribes, both of which are endemic (Stenocotini and Thymbrini). There are other groups of leafhoppers (e.g. Nirvaninae) which almost certainly, as well, form part of this faunal component.

Australian — South African — New Zealand — South American Association

The reason for placing this association after that with India is because it is supposed that it was retained for a considerable period after India and Australia had been separated from each other.

There are two groups of leafhoppers which are superficially alike and comprise small insects which are long and narrow and seed-like in appearance (Cephalelini, Ulopinae and Paradorydiini, Hecalinae). One of these, (the Cephalelini), is of particular interest as the adult insects occur in two forms, a winged form and a flightless one. So far as is known the Cephalelini feed only on rush-like plants belonging to the family Restioniaceae and they occur in South Africa, Australia, especially Tasmania, and New Zealand. As their food plants grow also in Chile they may well occur in South America also.

Endemic Australian Groups

The only major geographical area with more endemic groups of leafhoppers than Australia is the Neotropical region and the next in order of endemicity is Madagascar.

As well as the two endemic tribes which have already been mentioned, there occur in Australia two other distinctive tribes (the Trocnadini and Reuterellini which are derived from the Penthimiini by way of the Jassini), two sub-families of uncertain relationships (Austroagalloidinae, Tartessinae), and an endemic family, the Eurymelidae and this last will alone be discussed. The Eurymelidae which have been mentioned previously, since a fossil wing from the Triassic of Queensland has been tentatively ascribed to this family, may have formerly had a more widespread distribution and have become extinct elsewhere.

Eurymelids, which are comprised in three sub-families are gregarious insects and are ant-attended. The most generalised genera belong to a sub-family (Ipoinae) of drab insects which, as well as feeding on eucalypts (the dominant Australian trees) feed also on other trees, including *Casuarina*. One species, belonging to an Australian genus, has been recorded from New Caledonia, and a few, also belonging to Australian genera, from New Guinea.

Within Australia all described genera, apart from two, have representatives in both the eastern and western halves of the continent while the two in question *(Cornutipo* and *Cornutipoides)* are restricted to areas bordering the dry interior.

Insects in another sub-family (Pogonoscopinae) live in ants' nests and feed on the roots of eucalypts. Although largely restricted in distribution to south-western Australia, they have sparse representation also in the south of South Australia and in north-eastern Victoria.

The third sub-family (Eurymelinae) comprises leafhoppers with a bold colour pattern all of which feed on eucalypts. Most genera and species are widely distributed in both the east and west of the continent and an endemic species occurs in New Guinea. This sub-family represents a recent evolutionary development and its speciation is almost certainly associated with that of eucalypts, which only came into prominence in late Tertiary times. (GILL, 1957).

A consideration of the distribution of the Eurymelidae provides an indication of some of the salient geographical and climatic features of Australia during the Tertiary epoch.

Thus, it suggests an early or mid-Tertiary land contact with New Caledonia and late one with New Guinea. Furthermore, that prior to its present separation from eastern Australia by an arid barrier, Western Australia was isolated also at an earlier period and that between the two periods of separation there was one of environmental continuity.

The Late Indo-Malayan Invasion

There occur in New Guinea many Homoptera belonging to genera which are found also in the Oriental region and in Africa, but which, with a few exceptions, are absent from Australia (e.g. *Cosmoscarta, Leptaspis*, Cercopoidea). The exceptions are a few species which are limited in their occurrence within Australia to eastern Australia,

PLATE I

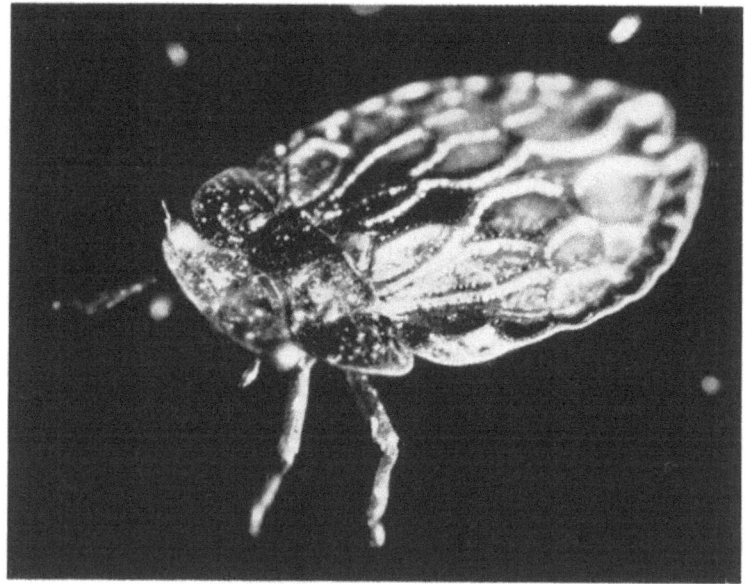

Fig. 2. A Peloridiid bug from Tasmania *(Hemiodoecus fidelis)*.

more particularly Queensland. Among leafhoppers, species in the genera which come into this category *(Coelidia, Drabescus, Vulturnus,* etc.) have the same distributional occurrence as those plants which undoubtedly also entered Australia at the close of the Tertiary period. It is presumed that these leafhopper genera either evolved elsewhere during the Tertiary, after Australia had become isolated, or else had formerly occurred in a region which was not in direct land continuity with Australia.

*
* *

The foregoing account of the distribution of some Homoptera has as its basis the authors' own research and in this sense represents an original contribution. The geographical and chronological pattern suggested, is however one which has been repeated in scores of publications dealing with completely different groups of organisms, (e.g. HARRISON, 1926, TILLYARD, 1926).

A late incursion of an Indo-Malayan fauna and flora into northeastern Australia is generally accepted, likewise the supposition that Australia was isolated from other land areas during the greater part of the Tertiary. An earlier land association with India is also recognised and it is presumed to have preceded the time of the evolution of mammals or, at least their occurrence in India as a numerically important group.

In regard, however, to a possible land connection between Australia and other southern continents, there is no such general agreement and as this problem has remained the foremost one of plant and animal distribution since first drawn attention to by HOOKER (1847), it merits further consideration than the brief mention made of it on an earlier page.

Workers with vertebrate groups, as already stated, tend to ignore evidence for palaeography based on invertebrates, hence in order to avoid reverse criticism, before discussing some insects with a solely southern distribution, mention needs to be made of the distribution pattern of a few coldblooded vertebrates.

DARLINGTON (1957) has suggested that if there is such a thing as a general pattern of animal distribution the fresh-water fishes ought to show it perhaps better than any other animal, so it will be appropriate to begin by mentioning the Galaxiidae. This family of small fresh-water fishes has representatives in South Africa, Australia, New Caledonia, Tasmania, New Zealand and South America.

Among the Amphibia, the Leptodactylidae occur only in South Africa, Australia and South America while in the Reptilia, extinct terrestrial turtles in the Family Meiolaniidae, are known as Tertiary fossils from South America, Lord Howe Island, eastern Australia and Walpole Island, close to New Caledonia.

DARLINGTON has sought to explain these distributional occurrences in the following way. In regard to the Galaxiidae he states that the family is considered to be of marine origin just completing the process of establishing itself in fresh-water. A presumed Leptodactylid frog is known from Eocene deposits in India, hence he infers the occurrence of this family formerly in the northern hemisphere, while he would explain the distribution of *Meiolania* by supposing, that these animals though apparently undoubtedly terrestrial, were able nevertheless to cross considerable ocean gaps.

Before the significance of the distribution of these various vertebrates is discussed, mention will be made of an assemblage of insects, which are bound together as a group by environmental factors.

Insects with aquatic, or marsh-inhabiting larvae and with a southern distribution

The various insects which are mentioned below are all associated with a fresh-water or near fresh-water environment.

Mayflies (Ephemeroptera) are delicate insects seldom found at any great distances from fresh-water in which their immature stages are passed. In the sub-family Leptophlebiidae there is a genus *Atalonella* with species that have been recorded only from Australia, Tasmania and Chile and another genus, *Atalophlebia* with a similar distribution but occurring also in New Zealand.

Dragonflies (Odonata) are another group in which the nymphs live in water, but they differ from Mayflies in being sturdy, swift-flying insects. Representatives of the archaic sub-family Petaliinae, the nymphs of which inhabit waterfalls, have the same distribution as the Mayfly genus *Atalonella*.

Stoneflies (Plecoptera) which are weak fliers, likewise have aquatic nymphs. One Family, the Eustheniidae, has representatives confined to Australia, Tasmania, New Zealand and South America. Then there is a sub-family (Notonemourinae) known only from Australia, Tasmania, New Zealand, Tierra del Fuego and the Cape Province of South Africa.

In another Order, the Megaloptera, (Dobson Flies), with aquatic larvae a genus *(Archichauliodes)* in the Family Corydalidae is confined to Australia, Tasmania, New Zealand and Chile, while in the Order Trichoptera (Caddis Flies), the genus *Hudsonema* in the family Leptoceridae, has a similar distribution.

Among the Diptera the larvae of the dipterous Family Blepharoceridae live in the cascades of waterfalls and under stones in fast flowing streams, and species belonging to the genus *Edwardsina* are confined to eastern Australia, Tasmania and cool, temperate South America. In the family Simuliidae there are two genera *(Cnephia* and *Austrosimulium)* with a similar distribution. This Order is notable for the very large number of comprised genera which have a solely southern distribution.

With regard to groups in which the larvae, though not aquatic yet occur in a wet environment, insects in the neuropterous subfamilies Stenosmylinae and Calosmylinae are restricted to Australia, Tasmania, New Zealand and Chile. In the Order Mecoptera (Scorpion Flies), the larvae of insects in the family Nannochoristidae are not known with certainty, but the adult insects haunt an environment characterised by the presence of alpine tarns and the family has the same distribution as those mentioned immediately above.

The pattern of distribution of the insects belonging to the eight different Orders mentioned above is identical with that of the families of vertebrates previously referred to and as well with that of numerous other invertebrates in several other classes. The inference follows that the distribution of both the vertebrates and the invertebrates has come about as a result of identical causes. It would be unrealistic to suppose that all these various insects, many of which never travel far from fresh water, could have been transported between the southern continents by adventitious means. This is particularly so when it is remembered that they occur in an environment which is dominated by plants with a similar pattern of distribution.

Accordingly, the only possible alternative explanation is that they reached their present areas of occurrence by passage over land and not over sea. This being so there is no necessity for supposing that fresh-water Galaxiid fishes independently forsook a marine existence in widely separated parts of the world or that the terrestrial Meiolaniids ever entered the sea. If it is accepted that the various, and very numerous components of the "Antarctic" fauna and flora increased their range by spreading over land surfaces then it means either that the southern continents were at some time, and as late as the mid-Mesozoic, in close or direct contact with each other, or else that all the plants and animals, which are now restricted in distribution to the southern continents, were formerly universally distributed, and were able to live under a wide range of climatic conditions, since passage of the tropics would be involved.

Mention has already been made of the supposition based on fossil evidence that the Leptodactylid frogs formerly inhabited India (a land area formerly associated with the southern continents), and ANDER (1942) has shown that several insects now confined to the southern hemisphere formerly also lived in Europe. These records may mean no more than that a northward extension of certain southern groups formerly took place. It is known that during the past 100 million years, great climatic changes have taken place in the northern hemisphere, as also in the southern one, but there is no evidence to suggest, apart from the Pleistocence glaciation in the north, that anything in the nature of a widespread cataclysm has occurred recently throughout the former. In fact, relict species from early Mesozoic times are still represented as abundantly in the northern as in the southern hemisphere. For example in the Homoptera one may mention the most archaic of all Cicadelloids, *Darthula hardwickii* (Aetalionidae) which occurs in south-eastern India and the Yunnan Province of China and also the family Hylicidae.

Hylicids, which are probably derived from the Prosbolidae (a family of Permian Homoptera which are known as fossils from northern Russia), are now widely distributed in the Oriental region and have sparse representation in tropical Africa. The reason for their absence from Australia, even though they were undoubtedly in existence prior to its Tertiary isolation, is very possibly because there was no land access between India and Angaraland (north-east Asia) until after the closing of the Tethys sea.

It has frequently been suggested that the explanation for the occurrence solely in the southern hemisphere of numerous ancient forms of life may be because they have been exterminated elsewhere as a result of competition with more recently evolved forms. This would imply that the old forms lack qualities essential for "success" but the evidence of their very survival contradicts this assumption. Furthermore they live together, in many instances in ecological balance, with more recently evolved representatives of the groups to which they belong.

Those who are convinced of the reality of former land connections between the southern continents are often criticised on the grounds that they neglect to take into account the long period of time that has elapsed since the end of the Mesozoic era.

It is known that even the most isolated oceanic islands carry an insect fauna that must have reached them by adventitious means, hence it is supposed that when the time element is considered, such means of distribution can readily explain the occurrence of related insects in the southern continents.

The insect faunas of oceanic islands are, however, of a very restricted nature and among leafhoppers comprise particularly forms which lack the need for specific food plants and are hence readily capable of establishment. Thus, species belonging to the same few cosmopolitan genera (e.g. *Batrachomorphus, Macropsis*) are to be found on nearly all islands, irrespective of whether they are in the Atlantic or Pacific oceans, but the vast majority of genera and there are over 1000 in the family Cicadellidae alone, lack representation.

There are several groups of insects with a southern distribution which, not only comprise species which are predominantly flightless, but as well have such specialised physical requirements, that they could not possibly be carried long distances in the upper air or survive on rafts of floating vegetation.

Among these is a family of small and ancient Homoptera, belonging to the Family Peloridiidae. These bugs usually lack hind wings and hence cannot fly and they live in wet moss which needs to be in a near-saturated condition to enable their survival.

Species belonging to several genera have been recorded from Chile, Patagonia, Tierra del Fuego and New Zealand, Tasmania, Lord Howe Island and from certain restricted areas on the mainland of Australia where the vegetation is dominated by the Southern Beech *(Nothofagus)*.

Apart from the facts of their distribution, these insects are of particular interest as they have retained on the first segment of their thorax flap-like lateral projections (Fig. 2) which are homologous with wings (EVANS, 1939). Similar flaps are known to have occurred on several Palaeozoic and Mesozoic Homoptera. Their close restriction to a moist environment suggests not only that there must have been means of direct land access between the various places they now inhabit, but also that there must have been climatic and environmental continuity.

Within Australia, as apart from Tasmania, while they are now known only from a few localities in southern Victoria and the McPherson Ranges in southern Queensland, doubtless they were formerly widespread when the climate was wetter and *Nothofagus* covered a large part of the continent (GILL, 1952, PIKE, 1954).

If they had ever inhabited lands lying in the northern hemisphere then their continued existence in them would be expected since their ecological requirements are of widespread occurrence.

The evidence provided by the distribution of insects is thus overwhelmingly in favour of the existence in mid or late Mesozoic times of a large land area of which all the southern continents formed a part. The problem of how this supposed land mass became separated into its present components is not a biological one and hence needs no discussion here.

This Chapter began with mention of the problem of the direction from which marsupials entered Australia. It is appropriate to conclude it by an examination of the contribution which a knowledge of insect distribution can make to its solution.

SIMPSON (1948) states that the diversity of marsupials in the early Tertiary clearly indicates the great antiquity of the group and that, not only must they date from the Cretaceous period, but also perhaps from well back into the Cretaceous.

GREGORY (1944) after mentioning that opossums of the Cretaceous period will have spread from North into South America to produce the several South American forms, suggests that other marsupials may have migrated into Australia from northern Asia during the Upper Cretaceous.

The evidence furnished by insect distribution supports an early connection between Australia and India and it is probable that this was in existence during the Jurassic and may have extended into the Lower Cretaceous.

It supports also a land connection with the Indo-Malayan region at the close of the Tertiary. It lends no support to a land connection with Asia during the Upper Cretaceous; moreover at the time of the earlier link with India it is presumed that the Tethys Sea separated the sub-continent from Angaraland.

On the other hand, there is a great deal of evidence based on insect distribution which suggests that South America and Australia may have been in land continuity, by way of Antarctica, during the very period when marsupials are believed to have entered Australia. This being so, surely it is more logical to suppose that marsupials may have entered Australia from the south and not from the north.

The absence of marsupials from South Africa, and New Zealand is sometimes used to support arguments in favour of a northern origin. The faunal and floral links which South Africa has with Australia though abundant are very much fewer than those shared by Australia and South America. This suggests that after India, Africa may have been the next land area to sever its connection with Gondwanaland and presumably the severance took place before the advent of marsupials.

So far as marsupial absence from New Zealand is concerned, nothing can be deduced from this; many groups of insects, of otherwise almost universal occurrence, also lack representation in New Zealand; but it is the presence in and not the absence from land areas of organisms which calls for reasoned explanations.

Acknowledgment is made to Mr. D. E. KIMMINS, Dr. I. MACKERRAS and Miss C. LONGFIELD for information relating to the distribution of certain groups of insects mentioned and to Dr. MACKERRAS and Mr. R. H. TEDFORD for helpful comments.

REFERENCES

ABBIE, A. A., 1941. Marsupials and the evolution of mammals. *Aust. J. Sci.* **1**, 77.
ANDER, K., 1942. Die Insekten-Fauna des Baltischen Bernsteins nebst damit verknüpften zoogeographischen Problemen. *Acta Univ. Lundensis* **38** (4), (N.F.).
BEKKER-MIGDISOVA, H. F., 1949. Mesozoic Homoptera from Central Asia. *Trav. Inst. Paléozool. Acad. Sci. U.R.S.S.* **22,** 1.
DARLINGTON, P. J., 1957. Zoogeography. *John Wiley & Sons Inc.* N.Y.
EVANS, J. W., 1939. The morphology of the thorax of the Peloridiidae. *Proc. R. ent. Soc. Lond.* (B) **8,** 143.
EVANS, J. W., 1940. The morphology of *Tettigarcta tomentosa*. *Pap. & Proc. R. Soc. Tasm.* **1940,** 35.

EVANS, J. W., 1946. A natural classification of leafhoppers. (Pt. 2) *Trans. R. ent. Soc. Lond.* **97,** *39.*
EVANS, J. W., 1947. A natural classification of leafhoppers. (Pt. 3) *Trans. R. ent. Soc. Lond.* **98,** *108.*
EVANS, J. W., 1956. Palaeozoic and Mesozoic Hemiptera. *Aust. J. Zool.* **4,** *165.*
EVANS, J. W., 1957a. New Upper Permian Homoptera from the Belmont Beds. *Rec. Aust. Mus.* **14,** *113.*
EVANS, J. W., 1957b. Some aspects of the morphology and interrelationships of extinct and recent Homoptera. *Trans. R. ent. Soc.* **109,** *275.*
GILL, E., 1952. Range in time of the Australian Tertiary Flora. *Aust. J. Sci.* **15,** *47.*
GILL, E., 1957. Whence Australia's mantle of green? *Aust. Mus. Mag.* **12,** *217.*
GREGORY, W. A., 1944. Australia - the story of a continent. *Nat. Hist.* **53,** *365.*
HARRISON, L., 1926. The composition and origin of the Australian fauna with special reference to the Wegener hypothesis. *Rept. Aust. Ass. Adv. Sci.* **18,** *322.*
HOOKER, J. D., 1847. *Flora Antarctica.* **2,** *211.*
MACFARLANE, J. M., 1923. The evolution and distribution of fishes. *Macmillan Co.,* N.Y.
PIKE, K. M., 1954. Some dicotyledonous pollen types from Cainozoic deposits in the Australian region. *Aust. J. Bot.* **2,** *197.*
SIMPSON, G., 1948. The beginning of the age of mammals in South America. *Bull. Amer. Mus. Nat. Hist.* **91,** (1).
SMITH, G., 1909. A naturalist in Tasmania. *Clarendon Press.* Oxford.
TILLYARD, R. J., 1926. Insects of Australia and New Zealand. *Angus & Robertson,* Sydney.

X
ZOOGEOGRAPHICAL ASPECTS OF THE AUSTRALIAN DIPTEROFAUNA

by

S. J. PARAMONOV
(C.S.I.R.O., Canberra)

It is not possible at present to give an exact picture of the Australian fauna of Diptera from the zoogeographical aspect, because our knowledge of it is very far from complete. Many families are inadequately studied. Papers published by the author make this very evident: for example, in 1926 TILLYARD knew only 1 genus, with 5 species, of Apioceridae, a family very well developed in Australia; in 1953 the author listed 2 genera with 64 and 6 species respectively. Not only were new species discovered, but new groups of them. Some families which were unknown to TILLYARD, for example the Thyreophoridae and Lonchopteridae, were recorded later. The position is altered by each new family revision.

Most of Australia is practically "terra incognita", especially its central part. In all Australian museums and collections there are only odd specimens from this area. Another obstacle is our poor knowledge of the Dipterofaunas of surrounding areas, especially of the islands eastwards from the mainland of Australia; we cannot compare the Australian fauna with them. Two short visits by the author to Lord Howe Island produced evidence that the relationship between the fauna of this island and that of the mainland is very complex, and only a study of both can solve the many problems that arise. The author must stress therefore, that the picture given below is only of a preliminary nature.

The main points of the zoogeography of Australian Diptera

It is advisable to consider first the following salient features which characterise the Australian dipterous fauna as a whole. We will also consider subsequently the peculiarities of different families, subfamilies and genera occurring in Australia, and in conclusion, see what faunal divisions can be established from a study of Australian animals, particularly Diptera.

(1) Each fauna can be characterised not only by the presence and numerical representation of different families and genera, but also by their absence. In this regard there is no impassable gap between the Dipterofauna of Australia and the faunas of other zoogeographical regions. Most of the families distributed throughout the world are

represented, and there are practically no autochthonous families. The Australian dipterofauna as a whole cannot be regarded as something very archaic, strongly demarcated from that of other continents, as is the marsupial fauna.

Although exhibiting many distinctive characters, the dipterofauna of Australia shows a close relationship to that of many other regions. This is because flies are, as a rule, strong fliers, and are therefore not subject to marked isolation. However, the dipterofauna of Australia has many endemic genera, and a great number of archaic groups (see details below).

(2) Recently the author found, in the Northern Territory, a new family of Acalyptrata, which he has called the Echiniidae PARAMONOV (in press). It has a two-segmented abdomen and very peculiar wing venation, but there is no evidence that this family is an autochthonous element of the fauna. Probably it is an element of the faunas lying to the north, just as other forms, which were recorded first from Australia, have their probable centre of origin, and main area of distribution, beyond of Australia. We must therefore accept the tentative conclusion that, in Australia, there are no endemic families of Diptera.

(3) There are many endemic genera in Australia, but only a few are without close relationship to genera from other regions. The oestrid genus *Tracheomyia*, with one species, almost certainly occurs only in Australia because the larvae live in the tracheal system of kangaroos and wallabies, a quite exceptional habitat for fly larvae. Another genus, *Exeretoneura* (Nemestrinidae), with a number of species, is also widely separated from other genera of the family, and its systematic position is in doubt.

(4) There are families and subfamilies of archaic character that are especially well represented in Australia compared to other zoogeographical regions. For example, the family Mydaidae has 2 genera *(Diochlistus* GERST. with 11, and *Miltinus* GERST. with 19 described species) and it is evident that the actual number of species is distinctly greater. The family Apioceridae has 2 genera *(Neorhaphiomydas* NOR. with 6 species, and *Apiocera* WESTW. with 64 species), whereas in North America it is represented by 2 genera with 32 species, and in South Africa by 1 genus of 2 species. Various *Apiocera* live under widely different conditions; they may be found from the tops of mountains right to sea level, where highly specialized forms occur on sandy beaches.

The genus *Comptosia* (Lomatiinae, Bombyliidae) is represented by about 150 species, descriptions of most of which are in preparation by the author. It includes species with 5, 4, 3, or 2 submarginal cells, showing a line of evolution from 5 or 4 cells to more recent forms with 2 submarginal cells. Among the latter group there are species practically inseparable from the widely spread genus *Lomatia* MEIG.

(Europe, Asia, Africa). Some of the Australian species show a very archaic venational character: a large number of short, additional veinlets placed perpendicular to the usual longitudinal veins which may be regarded as remains of the net-like venation of ancestral forms. In the genus *Comptosia* we can see all gradations in the evolution of venation, but the archaic forms are distinctly predominant.

(5) Of special interest is the genus *Rutilia* (Tachinidae) which reaches its maximum development in Australia where it probably originated. There are about 100 species whereas in surrounding areas there have been recorded: 1 species in India, 1 in the Philippines, 2 in Samoa, and 1 in Fiji. A group which can be also regarded as a genus, called *Formosia*, is widely distributed on the islands of the Malayan Archipelago. In Australia the numerous species of *Rutilia* are very common and are usually extremely abundant on *Leptospermum* and *Eucalyptus* flowers. There are also some closely related genera with a considerable number of species.

A study of the genus *Rutilia* shows that it is in a stage of active speciation, the species representing all gradations in evolution. There are no "missing links", and the species can be arranged in tabular form, in vertical and horizontal rows, in which there are few gaps. This net is so complete that the author was able to predict the existence of certain species with certain combinations of characters and later find them in nature. Each species occupies a special narrow biological niche, e.g. an altitudinal zone in the mountains, and each preserves its own characters very well. This genus is therefore an extremely valuable subject for the study of the origin of species.

(6) It is beyond doubt that the general fauna of Western Australia (especially its south-western corner) is markedly distinct from that of the south-east of the continent. These two parts of Australia are like two islands having different faunas, being only later united as geological evidence indicates; however, there is no marked distinction between their dipterofaunas. So far as plants are concerned it is well established that if these regions have common genera, the south-west is, as a rule, richer in species than is the south-east.

(7) Because of the large areas of ocean to the west, south, and east, the Australian fauna naturally does not exhibit close relationship to the faunas of countries lying in these directions. However, an important influence has occurred from the north, and both Malayan and Papuan elements have invaded the northern Australian Coast, from Darwin to Cairns. These tropical elements are also distributed along the Pacific coast nearly to Brisbane, and also occur in isolated localities further southwards.

(8) Some genera (e.g.: *Pelecorhynchus* (Tabanidae), *Dasyomma* (Leptidae), *Comptosia* (near *Lyophlaeba*) and *Dolichomyia* (Bom-

byliidae), *Chiromyza* (Stratiomyidae), *Trichophthalma* (Nemestrinidae), *Bathypogon, Deromyia* (Asilidae), *Ectinorrhynchus, Anabarrhynchus* (Therevidae), etc.) show very close relationship to Chilean and Patagonian forms. Even an experienced eye may sometimes find difficulty in detecting a specific difference between certain species from Australia and South America. This suggests that there may have been direct contact between these areas in the past. This hypothesis is supported by botanical data, although the Western Australian vegetation is distinctly richer than that of Eastern Australia in species and shows closer affinities with the South American flora.

(9) There are some recent immigrants (postdating European settlement) *(Lucilia sericata, L. cuprina, Musca domestica,* etc.), but generally speaking the composition of the Australian dipterofauna has not changed markedly under the influence of recent immigrants. Only *L. cuprina* WIED., probably introduced from South Africa with sheep, plays an important role. There are also some introduced New Zealand species, for example, *Calliphora hortona* WALK. (Calliphoridae), distributed discontinuously along the eastern seaboard, but their influence is unimportant. The Lonchopterids also are represented only by introduced species. All oestrids (except *Tracheomyia)* are also new immigrants.

(10) Although the mountains of Australia attain a maximum height of only 7000 ft. there is a marked altitudinal zonation in the distribution of species of some genera: there are low-land, middle zone and montain species. It is of special interest that in some families (the genus *Rutilia* is an excellent example) species replace one another at different altitudes. There are successions of 4—5 species from sea-level to the mountain summits, each occupying a particular range of altitude, despite the fact they are strong enough fliers to reach both extreme points easily.

It is reasonable to assume that these chains of species have a common origin, being derived either from a mountain or a lowland species. Closer study shows that the evolution of these chains of species was probably caused by past geological events. We can suppose that, during glaciation, the preglacial species were pushed northwards, and new species evolved under cold climatic conditions; also the preglacial species more adapted to cooler conditions, now became more widely distributed. At the end of the glacial period the warmth-loving species returned to their former habitats, but on the higher altitudes they met the closely related species adapted to a much cooler climate and formed a vertically distributed chain of species, physiologically and ecologically isolated from one another.

In central Australia, as well as in the western part, the ranges rise to 4—5000 ft. However, they have not retained the species of the glacial period because these have been exterminated by the dry climate and one finds only the desert representatives where one would except to find the mountain forms.

(11) In Eastern Australia a very interesting phenomenon may be observed: in mountain zones higher than 3000 ft. a typical Tasmanian fauna occurs, not as isolated representatives, but in a complex of species. If we find a Tasmanian Asilid we can be sure that we will also find nearby Tasmanian Stratiomyiids, Therevids, etc. In effect the Tasmanian fauna is distributed on the mainland island-wise at altitudes higher than 3000 ft., from the Grampians to the New England Range. Unfortunately we have no data about the mountainous areas in Queensland. In Tasmania these forms occur at much lower altitudes, often down to sea-level. This distribution of species probably also dates back to the glacial period in the past.

(12) The affinity with the fauna of New Zealand is very slight: some families are quite absent from New Zealand, others are very poorly represented; for example, the Bombyliidae has only one species in New Zealand — *Tillyardomyia gracilis*. This marked difference is probably due to the extermination in New Zealand, by glaciation, of a great number of genera (and families) formerly common to Australia, and principally of desert ecological habitats. Probably cold and high humidity caused their complete extinction (see PARAMONOV, 1955).

(13) The relationship with South America involves not only genera of the desert habitat (some Bombyliidae, Apioceridae, Asilidae, etc.) but also the genera adapted to lower temperatures and higher humidity, and ecologically linked with shrubs and forests *(Pelecorhynchus*, some Therevidae, etc.).

(14) The development of the Australian fauna has been influenced by three major factors:
- (a) It was isolated from the land masses of the Afreurasian continent probably during the Cretaceous, and since that epoch, has not been strongly influenced by invasion from other countries; it has therefore preserved very old elements and developed many peculiar forms.
- (b) The development of the fauna occurred at a steady pace, unaffected by environmental cataclysms, in comparison, for example, with Europe, where the glacial period caused very great changes in flora and fauna. As a result its fauna is: 1) rich in archaic forms, and 2) not split into small well-isolated groups.
- (c) The climate of Australia at all latitudes has always been only subtropical or temperate. Glaciation covered only a comparatively small part of the continent and had no overall catastrophic effects on the fauna. There is also reason to think that Australia, being isolated, never had a purely tropical fauna.

Although the glaciation in Australia was not very severe in the opinion of most authors, its influence was underestimated by earlier workers on the faunal and floral divisions of Australia. It did play an especially important role in determining the composition of the fauna and flora of the "Bassian" division.

(15) As will be seen later, there is such a close relationship between the representatives of certain old groups in the faunas of Australia and South America, that the author is convinced that an ancient geological connection must have existed. This connection was probably via the Antartic continent which in the past had quite a different climate. We propose to call these elements in Australian fauna "paleoantarcts".

The peculiarities of Australian families of Diptera
(Distribution and systematic position)

NEMATOCERA

1. Trichoceridae — 4 spp. (TILLYARD includes this family among Tipulidae).

This archaic family is absent from Africa and tropical America.

2. Limnobiidae —

This family has a world-wide distribution. The number of species is included under Tipulidae. The dominant genus in Australia and New Zealand is *Dicranomyia*.

3. Tipulidae — 691 spp. (TILLYARD: Aus. 250 spp.; N.Z. 500 spp.).

The genera *Semnotes*, *Longurio* and *Clytocosmus* have no analogues in other regions.

4. Anisopodidae — 4 spp. (TILLYARD: Aus. 2; N.Z. 3).

This family of about 100 species occurs throughout the world, but is unknown from Madagascar. It is poorly represented in Australia. A single species of *Olbiogaster* occurs on Lord Howe Island.

5. Mycetobiidae —

A family of about 30 species distributed throughout the world, but absent from Australia.

6. Mycetophilidae (including Fungivoridae, Lycoriidae, Sciaridae) — 136 spp. (TILLYARD: Aus. 100; N.Z. 225).

Although the Mycetophilidae as a whole is much better represented in New Zealand than in Australia, the Sciarinae is much better represented in Australia. There are some endemic Australian genera, such as: *Antriadophila* SKUSE, *Heteropterus* SKUSE, *Pseudoplatyura* SKUSE, *Stenophragma* SKUSE, *Lygistorrhina* SKUSE (some authors accept this genus as a family - Lygistorrhinidae), *Clastobasis* SKUSE, *Acrodicrania* SKUSE, *Aneura* MARSH., *Synplasta* SKUSE (2 spp., and a third species known from Baltic amber), and *Delopsis* SKUSE.

7. Cecidomyiidae — 112 spp. (TILLYARD: Aus. 100; N.Z. 25).

There are some endemic Australian genera: *Gonioclema* SKUSE, *Chastomera* SKUSE, *Necrophlebia* SKUSE, represented by single species.

8. Bibionidae — 16 spp. (TILLYARD: Aus. 15; N.Z. 6).

9. Scatopsidae — 5 spp. (TILLYARD: Aus. 1; N.Z. 5).

This family contains about 150 species, mostly holarctic; it is very poorly represented in Australia.

10. Simuliidae — 26 spp. (TILLYARD: Aus. 14; N.Z. 6).

About 600 species are known in the world. The subgenus *Austrosimulium* TONN. is dominant in Australia.

11. Ceratopogonidae — 51 spp. (TILLYARD includes this family under Chironomidae).

About 600 species known, of which about 500 are holarctic.

12. Chironomidae — 75 spp. (TILLYARD, including Ceratopogonidae: Aus. 75; N.Z. 75).

There are about 3000 known species, mostly occurring in the northern hemisphere.

13. Culicidae — 168 spp. (TILLYARD: Aus. 100; N.Z. 8).

There are more than 2000 species in the world, mostly in tropical areas. There are no groups peculiar to Australia.

14. Dixidae — 7 spp. (TILLYARD: Aus. 6; N.Z. 7).

150 species are known throughout the world.

15. Tanyderidae — 6 spp. (TILLYARD: Aus. 1; N.Z. 4).

About 20 species are described in the entire world.

16. Thaumaleidae — 8 spp. (TILLYARD: Aus. 7; N.Z. 2).

This family is recorded from Australia by TILLYARD. Eight undescribed species are preserved in the Museum of the Division of Entomology, C.S.I.R.O., Canberra. About 50 species are known from the whole world.

17. Ptichopteridae —

About 50 species are known from the holarctic region, Africa and South America, but the family has not yet been recorded from Australia.

18. Psychodidae — 56 spp. (TILLYARD: Aus. 30; N.Z. 40).

The number of species given by TILLYARD probably includes undescribed species.

19. Blepharoceridae — 11 spp. (TILLYARD: Aus. 11; N.Z. 8).

About 75 species are known throughout the world.

Total: In 1956—1376 described species: (in 1926 — 712 species, including some undescribed species).

BRACHYCERA

20. Stratiomyiidae — 108 spp. (TILLYARD: Aus. 44; N.Z. 24).

This old family is very well represented in Australia. TILLYARD, in 1926, listed 44 species but the real number is distinctly higher. It is interesting to note that typical *Stratiomyia* are extremely poorly represented in Australia, having been found only in the Cape York area. It looks as if this group had its origin in the Northern hemisphere. There are numerous endemic genera but the systematics of the family are not advanced enough to permit far-reaching conclusions. The genus *Chiromyza* occurs also in South America. Very distinctive is the genus *Boreoides* which has wingless females, and a head structure quite unique among the Diptera: in the lower part of the face there are two small, symmetrical tunnels through the head, through which one can pass an entomological pin; the purpose of these openings is unknown.

21. Leptidae — 20 spp. (TILLYARD: Aus. 11; N.Z. 0).

This small family is poorly represented in Australia (in New Zealand there is only 1 species) but is represented by 5 genera. The genus *Dasyomma* occurs only in Australia and South America. The author knows of some undescribed species, and the family is richer than previously believed.

22. Tabanidae — 307 spp. (TILLYARD: Aus. 200; N.Z. 15).

MACKERRAS, 1954, separates the family into 4 subfamilies (excluding the genus *Pelecorhynchus* from the family); only the Scepsidinae is absent from Australia. The three subfamilies which are present in Australia are very old and rich in species.

The subfamily Tabaninae has 3 tribes: the large and recently evolved tribe Tabanini is distributed very widely throughout the world; it is absent from Tasmania, New Zealand and the southern part of South America. The tribe Haematopotini has not yet reached Australia, South America, Tasmania, New Zealand and Madagascar; the tribe Diachlorini is the oldest, being represented in Australia, Tasmania, New Zealand, Madagascar, Africa and America.

The subfamily Chrysopinae also has 3 tribes: the tribe Chrysopini is evidently now developing vigorously. It is absent from the southern part of South America, New Zealand, Tasmania, and in Australia is represented only in Queensland. The tribe Rhinomyzini is present only in Africa and south-eastern Asia. The tribe Bouvieromyiini is evidently an old one with most of its representatives in the southern hemisphere.

The subfamily Pangoniinae has also 3 tribes: the ancient tribe Philolichini is absent from Australia; the tribes Pangoniini and Scionini are mostly represented in the southern hemisphere.

Pelecorhynchinae. The unique genus *Pelecorhynchus* MACQ., containing in Australia about 30 species, is also represented in South America, by a half a dozen species. It is a typical "paleoantarct" in our terminology.

23. Nemestrinidae — 56 spp. (TILLYARD: Aus. 21; N.Z. 0).

This ancient family has 56 Australian species, distributed in 8 genera, a number very high in comparison with other faunas. The genera *Atriadops* WAND., *Nycterimyia* LICHTW., and *Nycterimorpha* LICHTW. are closely related to Malayan and New Guinean forms, but are absent from Western Australia and "Australia deserta" in general. The genus *Exeretoneura* MACQ. is endemic and has such exceptional characters that its position in the family is obscure. The endemic genus *Cyclopsidea* MACQ. probably occurs throughout Australia. The other three genera are distributed throughout Australia and also occur in other zoogeographical regions. The genus *Trichophthalma* WESTW. is remarkably well developed, having 45 described species. This genus is also well represented in Chile, some of the species resembling Australian species so closely that their separation is very difficult even for an experienced dipterist.

24. Acroceridae — 26 spp. (TILLYARD: Aus. 6; N.Z. 6).

This old group, parasites of spiders, is represented in Australia by 4 genera: *Leucopsina* WESTW. (1 species) is endemic; *Panops* LAMARCK (3 species) is found only in Australia and America, but with most species in Chile; *Pterodontia* GRAY (3 species), a genus represented only in Australia and America; the genus *Oncodes* LATR. with 19 species is widely distributed throughout the world. It is interesting to note that the whole subfamily Philopotinae, which is rather well represented in central and South America, and occurs in Europe. Africa and Japan, has not yet been recorded in Australia and is probably absent. The genera *Apsona* and *Helle* WESTW., endemic in New Zealand, are absent from Australia.

25. Mydaidae — 30 spp. (TILLYARD: Aus. 9; N.Z. 0).

This is one of the very old and primitive families. The two endemic genera, *Diochlistus* GERST. and *Miltinus* GERST., have 11 and 19 described species respectively, and probably more species awaiting discovery. The record of *Miltinus commoni* PAR. in New South Wales shows that the group of species around *M. maculipennis* WESTW., which was regarded as typical West Australian, is distributed throughout the entire "desert" Australia.

26. Asilidae — 231 spp. (TILLYARD: Aus. 160; N.Z. 15).

A family well represented in Australia. The subfamily Laphriinae is absent from New Zealand, and the arid areas in Australia. The very large flies - *Blepharotes* (35 mm. long) and *Phellus* are endemic genera. There are other endemic genera, but their close relationship to the genera of other zoogeographical regions is quite evident. The genus *Neoaratus* is very rich in species and specimens.

27. Bombyliidae — 180 spp. (TILLYARD: Aus. 80; N.Z. 1).

This very large family is well represented in Australia by nearly all its subfamilies, but the study of it is incomplete. TILLYARD (1926) listed only 80 species; in the work of ROBERTS, which covers all subfamilies except the Lomatiinae, 119 species are dealt with; the true number of species is at least three times as great. New Zealand has only 1 species belonging to an endemic genus, *Tillyardiomyia* TONN.

The opinion of MACKERRAS (1950) that *Lomatia* is a Lemurian element cannot be accepted, because species are also abundant in Europe and Asia, and in Africa the genus is represented by more highly specialised forms than in Australia. This author also writes that: "in Bombyliidae, *Comptosia* is southern, *Exoprosopa* cosmopolitan, but clearly part of the Indo-Malayan element so far as Australia is concerned." This statement is right in regard to Queensland, but generally speaking Australia contains species groups similar to those in the rest of the world, but naturally containing some endemic elements. *Comptosia*, and the extremely closely related *Lyophlaeba* from South America, is well represented in the southern hemisphere. *Exoprosopa* can be separated into at least 11 groups which authors variously regard as genera or subgenera. These groups are found in nearly all zoogeographical regions, and naturally have specific peculiarities in each of them.

There are some endemic Australian genera, but their status is not clear, and only revision of the World genera will clarify the position. It seems that the endemic genera are the residue of a very old fauna, now poorly represented. Although the Bombyliidae is very rich in species, the family is not represented by genera and species "*in statu nascendi*", as is the case among Therevidae, Rutiliini and some other groups in Australia. The Bombyliid fauna is very old and specialised, but apparently not undergoing rapid speciation. The Bombyliidae may be accepted as reasonably "stabilised" in Australia.

The discovery of more species may be expected because little investigation has been carried out in the arid central area of Australia, which is a classical habitat for Bombyliidae.

28. Therevidae — 108 spp. (TILLYARD: Aus. 56; N.Z. 10).

This family is extremely well represented in Australia, and there are many species and genera which are not yet described (PARAMONOV, in MS). TILLYARD (1926) recorded 56 species in this family, but the known forms now comprise two or three times this number. Most inhabit forest or shrub areas, but some groups also occur on sand dunes. The genera *Anabarrhynchus* MACQ. and *Ectinorrhynchus* MACQ. are common in Australia and South America (both also occur in New Zealand). The remainder of the genera are mostly endemic, represented by rather a large number of species and preserving many intermediate, annectant forms. The genus *Acraspiza* has a peculiar scutellum, which is long, narrow, and quite vertical, but in the genus *Acraspizella* (gen. nov. in Ms.) it is more in a declined position. There are also species representing all stages in the position of the scutellum, from the vertical to the normal horizontal. In regard to other characters we find a similar state of affairs: the existence of completely intergrading forms between two extremes. This family is apparently undergoing active evolution, and it is not easy to separate the genera. Some of the species are very abundant. Unfortunately Cape York and Western Australia have been poorly studied and the geographical relationships of many genera and species are not yet clear. From the theoretical point of view this is a very interesting family for concentrated study.

29. Apioceridae — 70 spp. (TILLYARD: Aus. 5; N.Z. 0).

The family is very archaic. It is very poorly represented in the Old World. There are two genera in Australia: *Neorhaphiomydas* NOR. (6 species) and *Apiocera* WESTW. (64 species), by far the richest fauna in the world. *Neorhaphiomydas* NOR. is very closely related to the Chilean *Megascelus* PHILIP., and occurs only in Western Australia, an example of related forms occurring in Western Australia and South America but unrepresented in Eastern Australia.

30. Scenopinidae — 8 spp. (TILLYARD: Aus. 2; N.Z. 0).

This small family which is comparatively very poor in species throughout the world, is represented in Australia by 5 genera. Two of them: *Scenopinula* PAR. and *Riekiella* PAR. are endemic; *Pseudomphrale* KROEB. recently recorded, shows a close relationship with Central Asian forms, and *Pseudatrichia* O.-S. with North American forms. The genus *Scenopinus* MEIG. is practically cosmopolitan. The recent discovery of two new genera and other species suggests that Australia probably has more undescribed species and genera, and may prove to possess the richest Scenopinid fauna in the world.

31. Empididae — 49 spp. (TILLYARD: Aus. 50; N.Z. 110).

TILLYARD (1926) gave the number of species in Australia as 50, and those of New Zealand as 110; this representation was due to the fact that Empididae were better studied in New Zealand than in Australia. Australian Empididae are too poorly known to be considered in zoogeographical discussions.

It is interesting to note, however, that the subfamily Ceratomerinae is restricted to Australia, New Zealand and South America, suggesting an Antarctic connection between these continents in the past.

32. Dolichopodidae — 112 spp. (TILLYARD: Aus. 20; N.Z. 45).

The position regarding the family Dolichopodidae is similar to that of the Empidae.

33. Lonchopteridae.

This family was not recorded by TILLYARD, 1926. In the C.S.I.R.O. collection, there are specimens from Tasmania and New South Wales. belonging probably to European species.

34. Platypezidae — 6 spp. (TILLYARD: Aus. 5; N.Z. 0).

35. Phoridae — 16 spp. (TILLYARD: Aus. 6; N.Z. 15).

There are probably more species in Australia, but there is no recent review of the Australian representatives.

36. Pipunculidae — 27 spp. (TILLYARD: Aus. 26; N.Z. 4).

This is a small family, but is comparatively well represented in Australia. All the described Australian species are from Queensland. Specimens from other Australian States are extremely few in all Australian museums.

37. Syrphidae — 167 spp. (TILLYARD: Aus. 60; N.Z. 30).

This very large family is poorly represented in Australia. Although the genus *Syrphus* is represented in Australia, the number of species is not comparable with that of Europe, Asia, etc., nor does one see an abundance of individuals as often as in Europe. Many genera, represented only in Queensland, are quite evidently immigrants from the North. However, most subfamilies are represented in Australia, and a considerable number of them have endemic species. Only further investigations can elucidate the reason for the poverty of this family in Australia.

Some genera, for example, *Cerioides* and *Microdon* are rather rich in species. The extremely large genus *Chilosia* MEIG. is represented in Australia by only a small number of species. It is also interesting to note that the genus *Volucella* GEOFFR., so richly represented throughout America and Europe, is absent from Australia.

38. Conopidae — 47 spp. (TILLYARD: Aus. 12; N.Z. 0).

This family is very well represented in Australia. Some endemic genera were described by KRÖBER. The studies of the author have also resulted in a further increase of the number. Some widely distributed genera are represented in Australia by very few species, for example, *Myopa* has only 2 species. It is evident that in Australia this family is very well developed and probably of ancient stock.

39. Pyrgotidae — 60 spp. (TILLYARD does not mention this family).

A very well represented family, containing 18 genera (7 new), with 63 species (3 of them from New Guinea). It is quite evident that this list is very far from complete; light traps bring more and more new forms in Canberra alone. Most of the genera are endemic. Their exact number is difficult to state because some of them probably also occur in other regions but have not yet been recorded, or may change their systematic status after a revision of the world genera. It is evident that this family is in speciating profusely in Australia (for example, the genus *Epicerella* is already represented by 27 species).

40. Ortalidae — 59 spp. (TILLYARD: Aus. 50; N.Z. 11).

This large family is well represented in Australia. The record of the genus *Herina* R.-D. by TILLYARD is probably wrong because MALLOCH in his paper on Australian Diptera did not record this genus for Australia; the author also failed to find it among all available Australian collections. A similar mistake of TILLYARD can be demonstrated by the fact that he recorded one *Herina* species from New Zealand, where the only ortalid recorded is *Zealandortalis interrupta* MALL.

The genera *Euprosopia*, *Duomyia* and *Lamprogaster* are very well represented by endemic species, but it is evident that the centre of dispersion lies to the north. This group of genera probably originated in the equatorial zone. The relationship of the ortalids of Australia cannot be unravelled without knowledge of those in surrounding areas, especially to the east.

41. Trypetidae — 60 spp. (TILLYARD: Aus. 32; N.Z. 4).

A group of genera around *Dacus* (fruit flies) contains many species. However, many of the genera are probably only equivalent to groups of species. The splitting and creation of subgenera theoretically and practically is not well based. The mediterranean fruit-fly *Ceratitis capitata* was introduced at the beginning of this century and has become a serious pest, especially in oranges.

42. Agromyzidae (including Phytomyzidae and Ochthiphilidae) 46 spp. (TILLYARD: Aus. 22; N.Z. 16).

This leaf-miner family is very poorly studied; in Australia there are many undescribed species of *Agromyza* and *Phytomyza*. It includes some very specialised forms, for example the genera *Cryptochaetum* and *Fergusonina*.

43. Lonchaeidae — 6 spp. (TILLYARD placed Lonchaeinae under Sapromyzidae).

Most of the species are closely related to those in the tropical and equatorial areas. This family includes the so-called Tomato-flies.

44. Tylidae and 45. Neriidae — 7 spp. (TILLYARD: Aus. 7; N.Z. 0).

These closely related families are poorly represented in Australia. The majority of the Australian species have been found in Queensland and are immigrants from the North. There has been no review of these families dealing with Australian species. Preliminary investigations show that besides the tropical elements from the North, there are undescribed endemic species, adapted to moderate climates, and living in humid valleys at an altitude of 3-4000 feet and higher.

46. Sepsidae — 8 spp. (TILLYARD: Aus. 6; N.Z. 0).

There are two endemic genera, but very closely related to the basic genus *Sepsis* FALL.

47. Piophilidae.

Two introduced species were found in Australia.

48. Thyreophoridae - 8 spp.

Until 1954 it was thought that this family was not represented in Australia. The author has shown that there are 5 species of the genus *Chaetopiophila* MALL., which can be regarded as purely Australian. It appears that the isolation of Australia was not as pronounced as was thought. The lateness of the record of this genus was due to the paper by OSTEN-SACKEN (1882) being overlooked, and to confusion with the family Piophilidae.

49. Psilidae — no species recorded. (TILLYARD: Aus. 2; N.Z. 0).

TILLYARD stated that an undescribed *Loxocera* species occurred in Australia. The second species is probably the introduced *Psilae rosae*, but this has not been confirmed

50. Tetanoceridae — (included in Sapromyzidae).

51. Neottiophilidae — 9 spp. (TILLYARD: Aus. 4; N.Z. 0). (TILLYARD includes *Tapeigaster* under Scatophagidae).

This family is represented in Australia only by the endemic genus *Tapeigaster* MACQ. with 9 species. The larvae develop in decaying mushrooms. It is curious that the genus is quite absent from New Zealand.

52. Sapromyzidae — 107 spp. (TILLYARD: Aus. 41; N.Z. 16).

This family is well represented in Australia by numerous genera and species; some of them are peculiar to Australia, but the difference from those in other zoogeographical areas is not very great, and the gap between different genera is often very narrow. There are many undescribed species.

53. Ochthiphidae — 5 species in Australia.

54. Coelopidae — 5—6 spp. (TILLYARD: Aus. 2; N.Z. 10).

This family (kelp-flies) is small but is usually represented by enormous numbers of specimens. In Australia there are 2 genera with a total of 5 or 6 species. The family requires a revision before it can be decided which species are endemic. An interesting fact is that rather a considerable number of specimens have been collected in light traps inland, very far from the sea-shore, suggesting that some representatives have adapted themselves to life in fresh-water.

55. Helomyzidae — 3 spp. (TILLYARD: Aus. 14; N.Z. 20).

The genera *Helomyza*, and *Huttonomyia*, were recorded erroneously by TILLYARD, There are two genera: *Diplogeomyza* (endemic) and *Pseudoleria* (Introduced).

56. Clusiidae (including Heteroneuridae) — 2 spp. (TILLYARD: Aus. 4; N.Z. 0).

57. Anthomyzidae — 1 sp. (TILLYARD includes this family under Geomyzidae).

58. Opomyzidae (including Geomyzidae) — 2 spp. (TILLYARD: Aus. 6; N.Z. 12).

(One *Opomyza* with several unplaced species, according to TILLYARD).

58a. Chiromyidae — 1 sp. *(Aphaniosoma)*

59. Drosophilidae — 47 spp. (TILLYARD: Aus. 24; N.Z. 13; TILLYARD includes *Asteia* here).

60. Asteidae — 3 spp. (TILLIARD; only mentioned the genus *Asteia* among Drosophilidae).

61. Borboridae (Cypselidae) — 7 spp. (TILLYARD: Aus. 6; N.Z. 12).

62. Ephydridae — 34 spp. (TILLYARD: Aus. 10; N.Z. 15).

Ephydra and *Hydrellia* occur in both countries; *Ectropa* and *Nothiphila* in Australia; *Clasiopa, Ehygrobia, Hyadina, Parahyadina, Scabella, Gymnopa* and *Parydra* in New Zealand.

63. Chloropidae (and Mindidae) — 188 spp. (TILLYARD: Aus. 41; N.Z. 15).

This large family is well represented in Australia, with some endemic genera. The representatives of the genus *Batrachomyia*, in the larval stage, live beneath the skin of certain Australian frogs.

64. **Milichiidae** — 6 spp. (TILLYARD does not mention this family).

65. **Echiniidae** — 1 sp. (PARAM., MS. see p. 165).

66. **Braulidae** —

1 introduced cosmopolitan species - *Braula coeca* NITZSCH, a parasite of honey-bees. TILLYARD has included it in Phoridae. Recorded from Tasmania.

67. **Hippoboscidae** — 18 spp. (TILLYARD: Aus. 5; N.Z. 1).

There are 2 endemic genera parasitic on kangaroos: *Ortholfersia* SPEIS. (4 spp.) and *Austrolfersia* BEQ. (1 sp.). The genera and species parasitic on birds are predominantly not endemic. The specificity of Australian parasites on birds is not so marked as that on marsupials, because their power of flight means that birds are less isolated from other countries than are marsupials. There are some undescribed genera and species in the Australian fauna but they are not very different from hippoboscids from other zoogeographical regions.

All hippoboscid parasites of domestic mammals in Australia were introduced with their hosts. It is interesting to note that up to the present no hippoboscid has been recorded on the Australian dingo, whereas *Hippobosca capensis* is widely spread on dogs in Africa, Europa and Asia. It may therefore be presumed that the dingo penetrated from a country where hippoboscids were absent as parasites on dogs.

68. **Streblidae and** 69. **Nycteribiidae** — 14 spp. (TILLYARD: Aus. 6; N.Z. 0)

Both families are parasitic on bats. There are some Australian species, but because bats are migratory, the Australian specificity is not clearcut. The fact that a species was described first from Australia cannot be accepted as proof that this species is purely Australian and originated there.

70. **Gasterophilidae** — bot-flies. 3 species, all introduced.

71. **Scatophagidae** — 1 sp.

The genus *Tapeigaster*, which was usually regarded as belonging to this family, belongs to the family Neottiophilidae. Only one species of Scatophagidae was recorded for Australia - *Scopeuma guérini* R.-D. *(Tomella)*, recorded also by FROGGATT (Australian Ins., 1907, p. 310), who gives some details about it in New South Wales, but it is uncertain whether FROGGATT correctly identified the insect.

The family Scatophagidae is typical of the Northern hemisphere and has not been recorded from South America.

72. **Muscidae (and Anthomyiidae)** — 171 spp. (TILLYARD: Anthomyiidae: Aus. 50, N.Z. 60; Muscidae: Aus. 60, N.Z. 8). TILLYARD also included Calliphoridae here.

This large family contains many introduced species or species common in the oriental areas. The genus *Helina* is very well represented in Australia, but it is also widely spread throughout the world. There are genera such as *Pygophora*, *Macrochaeta*, etc. peculiar to Australia, but in surrounding areas there are genera very closely related to the former. Summarising our data we can conclude that the muscid fauna of Australia is comparatively poor and possesses few characteristic forms. The general impression is that this continent has not produced many endemic forms, in contrast with the high endemicity of Calliphoridae and Tachinidae (see below).

73. Calliphoridae — 95 spp. (TILLYARD records under Muscidae).

The genus *Calliphora* is extremely rich in species, many of which are not yet described. The abundance of some representatives is enormous, and traps may be filled with hundreds of specimens in a few hours. Being separated into some very distinct groups, this genus also contains many species with extremely close relationships and with intermediate forms between them.

Practically no spot in Australia from the summits of mountains to the sea-shore is free from Calliphorids. It is a common practice in the country and in many of the large cities to cover windows with screens to protect the homes from invasion by blow-flies.

Besides the endemic species there are also some introduced from other zoogeographical areas (for example *C. erythrocephala* and *Lucilia cuprina*)

Australian Calliphoridae in their richness, diversity, and mixture of geologically old with comparatively recent forms, is a very characteristic element of the Australian fauna.

Some groups, for example, *C. hortona*, show a close relationship to the New Zealand fauna, which is the richest centre of this group. It is difficult to tell if the presence of this group at the sea-shore in Australia is the result of comparatively recent introduction by ships, or if it is only a relict of the old group, previously more widely spread (we have some records about this group in Tasmania, but it is not yet confirmed).

Northern Australia is occupied by oriental genera and species which have not penetrated very far southwards. The details will be published by the author in a monograph on Australian Calliphoridae (in preparation).

74. Sarcophagidae — 47 spp. (TILLYARD records under Tachinidae).

75. Tachinidae — 479 spp. (TILLYARD: Aus. 220; N.Z. 200; TILLYARD includes here also Sarcophagidae).

The Tachinid fauna of Australia contains many endemic genera. The tribe Rutiliini is especially well represented *(Rutilia, Amphibolia,* etc.*)* having more than 100 species. Although not very numerous, species of the genus *Microtopeza* are also very characteristic of Australia. There are some undescribed genera closely related to the very peculiar genus *Amphitropeza*.

The tribe Dexiini is very typical and well represented in the forest areas of Australia, only a few of the species being described. The tribe Ameniini is also very widely distributed in Australia, but probably most species are described, and only in northern and western Australia can the discovery of new species be expected.

The tribe Goniini is also very well represented (the genus *Tritaxys* includes a number of species not separated by previous dipterists). Knowledge of the Tachininae is in a chaotic position: there are some very widely spread, nearly cosmopolitan genera, but there are many genera endemic to Australia.

The tribe Cylindromiini is rather richly represented, and it is expected that many new species await description.

Omitting details it can be stated that the tachinid fauna of Australia is very characteristic, because it includes many endemic genera.

76. Oestridae — 2 spp. (TILLYARD: Aus. 1; N.Z. 0).

There is one endemic genus - *Tracheomyia* TOWNS. the larva of which infests the tracheae of kangaroos; the adult stage has been known only since 1953 (PARAMONOV) It is strange that Australia, having developed a very distinct genus with a quite exceptional larval habitat (tracheae), possesses only a single species - *T. macropi* (FROGG.). In view of the diversification of kangaroos, one would have expected further speciation of the parasite. Other oestrids have been introduced with their hosts, but the genus *Hypoderma* is known only in larval stages and we have no record of adult flies in Australia. The larvae were found only under the skin of animals arriving from overseas. Only the common *Oestrus ovis* has been acclimatised.

Notes: In some families discord in numbers between the completed up to 1956 present list and that of TILLYARD occurs owing to TILLYARD having often included undescribed species.

Concluding the review of the Australian fauna of Diptera it will be useful to give a census. The author's catalogue of the Australian Diptera (unpublished) shows that we have:

	TILLYARD 1926	End of 1956.
Nematocera	712 spp.	1376 spp.
Brachycera orthorrhapha	774 spp.	1568 spp.
Muscidae acalyptratae	208 spp.	706 spp.
Muscidae calyptratae	330 spp.	794 spp.
Total Diptera:	2124 spp.	4444 spp.

It is interesting that in 30 years there has been an increase of more than 100 % in the number of described species, during a period when Diptera were studied mostly by part-time dipterists. It is evident that the total of 4444 species represent only a portion of the real number of Australian Diptera.

Faunal division of the Australian Fauna
(with special reference to the Diptera)

It will be necessary to discuss some of the main points of the problem before turning to the examination of details.

(1) From a study of the works of zoologists and botanists we can state that the geographical distribution of plants and animals in Australia is not uniform. The great difference between the floras of Western and Eastern Australia has no analogy in the dipterous faunas of these parts. At least $\frac{3}{4}$ of Australia (including the south-western corner) is dominated by arid, desert animals, and there is no reason for regarding Western Australia as the counterpart of Eastern Australia.

(2) Australia can be divided into two parts or "subregions" (see fig. 1): (a) desert, eremian Australia with boundaries coinciding approximately with the average yearly rainfall up to 20 inches — Australia deserta (D), and (b) shrub and forest Australia with boundaries coinciding with average yearly rainfall over 20 inches (up to 75 inches) — Australia sylvatica (A, B, C).

(i) Australia deserta has approximately the same boundaries as "Eyrean subregion" of SPENCER 1896, but we cannot accept the term of SPENCER because: (1) the area of Lake Eyre cannot be regarded as a "corner stone" of this fauna; on the contrary, its fauna is only an impoverishment of the land westwards from it, (2) Lake Eyre is connected with a water fauna and we are dealing mostly with land fauna, (3) the term does not reflect the most important feature of its content, and (4) Lake Eyre area (sensu stricto) has its own peculiarities and cannot be confused with the rest of this extensive subregion.

The Eremian division is, due to uniformity of habitats, comparatively poor, and extremely poorly studied. It is practically a "terra incognita" in regard to insects. It will therefore be better

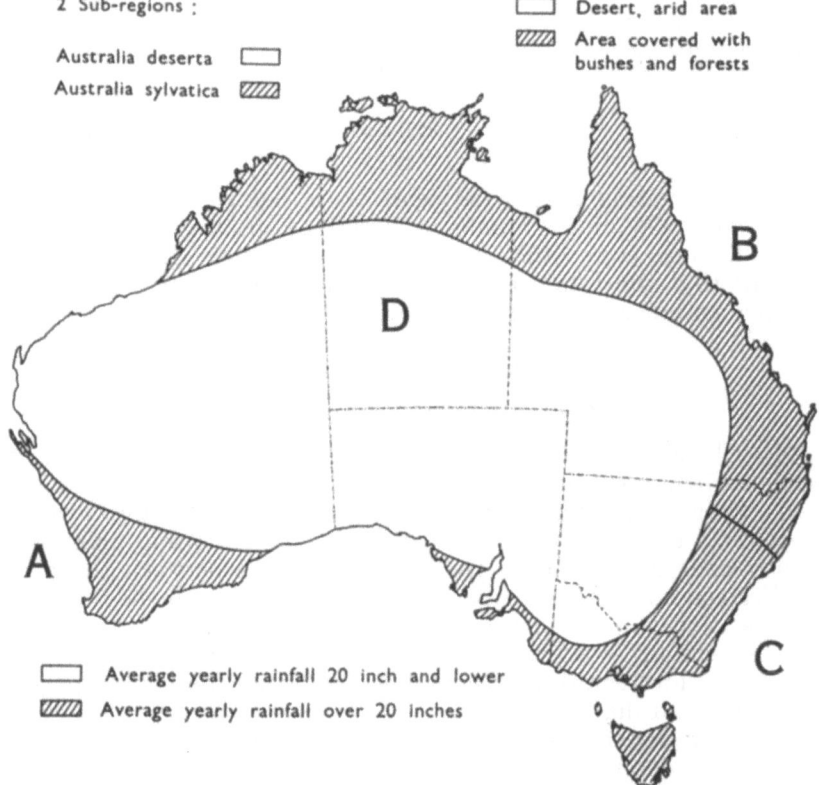

Fig. 1. Showing the main divisions of Australian fauna:
Australia deserta - D; A - Australia westralica
Australia sylvatica: B - Australia bororientalis
                        C - Australia merorientalis

to accept our ignorance of its fauna than to speculate about it.

(ii). Australia sylvatica includes practically the "Torresian subregion" and the "Bassian subregion" of SPENCER, plus "Hesperonotian" of NICHOLLS, 1933.

We can turn now to a more detailed analysis of this portion of the Australian fauna. Australia sylvatica can be separated easily into three provinces: (1) westralian — restricted to the south-western corner of Western Australia and approximately coinciding with the rainfall limit of 20 inches.

The Westralian division was established by TATE (1887) and named, very unfortunately, "Autochthonian". This term cannot be accepted because: (1) each country can have its autochthonous elements, not only this one; this limitation of the term can lead only to mis-understanding; (2) Eastern Australia has enough animal species to justify it also being called "autochthonian", whereas in West Australia the plants are especially rich in authochthonous elements. We propose to call it Westralian, the name signifying its geographical position. Geologically most of it is the oldest part of the continent. TATE (1887), from a study of the plants, thought that the flora arrived from the Austro-Malayan islands during or before the Cretaceous era, and spread over the whole of Australia.

We regard theoretical considerations with regard to the fauna very premature. The fossil remains are not helpful. We have no proof that there was a close relationship between the old fossils and the recent fauna. Most of the remains belong to Vertebrata, and they have little in common with the Evertebrata (especially Insecta) in evolution, habitats, etc.

The other two provinces are: (2) Torresian (SPENCER, 1896) and (3) Bassian (SPENCER, 1896). The former was established by HEDLEY (1893) under the name of Papuan. He wrote: "The types encountered by a traveller in tropical Queensland, or rather in that narrow belt of tropical Queensland hemmed in between the Cordillera and the Pacific, all wear a foreign aspect. Among mammals may be instanced the cuscus and tree kangaroo; among reptiles the crocodile, the *Rana* or true frog, and the tree snakes; among birds cassowary and rifle birds (Ptiloris); among butterflies *Ornithoptera;* among plants, the wild banana, orange, mangosteen, the rhododendron, the epiphytic orchids, and the palms; so that in the heart of a great Queensland "scrub", a naturalist could scarcely answer from his surroundings, whether he were in New Guinea or in Australia."

We regard SPENCER's terms more suitable because they avoid confusion between the true Papuan elements and the elements on the Australian continent, which would better be called "neopapuan"; on the other hand, not only the endemic Papuan elements but also elements from India, Malaya, and the Philippines, etc., can penetrate into Australia via Torres Strait. SPENCER's term indicates only the route of penetration, and is exact and unambiguous.

The "Torresian" and "Bassian" provinces together are practically the "Euronotian subregion" of TATE. Based on a study of the plants, he regarded its flora as reaching Tasmania from South America not later than the Miocene epoch; many of the original inhabitants, particularly on the east coast, probably disappeared before the invaders. HARRISON, 1928, thinks that the fauna of the Torresian province is of a different origin: "there is a Pliocene Pantropical (perhaps Indo-Pacific would be better), and a Pleistocene Indo-Malayan component." We regard such separation as premature; our knowledge of the fauna of this area is too scanty.

The Bassian province is inhabited by forms better adapted to lower temperature, lower humidity and savannah-type vegetation than are those from the Torresian province. Both provinces can be separated by a line slightly to the south of the Clarence River. Very probably, glaciation caused most change in Australia in this area, because this part contains higher mountains. Glaciation could have also caused the more marked impoverishment of the Tasmanian fauna than of that on the mainland. And as result of this we cannot now find in Tasmania the forms which penetrated through Tasmania from the Antarctic continent to the mainland.

The Bassian province can be separated into two districts: Tasmanian (Tasmania and surrounding islands) and Victorian (the rest of the province).

It is necessary to add that the faunal subdivision, sketched by us, is based on the principle of zoological geography, and not on the principle of geographical zoology (compare, for example, MACKERRAS 1950). We have characterised the various parts of Australia, i.e., its surface, by the different groups of animals (their presence and absence); geographical zoology is mainly concerned with the animals themselves, and with their geographical distribution. Our divisions are the natural complexes of animals and habitats with different history; in geographical zoology one studies the geographical distribution of groups of animals on the Earth.

The author thinks it necessary to explain why he rejects the terminology of other authors; he does so because: (1) no work uses the terminology of previous authors through disagreement with the limits of subregions, provinces, etc., (2) some of them are quite illogical, for example, "Authochthonous" (each country has its own autochthonous elements), (3) the terms are very often foreign and explain practically nothing by their names; what, for example, is meant by the term "Euronotian"? The use of such terms only overburdens our minds with new words. The author prefers a terminology understandable at once without special explanation: "Australia bororientalis", "Australia merorientalis", a.s.f.

Some proposed terms, such as "Torresian province" or "Bassian province" are not bad, but they express a wrong idea: one would think that the Torresian province is a province created by invasion or penetration of different forms by way of Torres Strait, but it has its own elements and is a mixture of forms from all directions. If there are forms penetrating into Australia by Torres Strait, there are also forms which penetrated into New Guinea from Australia by the same route. The process of penetration is not one-sided, but mutual, and we cannot regard the fauna of the Torresian province as a "mixtum compositum" created by Torres Strait; the province has its peculiar features which are caused not only by land connection in the direction of Torres Strait but also in other directions.

Therefore the author prefers to use a simple and evident term — A. bororientalis. It is necessary to know the fauna much better before establishing terms such as "Euronotian" and similar ones. The use of such terms at present is premature. They are erected on a too narrow basis.

Australia Deserta

Can be characterised largely by the absence of those families whose larvae are connected closely with water or with trees, or bushes. The dominant families are Bombyliidae and Asilidae, which are represented not only by numerous genera and species, but also by numerous individuals.

During the very short spring many families are well represented: Nemestrinidae, Calliphoridae, Muscidae, Tachinidae, etc. They are mostly associated with ephemeral flowers. Some weeks after the plants die off, the fauna of Diptera becomes very poor. Only *Bathypogon* (Asilidae) and *Exprosopa* (Bombyliidae) are found on stony or sandy soils lacking green plants.

Life is a little richer along the dissected streams and waterpools where some vegetation and water is preserved.

This subregion is nearly a "terra incognita", because no scientific institution exists in this territory and scientific expeditions are extremely rare, therefore the diptera collected are represented only by some odd specimens. Under such conditions it is difficult to know which endemic forms exist in this immense area. There are two possibilities: either the Dipterofauna of Australia deserta is only an impoverished fauna of the surrounding desert areas of A, B and C, or it contains many endemic forms, evolved there during millions of years of permanent isolation (the latter supposition is more probable).

The author considers that the organisation of a special institute at Alice Springs for study of the desert, its geology, botany, zoology, etc., is the only solution to this big problem which is connected with a number of purely practical problems. The central position of

Alice Springs, its accessibility by air, rail, and road, are good reasons for its selection as the site of such an institute, the main purpose of which would be the planned collection of material. Progress in science is impossible without material.

Australia Sylvatica (A — Australia westralica)

This part can be regarded as the western province forming with the Eastern part (C), a common southern province. Very often a species in one province is represented by a closely related species in the other. The difference in the faunas is much smaller than the difference in the floras.

Both these provinces, together with Australia deserta, represent the Australian fauna, *sensu stricto*, i.e., most of endemic Australian elements live there.

It appears that the majority of forms closely related to those of Patagonia and Chile, are preserved here.

The largest flies in Australia — *Phellus* (3 species) (Asilidae) are distributed in this area; there is a vague evidence of the occurrence of one species in the Mallee district (Victoria), but this is not yet confirmed. *Dakinomyia froggatti* DAK. & FORTH. also is a typical element. A second species (*D. claripennis* RIC.) is found in Queensland, but it differs strongly from the first.

The genera *Thereutria* Lw. and *Metalaphria* RIC., which are well represented in C (Australia merorientalis), are poorly represented here.

Among the numerous species of *Neoaratus* common in A and C, *N. hercules* WIED., which is very typical of C, has not yet been recorded in this province.

The tribe Ommatiini, which is well represented in B and C, is poorly represented here.

Up to the present date no *Pelecorhynchus* species has been found here. The Tabanids are represented mostly by species of *Scaptia*. The genus *Tabanus* (2 rare species here) undoubtedly originated in the Northern hemisphere.

Comptosia — species (Bombyliidae) are represented here in abundance (some dozens of species), including more archaic forms, but also more recent forms such as *Lomatia* which are widely distributed in Africa and Eurasia where the *Comptosia*-type species are absent.

Australia Sylvatica (B — Australia bororientalis).

This province is distinguished more sharply from A and C than the latter are from each other. It may be characterised by the presence of different families of oriental elements. From our present knowledge of the distribution of Diptera it appears that the oriental elements have penetrated into Australia from New Guinea via Torres Strait, because we know that the concentration of these elements is greatest in the Cape York area and that it diminishes to the west.

It is interesting to note that although the connection of New Guinea with Australia in the area of Torres Strait through a land bridge in the past has resulted in a mutual exchange of faunal elements, it appears that Oriental elements predominate in comparison with the Australian elements, i.e., that the movement of the fauna has mostly been in direction from North to South. Movement in the opposite direction was not so strong.

Among the Calliphorids the genera *Eucompsomyia* MALL., *Euphumosia* MALL., *Paratricyclea* VILL., and species: *Microcalliphora flavifrons* ALDR., *Chrysomyia incisuralis* MACQ., *Ch. megacephala* FABR. (in other parts of Australia — *Ch. bezziana* VILL.), *Ch. nigripes* AUB., *Lucilia porphyrina* WALK., *L. papuensis* MACQ., *Chlororhinia viridis* TOWNS., *Stomorrhinia pallida* MALL., are typical.

Among Tabanidae there is recorded the genus *Chrysops*, which is absent in the other areas of Australia; it is an element from the Northern hemisphere. It is interesting to note that *Haematopota* is quite absent from Australia and South America, but *Tabanus*, present in Australia, has not yet been recorded from Tasmania or New Zealand; the latter genus is absent also from the southern part of South America, being represented in most parts of this continent north of Chile and Patagonia — good evidence that this genus had its origin in the Northern hemisphere.

Australia Sylvatica (C- Australia merorientalis).

This area contains the elements most adapted to a cold climate, not only because of its geographical position but also because it includes the highest mountains in Australia (only slightly more than 7000 feet); it is also the best studied.

Heavily forested and containing a lot of streams and small rivers, it concentrates the greater part of those families in which the life-history is closely connected with water, trees and bushes. It contains also a wide range of ecological conditions: from sandy beaches at the sea-shore, savannah lands, dry sclerophyll and wet sclerophyll forests, to the subalpine zone at the summits of the mountains; it includes also the salt and fresh-water lakes, small parts of the desert areas, etc.

Tasmania is a natural unit of these areas, but the typical Tasmanian fauna if found on the mainland as a rule only in the zones over 3000 ft.

Here the elements of the temperate climate predominate; the tropical and subtropical elements are distributed mostly in "pockets" with more favourable conditions, resulting in a kind of "diffusion" from province B — Australia bororientalis.

Most of the species of the genera *Rutilia, Amphibolia, Senostoma, Chaetogaster*, etc. (Tachinidae) of the tribe Rutiliini are represented here; they are usually dominant on the flowers of *Leptospermum* and Eucalypts, the number of specimens being enormous. From the closely related genus *Formosia*, which is very well represented in equatorial and tropical areas (in B there are some species), only a few species penetrated from the north into this area.

The tribe Dexiini has most of its species in this area, and it is evident that a number of species are not yet described.

The greatest part of the tribe Ameniini is represented here also.

The *Cylindromyia* group also flourishes here, although some specific representatives are present in Queensland and West Australia. Many species, both described and undescribed, of the genus *Chaetophthalmus* can also be found in this area.

Typical of the Blow-flies present here are *Calliphora* species with bluish abdomens. They are predominant in the mountainous regions. The *C. stygia* group is represented by many species and by an enormous number of specimens. *Chrysomyia bezziana* VILLNV. is here predominant; *Chrysomyia albifacies* is the only species from the *Ch. albiceps* group which is distributed in the whole of Australia, but in this area it is extremely abundant. Among many thousands of specimens the author has not seen the true *Ch. albiceps;* this species occurs mostly in Africa. *Lucilia cuprina*, the most important species of blow-fly, was probably introduced.

Numerous species of *Pelecorhynchus* (Tabanidae) are concentrated in this area, some of them penetrate slightly into Queensland; another group of species is distributed in Patagonia and Chile; in Western Australia this genus has not yet been recorded. All species live in mountainous areas and are adapted to a comparatively cool, humid climate, and to life in forrested areas.

In this area we find the most numerous representatives of the Therevidae. There are many genera and species not yet described (a review of Australian Therevidae is in preparation), which provide evidence that this family is extremely well developed in Australia, not only numerically, but also by the presence of many intermediate forms. We can see a regular net of forms without "missing links", as in *Rutilia* (Tachinidae). For example, in the genus *Acrospiza* the scutellum is placed quite vertically, appearing as a finger, but we can also see it in a declinate position, again in the normal horizontal position, but with narrowed surface, and finally in the same position, but with a normal, rather wide surface. Lack of material prevents comparison of the faunas of the areas A, B, C, but it is hard to believe that in the provinces A and B the Therevids are better represented than in C.

Up to the present time the one Australian species of *Toxophora* (Bombyliidae) has been found only in this area.

The relationship of the Australian Dipterofauna to that of South America (Chile, Patagonia)

We will deal here in a very few lines with the relationship between Australia and South America (Chile-Patagonia), which is so evident that it cannot be omitted. Although the distance between these two countries is very great, some species are so closely related that only the experienced observer can separate them specifically, and this close relationship can be recognized not only in one, but in a number of families.

Tipulidae: *Phacelodocera* END.: *P. flabellifera* Lw. — Brazil; *P. tasmaniensis* ALEX. — Tasmania.

Tabanidae: Pangoniinae: *Ectenopsis* has been recognised so far only in Australia (but not Tasmania) and New Zealand; it has near relatives in South America. *Scaptia* is distributed in South America, Australia, Tasmania and New Zealand. *Dasybasis* is distributed in South America, Southern Africa, Australia (including Tasmania and also a small extension to the north of Australia), and New Zealand.

Pelecorhynchus is represented by rather numerous species in the mountainous areas of Eastern Australia, Tasmania and in Chile and Patagonia. This is one of the two groups of "paleoantarcts": one connected with arid, desert areas (for example *Comptosia*), another with the more humid, forested areas. It appears that the time and route of penetration were different, compared with the first group.

Cyrtidae: *Panops* is distributed in South and North America and in Australia.

Leptidae: *Dasyomma* is distributed in Australia and Chile.

Nemestrinidae: *Trichophthalma* is distributed in Chile and Australia.

Bombyliidae: *Comptosia* is found in Australia, and the extremely closely related *Lyophlaeba* in Chile and Patagonia.

Apioceridae: *Neorhaphiomydas* occurs in Australia, and the very closely related *Megascelus* in Chile.

Scenopinidae: *Pseudotrichia* is found in both countries, but also in North America.

Asilidae: *Bathypogon* (Australia, Chile, Columbia), *Deromyia* (Australia, Chile), *Aphestia* (Australia, Brazil, Mexico), *Astylum* (Australia, Tasmania, Venezuela), *Glaphyropyga* (Australia, Brazil).

Therevidae: *Anabarrhynchus* (Australia, Chile, Madagascar).

Syrphidae: *Ceratophya* (Australia, Brazil).

There are more genera and groups of species common to Australia and to the Chile-Patagonia area, but the author's only purpose is to show that the relationship exists without doubt, and that it embraces many families.

The relationship of the Australian Dipterofauna to that of Tasmania and of New Zealand

Tasmania.

TILLYARD (1926) thought that Tasmania remained a part of Australia until late Tertiary times, but was probably divided from and reunited with the mainland several times before its last very recent severance.

This statement is hardly applicable to the insect fauna. The ability of an insect to fly or to be carried by wind for very long distances, the presence of a group of islands between Tasmania and Australia forming a bridge between them, are factors nullifying the presence of the strait between Tasmania and Australia.

TILLYARD thought that Tasmania also received a great part of those insects of northern origin which worked their way at least as far south as Victoria and South Australia. Most of the tropical and subtropical groups are, however, absent, especially those representative of the latest stream of immigration from the north (but *Papilio macleayanus* and some of the large cicadas are penetrating into Tasmania).

It seems that an obstacle to penetration was not the strait but the ecological conditions and climate of Tasmania, to which TILLYARD did not give enough attention.

"Tasmania", said TILLYARD, "however, holds the largest share of the immigrants received through the Antarctic connections, and this gives its fauna its somewhat remote relationship with that of New Zealand", and he adds: "The great majority of insects found in Tasmania are closely related to those of southern Victoria, the faunas of these two regions being very similar."

The author cannot agree with these statements of TILLYARD. It is true that the fauna of Tasmania and of South Victoria are very similar, but the Tasmanian fauna is an **impoverished** fauna if compared with that of Victoria. If the greater part of the Victorian insect fauna penetrated from the Antarctic through Tasmania, we should find the "paleoantarcts" in more considerable numbers, than we in fact do.

Among "paleoantarcts" we can recognise two groups: one represented by species of the arid areas, desert forms (some *Comptosia* species), a second represented by forms more associated with forested areas (*Pelecorhynchus* species). The first group is poorly represented in Tasmania, the second better represented, but not well enough to show that a considerable part of the fauna was derived from a "paleoantarctic fauna".

In a few words, the history of the insect fauna of Tasmania is not such as is described by TILLYARD. Our knowledge of the Diptera of Tasmania is extremely poor; since 1847, practically nobody has collected intensively in Tasmania. Local Diptera are poorly represented in Tasmanian museums.

New Zealand.

In order to understand the composition of the Australian fauna it is necessary to know the role New Zealand played in it. TILLYARD thought that the insect fauna of New Zealand could rightly be considered as belonging to that of the Australian region, seeing that it originated from sources which also supplied other parts of that region.

New Zealand, however, had little or no access to the Gondwana faunas which first populated Australia, nor had it a sufficiently late connection with the north to receive the mass of representatives of the highest groups of insects which poured across into Australia. Nor, again, had it sufficient development of large continental areas to allow the formation of a striking autochthonous fauna of the xerophytic type such as we find in Australia itself.

These ideas of TILLYARD are in contradiction with some facts of geographical distribution of Diptera: the family Bombyliidae has in Australia some hundred species, but only one in New Zealand (an endemic genus). It is quite evident that New Zealand has had Bombyliids on a larger scale in earlier times, but now they are extinct. Not only have the presence of land bridges played an important part in determining the composition of the New Zealand and other faunas, but also changes of environment, climate, etc. have been important.

Table I.

Table showing the known sources of the present Australian and New Zealand insect Faunas.
(From: Report of the 16th Meeting, Australian Assoc. for the Advancement of Science, 1924: 413; by R. J. TILLYARD).

Faunal Element	Geological Period	Australia	N. Z.-land
1. Early Austro-Gondwanan	Upper Permian	Present	Absent
2. Late Austro-Gondwanan	Upper Triassic	Present	Absent
3. Early Austro-Malayan	Jurassic	Present	Present
4. Early Antarctic	Jurassic	Present	Present
5. West Australian	Cretaceous	Present	Absent
6. Middle Austro-Malayan	Upper Cretaceous to Early Tertiary	Present	Present
7. Penultimate Antarctic	Early Tertiary	Absent	Present
8. Last Antarctic	Middle Tertiary (? Miocene)	Present	Absent
9. Autochthonous (xerophytic)	Tertiary	Present dominant	Present slight
10. Late Austro-Malayan	Late Tertiary	Present	Absent

TILLYARD's scheme is both one-sided and simplified; as a paleontologist he is omitting the development of faunas in Holozön. Many species of northern origin (subtropical type) were introduced in the last 10—20.000 years by wind, sea-currents, by man, etc., which does not mean that they represent a remnant of a continent extending, for example, from New Hebrides to New Zealand. An absolute lack of fossil data does not permit generalisation; it makes the schemes of TILLYARD highly speculative and unsupported by facts.

It is necessary to add some critical remarks to the table of TILLYARD. Firstly, the statement that certain elements are "absent" does not mean that they have always been so. Such elements could have been eliminated by changes of climate. For example, the desert elements could be eliminated by the climate becoming more humid, and the desert becoming steppe-like, or even afforested.

The second wrong principle is the idea that the elements of a fauna always arrive from somewhere else and do not develop "in situ". This idea has been applied to all countries of the World. Different groups of animals are regarded as appearing as "Deux ex machina", suddenly, mysteriously and without further explanation.

The third wrong principle is the conception of many land bridges which arise, submerge, again arise, etc., and always serve as a route in one direction only. However, it is quite evident that a land bridge between continent A and continent B will always result in a mutual exchange of faunas.

The author hopes to develop a more detailed discussion of the problem of the origin of the Australian Dipterofauna, in a special work now in preparation.

REFERENCES

ANDERSON, C., 1925. The Australian Fauna. *Proc. Linn. Soc. N.S.W.*, **L**, *1—34*.
HARDY, G. H., 1922. The geographical distribution of genera belonging to the Diptera Brachycera of Australia. *Aust. Zool.* **ii** (4), *143—147*.
HARDY, G. H., 1944. Miscellaneous notes on Australian Diptera. X *Proc. Linn. Soc. N.S.W.*, **lxix** (3-4), *76—86*.
HARRISON, L., 1928. The composition and origin of the Australian Fauna. 18th Meeting Australas. Ass. Adv. Sci., **xviii**, *332—396* (large list of references).
HEDLEY, C., 1893. The Faunal Regions of Australia. *Rep. Australas. Ass. Adv. Sci.* **v**, *444—446*.
HEDLEY, C., 1912. The palaeographical Relation of Antarctica. *Proc. Linn. Soc. Lond.* 80. Reprinted in Smithsonian Institute, 1913, *443*.
IREDALE, T., 1928. The Avifauna districts of Australia. *Rep. Australas. Ass. Adv. Sci.*, **xix**, *244—245*.
LE SOUEF, A. S., 1928. The numbers and distribution of mammal species in the Australasian Region. *Rep. Australas. Ass. Adv. Sci.*, **xix**, *245—248*.
MACKERRAS, I. M. & FULLER, M. E., 1942. The genus *Pelecorhynchus*. *Proc. Linn. Soc. N.S. W.*, **lxvii** (1-2), *9—76*.

MACKERRAS, I. M., 1950. The Zoogeography of the Diptera. *Aust. J. Sci.*, **xii,** N. 5, *157—161.*
MACKERRAS, I. M., 1954. The classification and distribution of Tabanidae. *Aust. J. Zool.* **ii,** (3), *431—454.*
MACKERRAS, I. M., 1957. Tabanidae (Diptera) of New Zealand. *Trans. Roy. Soc. N.Z.*, **84** (3), *581—610.*
NICHOLLS, G. E., 1933. The composition and biogeographical Relations of the fauna of Western Australia. *Rep. Australas. Ass. Adv. Sci.*, **xxi,** *93—138.*
PARAMONOV, S. J., 1950. Review of Australian Mydaidae. Bull. No. 255. *Comm. Sci. Ind. Res. Org. Australia, 1—32.*
PARAMONOV, S. J., 1951. Note on Australian Streblidae. *Ann. Mag. Nat. Hist.* (12), **iv,** *752—760.*
PARAMONOV, S. J., 1953. The description of the adult *Tracheomyia macropi* Frogg. *Ibid.* (12), **vi,** *195—199.*
PARAMONOV, S. J., 1953. Review of Australian Nemestrinidae. *Aust. J. Zool.*, **i** (2), *242—290.*
PARAMONOV, S. J., 1953. Review of Australian Apioceridae. *Ibid.*, **i** (3), *449—536.*
PARAMONOV, S. J., 1954. Notes on some Hippoboscids (esp. *Ortholfersia*). *Ann. Mag. Nat. Hist.*, (12) **vii,** *283—292.*
PARAMONOV, S. J., 1954. On the family Thyreophoridae. *Ibid.*, (12) **vii,** *292—297.*
PARAMONOV, S. J., 1955. Review of the genus *Tapeigaster* Macq. (Neottiophilidae). *Ibid.*, (12) **viii,** *453—464.*
PARAMONOV, S. J., 1955. Review of Australian Scenopinidae. *Aust. J. Zool.*, **iii** (4), *634—653.*
PARAMONOV, S. J., 1955. New Zealand Cyrtidae and the Problem of the Pacific Island Fauna. Pacific Science, Jan., *16—25.*
PARAMONOV, S. J., 1957. The Review of Australian Acroceridae. *Aust. J. Zool.* **v** (4), *521—546.*
PARAMONOV, S. J., 1958. The Review of Australian Pyrgotidae. *Aust. J. Zool.* **vi** (1): *89—138.*
PARAMONOV, S. J., 1958. Many manuscripts.
SERVENTY, D. L. & WHITTELL, H. M., 1951. A handbook of the Birds of Western Australia (with the exception of the Kimberley division). *44—60.*
SPENCER, W. B., 1893. The Fauna and Zoological Relationship of Tasmania. *Rep. Australas. Ass. Adv. Sci.*, **iv,** *82—124.*
SPENCER, W. B., 1896. Report of the work of the Horn Sci. Expedition to Central Australia, **i,** *171—199.*
TATE, R., 1887. On the influence of physiographic changes in the distribution of life in Australia. *Rep. Australas. Ass. Adv. Sci.*, **i,** *312—325.*
TILLYARD, R. J., 1924. Origin of the Australian and New Zealand Insect Faunas. *Rep. Australas. Ass. Adv. Sci.*, **xvi,** *407—413.*
TILLYARD, R. J., 1926. The Insects of Australia and New Zealand, *1—560*, Sydney.

XI
THE ECOLOGY AND BIOGEOGRAPHY OF AUSTRALIAN GRASSHOPPERS AND LOCUSTS

by

K. H. L. KEY

(Division of Entomology, C.S.I.R.O., Canberra)

Over the past 20 years much important ecological work has been done on Australian locusts and grasshoppers of economic importance, chiefly by workers in the Division of Entomology, C.S.I.R.O., and the Waite Institute, University of Adelaide. This work has been recently reviewed (KEY 1958). The purpose of the present contribution is to discuss in a preliminary way the broad features of the ecology and biogeography of the grasshopper and locust fauna as a whole, drawing largely on hitherto unpublished observations of the author. Information on the economic species will be included only to the extent that it contributes to this aim.

In what follows, the term "grasshopper" is to be understood to cover the four species of Australian locusts* as well as the short-horned grasshoppers proper (i.e. all the Acridoidea), but not the long-horned grasshoppers (Tettigonioidea).

The Australian grasshopper fauna

The Australian grasshopper fauna exhibits the effects of the prolonged geographical isolation of the continent in the same way as many other groups of animals. Out of perhaps some 800 species, the great majority are representatives of two relatively primitive subfamilies: the endemic Morabinae, in the family Eumastacidae, and the Cyrtacanthacridinae (or Catantopinae), in the family Acrididae. The prominence of the Morabinae is shown by the fact that there are about 150 species in Australia (mostly undescribed), whereas the much larger continent of Africa has only about 100 species of Eumastacidae of all groups (JOHNSTON, 1956). Similarly, Cyrtacanthacridinae in Australia comprise about 81% of the Acrididae (estimate from SJÖSTEDT, 1935), but in Africa only about 38% (from JOHNSTON, 1956). This is related to the very poor representation of the Acridinae (including Oedipodinae), which total

---

* i.e., the Australian plague locust *(Chortoicetes terminifera)*, the chief economic species, swarming mainly in southern Australia; and the yellow-winged locust *(Gastrimargus musicus)*, the spur-throated locust *(Austracris guttulosa)*, and the migratory locust *(Locusta migratoria)*, all swarming more sporadically in northern Australia. See KEY (1938).

well under 50 species, as against over 700 in Africa: the Cyrtacanthacridinae and Acridinae are almost equally represented in Africa, whereas their ratio is about 7 : 1 in Australia.

At the generic level there is apparently over 90% endemism, although this figure may have to be revised when more is known of the fauna of islands immediately to the north of Australia. The few non-endemics are mostly large Old World genera (e.g. *Acrida, Gastrimargus*) represented by single species that are clearly recent invaders. Thus the bulk of the fauna is ancient and autochthonous. The two main groups, Morabinae and Cyrtacanthacridinae, have occupied every kind of habitat. In the process the former, although throwing up numerous species, have remained remarkably stereotyped in their general structure, while the latter have become highly diversified. The Acridinae, although poor in species, are often abundant as individuals and widely distributed; they include four of the six economic species.

Ecological characteristics of grasshoppers

Grasshoppers are strictly phytophagous insects and almost all species require green plant material. All stages except the eggs live on the outer foliage of plants or on the ground between plants — usually in situations exposed to the sun. The eggs are nearly always laid in soil, usually sparsely vegetated and hence exposed to the influences of sun and rain. Except to this extent, grasshoppers do not burrow in soil or plant material, or make nests, although a few species may at times hide under fallen leaves or partly bury themselves in loose sand.

Thus grasshoppers experience to an exceptional degree the direct impact of climate. The relation to climate depends upon two sets of physiological factors: the minimal requirements of the species for completion of its life-cycle (an adequate "growing season") and its thresholds of tolerance. The first involves the attainment of a minimal "temperature summation" over a period during which moisture is adequate for hatching and food production. Work on *Chortoicetes terminifera* has shown (KEY, 1942, 1945) that this may be used as the basis for a "climatic index" representing one aspect of the climatic favourableness of different seasons or regions. The second includes upper and lower temperature and moisture thresholds. Extremes of temperature usually act as direct mortality factors, and extremes of moisture may act in the same way against the eggs by leading to desiccation or asphyxiation. Against the active stages, drought exerts its effect through the food supply, and overmoist conditions through the bacterial and fugal diseases that they stimulate. There is interaction between the temperature and moisture thresholds (cf. KEY, 1942).

However, grasshoppers are by no means completely exposed to the gross climate. All species live in plant communities of one kind or another and these modify the climatic elements in varying degrees and produce microclimatic mosaics. Within its preferred plant community, a grasshopper will avoid injurious extremes and often attain optimal conditions, by virtue of its mobility (KEY, 1945, 1950; CLARK, 1947a). The mobility (or "vagility") of different species varies very greatly, from those which, like *Chortoicetes terminifera* and the other locusts, may cover hundreds of miles in a few days, to the apterous and sluggish Morabinae, where it is a matter of yards per generation.

Because of the difference in the requirements of the eggs and active stages of many — probably most — grasshopper species, we may distinguish between oviposition and food-shelter habitats (KEY, 1945; CLARK, 1947a, b; UVAROV, 1957). The mobility of the grasshopper enables it to move from the one habitat to the other as circumstances require; equally, the limits to which its mobility is subject determine the distance by which the two habitats can be separated and still remain effectively available. Thus we find that ecotones and mosaic habitats in which the scale of the patchwork is adjusted to the vagility of the species concerned are in general more favourable to grasshoppers (as to many other animals) than uniform habitats (KEY, 1945). Such mosaics are also much better fitted to buffer gross climatic fluctuations and ensure to a grasshopper favourable, or at least tolerable, conditions somewhere within them than are uniform habitats (KENNEDY, 1939; KEY, 1945, 1950).

Grasshoppers show numerous adaptations of behaviour, physiology, and structure, the function of which is evidently to enable them to withstand conditions which would otherwise be unfavourable. Movements within the temperature mosaic and changes of orientation in relation to incident radiation have been described by many authors (cf. CLARK, 1947a, on *Chortoicetes*). KEY & DAY (1954a, b) have described a physiological colour response in the Australian alpine grasshopper *Kosciuscola tristis*, which may facilitate its survival at high altitudes by enabling it to make better use of solar radiation. A few species frequenting the margins of water (*Bermiella acuta, Oxya gavisa, Loxilobus pulcher*) have swimming adaptations in the form of expanded hind tibiae and air-trapping chambers or hairs.

It is adaptations to drought, however, that are of most significance in the Australian environment. A few of the species inhabiting the arid regions may be classed as "drought-evading", as against the majority, which are "drought-enduring". They avoid coming to grips with the drought problem, by restricting themselves to the limited moist habitats associated with permanent or semi-permanent water. These species (e.g. *Bermius* spp., *Atractomorpha crenaticeps*) are, in fact, adapted to wet conditions and not to arid conditions at all. The drought-enduring species manage to survive prolonged periods of low moisture availability (usually combined with high temperatures and low shelter availability) by various adaptations of physiology, behaviour, and structure. Dormancy, ability to restrict water loss, ability to tolerate water loss, ability to survive on dew-soaked plant débris, special sheltering behaviour patterns, or a habit of feeding on the foliage of evergreen shrubs or trees, may be involved. Typically the capacity for drought resistance is well developed in only one life-cycle stage, either the egg or the adult, so that the problem of survival resolves itself very largely into the problem of ensuring that the insect will be in, or will rapidly enter, the appropriate resistant stage when the dry period commences. This problem tends to be solved in different ways according to the seasonal rainfall regime.

The species characteristic of very arid climates, where the limited rain is liable to fall in any season, usually pass the dry periods in the egg stage, the eggs remaining viable sometimes for years until sufficient moisture is available. They then hatch and the active stages, which may not be drought-resistant at all, develop very rapidly upon the resulting ephemeral plants and lay their eggs. In species characteristic of a mediterranean type of climate, low temperature thresholds for hatching and development permit the active stages to feed and grow during the moist spring, while the dry summer and usually also the cool winter are passed in the egg, often under the protection of a diapause (e.g. *Austroicetes cruciata* — see ANDREWARTHA, 1943, 1944; BIRCH & ANDREWARTHA, 1941). The typical life-cycle in climates with a markedly summer rainfall is one in which the dry, warm winter is passed through by the sexually immature adults. Onset of the rains leads to sexual maturation and oviposition, and development proceeds without interruption to the adult of the next generation. However, some species have a cycle similar to the mediterranean one, passing the dry season — in this case the winter — in the egg stage, probably under the protection of a diapause. Where an adequate rainfall is more evenly distributed over the year and temperatures are sufficiently high, we find species many of which are capable of passing through more than one annual generation. Both the egg and adult stages typically have some (but only moderate) capacity for drought resistance and there is no diapause. The number of generations depends on the length of the favourable period in a given year. *Chortoicetes terminifera* exemplifies this type of seasonal cycle (KEY, 1938, 1942).

The case of the drought-evading species is only one example of the general phenomenon whereby habitat selection by a grasshopper species may vary according to the climate in such a way as to compensate for its unfavourable features (cf. KEY, 1954). From this point of view it is the vegetation of the habitat that is chiefly important — more precisely, the structure of the vegetation, for structure is the relevant factor in the production of different microclimates. The species composition of vegetation (except in so far as this may be linked with structure) is of much less importance to grasshoppers. However, there is a broad division into predominantly dicotyledon feeders and predominantly grass feeders, the latter comprising, as might be expected, the phylogenetically advanced Acridinae, with individual specialised groups of Morabinae and Cyrtacanthacridinae. Although food preferences are probably quite strongly developed, restriction to one or a very few host plants is rare. Among such cases may be mentioned *Ecphantus quadrilobus*, largely or entirely restricted to species of *Sida* (Malvaceae), and the morabine genus *Warramunga*, restricted to the closely related grass genera *Triodia* and *Plectrachne*.

Life in the open exposes grasshoppers not only to the weather, but also to natural enemies, especially vertebrates. As a consequence most species show one or more of a wide range of procryptic mechanisms involving colour, form, and behaviour. These do not differ in principle from those known from other parts of the world. Certain of them have been described by KEY (1954) for the genera *Chortoicetes* and *Austroicetes*. We have leaf- and stick-like species, species resembling pebbles, colour patterns resembling gravelly soils or the light and shade of grassy habitats. We have genetic polymorphisms for pattern and colour, and physiological mechanisms for adjusting colour to that of the background. The large genus *Goniaea* comprises species associated with eucalypts and resembling dead eucalypt leaves. *Coryphistes* and its relatives resemble twigs of rough-barked trees or shrubs, such as many acacias. *Ecphantus quadrilobus* resembles, in colour, texture, and the shape of its pronotal crest, the leaves of *Sida* spp. A remarkable, pebble-like, new genus of Cyrtacanthacridinae from the arid shores of the Great Australian Bight has very variable foliaceous sculpturings on its enlarged pronotum that resemble with great accuracy, in both colour and form, the lichens that are so abundant in its natural habitat. Aposematic species are common only in the subfamily Pyrgomorphinae (genera *Monistria*, *Greyacris*, *Petasida*, and *Scutillya*).

The Australian environment in relation to grasshoppers

It is the interplay between the basic constitution of those grasshopper groups that have reached Australia, on the one hand, and the peculiarities of the Australian environment, on the other, that has been responsible for the evolutionary radiation of the grasshopper fauna. The further interactions of the species thus evolved with their environment constitutes the ecology of the fauna. Thus we must examine the Australian environment, especially as it affects grasshoppers. We may do this under six headings: climate, soil, topography, vegetation, animals, and human influences. These factors are by no means independent, but each has its characteristic direct as well as indirect effects on the grasshopper fauna.

Climate

All classifications of Australian climate, including those of KÖPPEN (see TAYLOR, 1932), THORNTHWAITE (1933), and DAVIDSON (1936), agree broadly in recognising **(1)** an interior hot and very arid region, in which the limited rain may fall at any season, **(2)** a hot semi-arid to subhumid region to the north and east, where there is a well marked dry season falling in the winter, **(3)** a corresponding cooler region to the south and east, where the dry season falls in the summer, and **(4)** a humid belt along the east coast, with outliers in Tasmania and the extreme south-west of Western Australia, where a regular dry season is absent or (in the south-west) brief. The last may conveniently be divided into **(a)** a northern section with higher temperatures and a summer preponderance of rainfall, and **(b)** a southern section with lower temperatures and a winter rainfall peak. The precise boundaries of the regions are not important from the present point of view. However, the isopleth for 9 months with $P/E > 0.3$ (see LEEPER, 1950, fig. 19) may be used for the inland limit of **(4)** and that for no month with $P/E > 0.5$ (see DAVIDSON, 1936) for the coastward limit of **(1)** — the latter corresponding roughly with the isopleth for 2 months with $P/E > 0.3$. The northern border of New South Wales may be used to separate **(2)** and **(3)** in the east, and also **(4) (a)** and **(b)**.

The grasshoppers inhabiting these climatic regions are differentiated with respect to their characteristic seasonal cycles broadly along the lines already indicated, i.e., region **(1)** is inhabited by species with a highly resistant egg and a rapid rate of development, **(2)** by univoltine species passing the winter in the adult or egg stage, **(3)** by univoltine species passing the late summer and usually also the winter in the egg stage, often with diapause, and **(4)** by species with no strongly drought-resistant stage, which may pass through several annual generations if conditions are sufficiently favourable. The distinctions are of course not sharp and exceptions may be found in every region. Within each region, the individual "growing season" requirements and tolerance thresholds of the different species delineate their potential distribution areas, within which the actual distributions are determined by the remaining environmental factors (cf. KEY, 1954, on *Austroicetes* spp.). Some versatile species, especially those of high vagility, occupy more than one region, often showing differences in habitat selection to compensate for the climatic differences.

Soil

Soil affects grasshoppers directly, being the environment of the eggs of almost all species as well as the substrate of the numerous geophilous ones. Indirect effects are also exercised, through the vegetation. The direct effects depend upon features of the surface 1—4 inches only, and from this point of view we may distinguish four soil classes: compact soils, self-mulching soils, sands, and stony soils. The compact soils (see KEY, 1945, p. 23) are the most widespread group. They afford a stable substrate to grasshoppers, and boring by ovipositing females produces a firm tunnel requiring no reinforcement. They are well drained and yet possess a fairly high waterholding capacity. The self-mulching soils are heavy friable clays and clay loams, developed on plains of fine-grained alluvium or *in situ* on basalt (KEY, 1945). When dry they develop deep wide fissures, as well as a nutty or flaky surface structure; when wet they are poorly aerated. Eggs laid in such soils run the risk of mechanical damage and desiccation in dry times and waterlogging in wet times. A few grasshoppers (e.g. *Perelytrana* spp., *Zabrala* spp., some species of *Peakesia* and *Cedarinia*) are most abundant on these soils or even confined to them (cf. CLARK, 1949), although there are indications that an admixture of near-compact soils with the purely self-mulching is still more favourable. Other species (e.g. *Austracris guttulosa, Locusta migratoria*) seem to tolerate them quite well, but are no more abundant than on other types. Most species avoid them. The "characteristic" species are mostly brachypterous; *Perelytrana* and *Zabrala*, at least, make use of the fissures as a refuge when pursued and possibly for other purposes (CLARK, 1949).

Sands provide an unstable medium for grasshopper eggs and an unstable substrate for the active stages. They are excessively drained, with a very low waterholding capacity. Several grasshopper species are confined to sand, the best examples being *Urnisiella* spp., *Urnisa* spp., *Ablectia rufescens*, and the tetrigid *Paratettix nigrescens*. The very widespread *Pycnostictus seriatus* is abundant on more stable sandy soils, and many of the species of compact soils favour or tolerate a surface layer of sand a few millimetres thick. *Urnisiella* and *Urnisa* possess two characteristic adaptations of sand-frequenting grasshoppers, namely a very long and slender second pair of legs and long spurs on the hind tibia. These features presumably assist the grasshopper in maintaining a foothold and are also used, at least in *Urnisiella*, in burying it beneath the surface (cf. KNIPPER & KEVAN, 1954). *Urnisiella* has the habit of running in rapid bursts on hot sand, the males waving their pale antennae about in a rather wasp-like manner. It is not known whether the Australian sand-loving species have a cemented egg-pod such as that described by KEY (1930) for the South African *Acrotylus deustus*.

Stony compact soils are characteristic of talus slopes and other situations in range country, especially in the more arid climates. The stony soils of the gibber tablelands, on the other hand, are often self-mulching or partly so. Whatever the nature of the soil itself, it is the presence of surface stone that confers on this class its characteristic feature in grasshopper ecology. The stones may range from large rocks to fine gravel and from angular, rough-textured, fragmented pieces to polished "gibbers"; they may be of almost any colour. Certain species (e.g. *Cuprascula corallipes*, *Aretza* spp., and an undescribed Tasmanian genus) are strictly saxicolous, living on and among rocks and often using the crevices between them as hiding places. *Perbellia picta* and *Macrazelota cervina* are scarcely less so. A larger number occur among stones and coarse gravel, and some of these show remarkable adaptations of form, colour, and behaviour, the effect of which is to produce a procryptic resemblance to the stones. The body tends to be flattened dorsoventrally and broadened. The pronotum may be greatly enlarged into a circular or oval shield, while the hind femora often have lamellate, pilose margins and can be closely appressed to the body. The face is usually vertical. Most of these species are sluggish, especially the females, and many are brachypterous or at least flightless *(Raniliella* spp., *Cratilopus* spp., *Exarhalltia obscura*, *Phanerocerus testudo*, some *Buforania* spp.). A less modified type is represented by *Qualetta maculata*.

Soils of these four classes are irregularly distributed over Australia. Self-mulching soils are most prominent on the larger alluvial plains, but also occur on a smaller scale along the course of many rivers; they are best developed in the eastern half of the continent. Sands are characteristic of much of the arid interior, but they also occur as lateritic sand-plains in betterwatered areas, as coastal dune-complexes, and along rivers.

Topography

Topography affects grasshoppers mainly indirectly, by contributing to the production of particular soils and vegetation. We may distinguish between an adequately drained level to undulating topography, wet depressions, rocky hills, and sand dunes. All but the first of these comprise essentially seral habitats, with immature soils and seral vegetation. Wet depressions also act directly in increasing the soil moisture available to eggs and the atmospheric moisture in the environment of the active stages. Topographical differences are usually the basic cause of habitat mosaics and also provide a basis for habitat selection in compensation for climatic differences.

Vegetation

Of all the environmental factors, vegetation is the most important in the ecology of grasshoppers, both because of its direct effects, especially as a mediator of the gross climate, and because it is so largely the integrated expression of all the other factors. The vegetation of Australia has a number of special features, such as the dominant position of the genus *Eucalyptus* and to a less extent *Acacia*, the absence of major coniferous and deciduous communities, and the remarkable characteristics of the grass genus *Triodia*, but on the whole its structure is related to climate and soil in the same general way as in other biogeographical regions. In the present discussion the concepts, terminology, and definitions of BEADLE & COSTIN (1952) will be largely adopted, especially their concept of structural "forms", applied to seral and disclimax as well as climax vegetation, in contrast to the synthetic category "formation".

There is no satisfactory vegetation map of Australia as a whole; various regions have been mapped in some detail, but under different systems of classification and nomenclature. However, the maps of PRESCOTT (1931) and WOOD (1950, fig. 25) will give some indication of the distribution of the major units.

Grasshoppers are chiefly inhabitants of the herbaceous stratum. The character of this, as of its microclimate, is greatly influenced by the presence of any higher strata, which modify particularly the degree of insolation and hence the light intensity, temperature, humidity, and evaporation rate. This conditioning of the habitat leads to the exclusion of many grasshopper species from the herbaceous stratum of communities in which higher strata are present (cf. CLARK, 1950). At the same time it admits a few other species which require just these conditions (KEY, 1954). The higher strata naturally have a more positive and direct influence on grasshoppers of thamnicolous or arboreal habit. Quite a number of species, including many Morabinae, live on shrubs 1—5 ft high, far fewer on taller shrubs and the lower branches of trees. Apart from such details as the possession of well developed tarsal arolia, it is not easy to see in what way these species are adapted to the phytophilous habit. It may be that the primary adaptation comprises physiological adjustments to reduced insolation and its consequences and that the phytophilous habit has followed on these.

Structural differences of any magnitude in vegetation nearly always entail differences in the associated grasshopper species, whereas floristic differences within the same structural form are often without effect. The over-riding consideration is the life-form and density of the higher strata (cf. CLARK, 1949). The following categories of vegetation, representing structural forms or groups of forms, may be recognised for purposes of grasshopper ecology: (1) rainforest, (2) sclerophyll forest and closed-canopy communities of woodland, mallee, and scrub, (3) heath, (4) saltbush, (5) grassland and other herbaceous communities, (6) fen and bog. To these should be added bare ground (absence of vegetation) which, although it does not occur over extensive areas, is an important local feature of some open communities. Continuity of the canopy of the dominant stratum is decisive for its effect on the lower strata and hence on the grasshoppers. Communities in which the dominants are well spaced (as in savannah woodland, savannah mallee, savannah scrub, and very open grassy heaths or saltbush stands, as well as savannahs with widely scattered trees or shrubs and alpine herbfield with scattered shrubs) are treated as ecotonal for present purposes.

Definitions of the forms listed above are given by BEADLE & COSTIN (1952). Rainforest is notable for its darkness and dampness. It is the very antithesis of the typical grasshopper habitat. An herbaceous stratum is often absent. The wooded communities comprised under (2) include a very wide range of types, differing in height of dominants and in the character of the subdominant and herbaceous strata, but agreeing in admitting much more light (including direct sunlight) than rainforest and in the fact that all grasshopper activities have to proceed under a closed leafy canopy. Grasses tend to be poorly represented except in wet sclerophyll forest. In heath and saltbush the dominants are much lower; grasshoppers can easily fly over their tops and thamnicolous species can climb all over them. The microphyllous and acidophilous heaths are sufficiently different from the succulent-leafed, alkaline, saltbush communities to harbour quite distinct grasshopper faunules. In both, the herbaceous stratum, especially grasses, may be poorly developed. Grassland and other herbaceous communities again vary widely in type: they may be tall or short, dense or sparse; they include the extraordinary hummock *(Triodia)* grasslands, which in some respects resemble heaths. Fen and bog communities are characterised by their extreme wetness: apart from this, they partake largely of the character of grassland or heath.

Ecotones between the structural groups listed bring together in one area the grasshopper species characteristic of the groups concerned and at the same time often increase the abundance of some of these species through the mosaic effect already discussed. Such ecotones may occur extensively between major communities, or locally within a major community in connection with local variations in soil or topography. There is no doubt that successional changes in the grasshoppers are as pronounced as those in the vegetation.

Animals

The effect of other animals on grasshoppers seems to be relatively unimportant. The chief vertebrate predators are apparently birds, among which the straw-necked ibis, *Threskiornis spinicollis*, is a well known grasshopper feeder. However, certain small dasyurid marsupials of nocturnal habit are also known to feed to a considerable extent on grasshoppers. There is the usual array of parasitic flies in the families Sarcophagidae, Tachinidae, and Nemestrinidae (cf. NOBLE, 1936). Scelionidae (Hymenoptera) (NOBLE, 1935) and at least one species of Bombyliidae (Diptera) (FULLER, 1938) attack the eggs of some species. Probably all species are subject to relatively benign parasitism by larval erythraeid mites, mainly species of *Leptus*, and some at least to parasitism by podapolipids. Ants have sometimes been noticed preying upon grasshopper hatchlings. It is possible that intensified predation by birds, ants, and spiders may have something to do with the absence of many species of grasshopper from wooded communities. However, we cannot make general statements about the population regulation of grasshoppers. Many grassland species fluctuate greatly in abundance under the influence of rainfall. Density-dependent effects of food shortage may be significant at dry times, while the disease epidemics that occur under over-moist conditions are presumably also density-dependent. Quite different factors are probably influential in thamnicolous species, which rarely reach high densities and do not suffer from food shortages.

Human Influences

The entry of European man upon the ecological scene in Australia has had profound effects, which have extended, mainly through vegetational changes, to the grasshoppers. It can safely be said that the relative abundance of different grasshoppers is now very different, at least in southern Australia, from what it was under virgin conditions. The plant communities that we see today are mainly grazing disclimaxes: their original composition and that of their grasshopper associates has to be determined by inference, with the help of small relict areas. Wooded areas of all kinds have been converted into artificial savannahs or savannah woodlands by felling and grasslands radically changed in species composition and structure (KEY, 1954). One of the most striking examples of the last is the transformation of an originally very widespread tall, dense, *Themeda-Poa* grassland (and herbaceous stratum in savannah woodland) into a much shorter and sparser *Stipa-Danthonia* community (PRYOR, 1954). This has been accompanied by the replacement of *Gastrimargus musicus* as the dominant grasshopper species by *Austroicetes* spp., *Phaulacridium vittatum*, and others. An even more striking, but less immediately evident change has been wrought in the status of the morabine *Moraba scurra* (WHITE, 1956).

Everywhere the trend has been towards more open communities, which, because of greater exposure, are necessarily more arid (cf.

UVAROV, 1957). Grass-frequenting species have been favoured, especially those characteristic of short, sparse, grass. Amongst these the Acridinae are disproportionately prominent. At the other end of the scale, the species of woody communities have been reduced in abundance and range. However, this trend is increasingly being given a new twist by the use of introduced pastures on the climax and disclimax grasslands. A good clover-grass sown pasture is an unfavourable habitat for grasshoppers of all species — probably because it is far too uniform, including neither bare ground for basking and oviposition, nor taller persistent clumps suitable for shelter.

Grasshoppers of the major habitat categories

Taxonomic knowledge of the Australian Acridoidea is still very incomplete. It is likely that less than half the existing species have been described: the C.S.I.R.O. collection alone includes representatives of more than 80 new genera. Certain regions, such as Cape York Peninsula and the Pilbara region of Western Australia, have scarcely been visited by collectors, and some others are almost as neglected. In these circumstances it would be premature to attempt to draw up definitive faunal lists or comprehensive lists of the species occurring in different habitats. However, collecting over the last 20 years, especially by C.S.I.R.O., has provided much information on habitats and distribution, so that preliminary lists of characteristic species of the major habitats can now be given.

In what follows, the six vegetation categories already discussed will be used as major habitat categories. The grasshoppers known to be characteristic of each will be listed, with indications of any special features of their occurrence, such as dependence upon a particular stratum or restriction to a particular climate. The species already mentioned as characteristic of sands and stony soils have not been included, since to these species the vegetation is of subordinate importance in comparison with soil features. This classification of habitats is suitable only for a very broad survey: a more discriminating one will be required when intensive work on the synecology of the Australian Acridoidea is undertaken.

Rainforest

Temperate rainforest does not appear to harbour any species of grasshopper. *Rectitropis exclusa* and *Loxilobus* sp. occur within monsoon rainforest in the Darwin area, and these species, with *Desmoptera truncatipennis* and *Desmopterella sundaica*, may also occur in tropical rainforest in Queensland. All other species known to be associated with rainforest seem to be confined to its ecotonal fringes. These include *Valanga irregularis*, *Stenocatantops angustifrons* (both these on shrubs and small trees), *Methiola picta*, *Cedarinia* spp. (both in the herbaceous stratum), and species of Tetrigidae on moist bare soil.

Sclerophyll Forest and Closed-canopy Woodland, Mallee, and Scrub.

Characteristic genera dependent on the tree and tall-shrub strata (but not confined to them) are *Goniaea* and *Pardillana* (both associated with eucalypts), *Pespulia* (associated with mallee eucalypts), *Coryphistes, Adreppus, Euophistes, Retuspia, Relatta, Macrolobalia, Typaya, Macrolopholia,* and *Capraxa* (associated with non-eucalypts, especially acacias: the last six mainly arid in distribution), and *Austracris* (mainly tropical).

Species dependent on the low-shrub stratum include *Cirphula* spp., *Peakesia brunnea* and *P. fulvipes, Apotropis* spp., *Monistria* spp., *Exarna includens,* and *Psednura* spp. (all occurring more characteristically in heath), *Percassa rugifrons* (montane and subalpine), *Stenocatantops angustifrons* (tropical), many species of *Moraba* and *Callitala,* and an undescribed species from Tasmania occurring in dense tangled undergrowth of *Bauera rubioides*.

Species of the herbaceous stratum are: *Rhitzala modesta* (shaded situations in wet sclerophyll forest), *Cryptobothrus chrysophorus, Rapsilla fusca, Epallia* spp. (these among fallen eucalypt leaves) *Macrotona* spp. (mostly associated with wiry tussock grasses and similar plants, especially *Triodia* hummocks), *Ecphantus quadrilobus* (associated with *Sida* spp. in arid and semi-arid regions), *Kosciuscola* spp. (montane and subalpine, associated with grasses, especially *Poa australis*), *Stenocatantops* spp., *Moraba curvicercus,* and *M. longirostris* (these three tropical, on tall grass), *Austracris guttulosa* (tall grass), *Praxibulus* spp., *Laxabilla* spp., *Froggattina australis,* and *Calephorops viridis* (short grass: first two southern, others mainly northern), *Curpilladia* spp., *Caloptilla* spp., *Zebratula flavonigra, Macrocara* spp., *Adlappa* spp. (all tropical), *Monistria* spp., *Peakesia* spp., *Cedarinia* spp., *Austroicetes frater, Phaulacridium vittatum, Moraba scurra* (last three southern), *Methiolopsis geniculata,* and miscellaneous Tetrigidae.

Heath

Characteristic of the shrub stratum are *Eumecistes gratiosus, Peakesia brunnea, P. fulvipes,* and *Exarna includens,* with species of *Cirphula, Coryphistes, Apotropis, Monistria, Macrotona, Psednura, Propsednura,* and many morabines. The sparse herbaceous stratum may include *Urnisa erythrocnemis* or *U. rugosa, Phaulacridium nanum,* and certain Tetrigidae (all these associated especially with bare sandy areas). We may consider along with heath the shrubby communities (not associated with bogs) within the sod tussock grassland and the alpine herbfield. These are occupied by *Percassa rugifrons*, *Monistria vinosa*, and, in Tasmania, *Tasmaniacris tasmaniensis*.

Saltbush

Characteristic species of the shrubs are the whole genus *Beplessia*, *Coryphistes* spp., *Macrolopholia* spp., and a few morabines. The usually sparse herbaceous stratum harbours *Austroicetes nullarborensis* (associated with grass), *Perelytrana* spp., *Zabrala* sp. (these mainly on self-mulching soils), *Caperrala* spp., and *Cratilopus* spp.

Grassland and other Herbaceous Communities

The following species are associated with medium to tall grasses: *Locusta migratoria*, *Austracris guttulosa*, *Heteropternis obscurella* (these mainly northern), *Chortoicetes terminifera*, *Aiolopus tamulus*, *Gastrimargus musicus*, *Acrida conica*, and *Caledia captiva*. The following occur mainly among short grasses: *Austroicetes tenuicornis*, *A. tricolor*, and *Calephorops viridis* (these tropical), *Austroicetes vulgaris*, *A. pusilla*, *Laxabilla* spp., *Praxibulus* spp., *Perala viridis* (all southern, humid), *Kosciuscola tristis*, *K. usitatus*, *K. cognatus*, *K. tasmanicus*, *Russalpia albertisi* (all alpine and subalpine), *Austroicetes cruciata*, *A. interioris*, and *A. arida* (arid to semi-arid regions), *Pycostictus seriatus*, *Collitera variegata* (sandy soils, latter in arid regions only), *Oedaleus australis*, and *Froggattina australis*.

A substantial proportion of dicotyledonous herbs in the community is probably important to the following species: *Moraba scurra*, *Brachyexarna lobipennis*, *Phaulacridium vittatum*, *Perunga ochracea*, *Perloccia evittata* (all southern species), *Perelytrana* sp., *Zabrala* spp., *Lagoonia scabronotum* (these on self-mulching soils, the last tropical), *Ecphantus quadrilobus* (on *Sida* spp.), *Fipurga crassa*, and *Cedarinia* spp. Most species of *Macrotona*, *Azelota*, and *Rusurplia* are associated with wiry tussock grasses, especially species of *Aristida*, or with hummock grassland of *Triodia* spp. or occasionally *Plectrachne* spp., and are arid or semi-arid in distribution. The whole of the morabine genus *Warramunga*, of which there are a number of undescribed species, is associated with hummock grassland, usually *Triodia*.

Fen and Bog

Few grasshoppers are found in these communities. *Bermiella acuta*, *Oxya gavisa*, *Tolgadia* spp. (these northern), *Bermius* spp., and *Schizobothrus flavovittatus* occur among tall grasses, rushes, and sedges, *Atractomorpha crenaticeps* on hygrophilous dicotyledons, and various Tetrigidae on bare damp soil, mud, or sand. *Monistria vinosa* and *Percassa rugifrons* may occur on the shrub components of subalpine bogs.

Grasshoppers and the biotic community

Grasshoppers occupy a very characteristic niche in the biotic community. They constitute the major group of invertebrate "grazers", corresponding, at a very different size level, to the grazing mammal. Indeed, it is a moot point whether, prior to the introduction of domestic stock into Australia, a greater proportion of herbaceous plant material was consumed by mammals or by grasshoppers. At the present time grasshoppers must surely be rated second only to the rabbit as grazing competitors of domestic stock. Most of the species of the herbaceous stratum participate in this competition — according to their relative abundance and the pastoral importance of their habitat — and not only the so-called "economic" species. They damage chiefly "natural" pasture — although this is nearly always disclimax in character. Sown pastures, as we have seen, discourage them, and some part of the increased carrying capacity resulting from pasture improvement must be due to reduced competition from grasshoppers. There is no doubt that the most satisfactory method of controlling these insects would be by control of the composition and structure of the herbaceous stratum, i.e. by pasture improvement and grazing control, wherever this is practicable having regard to climate, soil, and other relevant factors.

While these general points are clear, practically nothing is known in detail about the coactions of grasshoppers in the biotic community: their effect on the productivity and species composition of the vegetation, their place in food chains, etc. What detailed ecological knowledge we do have about the few "economic" species relates rather to the effects of the environment on the grasshoppers.

Biogeographic or faunal regions

Knowledge of the taxonomy and distribution of the Australian Acridoidea is still insufficient to permit a useful contribution from this group to discussions of Australian zoogeography, faunal origins, and faunal movements. However, certain points of view are suggested by the grasshopper data.

An autochthonous fauna free to spread over a continent will give rise to species adapted to the various existing environments. Where the environments are similar in their main features, this differentiation will involve relatively superficial changes, but at ecological "cliffs" it will need to be more fundamental and in general will take longer to achieve. Thus ecological cliffs constitute obstacles to faunal admixture, the faunules on each side of them spreading and differentiating readily within their respective territories but not mixing readily. Once two such faunules have effectively occupied the available habitats on each side of the cliff, invasion of one by members of the other will be rendered still more difficult. In this way distinct faunas may be expected to arise from a single autochthonous fauna by divergence, even in the absence of any real barriers*. Equally, invasion of a region by a truly exotic fauna can take place only at points ecologically similar to the region of origin. The invaders may spread and differentiate within this ecologically similar region, but will colonise other regions much more slowly and only after greater evolutionary differentiation.

If these arguments are valid, then we have not only to think of "fluid faunas" rather than "fixed regions", as urged by SERVENTY & WHITTELL (1948), following MAYR, but also to recognise the fundamental importance of ecology, especially in intraregional biogeography. The fact that the faunas of different subregions within Australia differ widely does not necessitate an assumption of different external origins. If we compare SPENCER's zoogeographical subregions, as modified by SERVENTY & WHITTELL (1948), with the climatic regions recognised here as fundamental in grasshopper ecology, we find that the Bassian subregion corresponds closely with our climatic region 4 (b), the Eyrean subregion with climatic regions 1 + 3, and the Torresian subregion (except at its south-eastern extremity, where it embraces our 4 (a)) with climatic region 2. In other words, the Bassian fauna is the fauna adapted to climate 4 or 4 (b), the Eyrean to climates 1 and 3, and the Torresian to climate 2, or 2 + 4 (a). The respective faunas are thus separated by the ecological cliffs (involving chiefly rainfall amount and distribution) separating these climates, although individual phylogenetic lines have scaled them and radiated on the other side.

The distribution and taxonomic relationships of the Australian Acridoidea certainly suggest relative discontinuities in the vicinity of SPENCER's boundaries. However, grasshoppers are more sensitive to the environment than SERVENTY & WHITTELL's birds (being poikilotherms of lower vagility), so that it is understandable that the significant ecological cliffs should be more numerous, and in particular that a discontinuity should exist also along the boundary between climates 1 and 3.

The bulk of the Australian grasshopper fauna seems to be autochthonous, but invasion has undoubtedly occurred from New Guinea, affecting especially the eastern part of the Torresian subregion (Cape York and the east Queensland coast). Species of wide tolerance, with an almost continent-wide distribution (e.g. *Gastrimargus musicus, Acrida conica, Aiolopus tamulus, Austracris guttulosa*), have

---

* Note that we are concerned here with the differentiation of faunas, not the origin of species.

also entered by this route or across the Timor Sea. There is at present no suggestion of southern faunal connections, except with New Zealand, where species of *Phaulacridium* closely related to the Australian *P. vittatum* occur.

Any consideration of Australian biogeography must take account of the climatic history of the continent. This has been characterised by great fluctuations in the average rainfall, with correlated expansions and contractions (or displacements) of the arid region, a mesic environment having more than once occupied the present arid centre. Biogeographical and evolutionary consequences of the most recent arid cycle have been considered by CROCKER & WOOD (1947) and the subject has been discussed with reference to birds by SERVENTY (e.g. 1951) and CONDON (1954). Expansion of the arid region at any period might be expected to push the more mesophilous grasshopper fauna towards the northern, southern, and eastern coasts and lead to a corresponding expansion and differentiation of a xerophilous fauna in the centre. This would tend to accentuate the adaptational divergence and discontinuity between the faunas now going under the names of "Bassian" and "Torresian". It is possible that the "Eyrean" fauna is more recent than the others and that the nearest approach to the pleistocene and pre-pleistocene faunas is provided by the grasshoppers of the central-eastern coast.

However, our knowledge on many essential points is so incomplete that any suggestions that may be made as to the actual causal sequences are little better than guesses. Thus there seems to be no agreement among paleoclimatologists as to whether the moisture distribution retained its present pattern during previous arid and humid cycles (in which case, if the Lake Eyre region was humid, Darwin, Perth, and Sydney would all have been more humid than they are now), or whether the system of moisture zones shifted northward or southward (in which case when Lake Eyre was humid either Darwin, or Perth and Sydney, might have been arid). This issue does not seem even to have been squarely faced. On the first interpretation, we need to know whether there was ever a time when no part of Australia was arid enough to support the characteristic xerophilous fauna. We need to know rather precisely, also, when such critical conditions last prevailed and whether the time elapsing to the present is adequate to account for the evolutionary changes postulated. On the second interpretation, we need to know how far the zones travelled and whether, and if so when, either the mesic or the xeric fauna would have been swept away against the northern or southern coastline. No firm answers have been given to any of these questions.

REFERENCES

ANDREWARTHA, H. G., 1943. Diapause in the eggs of *Austroicetes cruciata* Sauss. (Acrididae) with particular reference to the influence of temperature on the elimination of diapause. *Bull. ent. Res.* **34**, *1—17*.

ANDREWARTHA, H. G., 1944. The distribution of plagues of *Austroicetes cruciata* Sauss. (Acrididae) in Australia in relation to climate, vegetation and soil. *Trans. roy. Soc. S. Aust.* **68**, *315—26*.

BEADLE, N. C. W. & COSTIN, A. B., 1952. Ecological classification and nomenclature. *Proc. Linn. Soc. N.S.W.* **77**, *61—82*.

BIRCH, L. C. & ANDREWARTHA, H. G., 1941. The influence of weather on grasshopper plagues in South Australia. *J. Agric. S. Aust.* **45**, *95—100*.

CLARK, L. R., 1947a. An ecological study of the Australian Plague Locust (*Chortoicetes terminifera* Walk.) in the Bogan-Macquarie Outbreak Area, N.S.W. *Bull. Coun. sci. industr. Res. Aust.* no. 226, 71 pp.

CLARK, L. R., 1947b. Ecological observations on the Small Plague Grasshopper, *Austroicetes cruciata* (Sauss.), in the Trangie district, central western New South Wales. *Bull. Coun. sci. industr. Res. Aust.* no. 228, 26 pp.

CLARK, L. R., 1949. The habitats and community relations of the Acrididae of the Trangie district, central western New South Wales. *Bull. Comm. sci. industr. Res. Org.* no. 250, 20 pp.

CLARK, L. R., 1950. On the abundance of the Australian Plague Locust *Chortoicetes terminifera* (Walker) in relation to the presence of trees. *Aust. J. agric. Res.* **1**, *64—75*.

CONDON, H. T., 1954. Remarks on the evolution of Australian birds. *S. Aust. Ornith.* **21**, *17—27*.

CROCKER, R. L. & WOOD, J. G., 1947. Some historical influences on the development of the South Australian vegetation communities and their bearing on concepts and classification in ecology. *Trans. roy. Soc. S. Aust.* **71**, *91—136*.

DAVIDSON, J., 1936. Climate in relation to insect ecology in Australia. 3. Bioclimatic zones in Australia. *Trans. roy. Soc. S. Aust.* **60**, *88—92*.

FULLER, M. E., 1938. Notes on *Trichopsidea oestracea* (Nemestrinidae) and *Cyrtomorpha flaviscutellaris* (Bombyliidae) - two dipterous enemies of grasshoppers. *Proc. Linn. Soc. N.S.W.* **63**, *95—104*.

JOHNSTON, H. B., 1956. "Annotated Catalogue of African Grasshoppers". Cambridge: University Press: 833 pp.

KENNEDY, J. S., 1939. The behaviour of the desert locust (*Schistocerca gregaria* (Forsk.)) (Orthopt.) in an outbreak centre. *Trans. R. ent. Soc. Lond.* **89**, *385—542*.

KEY, K. H. L., 1930. Preliminary ecological notes on the Acrididae of the Cape Peninsula. *S. Afr. J. Sci.* **27**, *406—13*.

KEY, K. H. L., 1938. The regional and seasonal incidence of grasshopper plagues in Australia. *Bull. Coun. sci. industr. Res. Aust.* no. 117, 87 pp.

KEY, K. H. L., 1942. An analysis of the outbreaks of the Australian Plague Locust (*Chortoicetes terminifera* Walk.) during the seasons 1937-38 and 1938-39. *Bull. Coun. sci. industr. Res. Aust.* no. 146, 88 pp.

KEY, K. H. L., 1945. The general ecological characteristics of the outbreak areas and outbreak years of the Australian Plague Locust (*Chortoicetes terminifera* Walk.). *Bull. Coun. sci. industr. Res. Aust.* no. 186, 127 pp.

KEY, K. H. L., 1950. A critique on the phase theory of locusts. *Quart. Rev. Biol.* **25**, *363—407*.

KEY, K. H. L., 1954. "The Taxonomy, Phases, and Distribution of the Genera *Chortoicetes* Brunn. and *Austroicetes* Uv. (Orthoptera: Acrididae)." Canberra: *Comm. Sci. Industr. Res. Org.* : 237 pp.

KEY, K. H. L., 1958. Research on the Australian locust and grasshopper problems. *Proc. 10th. int. ent. Congr.* **3**, *63-7*.

KEY, K. H. L. & DAY, M. F., 1954a. A temperature-controlled physiological colour response in the grasshopper *Kosciuscola tristis* Sjöst. (Orthoptera: Acrididae). *Aust. J. Zool.* **2**, *309—39*.

KEY, K. M. L. & DAY. M. F., 1954b. The physiological mechanism of colour change in the grasshopper *Kosciuscola tristis* Sjöst. (Orthoptera: Acrididae). *Aust. J. Zool.* **2**, *340—63*.

KNIPPER, H. & KEVAN, D. K. McE., 1954. Über Flügelfärbung und Sicheingraben von *Acrotylus junodi* Schulthess (Orth. Acrid. Oedipodinae). *Ver. Überseemus. Bremen* (A) **2**, *213—26*.

LEEPER, G. W., 1950. Climates of Australia. "The Australian Environment" (Melbourne: Comm. Sci. Industr. Res. Org.), chap. 2 (*23—34*).

NOBLE, N. S., 1935. An egg parasite of the plague grasshopper. *Agric. Gaz. N.S.W.* **46**, *513—8*.

NOBLE, N. S., 1936. Fly parasites of grasshoppers. *Agric. Gaz. N.S.W.* **47**, *383—5*.

PRESCOTT, J. A., 1931. The soils of Australia in relation to vegetation and climate. *Bull. Coun. sci. industr. Res. Aust.* no. 52: 82 pp.

PRYOR, L. D., 1954. Plant communities. "Canberra A Nation's Capital" (Canberra: Aust. N. Zeal. Ass. Adv. Sci., ed. H. L. White), chap. 8 (pp. *162—177*).

SERVENTY, D. L., 1951. The evolution of the chestnut-shouldered wrens (*Malurus*). *Emu* **51**, *113—120*.

SERVENTY, D. L. & WHITTELL, H. M., 1951. "A Handbook of the Birds of Western Australia (with the exception of the Kimberley Division)." Perth: Paterson Brokensha Pty. Ltd.: 384 pp.

SJÖSTEDT, Y., 1935. Revision der australischen Acridoideen. 2. Monographie. *K. Sv. Vetensk. Handl.* (3) **15**, *1—191*.

TAYLOR, T. G., 1932. Climatology of Australia. "Handbuch der Klimatologie" (Berlin: Gebr. Borntraeger: ed. W. Köppen and R. Geiger), Vol. **4**, part S.

THORNTHWAITE, C. W., 1933. The climates of the earth. *Geogr. Rev.* **23**, *433—40*.

UVAROV, B. P., 1957. The aridity factor in the ecology of locusts and grasshoppers of the Old World. "Arid Zone Research: Human and Animal Ecology" (Paris: UNESCO), *164—98*.

WHITE, M. J. D., 1956. Adaptive chromosomal polymorphism in an Australian grasshopper. *Evolution* **10**, *298—313*.

WOOD, J. G., 1950. Vegetation of Australia. "The Australian Environment" (Melbourne: Comm. Sci. Industr. Res. Org.), chap. 6 (*77—96*).

XII
ASPECTS OF THE DISTRIBUTION AND ECOLOGY OF AUSTRALIAN TERMITES

by

J. H. CALABY & F. J. GAY

(Wildlife Survey Section, C.S.I.R.O., Canberra and Division of Entomology, C.S.I.R.O., Canberra)

The Australian fauna and its relationships

According to the most recent estimate of the world's living termites (EMERSON, 1955) approximately 2100 species, in 5 families and about 170 genera, are known. Of these about 150 species in 25 genera are Australian. A further 20 or more Australian species are known but not described, and a few new genera have yet to be characterized, including 3 of the 25 in EMERSON's figure. This is a comparatively rich fauna but is far exceeded by those of the Ethiopian region (approx. 700 species, 79 genera), Neotropical region (approx. 500 species, 56 genera) and Indo-Malayan region (approx. 430 species, 46 genera).

The Australian region is the only one which contains living representatives in all 5 families. Almost half of the genera are in the 4 primitive families but the region contains endemic genera in the two most highly specialised groups of termites, i.e. those with soldiers having asymmetrical snapping mandibles and those with nasute soldiers. Table I is a list of the Australian genera and a comparison with the other regions of the world. As can be seen from Table I, Australia has more genera in common with the Ethiopian than any other region, but if the relict Hodotermitidae are excluded, the Australian fauna is equally related to those of the Papuan, Indo-Malayan, and Ethiopian regions. It is almost certain that only the endemic genera in the Termitidae have arisen in Australia.

The most interesting genera from the point of view of distribution are the relict primitive genera in the Mastotermitidae and Hodotermitidae. The one Australian species of *Mastotermes*, morphologically the most primitive living termite, is the only surviving member of its family. Fossil species of the genus have been found in Tertiary rocks from the Eocene upwards, in Europe and North America, and other Tertiary fossil Mastotermitidae have been found in Australia, Europe, Asia, and South America. Why *Mastotermes* should persist in Australia is very puzzling. However, it occurs in large colonies, is a very destructive species and co-exists with an abundant fauna of higher termites. Three species of *Stolotermes*

Table I.

Families, genera, and numbers of species of living termites in the Australian fauna compared with their representation in the other zoogeographical regions

Families and genera	Number of Species in each Region							
	Australian	Papuan	Indo-Malayan	Ethiopian	Malagasy	Palaearctic	Nearctic	Neotropical
Mastotermitidae	1							
Mastotermes	1							
Kalotermitidae	23	48	53	45	27	6	17	109
Kalotermes	11	5	7	10	11	5	12	15
Procryptotermes	2	3		6	5		1	5
Cryptotermes	4	7	5	4	2		1	9
Neotermes	1	23	22	15	7	1	1	19
Glyptotermes	5	8	19	10	2			27
Hodotermitidae	6		2	5		12	3	1
Stolotermes	5			1				
Porotermes	1			1				1
Rhinotermitidae	16	31	59	12	3	9	8	28
Coptotermes	6	9	20	7	1	1		4
Heterotermes	7	1	7	1			2	10
Parrhinotermes	1	1	5					
Schedorhinotermes	2	7	17	3				
Termitidae	106	32	315	627	45	13	17	365
Ahamitermes	4							
Microcerotermes	11	6	20	42	19	5		10
Amitermes	27		3	36	1	5	8	11
Drepanotermes	2							
Termes	20	1	10	8	1			11
Paracapritermes	2							
Protocapritermes	1							
Nasutitermes	17	19	54	28	10			84
Tumulitermes	18							
Occasitermes	1							
"Subulitermes"	4							

The figures are taken from EMERSON's (1955) tables, except that those for the Australian region have been brought uptodate following recent work. The Australian figures do not include undescribed species of which our collection contains a number in the genera *Amitermes, Termes, Nasutitermes, Tumulitermes,* and *"Subulitermes"*. Included in the Australian totals are one species each of *Stolotermes* and *Kalotermes* which are endemic to New Zealand. No figures are given for *Subulitermes* in regions other than the Australian, as EMERSON (1955) has announced his intention of sub-dividing this genus. *Subulitermes* will then be restricted to South America and the Australian species will be placed in 3 new endemic genera.

occur in eastern and south-eastern mainland Australia, one in Tasmania, one in New Zealand, and one in South Africa, while *Porotermes* has one species in south-eastern Australia (including

Tasmania), one in South Africa, and one in Chile. It is likely that a revision of the Australian *Stolotermes* using adequate material would reduce the number of species recognised at present. This type of relict distribution is often cited as proof that the southern continents were formerly connected via Antarctica.

To those familiar with other animal groups, both vertebrate and invertebrate, it may come as a surprise that the Australian fauna is no more closely related to the Papuan than it is to the faunas of more distant regions. The chief reason for this anomaly is that the Papuan is the least explored for termites of all regions and no termite specialist has collected there. The known Papuan fauna lacks the conspicuous representation of species closely related to, or occasionally even identical with the Australian, that is found in other animal groups. However some of these surely await discovery in the savannahs and woodlands of southern New Guinea which are so like similar plant communities on Cape York Peninsula.

RAND & BRASS (1940, p. 358) state that "very large pinnacled termite nests, which may exceed 3 m in height, are a characteristic feature of the Mabadauan savannas" of southern New Guinea, and publish a photograph (plate 33) which includes a mound several feet high. A conspicuous object in a photograph of a southern New Guinea savannah woodland published by ARCHBOLD & RAND (1935, plate 45) is a large termite mound which must be about 6 ft high. Termites are not mentioned in this paper. None of the species recorded from New Guinea construct mounds which resemble the above, and they are probably species of *Nasutitermes* of an Australian group which build large mounds and gather and store food in them (e.g. *N. triodiae* (FROGGATT)), or *Amitermes*. It is almost certain that *Amitermes*, *Drepanotermes*, the large-mound-building food-storing group of *Nasutitermes* and *Tumulitermes* will be found in New Guinea. All but *Drepanotermes* have been collected on islands off Cape York. Four genera *(Prorhinotermes, Capritermes, Grallatotermes, Hospitalitermes)* which occur in the Papuan Region have not been found in Australia. The last 3 of these are moisture-loving genera which have apparently been unable to cross the dry woodlands of southern New Guinea to reach Australia. *Prorhinotermes* is a genus apparently dispersed in floating logs, and is almost entirely restricted to tropical islands which do not have the large competitive faunas found in continental areas.

For further information on the extra-Australian relationships of the termite fauna, the reader is referred to EMERSON (1955) from which most of the foregoing review has been abstracted.

Ecological factors in termite distribution

The environmental factors which have most effect on termite distribution in Australia as elsewhere are temperature, moisture, and vegetation. The part played by soil type in termite distribution is difficult to determine, as it is difficult to separate this factor from the vegetation one. Some examples of the operation of these factors are given in the following paragraphs and in the next Section.

There are approximately the same number of species north and south of the Tropic but there are more species of Termitidae north of the Tropic and more species of the primitive families south of the Tropic. The Kalotermitidae and Hodotermitidae are largely restricted to the wetter south-eastern and eastern parts, both temperate and tropical. They are mostly inhabitants of *Eucalyptus* forest and live on the wood of these trees. There is no tendency for these species to penetrate into semi-arid country, one possible exception being *Kalotermes condonensis* HILL which has been collected in a few low-rainfall areas. (The type-locality has an annual rainfall of about 12 in.) This species and one other, *K. hilli* EMERSON, which is endemic to the area, are the only Kalotermitidae recorded from south-western Australia, which has a dry summer and only a small area of wet sclerophyll forest. No Hodotermitidae have been found here. The species of *Stolotermes* live only in damp rotting logs or occasionally rot patches in trees in wet situations in the tropical east and temperate south-east.

Only four primitive species, *Stolotermes brunneicornis* (HAGEN), *Porotermes adamsoni* (FROGGATT), *Kalotermes convexus* (WALKER), and *K. improbus* HAGEN, (this latter species described in 1858 has not since been collected), occur in cool temperate Tasmania. The fauna of the cool south-eastern corner of the mainland is also an impoverished one largely made up of primitive species, but is somewhat greater than that of Tasmania.

The Australian rainforest is notable for the fact that it harbours very few termites, and so far as is known, only 4 species, *Stolotermes australicus* MJÖBERG, *S. queenslandicus* MJÖBERG, *Coptotermes dreghorni* HILL, and *Parrhinotermes queenslandicus* MJÖBERG are restricted to tropical rainforest. Little is known about most of these and the first-named is known only from the type series. Termites are, however, numerous in species and numbers in the nearby savannah woodlands. This depauperate rainforest fauna is doubtless due to the fact that this community is restricted to relatively small discontinuous areas. It has never been a large and important community with the conditions which could promote active speciation.

It is this impoverished rainforest fauna which provides the greatest contrast between the faunas of Australia and those of other tropical regions. In two weeks EMERSON (1949) collected over 50 species in a rainforest station in the Belgian Congo. He mentions also that he has collected over 100 species at Kartabo in the Guiana rainforest.

Certain soil types are avoided by some species of termites. They are virtually absent from the so-called black earths of inland north-eastern Australia which are clothed with certain grass species. The mounds of grass- and litter-harvesting species are however abundant on the sandy desert steppe soils adjacent to the heavy soils, It is thought that the physical characteristics of the heavy soils, which crack deeply and widely in dry times and waterlog during wet periods, are inimical to termite survival. (RATCLIFFE et al., 1952). *Mastotermes darwiniensis* FROGGATT is a tropical species the southern limit of which approximates to the tropic of Capricorn, both in coastal and inland localities. However, it is absent from rainforest and soils which once carried rainforest (RATCLIFFE et al., 1952). It also does not occur on the bauxite soils which occupy an extensive area of western Cape York Peninsula.

Aspects of the ecology of some Australian genera

In this Section it is not proposed to discuss the Isoptera genus by genus but rather to select a few representative genera which are successful and widespread and have especially interesting features in Australia. The genera treated are *Coptotermes*, *Ahamitermes*, *Amitermes*, *Nasutitermes*, and *Tumulitermes*.

Coptotermes

This tropicopolitan genus is represented in all Regions but the Nearctic, but is most abundant in the Indo-Malayan, Papuan, and Australian Regions. There are 6 species in Australia, some of which are notorious pests. Except for *C. dreghorni* HILL which occurs in a limited area of tropical rainforest, the genus is largely dependent on *Eucalyptus* spp. for food, and in fact *Coptotermes* are found in abundance only within eucalypt communities. This is best exemplified by *C. acinaciformis* (FROGGATT) for which all of Australia but the south-east corner and the rainforest areas is climatically suitable. In south-western Australia it is very common right up to the sharp boundary between the eucalypt formations and the mulga scrub. From here until one gets to the eucalypt savannah woodlands of northern Australia, where it again becomes abundant, it is very rare and found only along the water-courses in association with eucalypts. In the arid and semi-arid areas its place as the most destructive species is taken by *Schedorhinotermes intermedius actuosus* (HILL). The only eucalypt community from which *C. acinaciformis* appears to be absent is mallee growing on deep sand, probably because it cannot obtain clay which is essential for nest construction (CALABY & GAY, 1956).

The building of mound nests is an important aspect of the biology of 4 out of the 6 Australian species of *Coptotermes*, the only primitive genus in Australia which constructs such nests. Mound-building is almost if not entirely absent among the many species of the genus in other parts of the world. *C. brunneus* GAY which lives in a restricted area of semi-arid sub-tropical Western Australia is the only species which is found only in mounds. *C. lacteus* (FROGGATT) which occurs in the sclerophyll forests of south-eastern Australia is an obligatory mound-builder over almost all of its range, but at its northern limit in northern New South Wales and southern Queensland it is found in rotten logs in rainforest with no mounds in the vicinity (RATCLIFFE et al., 1952).

The two widely-ranging species of *Coptotermes* (*C. acinaciformis* and *C. frenchi* HILL) show geographical variation in their nesting behaviour and build mounds only in parts of their range. In tropical Queensland and the Northern Territory, *C. acinaciformis* builds characteristic large mounds and, at least on Cape York Peninsula, is an obligatory mound-builder. It is probable that the eucalypts of the northern Australian woodlands do not have the physical dimensions to support the large colonies of this species. In sub-tropical and temperate eastern Australia it nests within the trunks of trees and there is little or no indication on the outside of the trees that they are infested. The species is largely a tree-nester in south-western Australia but often the tree-trunks are buttressed by clay structures. Symmetrical domed mounds are also found in south-western Australia but they are rather uncommon and only found occasionally in the sclerophyll woodland and mallee.

C. frenchi occurs in the forests of eastern and south-eastern Australia and across the southern part of the continent in eucalypt communities. In the forests and savannah woodlands it nests strictly within tree-trunks. However from about Griffith, N.S.W. westward into Western Australia where it is found in semi-arid mallee and sclerophyll woodland, it builds symmetrical domed mounds. Whether it also nests in trees in these communities is not known — the one record of a tree nest was atypical in that the tree was dead and there was a clay structure round the base. Mounds are often found in the centres of mallee clumps. Two species, *C. dreghorni* which nests within trees in the north Queensland rainforest, and *C. michaelseni* SILVESTRI, do not build mounds. The latter species is restricted to the sandier soils of the west coast of south-western Australia, and makes subterranean nests.

There is a basic similarity about the mound nests of *Coptotermes* but there are considerable differences in detail. The mounds usually consist of three distinct sections sharply demarcated from each other. There is a hard clay outer wall, an "inner wall" or "inner section" consisting of lumps of carton made of partly-digested wood and faecal matter, and a "nursery" consisting of thin perforated carton lamellae regularly arranged into horizontal or concentric spherical layers. In this section the reproductives, eggs, and very small nymphs are normally found.

C. acinaciformis shows geographical variation in the structure of its mounds. Mounds built in the tropics are large — up to 6 ft or more in height. The outer wall, which is only a few inches thick at the top and up to a foot or so at the base, encloses a loosely packed mass of carton lumps. In the Northern Territory the nursery consists of some roughly horizontal cells in the centre of the nest at ground level (HILL, 1915), but on Cape York and the Atherton Tableland there does not appear to be a definite nursery region. The mounds built in south-western Australia are lower and have a very thick outer wall (2 ft to 2 ft 6 in.) The nursery is spherical, about 2 ft in diameter and is below ground level. There is no correlation between the distribution of this type of nest and that of the distinct morphological race (*C. a. raffrayi* (WASMANN)) which occurs in south-western Australia. The nursery in tree-nesting colonies in eastern Australia is of the spherical type.

The mounds of *C. lacteus* are as large as, or larger than those of tropical *C. acinaciformis* and have an outer wall a foot or more in thickness. The nursery is spherical and is near or above ground level. *C. frenchi* builds small mounds 1 to 2 ft in height which have a fairly thin outer wall and a spherical nursery about 9 in. diameter in an average mound. The inner section which encloses the nursery is solid carton traversed by numerous galleries. *C. brunneus* mounds are huge structures which may be over 9 ft high. The mound is largely a mass of clay or earth traversed by large galleries and often having a cavernous interior. The inner carton section is mostly underground and may go down 5 ft below the surface. The nursery consists of horizontal layers and is near the bottom of the inner section. In all species the food trees and logs are invaded by means of subterranean galleries which may radiate 100 yd from the nest.

The reasons why mound-building has assumed such an importance in *Coptotermes* in Australia are not known.

Perhaps mound nests enable the insects to better stabilize the living conditions in areas where there are long dry periods. The possibility that mounds are built by some species because of the lack of large trees to support big colonies, has already been mentioned. However there does not seem to be any reason why the nests should not be built underground, but in some areas at least (e.g. Cape York) the water-table is close to the surface during the wet season.

It is probable that competition plays a large part in determining nesting behaviour and micro-distribution in *Coptotermes*. In all areas where *Coptotermes* species are sympatric they are never found attacking the same log. Both *C. lacteus* and *C. frenchi* occur together over a large part of south-eastern Australia. *C. frenchi* nests in and attacks living trees while *C. lacteus* is a mound-builder and does not attack living wood. However both species eat fallen trees and logs. In the soil mosaic of the Western Australia mallee and sclerophyll woodland *C. acinaciformis* and *C. frenchi* occur sympatrically. *C. acinaciformis* largely nests in and attacks the big eucalypts which grow on the patches of heavier soils while *C. frenchi* builds mounds and generally attacks mallees. Both species eat fallen logs. Along the west coast of south-western Australia *C. acinaciformis* again usually nests within eucalypt trees in places where the soil has more clay while *C. michaelseni* prefers the sandier soils and is a subterranean nester. *C. frenchi* does not occur here — the ranges of this species and *C. michaelseni* are largely mutually exclusive.

Ahamitermes

From the points of view of biology and distribution, this endemic genus of 4 species is the most interesting of the Australian termites. The species are obligate inquilines in the nests of *Coptotermes* and their distributions of course depend on those of their hosts. The host-parasite relationship is species-specific. Three of the species are found in the nests of *C. acinaciformis* — *A. nidicola* MJÖBERG in north Queensland, *A. hillii* NICHOLLS in southern Western Australia, and *A. pumilus* (HILL) in New South Wales and southern Western Australia. The fourth species, *A. inclusus* GAY occurs in the nests of *C. brunneus*. The sole food of the genus is the carton composing the inner parts of the hosts' nests. The nests of *A. hillii* and *A. inclusus* are found in the inner carton sections of their hosts' nest while that of *A. nidicola* is found in the outer wall from which it invades adjacent parts of the carton section.

The nests of host and inquiline are blocked off from each other and the species are mutually hostile if accidentally brought together. *A. nidicola*, *A. hillii*, and *A. inclusus* have galleries running to the surface from which the alates are released at the appropriate time. Usually in *A. hillii* and rarely in *A. inclusus* these galleries terminate in specialised domes.

A. pumilus is considerably more specialised than its congeners in both morphology and biology. In fact its morphological distinctness particularly is so great that it will be transferred to a new genus. Its nest is found in the host's nursery where its very small workers eat out the thin lamellae composing the nursery. The nests of host and inquiline are quite separate and the members generally are mutually hostile. However the species releases its alates into the host's nest where they ascend up into the host's escape galleries and are released at the appropriate time with the host's alates by the host's workers. How the inquiline alates come to be accepted in the host's nest is unknown.

None of the species of *Ahamitermes* appears to prejudice the survival of the host colony or to outgrow its food supply. The hosts' populations and nests are very much larger than the inquilines' and it is probable that the normal growth of the hosts' nests more than makes up for the ravages of the inquiline species (CALABY, 1956).

Amitermes

This, the largest Australian genus, is widely distributed in the world and is notable among large genera for its success in desert regions. The Australian species eat wood or grass and litter. Only 6 of the 27 described species build mounds, and all of these are tropical except *A. obeuntis* SILVESTRI, a wood-eating non-food-storing species of south-western Australia. The 5 tropical mound-builders (except possibly for *A. scopulus* MJÖBERG about which little is known) feed on grass and litter and use the mound as a food store. The most interesting mound-builder is undoubtedly the well-known *A. meridionalis* (FROGGATT) which occurs in a limited area of the Northern Territory. The mounds are up to 10 ft or more in height, 8 ft broad, but only up to about 3 ft thick. They are narrowed from the middle towards each end and run up to a thin edge at the serrated apex. The long axis invariably points north and south. Several reasons have been advanced to explain these extraordinary structures, but that generally accepted is that of HILL (1942) who states, the "probable explanation appears to be the need for minimising the effects of rapid changes in temperature. Many kinds of tropical termites, including the species under notice, are unable to tolerate the excessive day temperatures in some parts of their nests and are compelled to seek more favourable conditions elsewhere, i.e. in galleries at the base of, or beneath the nests, returning to the temporarily vacated galleries when temperatures become satisfactory. In winter, when temperatures even in these latitudes often are comparatively low, there is a marked daily movement of the termites to the warmest part of the nest — to the eastern side in the morning and to the western side in the afternoon."

The building of meridional nests is not restricted to *A. meridionalis* as nests of this type are sometimes built by *A. laurensis* MJÖBERG and *A. vitiosus* HILL. These mounds are often only approximate in the direction of orientation, the long axis is not so much greater than the short axis, and also the broad sides are not so smooth as in *A. meridionalis* and are usually buttressed. These nests occur on ill-drained areas which incline to be swampy, similar to the habitat of *A. meridionalis*. The mounds of *A. laurensis* and *A. vitiosus* on sandy and better-drained soils are tall and circular in cross-section.

As far as is known *Amitermes* species do not gather food in the open but come to the surface from subterranean galleries and gather grass debris, seeds, litter, etc. beneath protective sheets of cemented soil specially constructed during damp weather. A common widely-distributed southern inland species, *A. neogermanus* HILL, does not build mounds but stores food in underground galleries. The food gathered beneath the thin protective sheets consists largely of grass and the weathered surface of wood and tree bark. An interesting species, as yet undescribed, occurs commonly in arid north-western Australia. It appears to live wholly on the dry or slightly moist dung of herbivores. The termites cover the dung with a layer of hard cemented earth, then take out the plant fibres and store them in underground chambers. This termite is usually found in horse dung, but also in kangaroo dung which presumably provided its food in pre-European times. Morphologically this species is closely related to a southern species, *A. capito* HILL, which is also a dry-country form but lives largely within eucalypt communities, feeds on wood and does not store food.

Nasutitermes

This large genus, widely distributed in the world, is one of the most successful in Australia. Of the higher termites it has had most success in penetrating the cool temperate south-eastern part of the continent. The Australian species can be divided into 3 distinct groups:

(i) a largely southern group, all subterranean wood-eaters, one or two species building earth-carton mounds,

(ii) a northern mostly-tropical group, which forage on the surface and store food in earth mounds, and

(iii) an eastern and somewhat northern coastal group of 2 species which build arboreal carton nests and eat wood. With the possible exception of one or two terrestrial wood-eating species, all the species so far found in the Papuan Region are arboreal-nesting wood-eaters. It is probable therefore that the ancestor of these two Australian species was a recent invader during Pleistocene land connections.

The best-known species is the southern wood-eating mound-builder *N. exitiosus* (HILL). Over most of its range its northern limit coincides with the boundary of the eucalypt communities. In eastern Australia its limits are determined by other factors, e.g. it does not occur in the wetter coastal country north of Newcastle nor on the cold upper parts of the southern highlands. An interesting example of a change in termite populations following a change in vegetation is given by this species. It is abundant in eucalypt forest but when this is replaced by pine plantations the colonies decline and gradually die out. This decline is not due to lack of food as abundant eucalypt logs and stumps are left when the forest is cleared. RATCLIFFE et al. (1952) and GAY (1957), who have reported this phenomenon, believe that the main factor responsible is probably the decrease in light intensity caused by the closed canopy of the pines. However it is known that pine plantations cause rapid changes in the soil (COSTIN, 1953) and it is possible that the changes in the soil make the habitat unsuitable.

Of the tropical mound-builders, *N. triodiae* (FROGGATT) calls for special mention. It builds huge mounds which may be columnar structures over 20 ft high. Nests of this magnitude are unique in the genus *Nasutitermes*.

Tumulitermes

This is the most successful of the endemic genera and is derived from *Nasutitermes*. It has speciated widely in the drier inland parts. A few species occur in south-western Australia, but there are none in south-eastern Australia. As far as is known all species are surface foragers and food storers. It seems reasonable therefore to suppose that the genus was derived from a northern form of *Nasutitermes*. Several species build mound nests, but like other Australian genera of Termitidae the majority of species live in subterranean nests. One of the most interesting forms is the type species *T. tumuli* (FROGGATT), of Western and Central Australia, the distribution of which is largely coincident with that of the mulga scrub. It penetrates into *Triodia* savannah woodland but can only be found there in the small patches of mulga growing on isolated areas of better soil. It does not depend on the mulga for food and it is possible that it is the soil type which determines its distribution. It harvests dried grass fragments and litter and processes it into small roughly spherical grey pellets for storage. Its small spire-shaped red mounds are common where they occur and are always built in the shade of a tree or bush.

Remarks on general ecology

Whether or not termite activity has much effect on soil fertility by breaking up the soil and redistributing organic matter is a debatable point. According to the most recent review (HARRIS, 1955) comparatively little increase in soil fertility results from termite activity in spite of the enormous quantities of vegetable matter consumed. This is because their utilization of the material is so complete. However the turning over of large amounts of soil in mound building and the construction of covered runways must have a beneficial effect on the soil, as would also the numerous tunnels which penetrate into the subsoil and allow the entry of water and air. It has been observed in semi-arid Australia that termite galleries sometimes penetrate masses of limestone, and that plant roots frequently follow deserted soil-packed galleries (RATCLIFFE et al., 1952).

With regard to forest trees, *Coptotermes* species are the most important pests. In parts of the eucalyptus forests of eastern Australia *C. acinaciformis* and *C. frenchi* nest in and feed on *Eucalyptus* trees. The central part of the trunk and branches is gradually eaten out. *C. frenchi* does more damage to trees than *C. acinaciformis* as the latter usually restricts its activity to a central pipe whereas the former tends to eat out annular galleries well out into otherwise sound wood. This damage is minimized to some extent by the normal practice of discarding the central four-inch square which is very prone to breakdown (RATCLIFFE et al., 1952). *C. acinaciformis* is not nearly so important in the forests of south-western Australia as two of the chief timber trees, jarrah (*Eucalyptus marginata* SMITH) and wandoo (*E. redunca* SCHAUER var. *elata* BENTHAM), have a fairly high resistance to termite attack, although the heartwood of both is attacked, particularly the latter. The most important forest pest in southern New South Wales and Victoria is *Porotermes adamsoni*. This species eats out a central pipe, round which a number of concentrically-arranged galleries extend into the sound wood. It has also been recorded as a minor pest in rainforest trees. *Neotermes insularis* (WALKER) damages forest trees of several species from southern Victoria to north Queensland but it is of lesser importance than the species mentioned above. Its attacks tend to be located in the upper parts of trees, which precludes their use for poles or long pieces of timber. The little known *Coptotermes dreghorni* has been recorded damaging rainforest trees but is of very minor importance.

Fruit and ornamental trees and vegetable crops of many varieties are destroyed by *Mastotermes darwiniensis*, and it is also an important pest of sugar cane. Damage to crops by species other than *Mastotermes* is of very little importance. *C. frenchi* has been recorded attacking potatoes in Victoria and also orchard trees and vines in inland New South Wales, while *C. acinaciformis* apparently damages fruit and almond trees around Adelaide (RATCLIFFE et al., 1952). *Neotermes insularis* occasionally attacks fruit and ornamental trees, and *Amitermes herbertensis* has been recorded as a pest of sugar cane (HILL, 1942).

Whether the many grass-feeding species of inland and northern Australia cause much loss to pasture is not known. However they do harvest a large amount of dry grass and cause extensive local denudation at times. An assessment of the possible losses caused by these termites must await further studies on the productivity and management of the semi-arid pastoral country.

REFERENCES

ARCHBOLD, R. & RAND, A. L., 1935. Results of the Archbold expeditions. No. 7. Summary of the 1933—1934 Papuan expedition. *Bull. Amer. Mus. Nat. Hist.* **68**, *527—79*.
CALABY, J. H., 1956. The distribution and biology of the genus *Ahamitermes* (Isoptera). *Aust. J. Zool.* **4**, *111—24*.
CALABY, J. H. & GAY, F. J., 1956. The distribution and biology of the genus *Coptotermes* (Isoptera) in Western Australia. *Aust. J. Zool.* **4**, *19—39*.
COSTIN, A. B., 1953. On coniferous forests in Australia. *Aust. Forestry.* **17**, *21—25*.
EMERSON, A. E., 1949. Termite studies in the Belgian Congo. *Deux. Rapp. Ann. I.R.S.A.C.* (Bruxelles) *149—59*.
EMERSON, A. E., 1955. Geographical origins and dispersions of termite genera. *Fieldiana: Zool.* **37**, *465—521*.
GAY, F. J., 1957. Termite attack on radiata pine timber. *Aust. Forestry* **21**, *86—91*.

HARRIS, W. V., 1955. Termites and the soil. pp. *62—72* in "Soil Zoology". (Butterworths Scientific Publications: London).
HILL, G. F., 1915. Northern Territory Termitidae. Pt. 1. *Proc. Linn. Soc. N.S.W.* **40,** *83—113*.
HILL, G. F., 1942. "Termites (Isoptera) from the Australian region". (C.S.I.R.O.: Melbourne). 479 pp.
RAND, A. L. & BRASS, L. J., 1940. Results of the Archbold expeditions. No. 29 Summary of the 1936—1937 New Guinea expedition. *Bull. Amer. Mus. Nat. Hist.* **77,** *341—80*.
RATCLIFFE, F. N., GAY, F. J., & GREAVES, T., 1952. "Australian termites. The biology recognition and economic importance of the common species". (C.S.I.R.O., Melbourne). 124 pp.

XIII
THE LAND AND FRESHWATER MOLLUSCA OF AUSTRALIA

by

DONALD F. MCMICHAEL and TOM IREDALE
(Australian Museum, Sydney)

Taxonomic study of the land and freshwater mollusca of Australia is still far from complete, while their ecology has, as yet, received practically no attention. Only one group, the freshwater mussels, has been studied sufficiently to allow the preparation of a monograph (MCMICHAEL & HISCOCK, 1958) and even here, one genus is still known from the shells alone. Many species are known from only a few shells from a single locality, and while a number of species have been studied anatomically enabling the probable relationships of the major groups to be established, the true affinities of many genera and species are still unknown. Before accurate zoogeographical conclusions can be reached, it is necessary to have a well established nomenclature, so that the present account is possibly somewhat premature.

However as a starting point for future research, IREDALE (1933—1945) has compiled lists of the named non-marine molluscs of Australia and of the land snails of Papua, Lord Howe and Norfolk Islands, from which some generalisations as to the nature and possible origin of the various components can be made and which serve as the basis for a discussion of distribution patterns. It must be emphasised however that many areas have not yet been properly collected and the smaller land snails have been largely overlooked, so that subsequent collecting may completely alter the pattern which is apparent now.

The non-marine molluscs, because of their comparatively limited means of dispersal and the wide variety of types are very suitable subjects for zoogeographical studies. It is therefore not surprising that three of the major contributors to Australian zoogeography (TATE, HEDLEY and IREDALE) have been malacologists. Previous workers have almost always assumed that molluscs are incapable of crossing even narrow sea-barriers without a land connection. While this may be true for the largest forms, it is realised today that the smaller forms are capable of being dispersed over long distances by cyclonic winds, by attachment to birds and other flying animals, and by rafting on drifting vegetation. Thus it is not necessary today to postulate land bridges to account for the molluscan faunas of oceanic islands (see MCMICHAEL, 1958) but these dispersal methods cannot explain all the distributional problems and the molluscs remain one of the key groups for zoogeographic study.

Nature, number and distribution

The Australian terrestrial molluscan fauna (including the freshwater forms) includes a wide variety of types which have radiated to fill the majority of the available ecological niches, and thus most of the well known holarctic families are represented by ecologically equivalent groups. The following tabulation of the fauna indicates briefly the nature of the groups represented and the extent of their development (as far as is known).

Freshwater Shells

Freshwater Mussels

All species which have been anatomically studied belong to the family Mutelidae (McMichael & Hiscock, 1958). One genus, *Lortiella* from north-west Australia has not yet been studied, and its shell form suggests that it may belong with the south-east Asian genus *Solenaia* in the family Unionidae. It is also known that all the New Zealand mussels and the majority of the New Guinea species belong to the family Mutelidae. This is of considerable zoogeographic significance and will be discussed below. Within the Australian region, the freshwater mussels exhibit a fairly wide range of shell form, recalling most of the developments of the northern families, Unionidae and Margaritiferidae, but there has been no significant anatomical modification. Over fifty names have been given to these variations, but only seventeen good species occur in Australia falling into seven genera, which contrasts greatly with the situation in north America, where several hundred species are found in an approximately equal area.

Other Freshwater Bivalves

The smaller bivalves, belonging to the families Sphaeriidae and Corbiculidae are abundant and widespread, occurring nearly everywhere in ponds and streams. They also show a good deal of morphological variation which has resulted in the naming of many species (seventeen *Corbiculina* and fourteen sphaeriids) but possibly only a few widespread species actually exist. *Corbiculina* is not known so far from south-western Australia.

Freshwater Pulmonate Gastropods

Five major groups of pulmonate snails occur in Australia, which are variously classified. The family arrangement of Hubendick

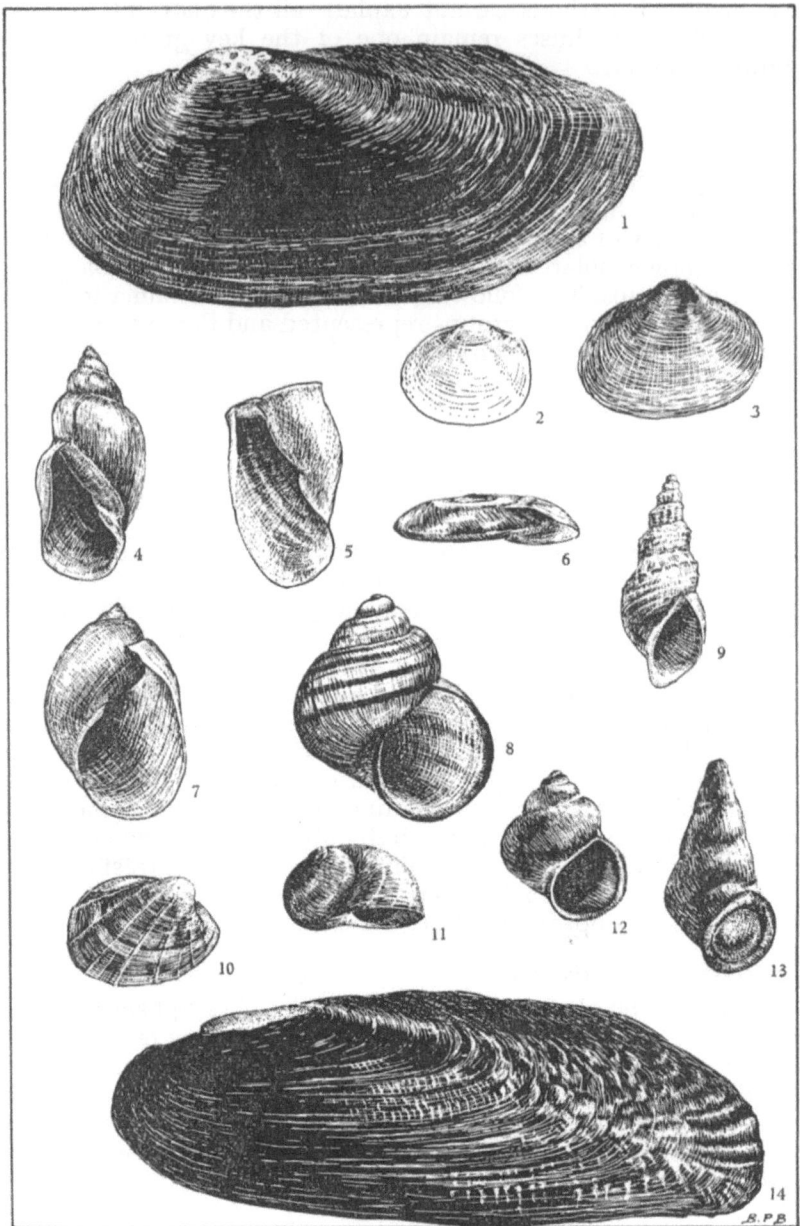

Plate 1. Freshwater Molluscs.

Fig 1. *Velesunio wilsonii* (LEA). Rockhampton, Q'ld. Mag. x $^3/_4$ (Family Mutelidae).
Fig. 2. *Sphaerinova macgillivrayi* (SMITH). Botany Swamps, Sydney, N.S.W. Mag. x 4. (Family Sphaeriidae).
Fig. 3. *Corbiculina australis* (DESHAYES). Wollondilly River, N.S.W. Mag. x 1½ (Family Corbiculidae)
Fig. 4. *Physastra gibbosa* (GOULD). Nepean River, N.S.W. Mag. x 2. (Family Planorbidae).
Fig. 5. *Amerianna carinata* (H. ADAMS). Burdekin River, Q'ld. Mag. x 3 (Family Planorbidae).
Fig. 6. *Segnitila australiensis* (SMITH). Ingleburn, N.S.W. Mag. x 4½. (Family Planorbidae).
Fig. 7. *Lymnaea lessoni* (DESHAYES). Narrabri, N.S.W. Mag. x 2. (Family Lymnaeidae).
Fig. 8. *Notopala essingtonensis* (FRAUENFELD). Townsville, Q'ld. Mag. x 2 (Family Viviparidae).
Fig. 9. *Plotiopsis balonnensis* (CONRAD). Lachlan River, N.S.W. Mag. x 1½ (Family Thiaridae).
Fig. 10. *Legrandia irvinae* (PETTERD). Great Lake, Tasmania. Mag. x 1½. (Family Ancylidae).
Fig. 11. *Larina strangei* (A. ADAMS). Burnett River, Q'ld. Mag. x 1¼ (Family Viviparidae).
Fig. 12. *Gabbia australis* TRYON. Armidale, N.S.W. Mag. x 4½. (Family Bithyniidae).
Fig. 13. *Tatea huonensis* (TENISON WOODS). Strahan, Tasmania. Mag. x 7½. (Family Hydrobiidae).
Fig. 14. *Cucumerunio novaehollandiae* (GRAY). Richmond River, N.S.W. Mag. x $^2/_3$. (Family Mutelidae).

(1948, 1951, 1955) is followed here. A number of species have been studied ecologically in connection with research on Liver Fluke and other Trematodes (BRADLEY, 1926; McKAY, 1926; CLUNIES ROSS & McKAY, 1929; JOHNSTON et al., 1937 et seq.). The Lymnaeidae are widespread, and over twenty species have been named, though HUBENDICK claims this is excessive. The species are apparently closely related to those of south-east Asia. They mostly have small, globose shells and fall into a least two distinct series *(Simlimnea* and *Lymnaea = Peplimnea)*. Two species, *Simlimnea brazieri* (SMITH*)* in New South Wales and *S. subaquatilis* (TATE) in southern Australia, are known to act as the intermediate host for the Sheep Liver Fluke, *Fasciola hepatica* L. The second major group of pulmonate snails are those which have *Physa*-like shells. Originally classified as Physidae, some species were later shown to be allied to the African genus *Bulinus* (= *Bullinus* auct.*)* so that all were later regarded as belonging to the family Bulinidae. However HUBENDICK (1948, 1955) has shown that at least one species is a true Physid, some belong in the genus *Bulinus* (sub-family Bulininae), some to the genera *Physastra* and *Amerianna* (sub-family Planorbinae) both the latter sub-families being placed in the family Planorbidae. These molluscs are common and widespread, with many local variations which have been named specifically. Further representatives of the Planorbinae are the small discoidal snails typical of the sub-family, which have been placed in five genera and about thirty species. Finally, a number of freshwater limpets (Ancylidae) are known, including two aberrant genera, the large Tasmanian *Legrandia*, and the capped forms known as *Problancylus*, which resemble the species of *Gundlachia* of Cuba and Central America, though they probably represent a separate development.

Freshwater Operculate Gastropods

These fall into four families. The Viviparidae includes ten species, classified in three genera. Two of these are rather similar to south-east Asian groups, but *Larina* with a single species from the south Queensland coastal district is very distinct morphologically, and has adopted an estuarine habitat. The family Hydrobiidae (= Paludestrinidae) is well developed in south-eastern Australia and Tasmania, where a number of species of the New Zealand genus *Potamopyrgus* occur. Some species of this family live in brackish water lakes in eastern Australia. The family Bithyniidae includes two genera and five species in Australia. The common group in eastern and central Australia is *Gabbia*, while a distinct genus *Hydrococcus* occurs in south-western Australia. The family Thiaridae is well represented in northern and eastern Australia, with five genera which are closely related to the tropical groups common in the Indo-Malayan region. Finally mention should be made of the Coxiellidae, which occur in the estuarine lagoons and salt lakes of southern and western Australia. They are probably a specialised development from the littoral family Acmeidae.

Land Snails

Operculate groups

Representatives of five families (Hydrocenidae, Helicinidae, Cyclophoridae, Pupinidae and Diplommatinidae) occur in northern Australia. Only the Pupinidae are represented by more than a few species, this family including nine genera and twenty-four species. The operculates make up only 6% of the total land snail fauna, whereas in New Guinea they represent about 25% of the fauna.

Pulmonate groups

Among the smaller families, the Tornatellinidae are represented by four species, while there are two native species of Subulinidae (one in coastal Queensland, the other in Central Australia) apart from the introduced species *Subulina octona* (BRUG.) and *Opeas gracile* (HUTTON). The Succineidae includes a dozen species occurring all over Australia. The pupoid snails (family Vertiginidae sensu lat.) are apparently well represented, about forty species having been named so far, though these have been largely overlooked by collectors. Probably the most remarkable Australian group is the endemic family Bothriembryontidae. It is best developed in south-western Australia, where thirty-three species have been named, while a few species occur in South Australia, with one surprisingly penetrating into the arid region of Central Australia. One possible relative occurs in Tasmania. The shell form is bulimoid, and PILSBRY (1946) has confirmed this anatomically, the South American Bulimulidae being the nearest relatives. The monotypic genus *Coelocion* occurs on the central Queensland coast, and has no known relatives. In shell form, it recalls some New Guinea groups, and PILSBRY has placed it with the Brazilian family Megaspiridae, but this seems very doubtful, and IREDALE (1937) has regarded it as a separate family Coelociontidae.

The "endodontid" snails are represented in Australia by three distinct groups, which have been called the families Laomidae, Charopidae and Flammulinidae. All three are based on New Zealand genera, the first and second occurring on Lord Howe and Norfolk Islands and all three in New Caledonia, while the Charopidae have a few representatives in New Guinea and the Indo-Malayan archipelago. About 140 species of Australian endodontids have been named to date, representing almost 20% of the total fauna, and probably many more remain unknown. Possibly related to these endodontids is a single species, *Stenopylis hemiclausa* (TATE) which has been

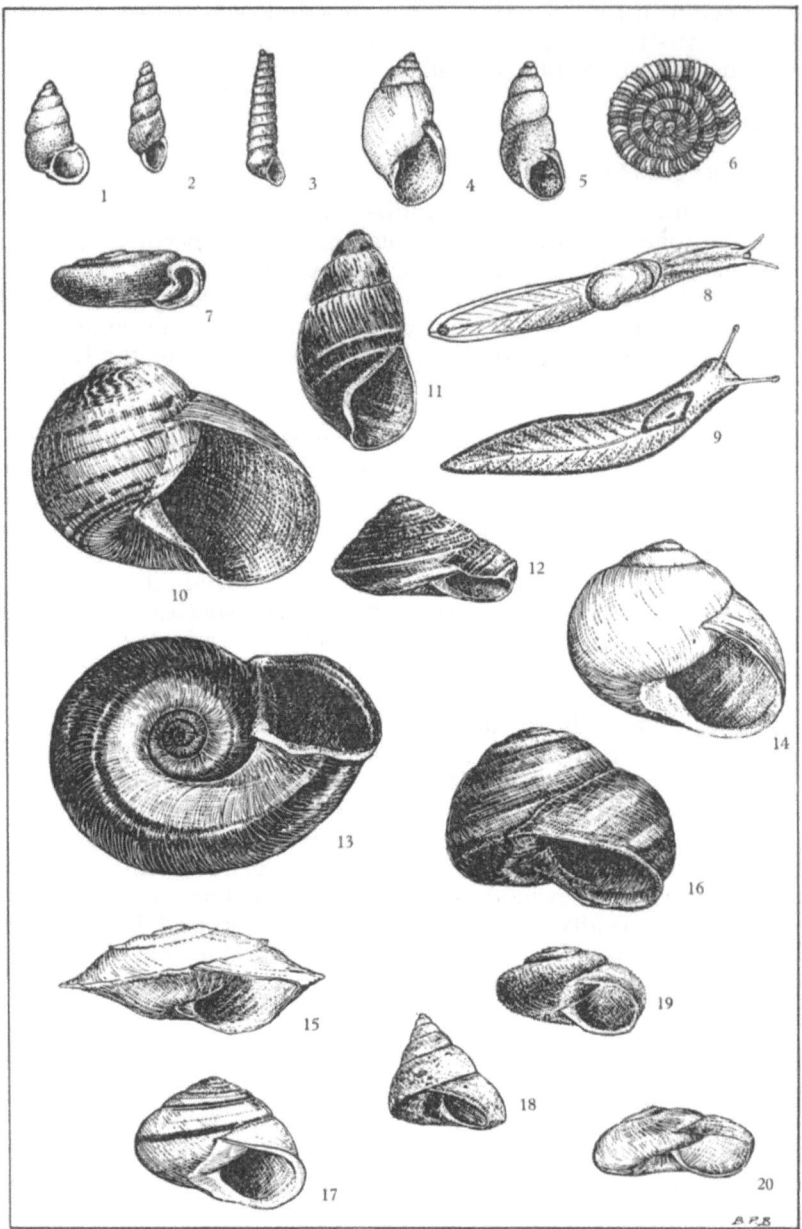

Plate 2. Land Mollusca.

Fig. 1. *Necopupina planilabris* (PFEIFFER). Port Curtis, Q'ld. Mag. x 1¼. (Family Pupinidae).
Fig. 2. *Eremopeas tuckeri* (PFEIFFER). Murray Id., Torres Strait. Mag. x 2 (Family Subulinidae).
Fig. 3. *Coelocion australis* (FORBES). North Pine River, Q'ld. Mag. x 1. (Family Coelociontidae).
Fig. 4. *Bothriembryon barretti* IREDALE. Nullarbor Plain, S.A. Mag. x 1. (Family Bothriembryontidae.)
Fig. 5. *Themapupa amolita* IREDALE. Broken Hill, N.S.W. Mag. x 5. (Family Vertiginidae).
Fig. 6. *Paralaoma tasmaniae* (COX) (After Cox). Mt. Wellington, Tas. Mag. x $5^3/_4$. (Family Laomidae).
Fig. 7. *Stenopylis hemiclausa* (TATE) (After ODHNER). Chillagoe, Q'ld. Mag. x $11^1/_3$ (Family Stenopylidae).
Fig. 8. *Vercularion freycinetti* (FERUSSAC). Lisarow, N.S.W. Mag. x $^2/_3$. (Family Helicarionidae).
Fig. 9. *Triboniophorus graeffei* HUMBERT. Sydney, N.S.W. Mag. x $^2/_3$. (Family Aneiteidae).
Fig. 10. *Hedleyella falconeri* (GRAY). Richmond River, N.S.W. Mag. x $^2/_3$. (Family Hedleyellidae).
Fig. 11. *Caryodes dufresnii* (LEACH). New Harbour, Tasmania. Mag. x 1. (Family Caryodidae).
Fig. 12. *Anoglypta launcestonensis* (REEVE). Tasmania. Mag. x 1. (Family Anoglyptidae).
Fig. 13. *Pedinogyra hayii* (GRIFFITH & PIDGEON). Kingaroy, Q'ld. Mag. x 1. (Family Pedinogyridae).
Fig. 14. *Xanthomelon pachystylum* (PFEIFFER). Chinchilla, Q'ld. Mag. x 1. (Family Xanthomelontidae).
Fig. 15. *Glyptorhagada kooringensis* (ANGAS) Waukaringa, S.A. Mag. x 2. (Family Xanthomelontidae).
Fig. 16. *Varohadra rainbirdi* (COX). Mt. Dryander, Q'ld. Mag. x 1. (Family Hadridae).
Fig. 17. *Rhagada convicta* (COX). Nichol Bay, W.A. Mag. x 1. (Family Rhagadidae)
Fig. 18. *Posorites bidwilli* (REEVE). Wide Bay, Q'ld. Mag. x 1. (Family Papuinidae).
Fig. 19. *Austrochloritis porteri* (COX). Richmond River, N.S.W. Mag. x $1^1/_4$. (Family Chloritidae).
Fig. 20. *Strangesta capillacea* (FERUSSAC). Sydney, N.S.W. Mag. x 1. (Family Paryphantidae).

given separate family rank. It is apparently related to some southeast Asian and New Guinea forms, and in Australia, the species has a remarkable distribution, occurring on the Queensland coast and in the MacDonnell Range in Central Australia.

The Australian zonitoid snails are mostly known from shells only, but BAKER (1938, 1940, 1941) has studied the group extensively using Pacific Islands material, and has suggested relationships for the Australian groups. BAKER showed that most of the Australian genera were "helicarionid" rather than "zonitid" but IREDALE used a series of family names which are followed here. The Nitoridae of eastern Australia includes only one genus with four species (which BAKER groups with the subfamily Sesarinae). A series of small shells are grouped as family Microcystidae, including thirteen species from eastern, southern and north-western Australia. BAKER refers these to the subfamily Helicarioninae. A single species from the central Queensland coast has been placed in the Macrochlamydidae, but BAKER considers it as subfamily Sesarinae. The dominant group however are the true Helicarionidae, with some fifteen species along the eastern Australian coast. A series of small shells are placed in the family Durgellidae, with relatives in New Guinea, but BAKER refers them to the widespread Pacific genus *Coneuplecta* (Helicarionidae, Euconulinae.) Finally, the family Cystopeltidae includes one genus of slug-like snails, with four species in Tasmania and the mountains of south-eastern Australia.

Four remarkable endemic groups (Caryodidae, Anoglyptidae, Hedleyellidae and Pedionogyridae) have been considered to belong with the primitive helicoids generally lumped as the Acavidae, or Acavacea, and anatomically they seem to be related. All four groups lay large, shelled eggs, but in shell characters, they diverge widely. *Anoglypta launcestonensis* (REEVE) and *Caryodes dufresnii* (LEACH) are each the sole representative of their families, both confined to Tasmania. The first possesses a conical helicoid shell, the second a bulimoid shell, and neither has any obvious relationship. The Pedinogyridae includes a single genus with five species, all large, discoidal shells occurring in southern Queensland. The Hedleyellidae includes six species which because of their great size variation have been allotted to four genera, all occurring in rainforest on mountain tops along the east coast.

The true Helicoids are by far the dominant element of the Australian land mollusca and fall naturally into six groups which together make up about 45% of the total fauna. The largest group, the family Hadridae includes some magnificent shells over three inches in diameter and height, and has been split into twenty-five generic groups including nearly 150 species, ranging all over the eastern half of Australia, including the arid central region. A specialised northern and central Australian group is the family Xanthomelontidae, with sixty species falling into ten genera. In the far northwest of the continent is a third group of helicoids the family Rhagadidae, which has representatives in the Lesser Sunda Islands. The remaining three families (Papuinidae, Chloritidae and Planispiridae) are found mainly in northern Australia, with some forms penetrating down the eastern coast into New South Wales. All three are closely related to groups occurring in New Guinea, though the Chloritidae includes 46 species with a number of endemic genera.

The family Paryphantidae, basically a New Zealand group, occurs in eastern Australia, and it may be related to the Streptaxidae of Africa and South America. The typical *Paryphanta* is confined to New Zealand, but two Australian species classified as *Victaphanta* are obvious relatives. A number of smaller shells are apparently related to the New Zealand genus *Rhytida*, and are among the commonest eastern Australian snails. They are carnivorous species, feeding on the other native snails as well as introduced forms. Ten genera and thirty-five species are recognised, the commonest groups being *Saladelos* (of small size) and *Strangesta* (some species reaching comparatively large size) which live together all along the eastern coast.

The list of Australian land molluscs concludes with three native slugs, though systematically these belong with the zonitids and helicarionids. Little is known of them, one species being comparatively common and referred to the family Aneiteidae *(Triboniophorus graeffei* HUMBERT). This species is olive green, with a red margin to the foot and a red triangle on the back, and lives in the comparatively dry scrub and sclerophyll forest of eastern Australia, occasionally penetrating into suburban gardens. The other two species of the genus *Prisma* (family Rathouisiidae) appear to be of northern origin. A few other species of slugs have been recorded, but are doubtfully Australian. The family Aneitiidae includes New Zealand, New Caledonian, New Hebridean and New Guinea representatives and related forms occur in Madagascar, East Africa, and India (HEDLEY, 1892).

In discussing the distribution of the non-marine mollusca it is best to consider the land and freshwater forms separately, as their distributions are governed to some extent by different factors, and the zoogeographic areas do not coincide. Thus while climate in the broad sense, especially rainfall is the prime factor in each case, the freshwater molluscs are in part affected by river systems and topography, while the land snails are apparently dependent mainly on the vegetational areas, as well as their historic source of origin.

Freshwater Faunal Regions

Comparatively little thought has been given to the distribution of freshwater animals in Australia in contrast to the terrestrial faunas. IREDALE & WHITLEY (1938) first outlined the fluvifaunal areas and gave names for the characteristic faunulae which inhabit them. IREDALE (1943) & WHITLEY (1947) later proposed slight modifications, and recently McMICHAEL & HISCOCK (1958) have suggested modifications based on study of the freshwater mussels. The areas and some of the characteristic forms are listed below.

(1) The Mitchellian Fluvifaunula, occupying the whole of the Murray-Darling River system and overlapping to the Victorian coast and to the central and southern Queensland coast is very distinctive. The Murray Cod and other fishes are characteristic while the whole region is the home of the two commonest Australian freshwater mussels, *Velesunio ambiguus* (PHILIPPI) & *Alathyria jacksoni* IREDALE. Also characteristic is the thiarid snail *Plotiopsis balonnensis* (CONRAD).

(2) The Lessonian Fluvifaunula, occupying the coastal rivers of south-eastern Australia and northern Tasmania, is characterised by the genus *Hyridella* among the mussels and an extensive development of hydrobiid gastropods.

(3) The Tobinian Fluvifaunula, occurring in southern Tasmania, is well distinguished by the absence of any freshwater mussels, and by the strong representation of hydrobiid and ancylid gastropods, the latter including the remarkable, large *Legrandia*, which lives in the high mountain lakes.

(4) The Krefftian Fluvifaunula, occupies the coastal rivers of southern Queensland from the Mary River, south to the Richmond and Clarence in northern New South Wales. There is some overlap here with the Lessonian to the south and also with the Jardinean to the north. The Lungfish, *Neoceratodus* is characteristic as is the mussel *Cucumerunio novaehollandiae* (GRAY) and also *Alathyria pertexta* IREDALE, the latter extending for some distance north and the former reaching as far south as the Hunter River. The aberrant viviparid gastropod genus *Larina* is also characteristic.

(5) The Jardinean Fluvifaunula, occupying the coastal rivers of eastern Queensland north of the Mary River has a characteristic fish fauna, but the molluscs are similar to those of the Leichhardtian.

(6) The Leichhardtian Fluvifaunula, extending westward from Queensland across the north of Australia to the Kimberley region has a fish fauna which links the area with southern New Guinea. However the molluscs, which extend to the Jardinean region, differ from the New Guinea forms. Characteristic of the combined Leichhardtian-Jardinean are the mussels *Velesunio wilsonii* (LEA) (which also extends southwards into Central Australia and the Lake Eyre drainage basin) and *Velesunio angasi* (SOWERBY). Thiarid gastropods and Viviparidae are dominant, while a peculiar development is the genus *Amerianna* among the Planorbinae.

(7) The Sturtian Fluvifaunula, occupying the rivers and lakes of

Central Australia, appears to be only a specialised development of the Leichardtian faunula, though the viviparid *Centrapala* may be characteristic. The fish fauna is not well known.

(8) The Vlaminghian Fluvifaunula, of south-western Australia, seems to have the most distinctive representatives, including a number of quite aberrant fishes, crustacea, and other freshwater forms. The mollusca include *Westralunio*, a genus whose only other representative occurs in the Fly River in New Guinea, and *Hydrococcus*, a distinctive Bithyniid gastropod.

It should be noted that McMICHAEL & HISCOCK (1958) separated the fluvifaunula of southern New Guinea as the Riechian, for although the fish fauna and the gastropoda agree with the Leichhardtian forms, the mussels are quite distinctive, and include three endemic genera.

The fluvifaunal areas here summarised agree more or less with the faunal areas which can be recognised from a study of the land mollusca. The few differences which exist are due to topographic differences which affect the drainage systems, but not the vegetational zones. Thus the sharp distinction between the Lessonian and the Mitchellian fluvifaunulas contrasts with the rather unified Euronotian faunula. Similarly the sharp break between the fluvifaunulae of northern and southern Tasmania (which may be due to the effects of glaciation) is not reflected in the land mollusca.

Terrestrial Faunal Regions

The zoogeographic areas at present recognised for terrestrial animals depend largely on the group studied. While some workers would not admit more than the primary divisions into Autochthonian, Euronotian and Eremian as first suggested by TATE (1887) and some groups show no distinction between the first named, workers on mollusca, however, have immediately recognised the distinction of a larger number of areas and these are confirmed by the studies of other workers in different zoological groups. Thus the districts outlined by SLOANE (1915) based on the distribution of beetles agree very closely with those recognised by IREDALE (1937) from land mollusca. IREDALE's scheme is merely an extension of the proposals of TATE (1887), HEDLEY (1894) & SPENCER (1896) in which the major areas are partly subdivided, through the recognition of distinctive aggregations of forms within portions of the faunal areas. The major characteristics of the areas with respect to their land molluscan faunas are set out below.

(1) The Autochthonian Faunula, occupies the south-west corner of Australia, which is probably the most distinctive biogeographic region in Australia, its most outstanding feature being the presence of the genus *Bothriembryon*, with over thirty species in the area, and about

five species elsewhere. These are the only large land snails occurring in south-western Australia, and they constitute about 50% of the total fauna of the region. A few of the species are adapted to the arid conditions which develop in the eastern part of the area, but the majority of the species are found in the well watered and forested country. Although a few Xanthomelontid species penetrate the area from the east, there are no endemic helicoids, and this niche has been filled by introduced helicoids, which flourish and have spread rapidly.

(2) The Eremian Faunula, occupying the arid and semi-arid central portion of the continent and most of South Australia, appears

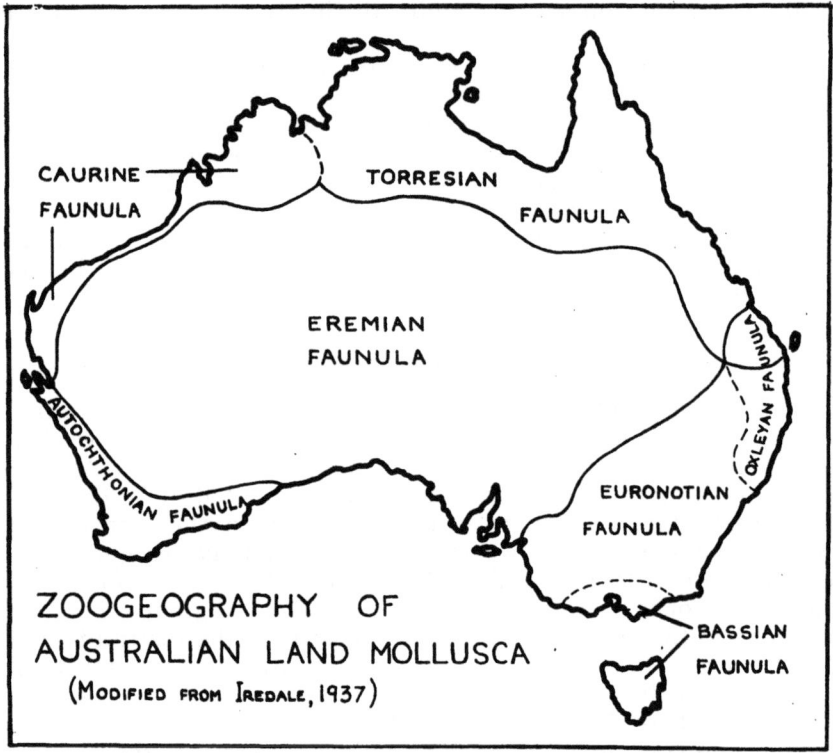

Map. 1. Zoogeographic Provinces of Australia, based on the distribution of terrestrial molluscs. Boundaries of major provinces are demarked by continuous lines. Broken lines indicate the subdivisions of these provinces, e.g. the Caurine faunula inhabits the western portion of the Torresian Province, while the Bassian and Oxleyan are the southern and northern portions of the Euronotian Province. The boundaries indicated should be considered as approximations only, and it should be noted that the Torresian and Euronotian (Oxleyan) faunulae overlap in the region of the Tropic of Capricorn on the coast of Queensland.

to have been derived from varied sources, with representatives of many groups characteristic of other areas, but adapted and modified for desert life. Helicoids, including about 40 Xanthomelontidae and 30 Hadridae dominate. The former includes the genus *Glyptorhagada* in which the shell has become sharply lenticular, and which apparently hides away in crannies in the rocks and soil during dry periods. One intriguing member of this faunula is *Stenopylis hemiclausa* (TATE) mentioned above as occurring elsewhere only on the coast of Queensland. Somewhat similar distribution is shown by the genus *Eremopeas*, with one species living in Central Australia (*E. interioris* TATE) and another in Queensland (*E. tuckeri* PFEIFFER). A number of smaller forms occur, mostly in the ranges of Central Australia which serve as refuges for these normally moisture-loving species. The occurrence of *Bothriembryon spenceri* (TATE) in Palm Valley is also intriguing, as the nearest record of this genus is two species on Yorkes Peninsula, and one (*Bothriembryon barrett* IREDALE) which lives on the Nullarbor Plain.

(3) The Euronotian Faunula, occupies the south-east portion of Australia from about the Tropic of Capricorn south to Tasmania. The northern section (extending south to about the Hunter River) has been separated as the Oxleyan Sub-Area, and has some distinctive species and genera, including the acavids *Pedinogyra* and *Hedleyella* and the Hadrid genus *Annakelea*. This corresponds closely with the distribution of the Krefftian Fluvifaunula. The southern section of the Euronotian, including all of Tasmania and southern Victoria has been differentiated as another Sub-Area (Maugean) inhabited by the Bassian Faunula, which is a specialised development of the Euronotian. The Bassian includes many endemic species and some genera, including the acavids, *Caryodes* and *Anoglypta*, the genus *Victaphanta* which appears to be relatively close to the true New Zealand *Paryphanta*, and a single species of *Bothriembryon*. Most characteristic of the Bassian however is the large number of endodontids, including over 30 Flammulinidae, 25 Charopidae and 15 Laomidae, representing over 50% of all Australian endodontid snails. The bulk of the Euronotian Faunula consists of about 24 Helicoids, including the Hadrid genus *Meridolum*, and about 20 zonitoids, more than half being Helicarionidae. A Bassian group which extends into the Euronotian proper is the slug-like *Cystopelta* two species being known from Tasmania, and others occurring on the high mountain tops in Victoria and New South Wales. Some degenerate members of the Hedleyellidae have a similar distribution on the mountain tops in New South Wales, but there is no Tasmanian representative.

(4) The Torresian Faunula, occupying eastern Queensland north of the Tropic of Capricorn and extending westward to the coastal region of Arnhem Land is the richest and most varied of the

Australian zoogeographic areas. A specialised western portion, occupying the Kimberley district of Western Australia and extending south down the Western Australian coast to Shark's Bay has been called the Caurine Faunula. The Torresian proper is characterised by the great predominance of Helicoids, over 160 species being recorded. The bulk of these are the family Hadridae, concentrated along the north coast of Queensland, with over 80 species distributed in five or six genera. Smaller numbers of Xanthomelontidae, Chloritidae, Planispiridae and Papuinidae occur, the last three groups being obviously of New Guinea origin. The Hadridae however are an endemic group, and include the largest snails in Australasia, the genera *Hadra*, *Varohadra* and *Bentosites* being characteristic. Also characteristic of this faunula are the operculate gastropods, all but three of the species occurring here. Again these groups are obvious invaders from New Guinea. In the western portion of the Torresian proper, the fauna is impoverished, and as yet little known, only about a dozen helicoids, including the large, typical Xanthomelontids being as yet recorded. No small shells are known, but these are well represented in eastern Queensland. The Caurine Faunula is rather sharply marked off from the true Torresian, by the presence of the family Rhagadidae as the dominant element, all forty species occurring in this area. A few Chloritids are the only other Helicoids, while the smaller forms are not well known.

This comparatively brief summary of the distribution of the molluscan fauna of Australia necessarily overlooks many interesting forms which further characterise the areas and faunulae. It is obvious to any worker familiar with the Australian mollusca that the zoogeographic provinces as outlined here exhibit very real differences, such that a random sample of species from any of them would suffice to identify their place of origin, not on the identity of the species, but on the nature and proportions of the groups represented. These differences are due to a large number of factors, including the climate, vegetational zones, and of course the adaptation of the species to the conditions prevailing. The differences are however partly historical, the recency and source of origin of particular groups affecting their present distribution. In the ensuing sections, the adaptations of the Australian non-marine mollusca to the environments in which they live will be discussed briefly, and some consideration given to the origins of the various groups.

Ecology and adaptations

Comparatively little is known about the ecology of the native molluscan fauna. Few laboratory or field studies have been made, but a number of field observations have been published, and personal knowledge enables some further discussion.

The freshwater molluscs show a complete range from those incapable of withstanding desiccation to any degree, to those fully adapted to drought conditions which can survive for months and even years without water. The former include most of the Hydrobiidae, Thiaridae and probably the Ancylidae among the gastropods, and the genera *Hyridella* and *Cucumerunio* among the mussels. All these groups are confined to the relatively well watered regions of the continent, occurring in east coast streams which are seldom subject to drought. The remainder of the freshwater forms show varying degrees of drought resistance, and some examples may be cited. The mussel *Velesunio ambiguus* (PHILIPPI) was recorded by ORTMANN (1920) as having survived for four months in a parcel mailed to Europe, while HISCOCK (personal communication) has reported that the species *Velesunio wilsonii* (LEA) which lives in the driest parts of Australia, wherever semi-permanent water holes occur, survived for three years in a box in the laboratory without water. This species was also undoubtedly the one referred to by SANGER (1883) who recorded that aboriginals dug the living animals from four to five feet below the bottom of Cooper's Creek, some eight months after the water had disappeared. McKAY (1926) experimentally dried populations of *Gabbia australis* TRYON (Bithyniidae) in the laboratory and found 25% still living after 203 days. The snails living in the streams and water holes of northern and Central Australia, such as *Amerianna, Notopala, Bulinus* and others, must be capable of similar drought resistance, as they appear very soon after rain fills the streams and ponds (though CLUNIES ROSS (1929) reported that *Simlimnea brazieri* would not tolerate desiccation). Mention should be made here of the Coxiellidae, which live in brackish lagoons and salt lakes in southern and western Australia. However as these may be of comparatively recent marine ancestry, this is not surprising.

Among the land mollusca, a similar range of adaptation is apparent. Some species, especially the larger species which inhabit the rain and sclerophyll forests of eastern Australia are very sensitive to dry conditions appearing only at night, and after rain, hiding away in moist places at other times. These species die quickly during dry spells, the populations apparently surviving through small numbers of individuals which find a safe refuge. Many species on the other hand are capable of surviving for years in a state of aestivation, emerging as soon as rain falls, breeding quickly and then returning to their hiding places in the crevices of rocks, around roots of trees and shrubs and deep in cracks in the soil. During these periods of aestivation, the snails form a calcareous epiphragm, which reaches nearly a quarter of an inch thick in some Eremian species. ROBSON (1913) recorded the sudden appearance of the Eremian species *Glyptorhagada silveri* (ANGAS) after rain in northern South Australia;

the species had been known from dead shells for many years but had never been seen alive previously. Similarly, *Bothriembryon barretti* IREDALE occurs as dead shells in countless numbers on the Nullarbor Plain, but is seldom seen alive. HARTLEY (1957) described the discarding of the epiphragm by *Bothriembryon barrett*, stating that the ground under each bush where the living animals were aestivating was covered with discarded epiphragma. A species inhabiting the drier areas of northern Queensland, *Chloritisanax banneri* (PFEIFFER) was found to be alive after nearly six years in a box at the Australian Museum, Sydney. The snails had eaten much of the soft cardboard from the inside of the box, but had survived without water. Even the native slug, *Triboniophorus graeffei* HUMBERT appears to be capable of drought resistance, for in 1950 it was recorded in large numbers at Mt. Kaputar, central New South Wales after a period of rain, though a road survey party some weeks earlier had seen no sign of molluscan life. MCLAUCHLAN (1951) has recorded some observations on the life cycle of *Strangesta capillacea* (FERRUSAC) (Paryphantidae) which indicate that the species is capable of entering a short period of facultative diapause. The normal incubation period is eight weeks, hatching depending on rain followed by warm winds, but the eggs will remain viable for up to four months if weather conditions are unsuitable.

However, the majority of the Australian land snails inhabit the moist forested regions of eastern Australia, where food and shelter are plentiful and where desiccation is seldom a problem. They are mostly microphagous, feeding on plant detritus, fungal hyphae and humus, but some of the larger species graze on new shoots, over-ripe fruits and other soft parts of living plants. Few native species attack living crop plants. Some freshwater snails (*Amerianna reevii* A. ADAMS & ANGAS) have been found attacking rice plants in the Northern Territory, but the majority of species feed on native aquatic plants. A few land snail groups, including the Paryphantidae, are carnivorous, feeding on other native snails, earthworms, soil crustaceans and insects. MCLAUCHLAN (1949, 1951) showed that *S. capillacea* feeds naturally on snails of the genera *Eremopeas, Paralaoma, Egilomen, Vercularion*, and will also attack the introduced *Helix aspersa* L.

Little is known about the effects of soil-type, food, topography and similar factors on the distribution of snails. MCLAUCHLAN (1951) has made a number of claims as a result of breeding experiments with some New South Wales species, but these require verification. In general, mollusca are more abundant in limestone areas, but this is not a limiting factor, as snails occur in the calcium deficient soils of the Hawkesbury Sandstone region. Freshwater mollusca are known to be limited by the acidity of the water, and do not occur in stream with a pH below about 5. While such factors do affect the distribution of the local population, they do not appear to be important in determining the range of the species.

A few groups of land snails have become arboreal. The Papuinidae, whose centre of distribution is New Guinea, are all tree-dwellers and are represented in Australia by four species, which are confined to the forested areas of Northern New South Wales and Queensland. Some species of *Bothriembryon* which live in the Jarrah forests of south-western Australia have become arboreal (*B. fuscus* THIELE, *B. jacksoni* IREDALE) as well as a rhagadid (*Parrhagada woodwardi* FULTON) and one genus of Succineidae. The latter *(Arborcinea)* lives under the loose bark of eucalypts.

The factors which control the size of terrestrial molluscan populations are primarily climatic. Drought and bushfires have severe effects as evidenced by the thousands of dead shells to be found following such events. McLAUCHLAN's experiments showed the importance of climatic factors in the hatching of land snail eggs. Predators therefore probably play a small part, but they include birds, rats, reptiles and carnivorous snails for land groups, and aquatic birds, water rats, tortoises and fishes for freshwater forms. Parasites are not known for land snails, while the trematodes which parasitise the aquatic snails do not seem to affect their hosts greatly.

Origin and relationships

As indicated previously, the relationship of many groups of Australian land molluscs is still speculative, but sufficient is known to make some comment on the probable sources of origin of the major groups. The fauna can be divided into a number of sections based on the geographic location of the nearest relatives of the families included.

Melanesian

This group includes those families with close relatives in New Guinea and the other Melanesian Islands and in the easterly islands of the Indo-Malayan archipelago (Celebes, Moluccas) but there is much less affinity with the northern Indo-Malayan islands (Java, Sumatra, Borneo). Nearly all are essentially Torresian or Leichhardtian: Hydrocenidae, Helicinidae, Cyclophoridae, Pupinidae, Diplommatinidae (land operculates) Tornatellinidae, Nitoridae, Microcystidae, Macrochlamydidae, Durgellidae, Subulinidae, Stenopylidae, Papuinidae, Chloritidae, Planispiridae, possibly Coelociontidae (land pulmonates) and Thiaridae (freshwater operculates). In a number of cases, the genera and occasionally species found in Australia are the same as those of Melanesia but some families show a rather high degree of endemism, especially the two most widespread, Chloritidae and Pupinidae in which nearly all the genera are endemic. It seems reasonable to assume that most of these families have originated comparatively recently through a succession of invasions from the North, probably during the intermittent connections of New Guinea and Australia across Torres Strait, but also by occasional rafting of colonisers on drifting vegetation. The distribution of two of the families, Subulinidae and Stenopylidae is intriguing. The former includes two native species of the genus *Eremopeas*, one of which lives in eastern Australia, the other *(E. interioris)* in Central Australia. This family includes the well known soil snails *Opeas* and *Subulina*, which are probably spread by cyclonic winds so the inland species may have originated in this manner. The family Stenopylidae however, includes only one genus and species, *Stenopylis hemiclausa* TATE, which lives on the coast of Queensland and in the MacDonnell Ranges of Central Australia. This distribution recalls that of the Cabbage Tree Palm *(Livistona)* and the cycad *Macrozamia*, which indicate that forest conditions once extended widely over eastern Australia, reaching as far as the central ranges.

Newer Asian

These families are mostly rather widely distributed groups, and the Australian representatives, while showing a certain degree of endemism, are not related specifically to the New Guinea forms, but rather to the whole south-east Asian fauna. All are widely distributed in eastern Australia, while a few range over the whole continent: Succineidae, Vertiginidae, Helicarionidae, Aneitiidae, Rathouisiidae (land pulmonates) Corbiculidae, Sphaeriidae (freshwater bivalves) Viviparidae (freshwater operculates) Planorbidae (but not Bulininae) Lymnaeidae, Ancylidae (freshwater pulmonates). These groups can be assumed to have reached Australia from northern sources over a long period of time, and because of the ease with which most are dispersed, there has been no marked isolation to allow the development of very different forms. Only the Helicarionidae, and some southern Ancylidae show much divergence from the Asian representatives.

New Zealand

The families included here are well developed in New Zealand, and the Australian representatives are common in south-eastern Australia only. The groups also make up major elements of the faunas of Lord Howe and Norfolk Islands, and New Caledonia. There are no close relations anywhere else in the world, though the Hydrobiidae are distributed elsewhere and PILSBY (1946) claims that the endodontids appear to show South American affinities: Charopidae, Laomidae, Flammulinidae, Paryphantidae (land pulmonates) Hydrobiidae (freshwater operculates). The origin of these forms is puzzling, as they seem to have arisen in situ in New Zealand or south-eastern Australia, subsequent interchange accounting for the similarity between the two faunas. The Hydrobiidae are widely distributed throughout the world, but the genus *Potamopyrgus* and its Australian relatives are distinctive in this region. Of the land snails, the three endodontid families may be derived from South American groups, but this would be a very distant relationship. The Paryphantidae were associated by THIELE with the South American and African Streptaxidae, but the relationship if any is very distant.

South American and African

Only three groups show a definite relationship to the faunas of the southern continents, and in each case the affinity is marked: Bothriembryontidae (land pulmonates) Mutelidae (freshwater bivalves) Bulininae (freshwater pulmonates). The two freshwater groups are distributed all over the continent but the Bothriembryontidae are practically confined to south-western Australia. The freshwater mussels of South America, Australia and Africa form a group which is quite separable from the families of the northern hemisphere. MCMICHAEL & HISCOCK (1958) have accounted for this by assuming a common northern origin, with subsequent confinement to the southern continents, but an equally good case can be made for a southern distribution across a temperate antarctic land mass. The Bulininae are found in Africa, and in Australia, the Asian and South American continents being inhabited by Physidae and Planorbinae. It is difficult to account for this distribution. The Bothriembryontidae has a close relationship with the Bulimulidae of South America (PILSBRY, 1946) which it resembles both conchologically and anatomically. Its restriction to the south-west corner of the continent suggests that it is an old group which may once have been widely distributed in southern Australia, but has since been practically confined to its present range by climatic fluctuations. The Tasmanian representative of this group is only provisionally classified as a bothriembryontid, though a fossil species of *Bothriembryon* has been named from the Tasmanian Tertiary. Because of the complete absence of any bulimoid groups from northern Australia or the northern hemisphere, it seems only logical to postulate a southern origin for this family.

Older Asian

These might equally well be regarded as endemic Australian groups. The families are all endemic with the exception of the freshwater Bithyniidae and in the latter the two genera are endemic. The distribution is varied: Caryodidae, Anoglyptidae, Hedleyellidae, Pedinogyridae, Cystopeltidae, Hadridae, Xanthomelontidae,

Rhagadidae (land pulmonates) Bithyniidae (freshwater operculates). The first four families, grouped with the Acavacea by PILSBRY (1894) & THIELE (1929—31) do show anatomical relationship to the acavid molluscs of Ceylon, Madagascar and elsewhere. If there is a real relationship, it must be a very ancient one, for the Australian representatives possess very distinctive shell characters, and their distribution is such as to indicate that they are relicts of an ancient fauna. The three helicoid families, Hadridae, Xanthomelontidae and Rhagadidae have been related by PILSBRY, (1894) and others to the Camaenidae of Asia, together with the Chloritidae and Papuinidae. However all three families are characteristically Australian, and the absence of representatives outside Australia (except for two southern New Guinea hadrids and two rhagadids in the Sunda Islands, which can be assumed to be of recent origin from Australia) indicates that the groups have developed in isolation over a long period. Within Australia, it is probably that the Xanthomelontidae arose from the Hadridae by adaptation to desert conditions, while the Rhagadidae are probably ancient derivatives from the Hadridae which have been isolated in north-western Australia and have undergone considerable morphological divergence in shell form. The family Cystopeltidae is probably a degenerate zonitoid or helicarionid, but its distribution which parallels that of the "acavid" groups may indicate its age. Finally the Bithyniidae is represented by two endemic genera, one in Western Australia, the other eastern, which are only distantly related to the exotic members of the family, and no suggestions are offered as to their origin.

REFERENCES

BAKER, H. B., 1938-'41. Zonitid Snails from Pacific Islands, Parts 1, 2 & 3. *Bull. Bernice P. Bishop Mus.*, Nos. 158, 165, 166.

BRADLEY, B., 1926. Observations of the Water Snails of Monaro and New England, New South Wales, with Especial Reference to Their Cercaria Carrying Capacity. *Med. J. Aust.* February 9, 1926, 147—159.

HARTLEY, T., 1957. *Bothriembryon barretti* discards its Epiphragm. *J. Malacol. Soc. Aust.*, **1**, 33.

HEDLEY, C., 1892. An Enumeration of the Janellidae. *Trans. New Zeal. Inst.*, **25**, 156—162.

HEDLEY, C., 1894. The Faunal Regions of Australia. *Rep. Australas. Ass. Adv. Sci.*, **5**, 444—446.

HUBENDICK, B., 1948. Studies on *Bulinus*. *Ark. Zool.* **40A**, No. 16.

HUBENDICK, B., 1951. Recent Lymnaeidae. *K. Sv. Vet.-Ak. Handl.* (**4**), 3, No. 1.

HUBENDICK, B., 1955. On a Small Material of Freshwater Molluscs collected by Prof. T. Gislen in Australia. *Lunds Univ. Arskr.*, N.F., Avd. 2, **51**, No. 6.

IREDALE, T., 1933. Systematic Notes on Australian Land Shells. *Rec. Aust. Mus.*, **19**, *37—59*.

IREDALE, T., 1937. A Basic List of the Land Mollusca of Australia (Part 1). *Aust. Zool.*, **8**, *287—333*.

IREDALE, T., 1937. An Annotated Check List of the Land Shells of South and Central Australia. *S. Aust. Nat.*, **18**, *6—59*.

IREDALE, T., 1937. A Basic List of the Land Mollusca of Australia - Part 2. *Aust. Zool.*, **9,** *1—39,* pls. 1-3.
IREDALE, T., 1938. A Basic List of the Land Mollusca of Australia - Part 3. *Aust. Zool.*, **9,** *83—124,* pls. 12 & 13.
IREDALE, T., 1939. A Review of the Land Mollusca of Western Australia. *J. Roy. Soc. W. Aust.* **25,** *1—88.*
IREDALE, T., 1941. A Basic List of the Land Mollusca of Papua. *Aust. Zool.*, **10,** *51—94,* pls. 3 & 4.
IREDALE, T., 1943. A Basic List of the Fresh Water Mollusca of Australia. *Aust. Zool.*, **10,** *188—230.*
IREDALE, T., 1944. The Land Mollusca of Lord Howe Island. *Austr. Zool.*, **10,** *299—334,* pls. 17—20.
IREDALE, T., 1945. The Land Mollusca of Norfolk Island. *Aust. Zool.*, **11,** *46—71,* pls. 2—5.
IREDALE, T. & G. P. WHITLEY, 1938. The Fluvifaunulae of Australia. *S. Aust. Nat.*, **18,** *64—68.*
JOHNSTON, T. HARVEY & E. R. CLELAND, 1937. Larval Trematodes from Australian Terrestrial and Freshwater Molluscs. *Trans. Roy. Soc. S. Aust.*, **61,** *191—201* (and subsequent papers by these and other authors in the same *Transactions*).
McKAY, A. C., 1926. An Intermediate Host of *Fasciola Hepatica* in New South Wales. *J. Aust. Vet. Ass.*, **2,** No. 1: *9—14.*
MCLAUCHLAN, C. F., 1949. Experiments Upon Certain Snails Inhabiting Sydney Gardens. *Poc. Roy. Zool. Soc. N. S. W.* 1947-'48: *21—24.*
MCLAUCHLAN, C. F., 1951. Basic Work on the Life Cycle of Some Australian Snails. *Proc. Roy. Zool. Soc. N.S.W.*, 1949-'50: *26—36.*
MCMICHAEL, D. F., 1958. The Nature and Origin of the New Zealand Freshwater Mussel Faunas. *Trans. Roy. Soc. N. Zeal.* **85,** *427—432.*
MCMICHAEL, D. F. & I. D. HISCOCK, 1958. A Monograph of the Freshwater Mussels of Australasia. *Aust. J. mar. freshw. Res.*, **9,** No. 3, *372—508,* pls. 1-19.
ORTMANN, A. E., 1920. Über die Australische Muschelgattung *Hyridella. Arch. Molluskenk.*, **52,** *119—121.*
PILSBRY, H. A., 1894. Guide to the Study of the Helices. *Tryon's Man. of Conch,.* (2), Vol. **9.**
PILSBRY, H. A., 1946. Notes on the Anatomy of Australian and Galapagos Bulimulidae (Mollusca, Pulmonata). *Notulae Naturae (Acad. Nat. Si. Philadelphia),* No. 168.
ROBSON, G. C., 1913. Note on *Glyptorhagada silveri* (Angas). *P. Malacol. Soc. Lond.*, **10,** *265.*
ROSS, I. CLUNIES & A. C. McKAY, 1929. The Bionomics of *Fasciola Hepatica* in New South Wales and of the Intermediate Host *Limnea brazieri* (Smith). *Bull. C.S.I.R.O.* No. 43.
SANGER, E. B., 1883. The Freshwater Shells of Cooper's Creek, Central Australia. *American Nat.*, **17,** *1184—1185.*
SLOANE, T. G., 1915. On the Faunal Subregions of Australia. *Proc. Roy. Soc. Vict.*, (N.S.) **28,** *141—148.*
SPENCER, W. B., 1896. Summary of the Zoological, Botanical and Geological Results of the (Horn) Expedition. *Rept. Horn Exped.*, **1,** *139—199.*
TATE, R., 1887. On the Influence of Physiographic Changes in the Distribution of Life in Australia. *Rept. Australas. Ass. Adv. Sci.*, **1,** *312—325.*
THIELE, J., 1929-'31. *Handbuch der Systematischen Weichtierkunde* (Parts 1, 2 & 3) Gustav Fischer, Jena.
WHITLEY, G. P. 1947. The Fluvifaunulae of Australia with particular reference to Freshwater Fishes in Western Australia. *W. Aust. Nat.*, **1,** *49—53.*

XIV
THE AUSTRALIAN FRESHWATER CRUSTACEA

by

E. F. RIEK

(Division of Entomology, C.S.I.R.O., Canberra)

The Australian freshwater crustacean fauna consists of much the same elements as occur in other continental masses. There is only a small distinctive component consisting of the Syncarida and Phreatoicoidea. Though these two groups are not entirely restricted to Australia they are certainly best developed in this continent.

The Crustacea form an important and often conspicuous component of most freshwater biotas. Not only are the larger and more obvious Decapoda (crayfish, crabs and prawns) represented but also the smaller Amphipoda and Isopoda and often microscopic Entomostraca.

The Entomostraca which comprise Copepoda, Ostracoda, Cladocera and Conchostraca, are very widely and generally distributed and constitute by far the largest component of the Crustacea to be found in freshwaters. Most of the Entomostraca have resting eggs which tide the species over during unfavourable conditions. When conditions become favourable again, such eggs hatch and the individuals grow rapidly to maturity.

Such resting eggs do not occur in the amphipods, isopods or decapods, all of which must depend on other means of survival under unfavourable conditions.

Some of the most striking Entomostraca belong to the Conchostraca which embrace fairy, tadpole and clam-shrimps. This is the only group of Entomostraca without marine representation. They are mostly inhabitants of temporary pools and bodies of water which dry up completely so that resting eggs, which can withstand unusual heat, cold and prolonged dessication, constitute the sole means of tiding the species over from one season to the next. These phyllopods are effectively distributed through their resting eggs which may be blown about in dust or transported by birds and insects. So it is not surprising that the species are widely distributed. For example, it is possible that in Australia there is only a single species of *Apus* and one of *Lepidurus*, the two world-wide genera of tadpole shrimps. Resting eggs of a few species will hatch without drying but in most cases drying is obligatory for hatching. Phyllopods sometimes occur in ponds and lakes which never completely dry up but the active phyllopod population completely disappears for some period and then reappears when water conditions are again suitable following flooding or heavy rain. This would appear to be the case with some species of *Eulimnadia* (clam shrimps, with a carapace resembling a small bivalve shell) in the lakes of the southern interior. The Australian representatives of this group are in no way distinctive though the group as a whole is quite well represented.

The Cladocera or water-fleas bear translucent shells somewhat as in the clam shrimps but the individuals are generally much smaller. Because of their resistant resting eggs Cladocera are easily transported and the great majority are very widely distributed, some being truly cosmopolitan. They exhibit a cycle of abundance in temporary bodies of water reaching a peak as the waters commence to recede after flooding. Some species, such as *Daphne longispina* (O.F.M.) are common throughout the year in limnetic populations in large lakes and reservoirs.

The free-living Copepoda are almost universally distributed in the plankton, benthic and littoral regions of freshwater. Some genera, such as *Cyclops* and its relatives, are world-wide with some species cosmopolitan. In *Boeckella*, on the other hand, most species have a very limited distribution and the genus is almost confined to the southern hemisphere. It is common in Australia. Its place in the northern continents is taken by *Diaptomus*, the species of which also have a limited distribution. This variability in dispersal is easily explained for thick-walled resting eggs are produced only in some species though others have a resistant copepodid cyst stage as an aid to dispersal. The Cyclopidae are well represented in Australia and in the Calanoida there are species of *Diaptomus* as well as of the more typical southern hemisphere *Boeckella*.

The parasitic Copepoda are of rather limited occurrence in fresh waters. The long worm-like *Lernaea*, attached to the general body surface of fish, is the most conspicuous one. The larvae, of normal copepod form, become attached to the gills where they feed without much modification of body structure. Later they attach to the body surface and alter into long worm-like animals. The Australian species of this world-wide genus attacks the golden perch, *Plectoplites ambiguus* RICHARDSON and grows to a size of 12.5 mm. The adults cause ulcer-like lesions on the surface of the fish. A smaller, gill-inhabiting species of the widely distributed genus *Ergasilus* is also present on this fish. There is possibly one endemic species of the greatly flattened *Argulus*, or fish-louse, attacking the smaller fresh-water fish in the coastal zone of south-east Queensland.

The Ostracoda resemble minute mussels. Most of the genera and many of the species are cosmopolitan. Active and passive distribution are brought about much as in the Cladocera and the Copepoda. There is an interesting group of parasitic forms found in the branchial chamber of freshwater crayfish. In Australia they occur on crayfish of the genera *Astacopsis*, *Euastacus* and *Cherax*, from Tasmania to Queensland. Similar, though generically distinct, parasitic ostracods of the Entocytherinae occur on the North American crayfish. The Limnocytherinae, the other subfamily of the Cytheridae with freshwater representatives, is also present in the Australian fauna. The family Cytheridae, however, is predominantly a marine group.

Unlike the Entomostraca, the Amphipoda are not generally adapted to withstand drought and other adverse environmental conditions. The Gammaridae is well represented by the semi-burrowing and often blind and subterranean species of *Neoniphargus*, a genus closely allied to *Gammarus*. *Gammarus* and *Niphargus*, common northern hemisphere genera, are also present. And in West Australia there are the endemic *Uroctena* and *Protocrangonyx*. *Chiltonia* (family Talitridae) and *Pseudomoera* (family Pontogeneidae) are lowland forms occurring in standing water as well as in slow-flowing streams.

With the exception of the Phreatoicoidea there are few Australian freshwater isopods. *Austroargathona caridiphaga* RIEK is a parasitic form from coastal Queensland. During the immature stages it is temporarily attached to freshwater prawns of the families Palaemonidae and Atyidae. There are also one or two small Asellota of the family Janiridae in Victoria and Tasmania.

The Phreatoicoidea is a group of elongate subcylindrical or laterally compressed Isopoda with its headquarters in Australia. They are more typical of the cold mountain streams where they are found burrowing in the loose mud and sand on the bottom or hiding under stones. They occur also in the silt of swampy areas. But they also have a limited distribution in the tropical north of the continent, in the central arid regions and in the temperate south-west of West Australia. They form one of the most distinctive elements of the Australian freshwater crustacean fauna.

The Syncarida is the most distinctive group in the Australian fauna. The only other living syncarids are *Bathynella* and *Parabathynella* of Europe, including England, and Malaya.

Decapods are represented in the Australian freshwaters by crayfish, prawns and crabs. The freshwater crabs (family Potamonidae) are represented by species of the subfamily Gecarcinucinae, which is also widely distributed in south-east Asia. The palaemonid prawns all belong, with the exception of one species of *Palaemonetes*, to the world-wide genus *Macrobrachium* which includes both salt and freshwater species. Crayfish form a distinctive assemblage dealt with in some detail later in this chapter.

Adaptations

Few Australian freshwater Crustacea show special adaptation for survival under dry conditions with the exception of the Entomostraca. In this group as a whole it has been pointed out that the development of a resting egg which resists dessication is a very efficient means of survival under adverse conditions. The crayfish, particularly of the interior, have developed the burrowing habit to a marked degree. These species of the inland and some others seal the mud chimney above the burrow under dry conditions and survive in the artificial waterhole at the base of the burrow till the next rains. Some of the species, such as *Cherax dispar* RIEK of the permanent waters of the coastal streams, are very poor burrowers and are unable to survive an unusual drying out of the stream. The crabs also seal the entrance to their burrows with mud when the waterholes dry out and survive in the burrow till the next wet season.

None of the prawns shows any special adaptation for survival under dry conditions but one species of atyid prawns, *Caridina thermophila* RIEK, is able to survive in the relatively hot waters of an artesian bore. This adaptation is considered a recent development for such hot waters were unknown under natural conditions. The species is in no way modified for a subterranean existance but merely shows a marked temperature tolerance. There is the record of a phreatoicoid, *Phreatomerus latipes* (CHILTON) being "able to thrive in the steaming hot [and slightly salty] water issuing from deep artesian bores" (NICHOLLS, 1943).

Most of the Phreatoicoidea, however, live free on the bottom of cold streams and swamps but "a blind and wholly subterranean species occurs in the Darling Range of West Australia and semi-terrestrial burrowing forms in the Grampians and the wet Beech Forest of the Otways and still wetter West Coast region of Tasmania" (NICHOLLS, 1943). Many of the amphipods are subterranean and are to be found in surface waters only after heavy rains.

The Syncarida have possibly survived in Tasmania through adaptation to cold water conditions. They are now limited in Tasmania to the colder waters of the higher regions and to small swamps and mud-filled trickles on the west coast. At the lower levels they may possibly have been replaced by the atyid prawn, *Paratya tasmaniensis* RIEK. The only described syncarid of the mainland is a swamp-dwelling form living under condition unsuited to atyids and this too may account for its survival.

Some of the crayfish show adaptation towards a terrestrial existance. Spiny crayfish of the genus *Euastacus* wander overland during the summer in the wet rain-forests of the tropical east coast. The crayfish leave the water of the stream on occasion and travel for considerable distances, often as far as the watershed between two streams. At this time of the year the humidity of the environment is very high. This movement takes place during the day though the light intensity in the heavy rainforest is often low. When disturbed while roaming over the forest floor the crayfish emit a loud hissing not unlike that of a scarab beetle. The reason for this movement on land is not known but the species is a general scavenger and individuals may be seeking food which is often scarce following the frequent spates of this period of summer rains. Some species of smooth crayfish of the genus *Cherax* also leave the water and travel overland during the wet season or during heavy rain. In this way freshly dug dams and tanks become stocked in a very short time. This movement seems to take place at night or during dull weather and so is rarely observed.

Economics

Few of the freshwater Crustacea are directly utilised by man but many form important links in the food chain of fishes. This is particularly true of the Entomostraca, especially the Cladocera, which are also utilised by one species of duck. An unusually large ostracod found in some of the shallow lakes of Victoria is eaten directly in great numbers by large trout. At times, the stomachs of these trout may contain thousands of this relatively small crustacean. The Cladocera, which are also present and abundant but smaller, are almost completely ignored.

Crayfish and prawns are an important item in the diet of the larger native fish and of the introduced trout. Only a few of the larger species of crayfish and prawns are used directly as food and then only on a very small scale.

Some Crustacea are of importance because of the damage they cause. Crayfish (*Cherax albidus* CLARK) and crabs (*Liotelphusa leichardti* (MIERS)) burrow in the earth walls of bore drains. This causes seepage which results in reduced flow along the channel. They both cause similar damage to earth tanks and dams. Other large species of crayfish (*Euastacus* spp.) cause damage to the roughly cemented irrigation channels of the Murrumbidgee irrigation area.

Lernaea, the parasitic copepod of the golden perch, a highly regarded endemic food fish, causes ulcer-like lesions on the fish, which may occur in sufficient numbers to cause the death of the host.

Apus, the tadpole-shrimp, could become a pest in rice plantings in the Northern Territory. These shrimps attack the leaves of young plants and, when they are present in numbers, stir up the silt and so interfere with photosynthesis.

Distribution within Australia

Although it has been pointed out that the Australian freshwater crustacean fauna as a whole is similar to that of other continents the distribution of the various components within Australia is most interesting but not surprising considering the development of such a wide range of climatic conditions. The differences between the northern tropical half of the continent and the temperate south are most marked. Freshwater amphipods do not occur in the north but they abound in the colder waters of the highlands of southern Australia and Tasmania. They occur also in the south corner of Western Australia and in the coastal streams of south-east Australia. The Phreatoicoidea have a rather similar distribution but there is some extension of forms into the tropics. The two families of the Phreatoicoidea have very different distributions. The Phreatoicidae occur only in the Bassian region. The family is also represented in New Zealand. The Amphisopidae has a wider distribution occurring from tropical north Australia through the central arid regions to the temperate south-west of West Australia and to the subalpine areas of Victoria and Tasmania in the south-east, but is much more abundant in the southern part of this range. The family is recorded also from South Africa and India. The Syncarida are best developed in Tasmania but occur also in Victoria. *Anaspides tasmaniae* THOMPSON and *Paranaspides lacustris* SMITH occur in the very clear, cold waters of the high plateaux and peaks of Tasmania. *Anaspides* occurs in the south from Mt. Wellington, near Hobart on the east coast, through the inland peaks to the west coast mountains. *Paranaspides* occurs in the littoral zone of the Great Lake and in some of the surrounding small lakes and tarns. Although rather similar subalpine conditions occur in the Australian Alps on the mainland Syncarida are not known from the area. The other Tasmanian species *Micraspides calmani* NICHOLLS occurs in the ooze of shallow sphagnum swamps and in swampy ground in the high rainfall area of the west coast. The only described syncarid from the mainland is the swamp-dwelling *Koonunga* which occurs in regions of low light intensity in some of the swamps in Victoria.

The crayfish fall into two distinct groups. The *Engaeus* group occurs commonly in the northern half of Tasmania and in Victoria with an isolated species in the foothills at the northern end of the Australian Alps and another rather distinct species in the wallum swamps of south-east Queensland. The other group of crayfish is more widely distributed occurring over the whole of the continent from the highest subalpine areas to the driest interior and extends north into the highlands of New Guinea.

The freshwater crabs are distributed over the drier regions of Cape York and the inland of the continent reaching as far south as the Victorian border in the east and to the Kimberley region in the west. The Palaemonid prawns are more common in the north. *Macrobrachium* occurs in the coastal streams of the north and east coasts to almost as far south as Sydney and in the permanent waters of the inland but is unknown from south-west Australia where, however, a single species of *Palaemonetes* is adapted to the slightly brackish lakes of the coastal strip. The Atyid prawns have a distribution rather similar to that of the palaemonid prawns but they do occur in smaller bodies of water and have a wider range in the colder waters of the southern portion of the continent. They occur only at the lower levels in the cold mountain streams of the highlands of south-east Australia and in Tasmania they are restricted to the coastal streams. The genus *Caridina* occurs only in the north. It is common in the coastal streams of Queensland. *Paratya* is almost restricted to the southern half of the continent and to Tasmania but it does extend north into part of Queensland. It is absent from south-west Australia. The single species of *Atya* has a very sporadic distribution in the fast-flowing waters of the east coast streams.

The Entomostraca are rather generally distributed but fairy shrimps and tadpole shrimps of the genus *Apus* are typical of the dry interior. On rare occasions some of these species of the interior have been recorded from the east coast zone where the eggs have been blown by strong westerly dust storms. While *Apus* is typical of the drier regions the closely related *Lepidurus* is more widely distributed in the temperate coastal regions and occurs also in Tasmania.

Zoogeographical regions

There are few clearly defined zoogeographical regions based on a study of the freshwater Crustacea.

The most distinctive and interesting forms occur in the Bassian region of south-east Australia and Tasmania. The Syncarida are known only from this region. The *Engaeus* group of crayfish are almost restricted to the region. The Phreatoicoidea are most diverse and abundant in these waters. Freshwater amphipods are very common. There is a marked development and diversity in the *Euastacus* group of crayfish which, though widespread, is well adapted to the cold waters of this region.

The inland region has a distinctive fauna but it consists of only those species adapted to the severe climatic conditions prevailing, where surface water is quite restricted for a large part of the year or even for years at a time. Considered separately, few of the species are of special interest. Conchostraca, such as *Apus* and the fairy shrimps, are possibly the most typical elements. In the northern and eastern parts of the region there is the widespread crab, *Liotelphusa leichardti* (MIERS), the equally wide ranging crayfish, *Cherax albidus* CLARK and the prawn *Macrobrachium atactum sobrinum* RIEK. These species are restricted to this dry inland region though in the case of the prawn the form occurring there is not very different from that of the east coast streams.

There is a less clearly defined region in north-east Australia embracing Cape York and the tropical part of the east coast. Here the distinctive elements are the crabs and the atyid prawns of the genus *Caridina*. There is also a distinctive group of crayfish of the warm-water genus *Cherax*. The large species of palaemonid prawns are common in the northern waters but the smaller species of *Macrobrachium* have a wider range down the east coast. Many of the distinctive components of this region, such as the crabs and the large species of prawns, extend through the coastal region of north Australia to north-west Australia.

The south-west corner of Australia does not have quite the distinctive fauna in freshwater Crustacea that one might have expected. It has a well developed amphipod and phreatoicoid fauna with some endemic genera but these are mostly not very different from those occurring in south-east Australia. The crayfish are allied to those of north Australia and Cape York and freshwater prawns are virtually absent though the one genus, *Palaemonetes*, that does occur is not represented elsewhere in Australian freshwaters. However, this is a very wide ranged genus of prawns.

Origin of the Australian freshwater crustacean fauna

The freshwater crustacean fauna is clearly of multiple origin and has developed over a long period.

The Syncarida and Phreatoicoidea are very old components which have persisted, most probably, from the Permian period or even longer. They are more or less restricted now to the southern regions of the continent but were possibly much more widespread in the past.

The Australian Syncarida are closely related to Permian and Carboniferous marine fossils found in Europe and North America so that their entry into Australia could have taken place directly from the sea or as migration over land in a gradual dispersal through freshwater. The only other living Syncarida are *Bathynella* and *Parabathynella* which occur in springs and in the groundwater in Europe and Malaya. However, these forms have a number of quite distinctive characters and their relationships are not clear. The Phraetoicoidea are recorded from Australia, New Zealand, India and South Africa. It is possible that they were widespread in the early Mesozoic for there is a fossil species in the freshwater shales of the Australian Trias.

The crabs show clearly an entry from the north into Cape York and then radiation of one species over most of the inland of the continent. They are closely allied to the species of south-east Asia.

The palaemonid prawns are considered to have colonised the north and north-west very recently from juveniles carried by warm ocean currents as the more common species of this part of the continent are rather wide ranged in tropical regions. The freshwater species of the east coast most probably evolved from an estuarine species rather similar to that occurring there at the present time (*Macrobrachium danae* (HELLER)).

The atyid prawns show a relict element in *Atya striolata* McCULLOCH & McNEILL and possibly in *Paratya* but *Caridina* is considered to be a more recent introduction, spreading from the north in a number of species waves. It has possibly replaced *Paratya* to some extent. *Paratya* is now restricted to the colder southern waters though it does not occur in the coldest waters of the south-east highlands. The endemic *Caridinides* is considered to have evolved from *Caridina*.

The parasitic isopod *Austroargathona* is considered a relatively recent development from some marine corallanid. It was possibly introduced to freshwater in association with the *Macrobrachium* prawn host. Once established in fresh waters it has taken to parasitising atyid prawns as well as those of the family Palaemonidae.

The Australian amphipods of the widely distributed family Gammaridae are restricted to the southern part of the continent to where they would seem to have receded under changing climatic conditions. *Chiltonia* of the family Talitridae is closely allied to the South American *Hyalella* but both are possibly recent developments from marine forms.

In all there is the impression of a recession of many of the older established groups to the southern part of the continent but this is brought about most probably in the great majority of cases by changing climatic conditions rather than by pressure of new introductions from the north.

Some of the crayfish have shown a different reaction to changing climatic conditions. The genus *Euastacus* occurs from Victoria to north Queensland. In Victoria and New South Wales some of the species occur in the low lying reaches of the streams. As one moves north the species are to be found only at higher and higher altitudes so that in north Queensland they are rarely found below the 2,000 foot level. In the past the genus most probably had a more continuous distribution but with change to drier and hotter conditions the species have receded to the cooler, more permanent waters of the higher areas. This isolation has produced a large number of species each with a restricted distribution.

The relationship of the Australian freshwater crayfish to those of other regions is considered in some detail as it indicates how assumed relationships often disappear following detailed taxonomic study of a group.

Crayfish

The Australian freshwater crayfish fall into two very distinct groups. One group is typified by *Engaeus*, the other by *Euastacus* and *Cherax*. The *Engaeus* group consists of the small "land" crayfish. The *Euastacus* group includes both the spiny crayfish and the smooth crayfish or "yabbies".

The *Engaeus* group of genera has characters very similar to those of the South American *Parastacus* and clearly falls within the family Parastacidae. The *Euastacus* group is considered in a separate family, the Euastacidae, at present known from Australia, New Guinea and New Zealand, though it may also embrace the Madagascan *Astacoides*.

Crayfish of both these families have a similar reproductive system which differs from that of the Astacidae of the northern hemisphere.

The resemblance between the *Engaeus* group and the marine family Axiidae is considered to be more than a superficial one. The two families agree in general body conformation, including the shape and mode of action of the first periopods. In the Axiidae the first abdominal somite is reduced in size and in some species the pleura of this somite are very reduced as they are in all the Australian species of Parastacidae s. str. Although all female Axiidae, which I have examined, have at least a reduced pair of pleopods on the first abdominal somite, male specimens of certain species have the first abdominal somite devoid of appendages, as is the case in all specimens, male and female, of Parastacidae. In the Axiidae there is a transition from a condition where the appendages of the first abdominal somite are well developed in both sexes to one in which the male has no appendages and those of the female are very reduced. In the Parastacidae reduction, presumably, has proceeded further, the appendages being lost in both sexes. In the Axiidae the first and second periopods are chelate. The third periopods of a few species are subchelate. In the Parastacidae the first, second and third periopods are always chelate so that in this character the Parastacidae can be considered as forming the end point of the series occuring in the Axiidae. Some Axiidae have the exopodite of the uropod divided by an indistinct suture but in most species the uropods are not divided. In *Engaeus* the exopodite is usually divided but occasionally (in one species) the division is indistinct and in the related genus *Austroastacus* the exopodite is not divided. Both families exhibit variation in this character but the tendency is for the undivided exopodite of the Axiidae to become divided in the Parastacidae.

The branchio-cardiac grooves are more widely separated and are situated on the lateral cephalothorax in Axiidae while in the Parastacidae they are more approximated and are more or less parallel on the upper surface of the occupied cephalothorax. This change is correlated with an increase in the area by the gills in the Parastacidae associated with their freshwater habitat as against the marine habitat of the Axiidae.

A consideration of the above characters leads one to the conclusion that the freshwater family Parastacidae s. str. has been derived from a marine ancestral group not unlike the recent family Axiidae and it is possible that the two families arose from a common ancestral stock. The family Parastacidae s. str. is known to occur only in Australia and South America. The forms occurring in these two continents could have arisen independently by the invasion of freshwater by a marine ancestral stock or there may have been only one invasion and subsequent dispersal. I am inclined to the former view for, I maintain that even to-day some brackish water axiids have only to lose the reduced pleopod of the first abdominal segment and for the subchelate third periopods to become chelate and they would be considered within the Parastacidae.

The distribution of the species of the family in Australia supports this view of independent origins for the Australian and South American forms. In Australia the family is best represented in Victoria where there are many species and several genera. In Tasmania, separated only by the shallow Bass Strait, only species of the dominant genus *Engaeus* are present and these are restricted to the northern half of the island. The Tasmanian species all have the longer abdomen typical of the Axiidae. The only species common to Victoria and Tasmania is of this generalised type. As one progresses away from the Bass Strait area one finds that the species become progressively more unlike axiids in general appearance. The abdomen, for example, becomes greatly reduced. The only exception to this progressive change is the isolated Queensland *Tenuibranchiurus* which again is quite like the axiids. This distribution of all the Australian species with the exception of *Tenuibranchiurus* would be readily understood if one assumed a common marine ancestor for the Victorian and Tasmanian species, an ancestor occurring in the shallow seas of the Bass Strait area. The Tasmanian species are all of the generalised type which may indicate a more recent invasion of freshwater than is the case with the Victorian fauna. It is considered that *Tenuibranchiurus* has arisen from a quite independent marine stock to that which gave rise to the main group of the family in Australia. This is not to say that the marine ancestors could not have been the same or very similar species in both cases. The fact that the southern group has diversified into several genera and species may indicate a much earlier migration from brackish to freshwater conditions than is the case with *Tenuibranchiurus* in the north.

If this possibility of a dual origin of the Australian group is conceded then there is every reason to consider the possibility of a separate migration to freshwater for the South American *Parastacus*. If not, then the evidence from the family Parastacidae suggests dispersal in one direction between South America and Australia or, at most, some land connection in the past between the two but does not infer any association of these two areas with either New Zealand or Madagascar.

The *Euastacus* group is the dominant element in the Australian freshwater crayfish fauna. The Euastacidae are very similar in general appearance to the freshwater Astacidae and less so to the marine Homaridae. The Homaridae differ most from these two groups in retaining the dorsal longitudinal ecdysial suture of the cephalothorax. All can be derived without any great difficulty from a common ancestral form of Astacura most probably with this ecdysial suture and with appendages on the first abdominal somite.

Histriobdellid worms of the genus *Stratiodrilus* occur on most, if not all, genera of the family Euastacidae in Australia. The same genus occurs on *Astacoides* in Madagascar. The genus is not restricted to crayfish but occurs on other decapods (e.g. *Aeglea laevis* (LATREILLE) in South America). Other histriobdellid worms occur on the marine lobsters (Homaridae).

Parasitic ostracods of the subfamily Entocytherinae occur on all the major genera of Euastacidae in Australia. They are present also on the North American crayfish. The family Cytheridae, to which these ostracods belong, is typically a marine group though there is also a small subfamily, Limnocytherinae, found free-living in fresh waters.

It seems most probable that the freshwater crayfish of the family Euastacidae have evolved from an astacuran-stock already parasitised by a histriobdellid and an ostracod which survived the transfer to freshwater. Whether this took place independently in the three landmasses (Australia, New Zealand and Madagascar) or whether there was a single invasion of freshwater and subsequent dispersal it is difficult to say but there is every possibility that the family could have developed in salt water and only then invaded freshwaters subsequently dying out in the sea in much the same way that certain genera of the palaemonid prawns, for example *Macrobrachium*, appear to be doing at the present time.

The temnocephalid flat worms which lead an ectocommensal existence on the external surface or in the gill chambers of freshwater crayfish are often sighted in support of a common origin of the crayfish of the southern continents but members of the Temnocephalida occur on all groups of freshwater Crustacea and less often on turtles and snails. In Australia they are known from crayfish, from both families of prawns and from crabs. They occur not only in Australia, New Zealand and South America but also in India, the Balkan Peninsula and on various South Pacific islands.

The assumed common origin and close affinity of all the crayfish of the southern continents to one another with their parasitic histriobdellid worms and ostracods and commensal temnocephalids is seen, on closer examination, to be poorly based. Further, their present distribution could have been attained without the necessity of land connections between the southern continents.

REFERENCES

HASWELL, W. A., 1901. Notes on the fauna of the gill cavities of freshwater crayfishes. *Rept. Aust. Ass. Adv. Sci.* **8.**

HENRY, M., 1922. A monograph of the freshwater Entomostraca of New South Wales. Part. 1 Cladocera. *Proc. Linn. Soc. N.S.W.* **47,** *26—52.*

HENRY, M., 1922. A monograph of the freshwater Entomostraca of New South Wales. Part 2. Copepoda. *Proc. Linn. Soc. N.S.W* **47,** *551—570.*

HENRY, M., 1923. A monograph of the freshwater Entomostraca of New South Wales. Part 3. Ostracoda. *Proc. Linn. Soc. N.S.W.* **48,** *267—286.*

HENRY, M., 1924. A monograph of the freshwater Entomostraca of New South Wales. Part 4. Phyllopoda. *Proc. Linn. Soc. N.S.W.* **49,** *120—137.*

HOLTHUIS, L. B., 1950. The Decapoda of the Siboga - expedition. Part 10. The Palaemonidae, subfamily Palaemoninae.

MANTON, S. M., 1930. Notes on the habits and feeding mechanisms of Anaspides and Paranaspides (Crustacea, Syncarida). *Proc. zool. Soc. Lond.,* **791—800.**

NICHOLLS, G. E., 1943. The Phreatoicoidea Part. 1. *Pap. Proc. Roy. Soc. Tas. 1942: 1—145.*

NICHOLLS, G. E., 1943. The Phreatoicoidea. Part 2. *Pap. Proc. Roy. Soc. Tas. 1943: 1—157.*

RIEK, E. F., 1951a. The Australian freshwater crabs (Potamonidae). *Rec. Aust. Mus.* **22,** *351—357.*

RIEK, E. F., 1951b. The Australian freshwater prawns of the family Palaemonidae. *Rec. Aust. Mus.* **22,** *358—367.*

RIEK, E. F., 1951c. The freshwater crayfish (Family Parastacidae) of Queensland. *Rec. Aust. Mus.* **22,** *368—388.*

RIEK, E. F., 1953. The Australian freshwater prawns of the family Atyidae. *Rec. Aust. Mus.* **23,** *111—121.*

SHEARD, K., 1937. A catalogue of Australian Gammaridea. *Trans. Roy. Soc. S. Aust.* **61,** *17—29.*

SMITH, G. W., 1909. The freshwater crustacea of Tasmania with remarks on their geographical distribution. *Trans. Linn. Soc. Lond.* 2, **11,** *61—92.*

XV
LA PLACE DE L'AUSTRALIE MEDITERRANEENNE DANS L'ENSEMBLE DES PAYS MEDITERRANEENS DU VIEUX MONDE

(Remarques sur le climat méditerranéen de l'Australie)

par

Louis Emberger

(Université de Montpellier, France)

Le climat de l'Australie a déjà été très bien étudié par les savants australiens (J. Gentilli, G. W. Leeper, J. A. Prescott, les membres du Bureau de Météorologie australien, etc. . . .). Les pages qui suivent apportent le point de vue d'un botaniste ayant étudié pendant trente ans le bassin méditerranéen avec lequel le Sud de l'Australie a tant de ressemblances quant au milieu climatique. Elle n'ont pas la prétention d'être une monographie, même élémentaire.

Remarques Générales

Les climatologistes reconnaissent, en général, en Australie continentale[1] les climats suivants (J. Gentilli):

(1) Un climat des moussons occupant l'extrême N du continent (en gros, Terre d'Arnhem et peninsule du Queensland).

(2) Un climat tropical semi-aride, situé au S. du précédant.

(3) Un climat de la côte orientale septentrionale (Cooktown à Bundaberg) dominé par les alizés.

(4) Un climat subtropical de la côte orientale méridionale (Bundaberg à Gabo Island), caractérisé par l'influence des vents d'W.

(5) Un climat subtropical frais ⎫ dans le SE. de l'Australie;
(6) Un climat froid ⎭ Alpes australiennes.

(7) à l'W. des climats 3 à 5, il y a un climat subtropical sub-humide.

(8) Un climat subtropical à étés secs, dans le SW. du continent dans la presqu'île d'Eyre, et dans la région d'Adelaïde.

(9) Un climat frais à étés secs (région du Cap Northumberland).

(10) Un climat subtropical semi-aride en bordure septentrionale du climat 8 et à l'intérieur du climat 7.

(11) Un climat aride occupant un immense territoire à l'intérieur du continent.

---

[1] Tasmanie exclue.

Ces distinctions climatiques ont été faites d'après les principes classiques de la climatologie, en mettant l'accent sur l'origine causale des climats, déterminée, en premier lieu, par la circulation atmosphérique, c'est-à-dire par des considérations essentiellement d'ordre dynamique.

Le botaniste voit la climatologie autrement. Sa climatologie est statique. Peu lui importe si un climat est dû à un jeu de moussons ou à des alizés, à des masses d'air polaires ou équatoriales. Il veut connaître les facteurs et leur grandeur qui agissent sur les plantes. Il s'intéresse, par exemple, à la masse des précipitations dont les plantes disposent annuellement, à la répartition saisonnière des pluies, à l'existence et à la fréquence ou durée des gelées, etc....

Il s'ensuit qu'il étudie les climats d'un point de vue spécial, en biologiste; il fait de la bioclimatologie.

En examinant, sous cet angle, l'Australie, on constate que le continent est partagé entre 4 grands bioclimats:

(1) Dans la partie septentrionale du continent, au N. d'une ligne jalonnée, en gros, par Gladstone, Cue, Centre-Sud de l'Australie, région entre Brisbane et Newcastle, règne un bioclimat du type tropical, c'est-à-dire à pluviosité concentrée sur les mois les plus chauds de l'année et période de sécheresse nettement accusée coincidant avec l'hiver.

(2) Dans le SE. (région de Sydney jusqu'aux environs du Promontoire Wilson et sur les Alpes australiennes), il y a un climat tempéré, caractérisé par une pluviosité plus ou moins régulièrement répartie sur toutes les saisons.

(3) Dans tout le Sud de l'Australie, depuis les Alpes australiennes, jusqu'au littoral occidental et remontant jusqu'à Gladstone, le climat est du type méditerranéen, c'est-à-dire à pluviosité concentrée sur la saison froide de l'année.

(4) Le centre-sud est occupé par un climat sec, mais très particulier, que nous appelerons simplement climat aride de l'intérieur.

Chacun de ces climats existe sous différentes formes. Celles-ci ne sont pas encore précisées, mais on peut affirmer qu'il existe, par exemple, au moins, un climat tropical australien sec (pays d'Alice Springs) et un climat tropical humide (Innisfail). Le climat du SE. est également diversifié: vers l'intérieur, il devient de plus en plus sec, sans perdre ses caractères essentiels; en montagne, il y a une forme froide.

Remarques sur le climat aride de l'intérieur de l'Australie.

L'originalité du climat aride de l'intérieur réside du fait que la courbe des moyennes de pluviosité annuelle exprime un régime de pluies réparties à peu près également sur toutes les saisons.

Voici, à titre exemple, la courbe de Broken Hill (chiffres arrondis):

	P (mm)	
Décembre	21,5	
Janvier	14,8	Eté: 60,3 mm
Février	24,0	
Mars	14,0	
Avril	16,0	Automne: 53,8 mm
Mai	23,5	
Juin	23,0	
Juillet	18,0	Hiver: 56,8 mm
Août	15,8	
Septembre	16,9	
Octobre	21,0	Printemps: 59,4 mm
Novembre	21,5	
Total		230,3 mm

Mais, si l'on examine la pluviosité année par année, on constate que, suivant les influences qui dominent au cours de l'année envisagée, Broken Hill a un climat tropical aride ou un climat méditerranéen aride. La courbe des pluviosités moyennes annuelles de Broken Hill n'exprime donc pas fidèlement le climat de cette région. Certaines années, le climat y est méditerranéen; d'autres, il est tropical; il s'ensuit que la courbe, a, de ce fait, l'allure régulière que nous connaissons.

Pour les botanistes, le climat aride de l'intérieur de l'Australie a donc une originalité, importante à retenir, car la végétation qui vit sous un tel régime a naturellement une écologie particulière.

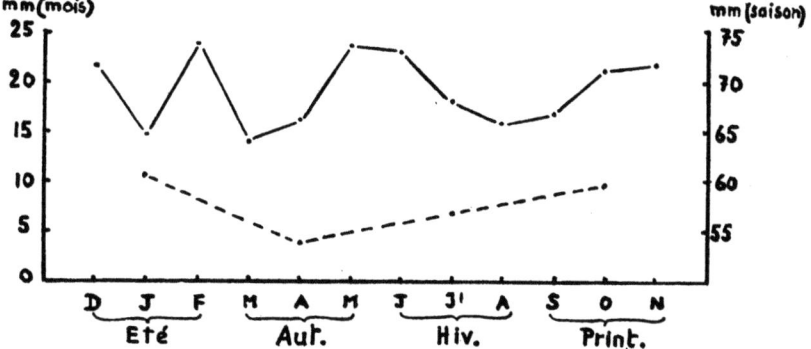

Fig. 1 – Courbes pluviométriques moyennes mensuelle (———) et saisonnière (-----) de Broken Hill (établies avec des documents pris dans Climatic averages Australia – Bureau of Meteorology, Melbourne, 1956).

Pour Broken Hill, les "années méditerranéennes" et "tropicales" doivent plus ou moins s'équilibrer, comme le montre la courbe moyenne mensuelle. Cependant, il y a une légère tendance générale "tropicale" mise en évidence par le graphique des pluies saisonnières (Fig. 1).

En 1956, lors de mon voyage en Australie, j'ai été frappé du nombre d'espèces annuelles méditerranéennes-européennes ou nord-africaines qui peuplaient la région de Broken Hill. J'ai appris que l'année avait été "méditerranéenne" et que durant les années "tropicales", le physionomie serait différente.

A Cook, situé très à l'W. de Broken Hill, sur le 130ème degré de longitude, c'est la tendance méditerranéenne qui se manifeste, quand on examine les pluies moyennes par saison.

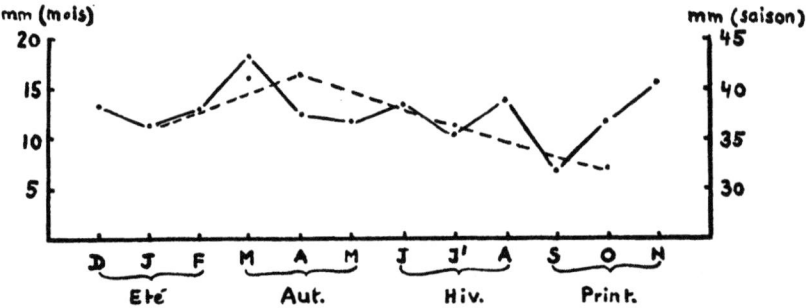

Fig. 2 – Courbes pluviométriques (———) et saisonnière (– – – –) de Cook (établies avec les documents pris dans Climatic averages Australia – Bureau of Meteorology, Melbourne, 1956).

Les répartitions mensuelle et saisonnière des pluies à Cook sont les suivantes (en mm):

Eté	Décembre	13.3	37.6
	Janvier	11.3	
	Février	13.0	
Automne	Mars	18.0	42.3
	Avril	12.5	
	Mai	11.8	
Hiver	Juin	13.3	37.6
	Juillet	10.3	
	Août	14.0	
Printemps	Septembre	6.8	33.8
	Octobre	11.5	
	Novembre	15.5	

Le Bioclimat Méditerranéen et la Végétation Méditerranéenne en Australie

Remarques sur le bioclimat méditerranéen

Le bioclimat du type méditerranéen nous retiendra davantage, car le territoire qu'il occupe fait l'objet du présent exposé.

En effet, il a un climat caractérisé par la concentration des pluies sur les mois froids ou frais, c'est à dire d'Automne, Hiver et Printemps, la saison chaude, l'été, avec parfois un empiètement sur le Printemps et l'Automne, est sec. C'est le climat qui règne tout autour de la Méditerranée.

Ce caractère de localisation saisonnière des pluies est le trait fondamental du climat méditerranéen.

Voici quelques exemples de pluviosités (en mm), prises en Australie méditerranéenne et comparées avec celles des stations situées autour de la Méditerranée (le minimum est souligné).

	Hiver	Printemps	Eté	Automne	Total annuel (mm)
Perth	510	161	34.0	195	900
Adelaide	200	130	79.0	121	530
St. Arnaud (Victoria)	160	119	87.0	106	472
Bugeaud (Algérie)	427	208	35.0	249	919
Marrakech (Maroc)	81	85	12.0	64	242
Monaco (France)	192	177	71.0	316	756
Montpellier (France)	181	192	123.0	258	754
Beyrouth (Liban)	564	158	3.0	176	901
Madrid (Espagne)	105	124	57.0	114	420
Rome (Italie)	253	198	67.0	310	828
Athènes (Grèce)	165	78	36.0	128	407
Kebili (Tunisie)	30	30	1.8	27	88.8

Le climat méditerranéen connait des variantes, suivant que le maximum pluviométrique a lieu en automne, en hiver, ou, parfois, au printemps, suivant l'importance de la saison sèche et les températures durant la saison la plus froide.

Nous ne pouvons pas détailler, ici, ces questions, et nous nous permettons de renvoyer le lecteur à des publications antérieures.

Disons, en un mot, qu'en nous basant sur l'ensemble de nos connaissances de ce climat, dans le monde entier, on peut appeler bioclimat méditerranéen, un climat qui est caractérisé par une saison pluvieuse coïncidant avec les mois froids (Automne à Printemps) de l'année les mois les plus chauds étant secs[1].

Au cours de nos recherches antérieures sur le bioclimat méditerranéen[2] nous avons montré que l'on peut distinguer, à l'intérieur du bioclimat général méditerranéen, un certain nombre de **sous-climats**, et chacun de ceux-ci comprend un certain nombre de **formes**.

Le tableau suivant schématise ces données:

Climat général	Sous-climats	Formes
Climat méditerranéen général	Climat médit. saharien	Forme froide Forme moyenne Forme chaude
	Climat médit. aride	Forme froide Forme moyenne Forme chaude
	Cl. méd. semi-aride	Forme froide Forme moyenne Forme chaude
	Cl. méd. subhumide	Forme froide Forme moyenne Forme chaude
	Cl. méd. humide	Forme froide Forme moyenne Forme chaude
	Cl. méd. perhumide (3)	Forme froide Forme moyenne Forme chaude
	Cl. méd. de haute montagne	Forme océanique Forme continentale

Une subdivision encore plus poussée, et biologiquement importante, fait intervenir, pour chaque **forme**, le **degré d'intensité de la période sèche** et le **régime de la pluviosité**.

Les divers sous-climats ont été définis à l'aide d'un **quotient pluviométrique** exprimant synthétiquement le climat méditerranéen, et de comparaisons des valeurs obtenues avec des observations sur le terrain en Afrique du Nord. Cette

---

[1]) La sécheresse des mois d'été doit être soulignée, car un régime pluviométrique méditerranéen, combiné avec des températures estivales trop basses pour créer une sécheresse ou ayant une pluviosité estivale relativement élevée, n'est pas **écologiquement** méditerranéen, bien qu'il le soit pluviométriquement.
[2]) loc. cit.
[3]) Nos observations récentes en Iran septentrional et en Australie nous ont conduit à distinguer ce sous-climat, dont nous ignorions jusqu'alors l'existence.

région possédant la gamme la plus complète des climats méditerranéens, elle a été prise, pour cette raison, comme étalon.

La limite du bioclimat méditerranéen est déterminée par l'effacement de la période de sècheresse estivale.

Le quotien pluviothermique est établi à l'aide des éléments climatologiques qui influent le plus sur la vie végétale, à savoir la pluviosité, la moyenne des maxima du mois la plus chaud (M) et la moyenne des minima du mois le plus froid (m), ces deux températures encadrant la vie végétale.

Le quotient que nous avons retenu comme exprimant le mieux le degré de sécheresse du climat méditerranéen est:

$$Q = \frac{P}{2\left[\frac{(M+m)}{2}(M-m)\right]} \times 100^{1)}$$

En principe plus Q est petit, plus le climat est sec.

Mais le quotient est, à lui seul, insuffisant. Deux stations peuvent avoir le même. Ballarat (Victoria, et Perth (Australie occ.) ont sensiblement le même quotient: 111,5 pour Ballarat et 112,5 pour Perth, mais à Ballarat, $m = 3.6°$, alors qu'à Perth, m est $= 8.9°$. m est donc un facteur différentiel très important: on en tient compte en utilisant un système d'axes de coordonnées portant sur l'abscisse m et, sur l'ordonnée, le quotien Q.

Les stations rapprochées dans ce graphique sont, enfin, différenciées par l'intensité de la sécheresse de la saison sèche qui résulte de la quantité plus ou moins grande des pluies tombant pendant la période sèche.

		Q	m	Pluviosité de la saison sèche (Eté)
Exemple:	*Ballarat*	111,5	3.6°	135
	Collie	110,5	4°	51

Le degré d'intensité de la sécheresse est très difficile à exprimer. Il n'existe encore aucune méthode qui soit à la fois exacte et pratiquement applicable à grande échelle pour la déterminer, bien que théoriquement le problème soit soluble.

Empiriquement, nous sommes arrivés à admettre qu'est sèche, en pays méditerranéen, une saison au cours de laquelle le rapport entre la pluviosité et la moyenne des maxima du mois le plus chaud (M) de la saison considérée est ≤ 7.

Plus ce rapport est petit, plus intense est la sécheresse.

Voici, à titre d'exemple, les données bioclimatiques d'un certain nombre de stations en application des principes énoncés.

PT = Pluie totale (annuelle).
QA = Quotient pluviothermique annuel.
PH/MH = Rapport entre pluviosité hivernale et moyenne des Maxima du mois le plus chaud de l'hiver.
PPR/MPR = Rapport entre pluviosité printanière et moyenne des Maxima du mois le plus chaud du printemps.

---

[1] En présence de températures négatives, on peut utiliser M et m à partir du 0 absolu ; la formule est alors $\dfrac{P}{\left(\dfrac{M+m}{2}\right)(M-m)} \times 1000$; les calculs sont un peu plus longs.

PE/ME = Rapport entre pluviosité estivale et moyenne des Maxima du mois le plus chaud de l'été.
PA/MA = Rapport entre pluviosité automnale et moyenne des Maxima du mois le plus chaud de l'automne.

	PT (mm)	QA	PH/MH	PPR/MPR	PE/ME	PA/MA
Perth	900	112.5	28.3	6.5	1.2	7.1
Adelaide	530	63.5	12.0	5.1	2.6	4.4
Albany	992	220.0	25.0	11.1	3.8	11.1
Karridale	1191	219	40.0	10.9	2.8	11.7
St.- Arnaud (Australie)	472	58.5	11.6	5.0	3.0	4.0
Bugeaud (Algérie)	919	136.5	54.0	10.7	1.3	10.6
Marrakech (Maroc)	242	23.9	4.2	3.0	0.3	1.9
Monaco	756	126.0	14.5	9.3	2.4	16.2
Montpellier	754	79.0	15 2	8.1	4.0	9.7
Beyrouth (Liban)	900	89	23.3	4.6	0.09	5.4
Kebili (Tunisie)	89	5	1.6	0.9	0.04	0.7

Cette conception du climat méditerranéen peut paraître trop large. Les limites qu'on est conduit à en tracer dépassent, en effet, beaucoup celles des frontières habituellement admises. Mais, elle est basée sur un ensemble complet dont les parties sont solidaires et inséparables. Dans une telle unité naturelle toute coupure serait artificielle, exactement comme les Papaveracées et les Fumariacées ne formant qu'une famille naturelle par enchaînement, dont les membres ne peuvent être séparés sans créer des divisions artificielles.

L'aire méditerranéenne en Australie.

Les bioclimats.

L'étendue donnée à ce climat sur le continent australien repose sur l'analyse des climats de tout le bassin méditerranéen (et même du Nouveau Monde), depuis Tanger à la Mer Caspienne et à l'Indus, incluant les Canaries, Madeire et le Portugal, et toute l'Afrique du Nord.

Sont donc méditerranéennes en Australie, les régions dont le climat comporte le régime pluviométrique spécifiquement méditerranéen à sécheresse estivale suffisamment accusée.

La frontière peut être jalonnée approximativement par les points suivants (Fig. 3):

Région de *Gladstone*, *W. de Cue*, région de *Coolgardie*, *N. de Balladonia* et d'*Eucla*, entre *Cook* et *Ceduna*, entre *Port Augusta* et

Yarramunda; de là, elle contourne le Golfe Spencer et passe légèrement au *N. de Mildura*, jusqu'à *Wagga-Wagga* et *Burrinjuck*, qui sont inclus; la ligne évite *Canberra*, passe dans le pays de *Bendigo-Seymour* et atteint *Ballarat*; de là, *Port Fairy*, qui en fait partie; la

Fig. 3. Les limites des climats méditerranéens en Australie.

pointe *d'Apollo Bay* et le *Wilson's Promotory* sont également méditerranéens météorologiquement. Mais, ces deux dernières stations ne le sont plus écologiquement malgré leur régime pluviométrique, car elles sont trop humides en été.

Les climats méditerranéens suivants existent en Australie (chaque station est suivie de son quotient pluviothermique annuel Q, de m et du rapport estival, les autres rapports saisonniers n'ont pas été indiqués).

Stations du climat méditerranéen aride:

	Q	m	PE/ME
Merredin	27.2	4.5°	1.1
Mildura	25.5	4.7°	1.7
Merbein	24.8	4.0°	1.7
Coolgardie	25.0	5.2°	1.6
Balladonia	25.3	5.0°	1.4

Stations du climat méditerranéen semi-aride:

	Q	m	PE/ME
Wagga-Wagga	52.3	3.2°	3.6
Nhill	45.0	3.7°	2.4
Yongala	41.0	2.3°	2.3
Longerenong	40.4	3.2°	2.4
Wycheproof	40.4	3.0°	2.7
Zara	36.0	3.6°	2.3
Rainbow	36.7	3.8°	2.5
Werai	35.2	3.0°	2.0
York	42.1	5.3°	1.1
Snowtown	41.4	5.3°	1.8
Katanning	40.0	5.5°	1.5
Hay	33.3	3.6°	2.3
Ouyen	31.0	4.3°	2.2
Deniliquin	40.2	4.4°	2.7
Narrogin	58.2	5.0°	1.3
St. Arnaud	58.5	3.7°	3.0
Northam	40.0	5.4°	1.1
Eucla	40.0	6.8°	1.6
Eyre	43.2	6.2°	1.4
Ceduna	36.8	6.5°	1.3
Chapman	42.8	7.4°	0.74
Geraldton	17.4	11°	0.74

Climat méditerranéen subhumide:

Bendigo	62.1	4.1°	3.1
Stawell	65.2	4.3°	3.2
Adelaïde (Ville)	63.5	7.4°	2.6
Adelaïde (Waite Inst.)	90.5	7.4°	3.4
Port Lincoln	78.0	8°	1.9

Climat méditerranéen humide:

Ballarat	111.5	3.6°	5.4
Collie	110.5	4.0°	1.7
Mt. Gambier	120	5.8	3.6
Mt Barker (Sw d'Australie)	120	5.5°	3.1
Donnybrook	119.2	5.3°	1.6
Dwellingup	174.0	4.7°	1.4
Beechworth	133	3.2°	6.2
Bridgetown	99	5.7°	1.7
Busselton	112	7.2°	1.3
Esperance	114	7.5°	2.3
Mandurah	108.7	8.2°	0.9
Perth	112.5	8.9°	1.2
Bunburry	118	8.3°	1.3
Kalamunda	126	8°	1.6
Fremantle	121.5	10.2°	1.1

Stations du climat méditerranéen perhumide:

Pemberton	202	5°	3.0
Albany	220	8°	3.8
Karridale	219	8.1°	2.8

Les climats méditerranéens, en Australie, sont typiquement développés dans le SW. et le S. du continent. Vers l'Est, ils se dégradent en forme de transition, par atténuation de la sécheresse estivale.

(exemples: *Wagga-Wagga, St. Arnaud, Stawell, Ballarat, Beechworth*).

Si l'on inscrit ces stations dans un système d'axes de coordonnées, l'abscisse portant m et, l'ordonnée Q, on constate que les stations y sont d'autant plus rapprochées que les bioclimats sont plus semblables.

C'est ainsi que *Mildura, Merbein, Balladonia* sont rapprochées, de même *Eucla* et *Eyre*, *York* et *Snowtown*, *Albany* et *Karridale*, etc....

L'homologation ou la différenciation des climats peut encore être plus fine, si l'on tient compte non seulement des quotients estivaux *(PE/ME)*, mais encore du quotient pour chaque saison, comme il a été dit plus haut (p. 266). On voit, par exemple, que bien que la résultante climatique générale soit la même pour *Albany* et *Karridale* ($Q = 220$ et 219), et que le climat estival ($PE/ME = 3.8$ et 2.8) ne diffère pas énormément, l'hiver est nettement plus humide à *Albany* qu'à *Karridale*, ce qui peut avoir une certaine importance écologique.

Bioclimats et végétation

Nous croyons que ces considérations bioclimatologiques ont un certain intérêt pour la phytogéographie et les applications de cette science à l'économie.

(1) Climat et végétation étant solidaires, il s'ensuit que l'aire générale de végétation méditerranéenne, unité phytogéographique suprême, se superpose exactement avec celle du climat méditerranéen.

(2) La même solidarité existe entre sous-climats ou formes. C'est ainsi qu'à chacun des 7 sous-climats méditerranéens correspond une unité de végétation subordonnée à l'aire générale. Ces unités sont les étages bioclimatiques de végétation méditerranéenne.

(3) Les types de végétation qui croissent sous des climats identiques sont équivalents, homologues, quelque soit la composition floristique. C'est ainsi que la végétation méditerranéenne de l'Australie est équivalente, homologue, de celle des autres pays du monde où règne le même climat.

En conséquence, l'aire générale de végétation méditerranéenne se compose de 7 étages bioclimatiques de végétation, ce que nous schématisons dans le tableau ci-dessous:

Etage bioclimatique de végétation	Bioclimat	
Aire de végétation méditerranéenne.	Méditerranéen saharien ⇆ Méditerranéen saharien Méditerranéen aride ⇆ Méditerranéen aride Méditerranéen semi-aride ⇆ Méditerranéen semi-aride Méditerranéen subhumide ⇆ Méditerranéen subhumide Méditerranéen humide ⇆ Méditerranéen humide Méditerranéen perhumide ⇆ Méditerranéen perhumide Méditerranéen de haute ⇆ Méditerranéen de haute montagne ⇆ montagne	Climat méditerranéen général.

A chaque forme de sous-climat correspond un sous-étage de végétation. Ainsi, on peut distinguer 3 sous-étages par sous-climat, un sous-étage froid, un sous-étage moyen, un sous-étage chaud, suivant les températures hivernales (voir p. 264).

(4) Les mêmes homologies peuvent être établies entre sous-climats et étages bioclimatiques de végétation, formes climatiques et sous-étages, etc.

(5) Le pays qui possède la gamme complète des climats méditerranéens, possède *ipso facto* la gamme complète des étages de végétation méditerranéens, c'est-à-dire des diverses modalités sous lesquelles la végétation méditerranéenne peut se présenter.

La végétation méditerranéenne australienne

Nous avons vu que l'Australie ne possède pas la gamme complète des climats méditerranéens. Les climats extrêmes, saharien et de haute montagne manquent. Seuls sont représentés les étages de végétation méditerranéenne aride, semi-aride, subhumide, humide et perhumide (voir la liste des stations pp. 267, 268).

De plus, aucun de ces climats distingués n'est représenté par une forme froide; seules les formes moyennes et chaudes y existent. Il en résulte deux conséquences:

(1) L'Australie ne possède pas une végétation méditerranéenne complète. Seuls sont représentés les étages bioclimatiques méditerranéens aride, semi-aride, subhumide, humide et perhumide (sous-étages moyens et chauds), caractérisés chacun par sa végétation propre, qui en est l'expression vivante.

(2) La végétation méditerranéenne australienne ne peut être comparée rationnellement en bloc avec celle des autres pays méditerranéens. On ne peut comparer que les étages, sous-étages... correspondants.

(3) Il est maintenant possible d'assigner à la végétation méditerranéenne australienne la place qu'elle occupe dans un ensemble plus vaste. Nous la comparerons, ici, seulement avec le bassin de notre Méditerranée.

Le tableau suivant, qui montre le répartition des étages de végétation, sera, croyons-nous, suggestif:

Etage bioclimatique de végétation méditerranéenne.

Pays	Saharien	aride	semi-aride	subhumide	humide	perhumide	de haute montagne
Maroc	+	+	+	+	+	+	+
Algérie	+	+	+	+	+	+	traces
Tunisie	+	+	+	+	+	0	0
Tripolitaine-Lybie	+	+	+	traces	0	0	0
Egypte	+	+	+	0	0	0	0
Liban	0	0	+	+	+	?	+
Israël	+	+	+	+	?	0	0
Jordanie	+	+	+	+	0	0	0
Syrie	+	+	+	+	+	0	0
Irak	+	+	+	+	+	?	+
Iran	+	+	+	+	+	+	+
Turquie	0	+	+	+	+	?	+
Balkans (notamment Grèce)	0	0	+	+	+	?	+
Italie	0	0	+	+	+	?	+
France	0	0	traces	+	+	0	0
Espagne	0	+	+	+	+	0	+
Portugal	0	0	+	+	+	+	0
Australie	0	+	+	+	+	+	0

Les homologies suivantes peuvent être établies; nous ne citerons que quelques exemples:

Avec le climat et l'étage de végétation méditerranéen aride d'Australie:

    Orléansville, Relizane (Algérie)
    Marrakech, Taroudant, Guercif (Maroc)
    Gafsa, Gabès, Kairouan (Tunisie) etc.

Avec le climat et l'étage semi-aride d'Australie:

    Constantine, Maillot, Mascare, Oran (Algérie)
    Oujda, Fès. Meknès, Casablanca (Maroc)
    Souk el Arba, Tunis (Tunisie)
    Campo Maior (Portugal)

Avec le climat et l'étage subhumide d'Australie:

    Tlemcen, Alger, Bougie (Algérie)
    Rabat, Mogador (Maroc)
    Bizerte, Beja (Tunisie)
    Lisbonne (Portugal)
    Beyrouth (Liban)

Avec le climat et l'étage humide d'Australie:
>Michelet, Djidjelli (Algérie)
>Tanger (Maroc)
>Ain Draham (Tunisie)
>Monaco (France)
>Coïmbra (Portugal)

Avec le climat de l'étage perhumide d'Australie:
>Ain el Ksar (Algérie)
>Guarda (Portugal)
>Maroc (certains points limités du Rif)
>Algérie (certains points limités)

Ces premières homologations établies, on procède à des distinctions plus subtiles, pour réunir séparément les localités et leur végétation correspondante de chaque sous-étage, ou suivant les quotients saisonniers, ainsi qu'il a été dit.

Dans le bassin méditerranéen occidental, par exemple, le climax de forêt à *Pistacia atlantica* est homologue du climax de l'étage méditerranéen aride d'Australie[1]. Par exemple, de *Mildura, Coolgardie, Balladonia*; les climax de *Tetraclinis articulata, Juniperus phoenicea, Pinus halepensis, Olea-Caratonia*, certains groupements à *Quercus Ilex*... sont homologues de la végétation de l'étage méditerranéen semi-aride australien, par exemple de *York, Snowtown, Narrogin, Werai*; le climax à *Quercus Ilex* ou de *Q. Suber* du Midi de la France, de l'Italie et des basses montagnes de l'Afrique du Nord est équivalent de la végétation de l'étage subhumide d'Australie, par exemple d'*Adelaïde*, de *Bendigo*; le climax de Chênes à feuilles caduques *(Quercus pubescens, Q. lusitanica* var., *Q. Mirbeckii*, certaines forêts de *Quercus Suber*, à *Cedrus atlantica, Abies maroccana*, sont homologues de la végétation de l'étage humide d'Australie, par exemple de *Dwellingup, Perth*; enfin certaines Cédraies *(Cedrus)* et certaines Chênaies *(Quercus Suber, Q. lusitanica, Q. Afaries*, de l'Afrique du Nord ou du Portugal, sont équivalentes de la végétation de l'étage de végétation perhumide australien, par exemple de celle de *Pemberton, Albany* et *Karridale*.

Conclusions

L'Australie méditerranéenne possède une gamme incomplète de climats méditerranéens et, par conséquent, une végétation méditerranéenne également incomplète. Seuls sont représentés les bioclimats méditerranéens aride, semi-aride, subhumide, humide et perhumide.

Manquent les bioclimats méditerranéens extrêmes, saharien et de haute montagne, qui existent autour de la Méditerranée, notam-

---
[1] en dehors des surfaces salées.

ment en Afrique du Nord et au Moyen-Orient. En contre-partie, seuls existent, en Australie méditerranéenne, les types de végétation méditerranéens correspondant aux bioclimats existant et représentant les 5 étages de végétation.

L'Australie méditerranéenne ne possède, semble-t-il, en aucun point des formes froides des divers types bioclimatiques méditerranéens distingués, ce qui diminue encore davantage la gamme des types de végétation présents.

En comparaison avec le bassin de la Méditerranée occidentale, sa méditerranée est donc nettement plus réduite, c'est-à-dire moins diversifiée.

La comparaison détaillée des bioclimats et des végétations des divers pays soumis au climat général méditerranéen permet des homologations intéressantes du point de vue phytogéographique.

Par sa valeur scientifique, sa précision et ses possibilités, la méthode exposée, est, pour les économistes chargés des plans de mise en valeur, une méthode de travail qui a croyons-nous avoir la plus grande importance.

XVI
THE VEGETATION OF WESTERN AUSTRALIA

by

C. A. GARDNER
(Govt. Botanist, Perth, Western Australia)

The State of Western Australia lies between 14 and 35 degrees South latitude. About two thirds lies to the south of the tropic of Capricorn, so that it comes within the tropical and South warm temperate zones. Climatically the greater part of the area comes under the influence of the trade winds, which, blowing from the south east or more correctly the east south east, over the continent-have a desiccating effect over much of the territory under consideration. The northern quarter of Western Australia experiences a submonsoonal summer rainfall alternating with a long period of winter drought, while the southern quarter, especially the south western quarter, experiences a winter rainfall due to the northern movement in winter of the thermal equator, and this in turn alternates with a long period of summer drought. Only in the extreme south west is there any amelioration of these conditions, and here the winter season is more prolonged.

It will thus be readily appreciated that three climatic influences are at work, resulting in three types of vegetation which, from their seasonal incidence of growth, their floristic constitution, and the temperature factor, result in what can be recognised as three phytogeographical provinces, — a summer wet, winter dry Northern Province; a winter wet, summer dry South Western Province; and a third, the Eremean Province, which is an area of low rainfall, receiving however, some marginal benefits from the adjacent provinces, but in its middle part, and especially the eastern portion bears the impress of marked aridity.

The vegetation of the Northern Province is profoundly affected by the large admixture of palaeotropic plants which enter into its composition, especially in those formations in which the soil retains moisture throughout the year namely the fluvial areas. These are mainly of Indo-Melanesian origin; many may have reached this country through the agency of water; others are not easily accounted for, but there remains a strong affinity to those lands which lie to the north. The closest connecting link is the Torres Strait, but there are, notwithstanding, conjunctive elements of a more direct linkage with India and Malaya exhibited by a few plants which are not generally present in tropical Australia. Some of these are difficult to account for, as for example *Adansonia,* found only in North Western

Australia, apart from its general distribution in Africa and Madagascar, and the same link with Madagascar is also evidenced through *Keraudrenia* and *Hibbertia*.

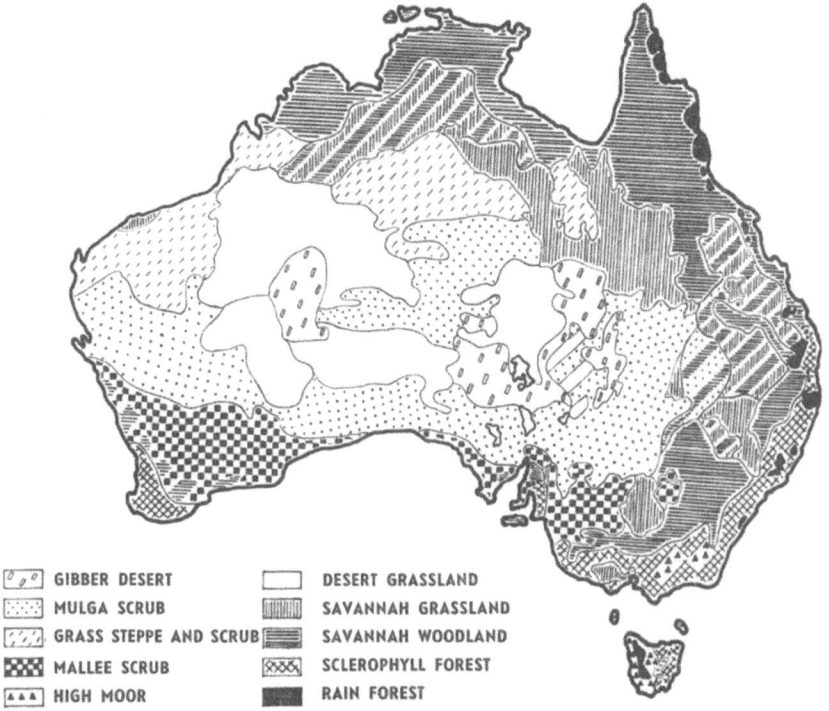

Fig. 1. Major vegetation associations in Australia (simplified) from the map of PRESCOTT (1931), as modified by WOODS (1949). See also Chapter II, pag. 25.

The South Western Province on the other hand, possesses strong examples of the Antarctic Element, typified by Proteaceae connecting Africa (Proteeae) South America (Grevilleoideae) while the Persoonioideae have representatives in Australia and New Zealand as well as Madagascar and South Africa. The South American linkage is also more firmly established through the conjuctive Centrolepidaceae, the Epacridaceae and Stylidiaceae, while the South African relationships are more strongly expressed through the Restionaceae, Liliaceae and Compositae. Three remarkable conjunctive genera linking South America and Australia are *Podocoma*, *Trichocline* (Compositae) and *Selliera* (Goodeniaceae).

The families endemic in Western Australia are Cephalotaceae (S.W. Australia), Byblidaceae (also the Northern Territory and Queensland in one species); and the Tremandraceae with three genera, one of which extends into Eastern Australia *(Tetratheca)*. On the other hand there are groups of a rank less than that of the family which are endemic, and most richly represented in South-Western Australia, e.g. the Xanthorrhoeeae and Calectasieae (Liliaceae), the Conostyleae (Amaryllidaceae), the Chloanthinae (Verbenaceae) the Angiantheae and the Chamaelaucieae (Myrteae) (Compositae), all of which are either confined to South Western Australia or most strongly represented there. The Goodeniaceae and the Stylidiaceae would almost come into this group but for the few plants which occur outside Australia. It is such groups that contribute so much to the Australian Element in the flora. The South West Province is by far the most richly endowed with this element, but it is by no means generally confined to the area. It is true that the Conostyleae are strictly endemic there, and the area confines the local representatives of the Epacridaceae and Restionaceae, but on the other hand, other groups extend into the other provinces, especially the Eremean Province, while there is evidence of a former traffic route connecting the Northern and South Western provinces by way of the Eremean littoral. On the other hand, the presence of *Pilostyles* (Rafflesiaceae) in South Western Australia is one of the most unaccountable features of the vegetation, especially when it relies on *Daviesia* (Podalyrieae) for its host.

Finally from the floristic aspect, an outstanding feature of the flora of the South West Province is the high degree of endemism amongst its constituent species. This is probably near 75% — a figure which compares favourably with that of the flora of the Cape Peninsula in South Africa.

The floristic character of the Eremean Province relies for its distinctiveness on the very high development of the leafless (phyllodineous) species of *Acacia*, and the high proportion of the species of *Eremophila* which are mainly restricted to the province, while the remaining genus of the family *Myoporum* (Myoporaceae) is largely littoral and relies on an physiologically dry soil for its environment. This family must be placed within the palaeotropic influences of the flora, but it is by far more richly represented in the Eremean Province than in any other part of the world.

The fact that with the exception of the south western portion of the South West Province the whole of Western Australia experiences a dry period of more than six months of the year accounts for several peculiarities in its plants. The adaptations to the drought factor are variously expressed: the leafless species of *Acacia* for example, which with a few south western and northern leafy species are a common feature of the landscape; the marked sclerophylly of the woody plants or the reduction to the ericoid form of leaf; the rich branching of most woody species with foliage only in the upper portions which suggest a light demanding habit; complete leaflessness in a number of plants, horizontal placing of the leafy areas, the sympodial form of branching, or, in the ericoid froms the funnel or umbrella-like crown formation which are suitable adaptations for reducing transpiration due to wind action, and the almost complete absence of horizontally placed foliage. All of these suggest a highly developed adaptation to the arid environment. It is only in the south western forest areas and the fluvial formation of the north that we find the mesophytic plants, and these occur as ombrophytes in the forest undergrowth.

A remarkable feature of the flora is to be seen in the adaptation of the plant to fire resistance. That fire has been a factor in moulding the flora is an inescapable fact. There can be no other explanation for the prodigious production of wood in the fruits of many species, particularly in certain Proteaceae, e.g. *Hakea, Mylomelum* and *Banksia*. Such fruits remain in an unopened condition on the plants for many years until they dehisce through the hygroscopicity occasioned by the death of the tissues, or through the direct influence of fire. The mechanisms by which the seeds are for a time retained in the follicles by processes on the seeds which cause them to adhere to the carpel wall until ultimately released by air movements, thus ensuring their deposition on unheated soil, or through the presence of the curious interseminal bilamellar plates which attain the same result in *Banksia* and *Dryandra* can leave no doubt as to the validity of this statement. Something of the kind may be seen in the indigenous Cupressaceae *(Callitris* and *Actinostrobus)* in *Casuarina* and a few Myrtaceae. Again the high proportion of "hard seeds" in the indigenous Leguminoseae serve the same purpose with less effort, and the writer maintains that the mallee habit in certain species of *Eucalyptus* is but another expression of the same adaptation to the influence of fire. It is well known that these mallees are never killed by fire, and that many of them produce their flowers and fruits in the second or third year after being burned.

In conformity with the climatic conditions the formations of the three provinces can broadly be classed as hereunder:

Northern Province: In the areas which receive adequate rainfall we find a type of monsoon woodland in which a large number of deciduous trees occur, especially *Terminalia, Brachychiton, Gossampinus, Cochlospermum* and others, often mixed with small palm woodlands. The river courses are fringed with a type of gallery forest which owes its existence to telluric water, and here are such character plants as several Rubiaceae — *Nauclea, Timonius, Randia*, large trees of *Ficus*, and others. Apart from this the country is principally savannah or savannah woodland, the conditions determining which appearing to be edaphic rather than climatic. For example the open grass country is most frequently that in which the soil is a rich loam often derived from the weathered limestone rocks, while in the woodland formations the soils are those derived from sandstones or quartzites, or from basalt. The former carry a greater variety of arborescent species than the latter which, usually occupying undulating country relatively impervious to water, is consequently much restricted in the number of species. In such country the predominence of *Eucalyptus tectifica* is characteristic, whilst *Themeda* constitutes the principal grass covering. The northern Kimberley Plateau is deeply eroded into gorges which support an interesting rich and varied flora, while the sandstone plateau by reason of its light soil retains little water, and its vegetation apart from a varied tree growth in which *Eucalyptus, Terminalia, Gardenia* etc. play a prominent role, is dominated by *Triodia* as a ground covering, thus emphasising the drought factor in the area, for *Triodia* is one of the character plants of the sandy desert soils.

The palaeotropic mangroves fringe the coast, the mangrove forests are found in the larger estuaries, often of considerable extent. In practically all formations *Eucalyptus* is present, but is far less important than in the South West Province. A number of species — particularly *Eucalyptus alba* and *E. brachyandra* are deciduous, totally or in part during the dry season, and *Adansonia Gregorii* with its gouty stems is one of the common feature plants of the sandstone savannahs in the north. In the drier areas, especially towards the coast in the southern parts of the Province we find a shrub and grass formation composed mainly of species of *Acacia*, but towards its northern limits of richer composition. This is known as "Pindan", and its grasses are mainly *Chrysopogon* and *Sorghum*.

Still further south within the Province there is a gradual change in the constituent flora: the palaeotropic element becomes less common, and the Australian Element more assertive. This is particularly true of *Acacia, Grevillea* and *Eucalyptus*. The last of the palaeotropic trees are species of *Bauhinia* and *Owenia* which remain to the latitude of 22 degrees, whilst *Erythrina* continues as far south as 26 degrees. Conditions which provide permanent moisture account for the presence of others such as *Sesbania grandiflora* and *Livistona Alfredii* which rely on moisture and the protection of other vegetation for their existence.

The savannahs of these southern areas still carry *Chrysopogon*, but *Eragrostis* becomes common in the loamy soils, while *Aristida* and *Triodia* replace *Sorghum* in the red sand.

The South West Province is dominated by forest and woodland formations, with intermediate areas of shrub heath which is edaphically controlled. In the lower South West corner of Western Australia under conditions of high seasonal rainfall the karri forest occurs, the trees of *Eucalyptus diversicolor* being amongst the tallest trees of the world. It finds its edaphic requirement in the gneissic soils of the hills, whereas lateritic areas carry *Eucalyptus marginata*, the jarrah. More sandy soil provides colonies of *Eucalyptus calophylla*, while the clay alluvial soils adjacent to streams have their stands of *Eucalyptus rudis* and *E. megacarpa*, the latter in somewhat lighter soils. In the wettest part *Eucalyptus Jacksonii*, the red tingle competes with the karri trees in the dense forest. The undergrowth is richly developed in the storied karri forest, large shrubs of the bipinnate-leaved *Acacia pentadenia* grow with such trees as *Banksia* and *Casuarina decussata*, and *Agonis* is not uncommon. The smaller shrubs are mainly mesophytic, and the ground covering is of still smaller shrubs of many species, amongst which *Boronia, Chorilaena, Dampiera, Leucopogon* and many species of *Lasiopetalum* and *Thomasia* are the principal. The littoral sandy country which separates the forest from the sea carries small trees of the jarrah *(Eucalyptus marginata)*, dwarf *Agonis* and a few other species, and a dense covering of low crowded shrubs forming heaths, while the wet swampy areas are rich in many plants, and these are particularly the environment required for many species of Proteaceae — notably *Banksia*, and the Myrtaceae which here are strongly represented by *Beaufortia* and *Callistemon*.

Northwards of the karri forest, and associated with the lateritic soils is the jarrah forest *(Eucalyptus marginata)* which again, like most of the forest areas consist of almost pure stands of the single predominant tree species, although the presence of lighter sandy soils is always indicated in the forest by the appearance of *Eucalyptus calophylla* the marri, or of clay or granite by the presence of the wandoo tree *(Eucalyptus redunca* var. *elata)*. The undergrowth, except in the wetter southern regions, stands in marked contrast to that of the karri forest because of the sclerophyllous nature of the shrubs, and the more open texture of both trees and shrubs. As we approach the inland margin the forest opens out and gives place to a narrow zone of savannah woodland in which the York gum *(Eucalyptus loxophleba)* is associated with *Acacia acuminata*. This open formation has a ground covering in which the grass flora is comparatively well developed and the shrubby plants are sparse. The predominance of Compositae in this zone as ephermerals lends a very attractive feature to the spring landscape.

The sand heaths are mosaiced into the woodland picture according to the prevailing soil. It is in these heaths that the Australian Element finds its highest expression. Any attempt to describe the bewildering number of small shrubs which constitute the picture, all of which are either ericoid or with broad sclerophyllous foliage, is impossible, the spring aspect of the landscape is also amazing in its kaleidoscopic effects. Here also the species of *Banksia* form an attractive feature, with species ranging from tall to diminutive shrubs embracing the greater number of the fifty endemic species of this genus.

Further inland we encounter the sclerophyllous woodland which is unique in temperate South Western Australia because of the mixture of the arborescent species. The commonest are *Eucalyptus salmonophloia* and *E. salubris*, *E. oleosa* and *E. gracilis* and a few other less common. The texture is open, and the ground flora sparser than in other formations.

The Eremean Province stands between the Northern and South West Provinces, and occupies that portion of Western Australia in which the annual rainfall is below eleven inches. It is thus the province in which the rainfall is less than the other two, and of short seasonal duration. In the north summer rainfall occurs, while in the south winter rainfall is experienced. The middle tract possesses a rainfall of no marked seasonal incidence, relying for its precipitation on extensions of the north summer rainfall (usually from cyclones which usually occur between January and April), while the southern area has its rainfall usually between the latter half of May and the end of September. On the average the wettest months lie between March and July, but it must be stressed that this rainfall is not of marked periodicity.

The province may thus be divided into three distinct zones. The northern zone carries extensive areas of *Triodia* steppe in which *Triodia* and *Plectrachne* are the common grasses, with areas of stunted open savannah in which *Eucalyptus*, *Hakea* and *Acacia* are the principal arborescent plants, while the stony ground carries shrub associations and these are also fairly extensive in the regions near the coast, *Acacia* being the dominant genus. The ground covering is mainly herbaceous, although scattered shrubs occur, and the genera *Ptilotus* and *Gomphrena*, together with *Amaranthus*, and many Aizoaceae are frequent. Amongst the shurbs *Hibiscus* and *Cassia* are common. *Triodia* and *Plectrachne* favour the sandy or stony soils, while in the loamy soils *Eragrostis* and *Cymbopogon* with *Sorghum* and *Eriachne* are the principal grasses.

The Southern zone retains many of the features of the sclerophyllous woodland zone of the South West Province, but is characterised by the greater number of species of *Eucalyptus* mixed in the woodland, the more open texture of the shrubs and trees, the total absence of Epacridaceae and Amaryllidaceae (e.g. *Anigosanthos* and *Conostylis*) and the importance of *Eremophila* (Myoporaceae) shrubby Compositae (*Olearia* and *Cratystylis* etc.), *Cassia* and numerous Chenopodiaceae, especially *Atriplex*, *Kochia* and *Bassia*, while the Myrtaceae and Proteaceae, although relatively common in the landscape are not as numerous as in the South West Province, and *Dryandra*, and all but two species of *Banksia* are absent. *Acacia* remains common, but its species are usually small spreading shrubs and characteristic of the loamy soils. *Dodonaea* becomes more important than in the South West. Amongst the species of *Eucalyp-*

tus the trees and shrubby mallees are mainly derived from the Oleosae and Dumosae, the latter in particular possessing many very closely related species.

The middle zone of the Eremean Province is the "Mulga Bush". This formation bears the impress of the thorn formations of tropical and warm temperate Africa, but in the place of *Acacia* species with thorns and bipinnate foliage, the Australian species are almost entirely shrubs or small trees with rich erect branching and usually vertical phyllodes. The principal species are mainly species of the section Juliflorae centring round *Acacia aneura*, and a group with narrow, usually clustered needle-like phyllodes characterised by *Acacia tetragonophylla*. Exceptional in growth form are *Acacia sibirica* which usually has horizontally spreading branches. The phyllodes of the mulga group have a close scurfy indumentum, and are important among the browsing plants of the area. Next in importance to *Acacia* is the genus *Eremophila* which has most of its species in the mulga country. The bewildering variation in growth forms in this genus, varying from the broad-leaved resinous-coated *Eremophila Fraseri* to the pinoid *E. abietina*, the magnificence of flowers of almost every colour, and the enlarging of the calyces of many provide much interest in this remarkably adaptable genus. *Cassia* is common, the species varying from pinnate-leaved species usually with a velvety tomentum (e.g. *C. desolata*), to the leafless forms found in *C. phyllodinia*. Other important genera are *Dodoneae, Solanum, Trichinium, Brachychiton* — the only deciduous tree in Western Australia outside the tropics, — *Callitris, Hibiscus,* and *Grevillea*. *Hakea* is represented mainly by arborescent forms. The number of lianas is greater here than in the South West Province, *Marsdenia, Pentatropis* and *Pandorea* being found in many spots. The ground flora is represented by scattered shrubs which show a preference for the shelter of the mulga bushes, and a number of hardy tussocky grasses, amongst which *Danthonia (D. bipartita) Eragrostis, Stipa* and *Neurachne* are the most common.

With suitable summer rains a large and varied annual grass ephemeral ground flora is brought into being, and there is a corresponding flowering season for many of the shrubs including the species of *Acacia*. On the other hand winter rainfall accounts for an amazing herbaceous wealth in which many species of everlastings (*Helipterum, Helichrysum, Waitzia, Cephalipterum* etc.), dominate the spring landscape with their vivid colours, interspersed with numerous species of Goodeniaceae and the ubiquitous *Erodium cygnorum* which is more at home in this formation than in the South West Province. Adequate winter rainfall promotes flowering in the trees and shrubs if this has not already occurred with the capricious previous summer rainfall or its absence, and it is this remarkable production of both a summer and a winter herbaceous ephemeral vegetation, together with the almost total absence of succulents which is the outstanding characteristic of the mulga country.

Salt pans occur scattered through this and the adjoining sclerophyllous woodland of the South West Province. The marginal flora of these is rich in *Arthrocnemum*, species of *Atriplex, Bassia, Eremophila*, sometimes *Melaleuca*, and *Casuarina cristata*, together with *Gunnia* (Aizoaceae) and less important genera. The succulent *Kochia Atkinsiana* is also a species which can tolerate conditions favoured by *Arthrocnemum* and *Atriplex*. These halophytic associations occupy the old watercourses which no longer find an outlet to the sea, or to large undrained depressions.

In the south east on the Nullarbor Plain, traversed by the Trans Australian Railway is an extensive tract of limestone soil within the Eremean Province. A total absence of trees, a paucity of the larger shrubs, and the predominance of small shrubs mainly belonging to the Chenopodiaceae characterise this area which from the dwarf habit of its constituents may be termed a salsolaceous steppe. The incidence of the rainfall is in general that of the mulga bush under somewhat cooler conditions, and a different edaphic requirement.

Finally mention should be made of the desert . . . The deserts of Western Australia are found towards the eastern margin of the State beyond the mulga and triodia steppe formations. There are no large sandy denuded areas but there are approximately parallel ridges of sand dunes clothed with the hardier elements of the Proteaceae and Myrtaceae, shrubby species of *Ptilotus* and a remarkable development of the genera *Newcastlia* and *Dicrastylis*, — shrubs which flower only infrequently, but well protected by their dense vestiture of interlocked woolly hairs. Stony plains occur with small scattered shrubs, mainly Chenopodiaceae, and the only tree of any size is the desert oak *(Casuarina Decaisneana)* which attains a stature of ten metres, and has pendulous branchlets. Strangely in this country there are scattered specimens of *Xanthorrhoea Thorntonii*, a southern element which has proved hardy enough to invade this hostile region.

XVII
PAST CLIMATIC FLUCTUATIONS AND THEIR INFLUENCE UPON AUSTRALIAN VEGETATION

by

R. L. CROCKER

(Dept. of Botany, University of Sydney)

Introduction

The developmental history of the flora of Australia is very poorly understood. The marked botanical discountinuity with neighbouring regions like New Zealand and New Guinea gives the Australian flora a certain uniqueness. This uniqueness, however, is at the same time contradicted by the close affinities of certain elements with the floras of both nearby and distant land masses. As a result the history of the Australian flora and its relationship to that of other regions has long been a subject of much interest and speculation. In addition to plant geographical relationships, there is the equally fascinating topic of the mechanism and causes of the changes which have taken place in the evolution of the Australian flora and vegetation. Since the early Tertiary there has been a remarkable change in the distribution of plants in Australia. The direction of this change, in the broad sense, must have been largely due to climatic causes because there has been only minor orogeny since the beginning of the Tertiary and this has been almost confined to simple vertical uplifts. These changes in climate and their influence upon the vegetation have not been sufficiently elucidated yet to make anything but an interim appraisal of the present state of knowledge and ideas possible here.

Tertiary and Quaternary Floras – The Paleobotanic Evidence

Lower Tertiary

Tertiary deposits in Australia, if we except the old Murravian Gulf and Bight deposits, were almost entirely terrestrial, and their exact position in the Tertiary sequence is difficult to define. Partly for that reason, and partly because of the limited paleobotanic record, especially in the northern half of the continent, it is not possible to speak of the Tertiary vegetation in great detail. Nevertheless enough is known to write confidently about some of the broader features.

The earlier paleobotanists (e.g. DEANE 1900, 1902, CHAPMAN 1921, 1937), working with macrofossils, chiefly leaf remains, decided that over a very wide area in the early Tertiary there occurred a somewhat broad-leafed mesic vegetation of *Cinnamomum, Laurus, Daphnandra, Nothofagus, Ficus,* admixed, more or less with *Eucalyptus* and *Casuarina* and Proteaceous genera allied to *Banksia, Grevillea, Persoonia,* and *Hakea.*

This picture of a predominantly mesic flora in the early Tertiary has been confirmed in general, but not in particular, by later palynological studies of COOKSON and her associates. These palynological studies have extended the range of distribution of *Nothofagus* (COOKSON, 1954) and many of the other genera recognised by earlier workers. For example, fossil pollens of many species of Proteaceae, most of which are now extinct, have been widely recorded, (COOKSON, 1950). So also have many other dicotyledonous families, including the Casuarinaceae, Santalaceae, Olacaceae and Myrtaceae. In addition many mesic southern gymnosperms like *Dacrydium, Phyllocladus* and *Araucaria* and *Agathis* (COOKSON & DUIGAN, 1951, COOKSON & PIKE, 1953, COOKSON, 1954) were also widely distributed at this time. It has also been postulated that, in a broad sense, there must have been something of a pan Australian flora in the Eocene and Oligocene (CROCKER & WOOD, 1946) — at least across the southern half of the continent.

Particular interest attaches to the Myrtaceae, especially the genus *Eucalyptus* because of its dominance today. But unfortunately the pollen morphology within this family does not permit assured identification of genera (PIKE, 1956) so that it has not been possible for the palynologists to confirm the earlier macrofossil identification of *Eucalyptus* in the Australian Oligocene.

One inference seems very well established in the early Tertiary paleobotanical evidence — the climate must have been humid. Not only was the period more humid, for mesophytic plants grew in some of what are today the most arid parts of Australia, but many of the plants had a more southerly (and more westerly) occurrence than today. For example *Podocarpus*, which occurred widely over southern Australia (South Australia, West Australia, Victoria, Tasmania, New South Wales and Southern Queensland) in the early Tertiary, has 3 representatives today — all confined to the subtropics or tropics of Northern Australia and New Guinea (COOKSON & PIKE, 1953). Another example is *Anacolosidites* (Olacaceae) which occurred in Victoria and South Australia in the Eocene, but which are now extinct in Australia, although having close affinities with living species in New Guinea and Fiji (COOKSON & PIKE, 1954). Still another example is the Cupanieae of the Sapindaceae which are now confined in Australia to the rain forests of the tropics and subtropics. Many other examples of this type of relationship between early Tertiary and present-day distribution could be given, so that the inference also seems entirely reasonable that not only was the climate much more humid in early Tertiary times but it was probably considerably warmer. This means that the paleo-Australian element, Proteaceae, Myrtaceae, Casuarinaceae etc., even though it may have been restricted to special soils, must have had subtropical affinities during the early Tertiary.

Upper Tertiary

Unfortunately knowledge of the flora of the upper Tertiary is much more scanty than for the lower Tertiary. It has been generally agreed by geologists that by mid-Tertiary Australia had been reduced to a very flat continent with widespread peneplanation. Late in the Tertiary the block faulting that elevated the main range systems of today were initiated.

Such paleobotanic evidence as does exist for this period suggests that the more mesic early Tertiary flora persisted with only gradual change into the upper Tertiary. *Eucalyptus*, to gauge by the sections which can be identified by their pollens, becomes more definite and widely distributed, at least in southern and eastern Australia, but many genera closely resembling current sub-tropical rain-forest genera have also been recorded. The pollen evidence suggests that it was here that *Acacia* and the Compositeae first became prominent, and that the genus *Dacrydium* and many of the families of the lower Tertiary, like Casuarinaceae, Haloragaceae, and Proteaceae were still widespread in southern Australia (COOKSON, 1954).

It seems then that the upper Tertiary too was a humid and probably warm period in Australia as a whole. Indeed sub-tropical conditions may have been widespread. This is rather strongly suggested by the widespread occurrence over much of the continent of what has been recognised as an old Upper Tertiary land-surface, on which lateritic or lateritised soils have been developed (PRESCOTT, 1931 and others).

Quaternary

Unfortunately there is virtually no knowledge of the Pleistocene and early Recent flora. Consequently a picture of the important and striking changes which have taken place between the Upper Tertiary and today has at this stage to be largely inferred by the evidence for environmental fluctuation from other sources. The Tertiary flora was apparently a mesic one. The present flora is predominantly a sclerophyllous and xeric one in which many of the earlier mesic elements have been restricted to small especially favoured locations (chiefly along the Eastern fringe of Australia) or have become extinct.

What are the nature of the environmental changes which might have brought this about?

Quaternary Environmental Changes

The Quaternary environmental changes of significance in plant distribution can be considered conveniently as orographic or climatic.

At the beginning of the Quaternary Australia was still a very flat country, despite the fact that locally, as adjacent to the S.E. and S.W. coasts, and in the Mt. Lofty-Flinders Range area in South Australia, the Tertiary peneplanation was modified by the vertical movements initiated about the end of the Miocene (DAVID & BROWNE, 1950). These uplifts are considered to have culminated in the Upper Pliocene or Pleistocene in the so-called Kosciusko period.

As a result of these crustal deformations new erosive cycles were initiated, and new soil parent materials exposed or deposited, and this changing pattern must have resulted in adjustments in the Australian flora. Adjustments would be necessary too as a result of modifications to the climatic pattern as determined by the new mountain ranges themselves. However, neither these edaphic nor orographic climatic effects were likely to have had any great importance over the continent as a whole. They would have been quite inadequate to explain the hiatus between the distribution of Tertiary forms and current distributions. Such a change can only be explained in terms of a major climatic shift (or shifts) on a continent-wide scale. What is the evidence for any such climatic change?

The climatic interpretation resulting from geological studies has been that the Pleistocene generally was much wetter than the present (HILLS, 1939, BROWNE 1945, DAVID & BROWNE, 1950, etc.). As south-east Australia experienced some minor mountain glaciation it is likely that there were also fluctuations in temperature, in harmony with world-wide glacial and interglacial phases of the Pleistocene. Since the Pleistocene there has been a marked decline in rainfall to arid conditions sometime in the early Recent. This is the general picture presented by the geologists, although some (e.g. WHITEHOUSE, 1940) have proposed minor variations to this general theme mainly in regard to the position in time of the arid conditions.

The main evidence for a wet Pleistocene rests in the distribution of alluvial deposits and the fossil evidence for a wide distribution in Australia at that time of giant herbivores, most of which are now extinct. While the principal evidence for relatively recent arid conditions is to be found in the distributions of truncated soils and aeolian sand systems (HILLS, 1939, CROCKER, 1946, DAVID & BROWNE, 1950). Many of these sand systems occur in regions which are now well vegetated (CROCKER, 1941, 1946) and are superimposed upon deposits of the later Pleistocene.

A great deal of interest has attached to the nature and extent of the arid period, largely because of its apparent nearness and its biological and pedalogical significance. The prevailing opinion has been that the dune systems of Australia (excepting the coastal series) belong to the one major period of aridity, although it has been appreciated that within this there may well have been fluctuations and interruptions (BROWNE, 1945, CROCKER, 1946, HILLS, 1939, etc.). Aridity was apparently somewhat greater than at present, but many of the effects of aridity had little to do with absolute aridity in terms of today's climate; they were relative to the wet phase which preceded the onset of dessication.

The exact location in time of the onset of aridity and its duration however is difficult to assess. A great deal of evidence has been produced (HILLS, 1939, BROWNE, 1945, CROCKER, 1946, etc.) to place it about the mid-Recent, and it has been proposed (GILL, 1955) that it was co-incident with the mid-Recent Xerothermic and Climatic Optimum periods of the Northern Hemisphere. KEBLE (1947) has proposed that it is logical to assume, having regard to Australia's latitudinal position, that there has always been a relatively arid zone sandwiched between the tropical northern and a wetter winter-rainfal southern zone. He suggests that the effect of glacial and interglacial conditions in the Pleistocene was to cause a latitudinal shift in these climatic belts along with modifications in their intensities. On theoretical grounds this seems a very reasonable hypothesis. If correct it implies a southerly shift of a central arid belt with each interglacial and a northerly swing corresponding to each of the glacials.

The suggestion of a southward march of deserts with the interglacials, raises the possibility that the sand systems refer not to a single period, but to successive periods. That is, they have had a multiple genesis and an evolution extending right back through the Pleistocene. BUTLER (1956) is inclined to such an interpretation from his studies of the loessial soil parent materials he calls parna in New South Wales. For the time being all that can be said with assurance is that there has been a mid-Recent period of aridity which led to extensive erosion and the build-up of large dune systems. It is likely that in part this involved the re-sorting and elaboration of existing (earlier) sand systems. The exact sequence of these events and their time of occurrence will be satisfactorily understood only after we have a much better knowledge of the Quaternary in Australia, and after sufficient critically chosen organic samples have been submitted to radio-carbon analysis.

Quaternary Climatic Change and the Australian Vegetation

As already explained the elimination and migration and restriction of many mesic components in the Tertiary flora must have been in response to climatic changes. It is likely that the slow changes culminating in the repetitive glacial and interglacial stages produced the main effects. It is apparent however that in this sequence a Quaternary arid phase, or phases, sudden in onset, must have had profound biological implications.

The consequences of aridity as far as the flora are concerned has been discussed by CROCKER & WOOD (1947). They assume only one major arid period. If there were more the result would have been a repetition and intensification of the effects they have discussed. These are briefly outlined as follows.

The onset of aridity was sudden and drastic. This is implied by the fact that the soils became freely exposed to wind erosion and were sufficiently unstable for the building up of the dune systems. Migration and reproduction of the more xeric members of the old flora was apparently not rapid enough for the purposes of maintaining soil cover and stability. Such a sudden change in climate might well have been basically the result of a sudden break-up of an antarctic ice barrier. Whatever its causes, widespread wind erosion of the order required could only have occurred if there was an extensive reduction in vegetative cover and considerable areas of virtually bare ground. This must have meant that an extensive destruction and some migration of the pre-arid flora preceded the initiation of the building or reorganisation of the sand systems. Successful migration was probably most frequently possible where the rainfall gradients were steepest, or where more humid niches were closely at hand. The mountain ranges, water courses and rivers were likely therefore to have been the main refuges of the relic flora during the arid phases. The slow spread of xeric species and the release of climatic pressure since the aridity has resulted in a recolonisation of the large relatively bare areas. The fact that the sand systems have, if we except parts of the Simpson Desert, almost all been completely colonised and stabilised since their formation seems eloquent testimony that some amelioration of climate has occurred. It is not considered however that this amelioration has been very great.

Considering the flatness of Australia and the broadness of the climatic zonation, it is to be expected that during the humid period, or periods, preceding the aridity, many floral elements were widely distributed. The onset of aridity in fragmenting these old populations is likely to have isolated in the refuges species relics with varying evolutionary potentials, and subjected them to varying selective pressures. Under the circumstances the species relics would be expected to diverge. Subsequently, with expansion on release of climatic pressure and opportunity for recombination and hybridization under conditions of low competition stress many of these divergences might be expected to be fixed and preserved within the flora. These effects would have been more complex and intensified if the oscillations and process had occurred many times.

The foregoing, it is believed, explains the main significance of the Quaternary climatic history to the Australian flora, for a history of retraction and expansion would undoubtedly profoundly alter the content, distribution of, and evolutionary directions within the flora. It is however a thesis which for its firm establishment still requires the collection of much more botanical evidence. Good evidence in support of the theory is thought to be provided by the large degree of ecologic or geographic speciation within the Australian flora, by disjunct vicarious pairs, by major species disjunctions and the occurrence of relic species. The genus *Eucalyptus* provides an interesting example. It contains some 522 species and 150 varieties as determined by the taxonomists (BLAKELY, 1955). A great number of these however can hybridize quite freely (PRYOR, 1957). Indeed it is possible that these might reduce in the strict genetic sense (coenospecies — that is, species which are maintained by a genetic barrier) — to as few as 5 or 10!!

Summary

Owing to peneplanation and an apparently humid climate, habitat diversity in Australia must have been limited in the Tertiary, and many plant genera and species apparently ranged widely. Over most of Australia favourable conditions persisted into the Pleistocene. Relatively humid times may indeed have persisted into the early Recent, although there is the possibility that some of the periods equivalent to the Pleistocene interglacials may have been dry and arid. It is impossible to be precise about this at present for the greater part of the Pleistocene in Australia is very poorly understood. At all events, sometime in the late Quaternary severe and sudden aridity, or aridities, profoundly affected the flora resulting in elimination of many units and retraction of others to the more favourable situations. Since that time there has been expansion. It is considered that these expansions and contractions during the Quaternary have left a profound impression upon the Australian flora: they were probably responsible for much species complexity and have led to the development of the current plant communities.

REFERENCES

BLAKELEY, W. F., 1955. A key to the genus *Eucalyptus*. For. Timber Bureau, Canberra.

BROWNE, W. R., 1945. An attempted post-Tertiary Chronology for Australia. *Pres. Add. Linn. Soc. N.S.W.* **70**.

BUTLER, B. E., 1956. Parna - an Aeolian Clay. *Aust. J. Sci.* **18** (5), 1956.

CHAPMAN, F. 1921. A sketch of the Geological History of Australian Plants - the Cainzoric Flora.

CHAPMAN, F., 1937. Descriptions of Tertiary Plant Remains. *Trans. Roy. Soc. S. Aust.* **61,** *1—16.*
COOKSON, I., 1945. Pollen Content of Tertiary Deposits. *Aust. J. Sci. Res. B.* **3** (2) *166—177.*
COOKSON I., 1954. The Occurrence of an older Tertiary Microflora in Western Australia. *Aust. J. Sci.* **17** (1), *37—38.*
COOKSON, I., & DUIGAN, S. L., 1951. Tertiary Araucariaceae from S.E. Australia. *Aust. J. Sci. Res.* **4** (4), *415—449.*
COOKSON, I. & PIKE, M. K., 1953. A Contribution to the Tertiary Occurrence of the Genus Dacrydium in the Australian Region. *Aust. J. Bot.* **1** (13), *474—484.*
COOKSON, I. & PIKE, M. K., 1954. The Fossil Occurrence of *Phyllocladus* and two other Podocarpaceous Types in Australia. *Aust. J. Bot.* **2** (1), *60—68.*
CROCKER, R. L., 1941. Notes on the Geology of South-East Australia with reference to late Climatic History. *Trans. Roy. Soc. S. Aust.* **65,** *103—107.*
CROCKER, R. L., 1946. Post Miocene Climatic and Geologic History and its Significance etc. Bull. 193, C.S.I.R.
CROCKER, R. L. & WOOD, J. G., 1947. Some Historical Influences on the Development of the South Australian Vegetation Communities. *Trans. Roy. Soc. S. Aust.* **71** (1) *91—136.*
DAVID, E. T. & BROWNE, W. R., 1950. The Geology of the Commonwealth of Australia Vol. 1. Edward Arnold, London.
DEANE, H., 1900. Observations on the Tertiary Flora of Australia. *Proc. Linn. Soc. N.S.W.* **25.**
DEANE, H. 1902. *Rec. Geol. Surv. N.S.W.* **1** (1), *14—20, 21—32.*
GILL, E., 1953. The Australian "Arid Period". *Aust. J. Sci.* **17** (6), *204—206.*
HILLS, E., 1939. The Physiography of North-Western Victoria. *Proc. Roy. Soc. Vict.* **51** (2), *297—320.*
KEBLE, R. A., 1947. Notes on the Australian Quaternary Climates and Migration. *Mem. Nat. Mus. Vict.* **15,** *28—80.*
PIKE, M. K., 1956. Pollen Morphology of Myrtaceae from the South-West Pacific Area. *Aust. J. Bot.* **4** (1), *13—53.*
PRYOR, L., 1957. Personal communication.
WHITEHOUSE, F. W., 1940. Studies on the late Geological History of Queensland. *Univ. of Q. Papers, Geology* **2,** *N.S.* 1().

XVIII
THE PHYTOGEOGRAPHY OF AUSTRALIA (IN RELATION TO RADIATION OF EUCALYPTUS, ACACIA, ETC.)

by

J. G. WOOD

(Professor of Botany, University of Adelaide, South Australia)

Introduction

The most characteristic feature of present day Australian climatology is the arid centre surrounded by belts of progressively wetter country. The arid zone, bioclimatically determined, is bounded approximately by the 10 inch annual rainfall isohyet in the south, the 25 inch isohyet in the north, the 15 inch isohyet in New South Wales and the 20 inch isohyet in Queensland; it occupies about four-fifths of the Continent.

The length of the growing season follows a pattern similar to that of rainfall; so do the Zonal Soil Groups, a sequence being Desert Sandhills, Desert Loams, Brown Soils of light texture, Red Brown Earths and some Black Earths, Podsolized Soils and Red Loams. These soils show a broad, though not absolute correlation with the following sequence of Vegetation Formations: Hummock Grassland, Low Layered Woodland, Woodland, Forest and Rain Forest. The sequence is not determined by climate or by length of growing season alone but also by nutrient status of the soils; and the sequence of plant Formations and Zonal Soil Groups is replaced in southern Australia, especially in the Woodland zone by polygenetic soils, with characters not solely dependent on present climate and parent material, and which carry other plant communities, mostly those of Mallee and Mallee-Heath. Descriptions and maps illustrating the approximate limits of the Australian plant formations have been published by WOOD (1949) WILLIAMS (1955) and WOOD & WILLIAMS (1958).

The vegetation of present day Australia is a mosaic of types. Rain Forest, now restricted, was once probably widespread in the early Tertiary. Floristically the northern rain forests and their allies are clearly akin to those of Malaysia, but are very old members of the Australian flora. They are not considered here but only the strongly developed element of the flora which gives a characteristic landscape to most of Australia and which is mainly restricted to it. The rain forest types have been discussed by HERBERT (1950).

The woodlands and forests are dominated by various species of

Eucalyptus. Of the 600 species of *Eucalyptus* listed by BLAKELY (1934) not more than a dozen or two occur within the arid zone, either along watercourses or in specialized habitats, and none occurs in Rain Forests.

The typically Australian phyllodineous section of the genus *Acacia*, like *Eucalyptus*, is barely represented in Rain Forest; many species occur in association with *Eucalyptus* in forest and woodland, but in addition, and unlike *Eucalyptus*, some species are widespread in arid Australia and are dominants in various arid woodlands or scrubs (e.g. *Acacia aneura* (mulga), *A. sowdenii* (myall), *A. cambagei* (gidea) as well as other species in sandhill scrubs.

Besides *Eucalyptus* and *Acacia* there are ten genera which possess the greatest number of species in Australia and which are usually referred to as "typically Australian genera". They are the genera *Grevillea, Styphelia, Melaleuca, Candollea, Goodenia, Hakea, Hibbertia, Pultenaea, Eremophila* and *Schoenus*. Of these, *Eremophila*, like some species of *Acacia*, is important in the vegetation of rocky hills and scrubs in arid Australia. The remaining genera, as pointed out by HERBERT (1935) are essentially forest genera, associated with *Eucalyptus* species as dominants. Furthermore, most of these are sclerophyllous shrubs or undershrubs in habit.

At the present day sclerophyllous shrubs belonging to these genera occur in dense assemblages in the higher rainfall areas of southern and south-eastern Australia either alone in a heath formation, as an understory to mallee in mallee-heath formation, to woodland trees in dry sclerophyll woodland or to forest trees in sclerophyll forest formations. The soils in which they occur are acid to neutral in reaction, are consistently low in phosphorus and nitrogen (WOOD, 1939; SPECHT & PERRY, 1948; SPECHT, 1951; BEADLE, 1953 and 1954; SPECHT & RAYSON, 1957) and often in the micronutrient elements copper, zinc and molybdenum. They are formed wherever the parent rock is initially low in plant nutrients e.g. from sandstones and quartzites, on sandplains derived from redistribution of the A horizon of fossil soils or dunes and wherever the climate has caused a considerable degree of leaching of nutrients from developing soil. Furthermore, throughout Australia in high rainfall areas with long growing periods, sharp boundaries are frequently seen between sclerophyllous forests on soils, low in plant nutrients, derived from siliceous rocks etc. and woodland with a herbaceous understorey occurring on more fertile soil derived from argillaceous rocks. (PIDGEON, 1937; CROCKER, 1944; SPECHT & PERRY, 1948; BEADLE, 1954).

It will be clear that the three categories discussed above differ in their present day distribution and habitat requirements. The history of the sclerophyllous shrub genera, of the genus *Eucalyptus* and of the genus *Acacia* will therefore first be traced separately and factors influencing their present day distribution then considered.

Any attempt to describe radiation of species presupposes knowledge of centres from which dispersal occurred; in Australia we start with a dilemma — the fragmentary nature of the Tertiary record makes it impossible to determine centres of origin of the genera.

The Sclerophyllous Shrub Genera

Work described in the chapter of Prof. CROCKER has provided evidence for the widespread occurrence, throughout southern Australia* during the Early Tertiary of mesic *Nothofagus* — coniferous forests, with which were associated various genera of the Proteaceae among which *Banksia* was especially prominent.

Of the typically Australian genera described from the early Tertiary, those of the Proteaceae preponderate; DUIGAN (1950) has reported 75 species of different genera of this family derived from wood, leaf and fossil pollen. The fossil occurrence of other Australian sclerophyllous genera in the early Tertiary are more fragmentary but present (DUIGAN, 1950).

The sclerophyllous Australian genera continued to occur in the Miocene and post-Miocene fossil floras. As will be shown later, so did the genera *Eucalyptus* and *Acacia* both of which showed very great post-Miocene development and the three categories occurred together in plant communities which replaced the *Nothofagus*-coniferous forest throughout Southern Australia.

The Miocene, therefore, was an important period in Australian plant geography — indeed in its results one of the two most important periods. It marks the end of a period of great stability when Australia was apparently reduced to a peneplain (DAVID, 1932) with edaphic conditions relatively uniform, the climate mild and its zonation broad and with no major physical barriers to migration of vegetation. The fossil record, so far as it goes, suggests the occurrence of a pan-Australian flora.

The long period of post-Cretaceous stability was broken in the Miocene by epeirogenic earth movements, which initiated the breakup of the old peneplain, which had their greatest effect in southeastern Australia (the Kosciusko upflift) and caused the uplift of the Great Dividing Range, the Mount Lofty-Flinders Range System and formed the Gulf Region in South Australia. At the same time marine inundation (which had withdrawn by the Pliocene) caused

---

* Throughout the text the term "Southern Australia" refers to the area extending from southern Western Australia, through South Australia to Victoria and extending into southern New South Wales.

large deposits of limestone inland from the Head of the Bight. These changes resulted in great habitat diversity and, as will be shown, profoundly influenced the development of the post-Miocene flora.

HOOKER (1860) first pointed out the present day richness of the flora of south-western Australia in the characteristic Australian families and genera, approximately six-sevenths of which then attained their maximum development in that region, the remaining one-seventh reaching maximum development in south-eastern Australia with none showing great development in the tropics. He suggested that Western Australia was the centrum of the Australian flora from which migration had proceeded. WOOD (1930, 1937) showed that the present day flora of South Australia, which is poor in endemics, was derivative and intermediate in character between that of east and west and he also demonstrated that the Gulf Region of South Australian has acted as a barrier to migration, the area to the west of the Gulfs being richer in species of western origin whilst that to the east of the Gulfs was richer in species of eastern affinities.

CROCKER & WOOD (1947) pointed out that since HOOKER's day no convincing arguments have been advanced to support his suggestion that Western Australia was the centrum of the Australian element, although it is established as a centre of dispersal. The centre of origin of the Australian element, if it had a common one, is unknown. They suggest that the Miocene earth movements and inundations were sufficient to isolate the Australian element of the flora in two widely separated parts of the continent and this isolation has been maintained since the Pliocene, partly by a southward migration of the flora in southern Australia in response to climatic changes which were probably responsible for the disappearance of the more mesic *Nothofagus*-coniferous vegetation except in suitably mountainous areas of south-eastern Australia and partly by the edaphic barrier of the soil type developed on the Miocene limestone. The flora of the south-west was one of acid, sandy soils low in plant nutrients and even if no climatic bar to migration was present (and such was probably the case during the Pleistocene) the limestone soil of the Bight area would have prevented migration. With age and continued isolation an endemic flora developed in both south-western and south-eastern Australia, in large measure on the lateritic soils formed during the Pliocene and Pleistocene.

Additional evidence that the present day Australian sclerophyllous genera developed in a warmer climate than now prevails has been supplied by the work of SPECHT & RAYSON (1957) who determined the growth rhythm of the major components of a heath community in South Australia by measuring the monthly dry weight increment of current terminal shoots of the larger sclerophyllous species. This community consists of sclerophyllous shrubs belonging to the Australian Sections of the families Proteaceae, Leguminoseae,

Epacridaceae, Rutaceae and Casuarinaceae and occurs on very infertile sands in a typically Mediterranean-type climate with cool, wet winters and hot, dry summers. The species all show a period of maximum growth during summer, i.e. from December to March; growth is initiated when the mean air temperature rises above 65° F in December and is inhibited when the mean air temperature falls below 65° F in autumn. Growth occurs at a time of water stress; adequate water resources are usually available in the deep sands until January, thereafter growth depends on water reserves available in the B horizon. It is noteworthy that in coastal Queensland, where there is a marked coincidence of summer rainfall and high air temperatures, growth of heath plants of the same genera, but different species, is vigorous and coincides with growth of heath plants in South Australia. In southern Australia, therefore, growth is not initiated during spring as is the case in analogous sclerophyllous communities oversea (e.g. the chapparal of California and macchia of the Mediterranean region). On the contrary, these typically Australian species exhibit a growth rhythm which is markedly out of phase with the present climatic cycle and strongly suggests an origin in a warmer, wetter climate.

The Genus Eucalyptus

The present day distribution of the genus *Eucalyptus* is wider than that of the genera of sclerophyllous undershrubs of typically Australian character so far considered. As a genus its potential habitat requirements are wide; it occurs as dominant in woodlands, with grassy and herbaceous undergrowth, growing in soils relatively rich in plant nutrients as well as in sclerophyllous forests in soils low in plant nutrients. Furthermore, unlike the sclerophyllous shrubs, species of *Eucalyptus*, mainly restricted to the region concerned, occur as forest or woodland dominants over much of northern tropical Australia. The genus is absent from Rain Forests and occurs only extremely rarely in arid communities. The distribution of individual species of the genus have been mapped by BLAKE (1953) and CARTER (1946).

Eucalyptus is an essentially Australian genus. At the present day, six species belonging to four distinct series are common to both Australia and New Guinea and BLAKE (1953) pointed out that these and the groups to which they belong must have been differentiated before the separation of Australia and New Guinea in late Pleistocene times. One Australian species *(E. alba)* and one non-Australian species (*E. deglupta*, BL.) extends through New Guinea to islands of Indonesia east of WALLACE's Line. No eucalypt occurs in New Zealand or New Caledonia though members of the Metrosidereae do; it has been suggested that genera like *Leptospermum* and *Metros-*

ideros must have arisen whilst Australia and New Zealand were connected — probably not later than the Cretaceous — whilst the Eucalyptineae may have arisen at a later date. However HERBERT (1935) has pointed out that the New Zealand forests were and still are Rain Forests and that on climatic considerations alone *Eucalyptus* species were unlikely to survive.

The scarcity and unreliability of the fossil record makes highly conjectural any attempt to reconstruct the early history of the genus. UNGER (1865) and ETTINGSHAUSEN (1888) referred leaf impressions from Tertiary sources in both Europe and North America to the genus *Eucalyptus* and the latter propounded the concept of a cosmopolitan Tertiary flora. Criticism (DEANE, 1897, MAIDEN, 1922, HERBERT 1929) has suggested strongly that there is no valid evidence that *Eucalyptus* existed outside Australia in Tertiary times.

The Tertiary fossil record within Australia is also meagre. DUIGAN (1950) listed 27 species of *Eucalyptus*, which are supported by descriptions or illustrations, from the Tertiary. These occurred in various sites in South Australia, Tasmania, Victoria, New South Wales and Queensland in beds of from Eocene to Miocene, but chiefly Miocene age. Of the species recorded all but two (from wood fragments) have been described from leaf impressions. These records are generally accepted as valid, though BLAKE (1953) pointed out that some of the leaves resemble those of members of the section *Corymbosae* and that the leaves of this and of some allied groups resemble so closely those of living species of *Angophora*, *Eugenia* and other diverse genera as to render hazardous determinations of single leaves. COOKSON (1954) has described fossil eucalyptoid pollen associated with coniferous sporomorphs and pollen grains of Casuarinaceae, Compositae, Proteaceae, Haloragaceae and ferns from beds of Lower Miocene to Upper Pliocene ages suggesting the occurrence then of mixed conifer-*Eucalyptus* forest and perhaps also more open eucalypt forest. The fragmentary fossil record suggests, like that for the sclerophyllous species, a widespread occurrence of *Eucalyptus* during the early Tertiary, but increasing from the Miocene onwards.

Other suggestions concerning the history of the genus have been based on deductions from its present day distribution. BLAKE (1953) in his work on northern Australian eucalypts pointed out that, unlike southern Australia, in northern Australia there has been relatively little earth movement since the Cretaceous, though lateritization occurred probably during the Pliocene. There have been no serious barriers to migration such as occurred in the south and climatic changes subsequent to the Miocene appear to have affected the habitat chiefly by erosion of the substrate. Old species have had time and opportunity to become established over large areas and BLAKE pointed out that several species are so wide spread over northern Australia as to suggest a definite vegetation region, a view supported by the similar pattern of distribution of several genera of grasses (Andropogonineae) and such genera as *Brachychiton, Ficus, Fimbristylis* and the family Combretaceae. Other *Eucalyptus* species are more restricted. The 42 species recorded belong to 11 series of the genus and geographic distribution and taxonomic relationship did not show any consistent relationship with one another.

BLAKE (1953) discussed diverse views which have been held about the origin of the genus and also the generally accepted view that it arose and developed in the northern part of Australia — views based on distribution of various sections of the genus and on supposedly primitive characteristics. Much of this is conjectural and the notion of the northern Australian centrum cannot be substantiated. If the leaf fossils are accepted as valid the Corymbosae, now found predominantly in northern Australia, occurred in southern Australia and Tasmania during the early Tertiary.

It has been suggested (BURBIDGE, 1952) that one plant form of *Eucalyptus*, the mallee habit, is of late development. In mallees a number of woody stems, carrying a scant canopy of foliage, arise from an underground root-stock. They occur in several sections of the genus, especially the *Dumosae* and *Subulatae* several members of which are important dominants in Mallee Associations in the semi-arid zone of southern Australia. From present day distribution data and from specialized morphological features, BURBIDGE considered that the mallee form represents a secondary development in the genus and that the chief centre of distribution was in the Southern Eremaea region of Western Australia.

The view that some differentiation of the genus must have occurred whilst Australia had a relatively homogeneous topography and climate during the early Tertiary is in harmony with the evidence from all parts of Australia. HERBERT's (1929) view seems the most likely, namely that the genus *Eucalyptus* is the present day development of the fringe of species which remained around the coast after the destruction of the greater number of species in the central part of the continent in response to the drier climatic conditions which destroyed or restricted the *Nothofagus* - coniferous forest and probably also Rain Forest.

The Genus Acacia

This genus contains about 700 species of which about 600 are found in Australia. The fossil record, however, is so fragmentary as to preclude adequate discussion of its history in Australia. DUIGAN (1950) records only one fossil species from a leaf impression from Tasmania and five doubtful determinations from Victoria and

Queensland varying in age from doubtfully Eocene to Miocene. COOKSON (1954) has described two *Acacia* sporomorphs from four Victorian deposits, the earliest being middle-Miocene. She commented on the marked absence of *Acacia* from the large number of early Tertiary beds of *Nothofagus* - conifer forest which she had examined and expressed the opinion that it did not become an "integral part" of the Australian flora until after the Lower Miocene period when it was accompanied by pollen grains of genera not occurring in the early Tertiary sediments. Its association with eucalyptoid pollen in the late Miocene suggests the beginning of the common association of *Acacia* and *Eucalyptus*.

ANDREWS (1914) studied the present day distribution of the two series of phyllodineous *Acacia*'s (the Uninerves and Plurinerves) and suggested that they originated in North Australia and later occupied temperate parts of the continent. No recent surveys of the genus have been made. It is probable that its history has been similar to that of *Eucalyptus* and that the present day genus developed from the fringe of species left after break up of an early pan-Australian distribution with increasing aridity. Some species of *Acacia*, however, are more tolerant of arid conditions than any *Eucalyptus* species and play an important role in arid plant communities.

Distribution During Late Tertiary – Recent

The picture which has emerged from the above discussion indicates widespread distribution during the early Tertiary of typically Australian genera, especially of the Proteaceae but including *Eucalyptus* and probably *Acacia* and probably equally widespread beech forests and rain forests for which the fossil evidence is slight but that from residuals stronger (HERBERT, 1950); isolation of south-eastern and south-western Australia with similar climate during the Miocene, relative climatic isolation of south and north Australia; and subsequent development of endemism in genera in several centres. The mid to late Miocene earth movements resulted in great diversity of habitats and in local climates and resulted in the development of different plant communities, each consisting of species whose potential environmental requirements were approximately equal. There occurred sclerophyll forest communities on soils poor in nutrients and woodlands on more fertile soils, all *Eucalyptus* dominated. In northern Australia however, the diversity of environment was much less.

However the late Tertiary pattern of distribution of vegetation, except in broad outlines, was not that of the present day but has been influenced by changes, especially in southern Australia, which are of recent origin.

CROCKER & WOOD (1947) postulated a mid-Recent period of

aridity based on pedological, geological and palaeo-ecological evidence, a period of climatic stress which formed the major dune and dune-sheet systems of central and southern Australia. It was suggested that a decrease in rainfall, catastrophic from a biological point of view, initiated the aridity, placed stress on the pre-arid flora much of which was destroyed except in more favoured refuges where habitat diversity was greatest (e.g. mountains, rivers, etc.). The impact of such stress, due to prolonged droughts, especially on trees, is seen to occur today in arid Australia (WOOD, 1936). With loss of vegetation the A horizons of soils on southern Australian calcareous dunes were stripped to form sand sheet and dune systems and, under prevailing winds, the present mallee areas received quantities of calcium carbonate as loess. Higher rainfall subsequent to the aridity initiated migration and re-colonization from refuges, the distribution of the early elements being determined, within their climatic tolerances, chiefly by the edaphic environment, much of which was new. The present day plant communities are the result of such re-colonization of large virtually bare areas especially in the drier regions. One result in the arid regions was a blending of Indo-Melanesian elements of the flora from the north with the Australian from the south and of these *Acacia* species play a prominent part as dominants. The southern flora contains essentially a pre-arid element, though with different distributions.

A fact to be stressed is that the Australian plant communities are young and that their distribution within a climatic zone has been determined chiefly by edaphic factors, though other factors including individual dispersal capacity, chance dispersals, location of surviving centres, degrees of biotypical differentiation and barriers have also played parts.

The theory accounts for the discontinuous distribution in southern Australia of some species of *Eucalyptus*, *Acacia* and *Casuarina*, examples of which are given by CROCKER & WOOD (1947). It also accounts for the occurrence of Mallee and Mallee-heath communities in areas of southern Australia where woodland would be expected if the distribution were controlled by climate alone; in these areas instead of the Zonal Soil Type (Red Brown Earth) are found either the specialized calcareous mallee soils developed on a great variety of parent rock or sand sheets, each with their distinctive *Eucalyptus*-dominated plant communities.

Though young, the plant communities are relatively stable. They illustrate a fundamental fact: that the basis underlying ecology is a physiological one, concerned with individual tolerances of species, and that in the case of species growing together in a community the potential environment of the individual species overlaps the potential environment.

As a result of their past history — especially the post-Miocene isolation of centres and the post-aridity colonization — the dominants and the floristic compostition of communities are diverse. For this reason Australian plant communities appear most complex to non-Australian observers. One example will suffice: The Dry Sclerophyll Forests of Australia all look much alike; they occur in similar climatic zones on poor soils; the communities are similar in structure and the facies of the species themselves are similar. They are dominated by *Eucalyptus* species; for example in Western Australia by *E. marginata* and *E. calophylla*, in South Australia by *E. obliqua* and *E. baxteri*, in Victoria by *E. obliqua* and *E. viminalis*, in parts of New South Wales by *E. macrorrhynca* and *E. rossii*, in Queensland by *E. intermedia*, *E. micrantha* and *E. acmenioides*. Further, examination reveals that few of the associated sclerophyllous shrub species occur as characteristic plants in more than one of the communities dominated by the eucalypt named; the plants in the understories belong to the same genera — of *Proteaceae, Leguminoseae, Myrtaceae, Epacridaceae* etc — but are closely allied vicarious species-pairs; examples for different communities have been listed by WOOD (1949). Similar facts hold for other Formations.

Furthermore, species of *Eucalyptus* vary greatly in their environmental tolerances, some are "wides" but many are extremely sensitive to changes in the micro-habitat. SPECHT & PERRY (1948), for example, have shown within a restricted area that the distribution of the *Eucalyptus* species is controlled by moisture relationships within the soil and by nutrient status of the soil. The result is that a slight change in the micro-habitat results in a change not in subordinate strata as is usually the case in Northern Hemisphere communities, but in a change in the dominant *Eucalyptus* species, one species locally replacing another.

Many of the important sclerophyllous species are also sensitive to slight changes in the micro-habitat. RAYSON (1957) has described the individual growth tolerances of plants on a single sand dune in a sclerophyllous heath community and showed the changes in species composition and frequency which occur as a result of essentially unidirectional winds and dune topography which influence amount of rainfall and hours of sunlight in different parts of the dune.

In general the present day distribution of *Eucalyptus, Acacia* and the sclerophyllous genera and their grouping into plant communities is the result of past history associated with often extreme sensitivity to changes in the micro-habitat, the boundaries of which are not usually abrupt but present a varying continuum and the character of the communities depends primarily upon the growth tolerances of the species. The most important work confronting Australian ecologists is the determination of these tolerances.

REFERENCES

ANDREWS, E. C., 1914. The Development and Distribution of the Natural Order Leguminoseae. *Proc. Roy. Soc. N.S.W.* **48**, *337—407*.

BEADLE, N. C. W., 1953. The Edaphic Factor in Plant Ecology. *Ecology* **34**, *426—428*.

BEADLE, N. C. W., 1954. Soil Phosphate and the Delimitation of Plant Communities in Eastern Australia. *Ecology* **35**, *370—375*.

BLAKE, S. T., 1953. Botanical Contributions to the Northern Australia Regional Survey 1. Studies on Northern Australian Species of *Eucalyptus*. *Aust. J. Bot.* **1**, *185—352*.

BLAKELY, W. F., 1934. "A Key to the Eucalypts". Sydney.

BURBIDGE, N. T., 1952. The significance of the Mallee Habit in *Eucalyptus*. *Proc. Roy. Soc. Qld.* **62**, *73—78*.

CARTER, C. E., 1946. The Distribution of the more Important Timber Trees of the Genus *Eucalyptus*. Canberra Forestry Bureau, Atlas No .1.

COOKSON, I. B., 1954. The Cainozoic occurrence of *Acacia* in Australia. *Aust. J. Bot.* **2**, *52—59*.

CROCKER, R. L., 1944. Soil and Vegetation Relationships in the Lower South East of South Australia. *Trans. Roy. Soc. S. Aust.* **68**, *144—172*.

CROCKER, R. L. & WOOD, J. G., 1947. Some Historical Influences on the Development of the South Australian Vegetation Communities and their Bearing on Concepts and Classification in Ecology. *Trans. Roy. Soc. S. Aust.* **71**, *91—136*.

DAVID, T. E., 1932. Explanatory Notes to accompany a New Geological Map of Australia. *Aust. Med. Pub. Co. Sydney*.

DEANE, H., 1897. Presidential Address. *Proc. Roy. Soc. N.S. W.* **21**, *811—854*.

DUIGAN, S. L., 1950. A catalogue of the Australian Tertiary Flora. *Proc. Roy. Soc. Vict.* **63**, *41—56*.

ETTINGSHAUSEN, C. V., 1888. Cosmopolitan Flora of Tertiary Australia. *Mem. Geol. Soc. N.S.W.* Palaeontology. No. 2.

HERBERT, D. A., 1929. The Major Factors in the Present Distribution of the Genus *Eucalyptus*. *Proc. Roy. Soc. Qld.* **40**, *165—193*.

HERBERT, D. A., 1935. The Climatic Sifting of Australian Vegetation. *Rep. Aust. and N. Zeal. Ass. Adv. Sci.* **22**, *349—367*.

HOOKER, J. D., 1860. The Botany of the Antarctic Voyage of H.M. Discovery Ships "Erebus" and "Terror". Introductory Essay to the Flora of Tasmania. Lovell Reeve. London.

MAIDEN, J. H., 1922. A Critical Revision of the Genus *Eucalyptus*. Vol. 6 part 8. Govt. Printer, Sydney.

PIDGEON, I. M., 1937. The Ecology of the Central Coastal Area of New South Wales. 1. The Environment and General Features of the Vegetation. *Proc. Linn. Soc. N.S.W.* **62**, *315—340*.

RAYSON, P., 1957. Dark Island Heath (Ninety Mile Plain, South Australia). 2. The Effects of Microtopography on Climate, Soils and Vegetation. *Aust. J. Bot.* **5**, *86—102*.

SPECHT, R. L., 1951. A Reconnaisance Survey of the Soils and Vegetation of the Hundreds of Tatiara, Wirrega and Stirling, County Buckingham, South Australia. *Trans. Roy. Soc. S. Aust.* **74**, *79—107*.

SPECHT, R. L. & RAYSON, P., 1957. Dark Island Heath (Ninety Mile Plain, South Australia). 1. Definition of the Ecosystem. *Aust. J. Bot.* **5**, *52—85*.

SPECHT, R. L., & PERRY, R. A., 1948. Plant Ecology of Part of the Mt. Lofty Ranges. *Trans. Roy. Soc. S. Aust.* **72**, *91—132*.

UNGER, F., 1865. New Holland in Europe. *J. Bot.* **3**, *39*.

WILLIAMS, R. L., 1955. Vegetation Regions. Commentary accompanying Map. Dept. of National Development. Canberra.

WOOD, J. G., 1930. An Analysis of the Vegetation of Kangaroo Is. and the adjacent Peninsulas. *Trans. Roy. Soc. S. Aust.* **54**, *105—139*.

WOOD, J. G., 1936. Regeneration of the Vegetation of the Koonamore Vegetation Reserve, 1926 to 1936. *Trans. Roy. Soc. S. Aust.* **60,** *96—111.*
WOOD, J. G., 1937. The Vegetation of South Australia. Govt. Printer, Adelaide.
WOOD, J. G., 1939. Ecological Concepts and Nomenclature. *Trans. Roy. Soc. S. Aust.* **63,** *215—223.*
WOOD, J. G., 1949. The Australian Environment. Chap. VI. Vegetation of Australia. 1st. Edition. C.S.I.R.O., Melbourne.
WOOD, J. G. & WILLIAMS, R. J., 1958. The Australian Environment. Chap. VI. Vegetation of Australia. Revised Edition. C.S.I.R.O., Melbourne (in press).

XIX
SOME ASPECTS OF SOIL ECOLOGY

by

G. A. STEWART

(Regional Survey Section, Division of Land Research and Regional Survey, C.S.I.R.O., Canberra, A.C.T.)

In general Australian soils are of low fertility. This is attributable to two major factors:

(a) the geochemical nature of the rocks of the Australian continent — most are apparently low in phosphates and some appear to lack adequate amounts of some of the minor elements.

(b) the strongly weathered nature of most soil materials due to either their formation from sedimentary rocks that contain only small amounts of weatherable minerals, or the long continued leeching of soils of old land surfaces from as far back as mid Tertiary, or their formation from superficial deposits that have passed through several cycles of soil formation. Quaternary vulcanism and Pleistocene glaciation — important sources of relatively unweathered parent materials in other continents — were both of minor extent in Australia. All of these characteristics are related to the low relief and geological stability of the continent.

This low fertility level has imposed limitations on the flora, its most marked expression being the depauperate heathy vegetation of the sandplains of Western Australia and the "wet deserts" of south-eastern South Australia. In the wheat-belt of Western Australia the type of vegetation was found to be a good indication of the fertility of the soil and was used by the Lands Department as a means of classifying land suitable for farming development.

Australia is the most arid of all the continents. Approximately two-thirds of it receives less than 15 in (400 mm.) of rainfall annually and the average rainfall for the continent is only two-thirds of that in America, Europe and Asia, half of that in Africa, and one-third of that in South America.

More than half of Australia is less than 1000 ft (300 m.) above sea level and less than 10% exceeds 2000 ft (600 m.) while the highest point is only 7,300 ft (2,190 m.).

Geologically, Australia is an old and relatively stable land mass with no active volcanoes and only relatively minor land movements since the rather gentle warping and monoclinal folding in the late Tertiary time. Neglecting the minor discontinuous belt of coastal lowlands the continent consists of three broad topographic regions.

The Great Western Plateau, which covers the western half

of the continent, consists essentially of a Precambrian shield which has remained a land mass for much of geological time. The basement of lower Precambrian granite and strongly contorted metamorphic rocks is partially overlain by Upper Precambrian and Lower Palaeozoic sediments and smaller areas of younger sediments, most of which have been protected from appreciable deformation by the rigidity of the old shield. Roughly three-fifths of the plateau is 1000 to 2000 ft (300 to 600 m.) above sea level, the highest peaks in central Australia rising to almost 5000 ft (1,500 m.).

The Central-Eastern Lowlands extend southward from the Gulf of Carpentaria to the Southern Ocean. They consist essentially of gently undulating to flat plains formed on undeformed relatively young rocks — mostly Mezozoic sediments in the north and Tertiary sediments and Quaternary alluvia in the south. They do not exceed 800 ft (240 m.) above sea level except along their eastern margin where they merge gradually into the third topographic zone.

The Eastern Highlands extend along the complete eastern coast of Australia and include Tasmania. The geology is rather complex, but the outstanding feature is the number of well-marked depositional phases and orogenies of the Palaeozoic Era. The last major orogeny was in Upper Palaeozoic time but this was not universal throughout the zone and virtually undeformed rocks ranging in age from Upper Palaeozioc to Cainozoic now overlie the contorted older rocks. Quaternary volcanics are restricted to a number of small areas in the Eastern Highlands, ranging from northern Queensland to western Victoria.

The stability of the Australian continent is in marked contrast to the two nearby land masses, New Guinea and New Zealand, where relatively young sediments have been greatly contorted and post-Tertiary volcanic activity is widespread. This stability had has a marked influence on the nature and distribution of soils in Australia. The major example of this is the widespread peneplained surface, believed to be of mid-Tertiary age, with deeply weathered relict soils generally containing ferruginous concretions. In places the peneplain has been truncated and dissected and a younger cycle of soils has formed on the exposed strongly weathered and unweathered underlying rocks. Evidence of the deeply weathered materials of the Tertiary peneplain are most common on the Great Western Plateau and are least common on the Eastern Highlands and the southern part of the Central Lowlands.

Unlike the parts of northern America and northern Europe where much of the early development of ecology and soil science took place only small areas (at the higher levels of southeastern Australia and Tasmania) underwent glaciation in the Pleistocene. Glaciation has impact on soils in two ways — the removal of old soils by moving ice and the provision of large amounts of mechanically pulverised but chemically relatively unaltered materials such as till, moraines and loess, i.e. the provision of an environment in which the soils are relatively young. However, in southern Australia, the glacial period is believed to have profoundly affected landscapes and soils. In the southern part of South Australia (CROCKER, 1946) the lowering of sea level during the ice ages resulted in windpiling of calcareous sandy material from the exposed shore lines into dunes and dune sheets while finer materials were carried inland and deposited as loess. With the onset of soil formation the leaching of carbonates in the calcareous dunes resulted in the formation of sandy siliceous surface horizons overlying cemented calcareous horizons. Subsequent wind erosion during a period of marked aridity in the Early Recent resulted in piling of the sandy topsoil into dunes and sandsheets and exposure of the hardened calcareous travertine layer and further loess deposited inland. The soil pattern has been largely determined by these superficial deposits, which in general are highly weathered, regardless of the nature of the underlying rocks.

Further east in the riverine plain of south-eastern Australia, BUTLER (1956) has found extensive evidence of clayey, calcareous, wind borne deposits that he has called parna as they have much higher contents of clay (apparently transported as aggregates) and much lower content of primary minerals than loess. Several layers of parna have been identified and their origin has been ascribed to wind transport from sand dune areas to the west during periods of aridity in the interglacial periods, interspersed with periods of ecological stability and soil formation during the glacial periods.

A number of soil layers, providing evidence for several periods of stability interspersed with periods of instability, have also been found by WALKER (1957) and VAN DIJK (1957) in the southeast of New South Wales, but parna was not found in this area.

Less is known of the post Tertiary landscape history in other parts of Australia. In central Australia the mid-Tertiary peneplain surface has been largely dissected, some calcareous sediments and alluvia were laid down, but the major characteristics of much of the landscape have been determined by aeolian influences, probably during the arid period of the Early Recent. These produced the extensive sief dune systems of the Simpson and Great Sandy Deserts, the irregular dunes to the southeast of Alice Springs, N.T., and the very extensive riverless sandplains in which wind movement was sufficient to obliterate previous drainage systems but insufficient to form dunes. No evidence is yet known of loess or parna derived from this wind action but further to the northeast, on the Barkly Tableland, calcareous earthy material appears to have been windcarried some 30 miles southward from drying internal drainage basins.

Northward of the 25 in (625 mm.) rainfall isohyet no soil features providing evidence of periods of landscape stability and instability are known except that the mid Tertiary peneplain has been widely dissected and there are at least 2 ages of coastal alluvia that have been correlated with changes in sea level in later Pleistocene and Recent time.

In southern Australia rainwater contains appreciable quantities of sodium chloride, probably picked up as ocean spray by the westerly winds (e.g. TEAKLE, 1937). This cyclic salt is thought to be the source of salt for many of the pans of southern Australia and, where total rainfall and rainfall intensity is low (PRESCOTT, 1931), for the high salt content of many soils. DOWNES (1954) also attributes the occurrence of solonetzic and solodic soils in parts of southeastern Australia to the access of cyclic salts in rain.

In northern Australia, however, the rain must be virtually free of salts as internal drainage basins of the Barkly Tableland that are considered to date back to mid Tertiary have soils that are low in chlorides in their terminal distributory areas.

In some arid parts of southern Australia another factor that probably contributes to high salt contents and solonetzic features is the continual return of salts to the surface through leaf and stem residues of *Atriplex* spp. These contain 15—20% of sodium chloride on a dry weight basis, irrespective of the salt content of the soils in which they grow (BEADLE, WHALLEY & GIBSON, 1957).

In general Australian soils are not well supplied with plant nutrients. On almost all soils crops respond markedly to phosphate fertiliser and some soils give virtually no growth of introduced crops without it.

WILD (1958) has recently surveyed the known data on phosphate levels in Australian soils and confirmed their low content. The contributions of parent material, leaching, and drainage were discussed and the very low phosphate content over much of Australia was attributed to the prolonged leaching and poor drainage on the extensive Tertiary peneplain. This thesis cannot be supported for much of the northern part of the Great Western Plateau where soils from the undissected old peneplain have virtually the same phosphate content as similar soils of younger land surfaces. Excluding soils rich in organic matter the most juvenile soils sampled do not exceed 0.08% P_2O_5 (boiling HCl extraction) and some highly leached sandy soils have less than 0.01% P_2O_5. The phosphate content of the soil parent materials is not known but it would appear that an initial low phosphate content in the parent material has been accentuated by soil leaching. This may also be true for much of southern and eastern Australia as many young alluvial soils show responses to phosphate fertiliser and much more marked responses occur on strongly leached soils such as lateritic podsolic, red and yellow podzolic, and solonised soils.

Soil nitrogen is also low in general. Nitrogen is essentially a part of the soil organic matter which is dependent to a large degree on the associated vegetation. Also, it appears that at least carbon, nitrogen, phosphorus and sulphur tend to occur in constant proportions in organic matter (WILLIAMS & DONALD, 1957) and deficiencies in phosphorus and sulphur may prevent the buildup of organic matter and nitrogen.

Potash deficiency is very uncommon in virgin soils, but responses to this element are occurring more widely as soils of more humid areas are more intensively farmed and adequately fertilised with other elements, e.g. HOSKING (1956).

Australia is renowned for the extreme responses that have been obtained with a wide range of minor elements including copper, zinc, manganese, molybdenum, cobalt, sulphur and boron. These appear to be partly geochemical and partly pedological. For example (STEPHENS, 1951), most of the soils associated with the previously mentioned windsorted sand and windborne loess in southeastern South Australia appear to be deficient in copper and zinc. From this it is implied that the parent materials were apparently low in these elements. However the deficiencies are most marked on highly calcareous or strongly leached soils, indicating pedological accentuation of the deficiencies under those conditions.

Zonality in the classical Russian and American sense in soils, can be developed only where very similar parent materials have been exposed to soil forming processes under good drainage over a similar time under a range of climatic conditions, and these requirements are generally lacking in Australia. The major limitation is inheritance of some features from earlier weathering cycles and as more becomes known of our soils more of them are included in this category of polygenetic soils.

A soil characteristic that appears to be more widely developed in Australia than most other countries is a marked, and in many cases sharply defined, increase in the texture of the subsoil. There is no universal explanation for this — in part it appears to be due to downward leaching of clay under "podzolic" conditions, in part due to downward leaching under the influence of sodium, i.e. solonization, and in part to the deposition of younger, coarser materials over older clayey soils. Another possible explanation is a biotic one — the movement to the surface of the finer soil fractions by soil fauna and their removal by erosion. This factor is believed to be important in some rainforest soils in Nigeria (H. VINE, personal communication). This could also be important in parts of southeastern Australia where some kinds of ants build a rim of clayey subsoil material around the entrance to their hole. Over a long period of time the amount of subsoil brought to the surface must be very considerable yet there is no accumulation of this material at the surface i.e. it is apparently removed by erosion. Northern

308

Australian termites, whose termitaria generally appear to contain much subsoil material, may also produce the same effects.

Soil science is a comparatively young science, having commenced only in the late nineteenth century, and as yet no widely acceptable systems of soil classification and nomenclature have been developed. Much effort is being expended in various lines of attack on these requirements at the present time but at the moment the position can only be described as chaotic and equivalent to the pre-Linnean period of botany. This situation makes it difficult to draw out relationships of the soils of different countries. Also the soil map

Fig. 1. Soil Map of Australia (Simplified from that of PRESCOTT, 1944). See also Chapter II, page 24.

- Stony deserts
- Table lands and ranges
- Desert sandhills and sandplains
- Low country subject to periodic flooding: tidal marshes and deltaic formations
- Desert loams
- Brown soils of light texture
- Grey and Brown soils of heavy texture
- Mallee soils and sandhills, solonetz soils
- Podsols, residual podsols and lateritic sandplain, red loams
- Red-brown earths and terra rossas
- Rendzinas and black earths
- High moor peat

of Australia reproduced here is that of PRESCOTT (1944) and, while it was a major pioneering achievement based on limited field data, extensive field work and changes in soil nomenclature since that time make it difficult to use for general soil geography.

In the following discussion of the major groups of Australian soils and their overseas equivalents the nomenclature follows that of STEPHENS (1956) unless indicated otherwise, and major amendments to the soil map, principally for northern and central Australia, are given.

Podzols and Groundwater Podzols occur in the wetter parts of southern and eastern Australia and are mostly associated with relatively small areas of highly arenaceous parent materials. In Europe and America these soils were thought to be characteristically cool temperate but they occur almost as far north as the Tropic of Capricorn in Australia and in recent years have been identified also in tropical Africa, northern South America and Australian New Guinea.

The major soils of the zone mapped as podzols by PRESCOTT are Red and Yellow Podzolic soils similar to those in eastern United States. In Australian podzol and podzolic soils the A_{00} and A_0 horizons are much less developed under the sclerophyllous or evergreen vegetation than under the deciduous vegetation of northern America and Europe (STEPHENS, 1950). DOWNES (1954) considers that many of the texture-contrast soils in the south-eastern part of the podzol zone are solodic, and solodic and solonetzic soils also occur in the far north of this zone near Townsville.

Terra Rossa soils, similar to those of Mediterranean Europe, occur over small areas, mainly along parts of the coastal fringe in South Australia, on hard calcareous rocks. In some places they occur in complexes with the dark coloured Rendzinas. Apparently Rendzinas form where the ratio between organic matter and free iron oxides is high and Terra Rossa where the ratio is low (STACE, 1956).

Chernozems and the related Prairie and Chestnut soils, the fertile black soils of northern America and Europe, are of very minor extent in Australia and occur under climatic conditions that are distinctly different from those in the Chernozem areas of the northern Hemisphere.

In the past the Black Earths of semi-humid eastern Australia have frequently (e.g. HALLSWORTH & COSTIN, 1950) been correlated with Chernozems but they are now recognised as belonging to the dark coloured cracking clay soils, Grumusols, which are widely known in southern U.S.A., Africa, India and Indonesia. Grumusols also include the Australian Grey and Brown Soils of Heavy Texture which occur extensively in the semi-arid parts of northern and eastern Australia. Similar soils are extensive to the south of Khartoum, Sudan. They are lighter coloured (ranging from grey to olive, brown and red-brown) but otherwise are very similar to Black Earths. As well as the areas mapped by PRESCOTT the Stony Deserts mapped in Queensland are probably mostly Grey and Brown Soils of Heavy Texture with a pavement of stones on the surface.

All of these Grumusols are characterised by a relatively uniform, plastic clay soil that cracks severely on drying. It is believed that "self-ploughing" is a dominating soil-forming process in these soils. The cracks, which represent the shrinking in volume on drying, become partially filled with materials washed in from the surface by rains and as the whole soil mass becomes wet it cannot expand to its former shape and volume and must expand upwards. The repetition of this cycle results in thorough churning and mixing of the soil.

The Red-Brown Earths of the semi-arid to semi-humid parts of southern and eastern Australia are characterised by a brown loamy surface overlying a red to red-brown clay subsoil with lime in its lower part. They are similar to some of the Reddish Chestnut soils of U.S.A., but in general the contrast between surface and subsoil is less marked in Reddish Chestnut soils. In most Red-brown Earths exchangeable sodium is low and clay eluviation from surface to subsoil is attributed to "podzolic" leaching. The red-brown Earths mapped by PRESCOTT in far northern Australia (Cape York Peninsula and the northern part of the Northern Territory) are now classed as Red Earths and Yellow Earths.

As their arid margin in South Australia the Red-brown Earths merge into Desert Loams which are widespread in the central and southeastern arid zone, and normally support saltbush steppe, with many salt accumulating species. Desert Loams have a marked bleached horizon at the base of the sandy or loamy surface horizon and a dark red angular blocky or massive clay subsoil with gypsum and/or lime in the lower parts. They have high exchangeable sodium and high soluble salts in the subsoil but have not developed the domed columnar structure typical of solonetzic soils, possibly due to the aridity of their environment (mean annual rainfall 5 to 10 in.). To date no similar soils are known from other continents.

As acknowledged by PRESCOTT (1944), his Desert Loam mapping unit is a complex one. Soils now recognised as Desert Loams are probably most extensive in the areas mapped as Desert Loam in South Australia and the southern part of that mapped in the Northern Territory, together with the areas of Stony Deserts mapped in South Australia. Grey-brown and Red Calcareous Desert soils occupy virtually all the Nullabor Plain and are a lesser component of other areas mapped as Desert Loam. The areas mapped as Desert Loam in Queensland, New South Wales, Western Australia and to the north of Alice Springs, Northern Territory, are mostly Red Earths.

The Solonised Brown soils of the southern margin of the arid zone are greyish brown to reddish brown sandy soils with a well-marked accumulation of lime and soluble salts in the subsoil. They are now considered to be "salinised" rather than "solonised" and it is proposed (NORTHCOTE, 1956) to call them Mallisols after their typical vegetation — "mallee" eucalypts, which are characterised by marked branching at ground level from an enlarged root. The Mallisols are believed to have formed on wind resorted siliceous sands that received accessions of calcareous loess or parna, and cyclic salt has probably contributed to their salt content. The Red and Grey Desert soils of southwestern United States have much in common with Mallee soils but their parent materials are generally alluvial deposits that still contain appreciable quantities of weatherable minerals.

Grey-brown and Red Calcareous Desert soils occur on calcareous rocks exposed on young land surfaces in the arid zone (extending from 10 in rainfall in the south to 25 in rainfall in the north). They have shallow weakly developed profiles that appear to be the arid equivalent of the Rendzina. Readily soluble salts have been leached out and carbonates have been at least partially leached but otherwise profile development is very weak. No counterpart to these soils is known in overseas countries.

By far the most extensive areas in the arid zone of Australia are occupied by soils having Red Earth profile form. They are characterised by brown A horizons of sandy to loamy texture that merge gradually into redder, generally finer textured, B horizons. They have massive but porous structure throughout and are moderately acid to weakly alkaline. This profile form may overlie a wide variety of substrates including ferruginous, siliceous, calcareous, or alluvial layers, or a wide range of weathered igneous and sedimentary rocks. They occur on land surfaces ranging in age from probable mid Tertiary to young surfaces with relatively immature soils. The latter appear to be in equilibrium with their current environment. At first sight it appears surprising to find moderately leached soil where rainfall is under 12 in (300 mm), but this may partly be attributed to the nature of the rainfall (a high proportion in thunderstorms of relatively high intensity) and, in some sites, to the accession of runoff from neighbouring hills and rock outcrops.

The most extensive soils of this profile form in the arid zone are the sandy soils of Desert Sandplains and Desert Sandhills of PRESCOTT (1944) and STEPHENS (1956), excluding the unfixed crests of the dunes which show virtually no evidence of profile development, i.e. are regosols. As well as the areas mapped by PRESCOTT (1944) they occur in much of the most northerly mapped Desert Loam in the Northern Territory. They are very similar to the Kalahari sands of southwestern Africa, and the stabilised sands on the southern margin of the Sahara in French West Africa. Parts of the Kalahari sands in Angola and extending northward have apparently had a marked change in climate since their deposition as they now support tall, closed woodland communities. They have their counterpart in the tall eucalypt woodlands on sandridges near Derby, Western Australia (rainfall 670 mm.) which is the northwestern tip of the extensive sandhill system of the Great Sandy Desert. In general, however, the sandy soils support tussock perennial grasses (*Triodia* spp. and *Plectrachne* spp.) with varying amounts of leguminous and myrtaceous shrubs.

Finer textured Red Earths, some of which are calcareous, occur around the main mountain ranges in central Australia and also to some extent in sandplains and interdune corridors. The Red and Brown Hardpan soils of Western Australia appear to have a finer textured Red Earth profile over a hardened siliceous layer.

In monsoonal northern Australia there are extensive areas of Red Earth soils. The coarse textured ones have been called Brown Soils of Light Texture by PRESCOTT (1944) and STEPHENS (1956), but areas mapped in this category in eastern Australia also include finer textured Red Earths.

Finer textured Red Earths in northern Australia are classified as Lateritic Red Earths by STEPHENS (1956) and were mapped as Red-brown Earths or included in Tablelands and Ranges by PRESCOTT (1944). They occur on a wide range of parent materials on both old and young land surfaces.

Yellow soils with very similar profile forms (Yellow Podzolic soils — STEWART, 1954) also occur widely in monsoonal Australia and, to a minor extent in the northern part of the arid zone. They are now generally called Yellow Earths.

Both coarse-textured and finer-textured Red Earths occur extensively in the wet-dry parts of eastern and western Africa and fine-textured Red Earths occur in Indonesia and Ceylon.

Lateritic Podzolic soils, which are equivalent to the American Groundwater Laterite, are the major soils of preserved parts of the Tertiary land surface in southern Australia but in northern and eastern Australia the old land surface also has extensive areas of Red Earths and Yellow Earths overlying laterite. PRESCOTT's (1944) Residual Podzols and Lateritic Sandplain was meant to include these old surfaces and there are considerable unmapped areas in the Northern Territory, e.g. on Melville Island, to the south and east of Darwin, and in the vicinity of Daly Waters.

In the 35 to 60 in (870—1600 mm) rainfall parts of northern Australia near Darwin, Lateritic Podzolic soils also occur on post Tertiary land surfaces on granites and arenaceous sedimentary or metamorphic rocks.

Krasnozems, frequently called "red loams", are deep friable red clayey soils, generally formed on more basic igneous and metamorphic rocks. They occur over relatively small areas in the more humid parts of eastern Australia, from Tasmania in the south to northern Queensland. Some of them are equivalent to Low Humic Latosols of Hawaii but overseas equivalents for some of the other soils included in this category, e.g. those with very high exchange capacities, are not known.

Halomorphic soils of solonchak, solonetz and solod forms are known from many parts of Australia, particularly in the south where their occurrence has been linked with the previously mentioned accession of cyclic salts in rainwater.

Skeletal soils of different forms are widely distributed throughout the continent but are most common, in association with rock outcrops, in the Tablelands and Ranges of PRESCOTT (1944).

The Tidal Marshes and Deltaic Formations of PRESCOTT (1944) are predominantly saline alluvial soils with small areas of sand dunes and non-saline alluvial soils, and, near Darwin, extensive areas of cracking clay soils liable to wet season flooding (Wildman Family — STEWART 1954) that appear to be hydromorphic Grumusols.

PRESCOTT (1944) notes that his Low Country subject to Periodic Flooding is a vague geographic unit and its mapping is rather inaccurate. It is meant to include gleyed Meadow Podzolic soils, together with some Yellow Earth, Yellow Podzolic, Soloth and Solodised Solonetz soils.

REFERENCES

BEADLE, N. C. W., WHALLEY, R. D. B. & GIBSON, J. B., 1957. Studies in halophytes. II. Analytical data on the mineral constituents of three species of *Atriplex* and their accompanying soils in Australia. *Ecology* 38, 340—344.
BUTLER, B. E., 1956. Parna - an aeolian clay. *Aust. J. Sci.* 18, 145—151.
CROCKER, R. L., 1946. Post-Miocene climatic and geologic history and its significance in relation to the major soil groups of South Australia. *C. S. I. R. Aust. Bull.* No. 193.
DOWNES, R. G., 1954. Cyclic salt as a dominant factor in the genesis of soils in southeastern Australia. *Aust. J. Agric. Res.* 5, 448—64.
HALLSWORTH, E. G., & COSTIN A. B., 1950. Soil classification. *J. Aust. Inst. Agric. Sci.* 16, (3): 84.
HOSKING, W. J., 1956. What is Potash? *J. Dept. Agric. Vic.* 54, 241—5.
NORTHCOTE, K. H., 1956. The solonised brown (Mallee) soil group of southeastern Australia. Sixième Congrès de la Science du Sol, Paris, 1956, 2, 9—19.
PRESCOTT, J. A., 1931. The soil of Australia in relation to vegetation and climate. *C.S.I.R. Aust. Bull.* No. 52.
PRESCOTT, J. A., 1944. A soil map of Australia. *C. S. I. R. Aust. Bull.* No. 177.
STACE, H. C. T., 1956. Chemical characteristics of Terra Rossas and Rendzinas of South Australia. *J. Soil. Sci.* 7, 280—293.
STEPHENS, C. G., 1950. Comparative morphology and genetic relationships of Australian, North American and European soils. *J. Soil. Sci.* 1, 123—49.

STEPHENS, C. G., 1951. The influence of pedological age and parent material on the nutritional status of Australian soils. Brit. Commonw. Sci. Offic. Conf., Spec. Conf. in Agric., Aust. 1949. Proc, *51—7*.

STEPHENS, C. G., 1956. A manual of Australian soils. 2nd. edition. Comm. Sci. Industr. Res. Org. Melbourne, Australia.

STEWART, G. A., 1954. The soils of monsoonal Australia. *Trans. V. Int. Cong. Soil. Sci. Leopoldville* **4**, *101—8*.

TEAKLE, L. J. H., 1937. The salt (sodium chloride) content of rainwater. *J. Dept. Agric. W. Aust.* **14,** *115*.

VAN DIJK, D. C., 1957. Soil pattern in relation to erosional history on the Southern Tablelands of N.S.W. Second Aust. Conf. in Soil Science, Melbourne, 1957.

WALKER, P. H., 1957. The occurrence of soil layers related to topography at Nowra. Second Aust. Conf. in Soil Science, Melbourne, 1957.

WILD, A., 1958. The phosphate content of Australian soils. *Aust. J. Agric. Res.* **9**, *193—204.*

WILLIAMS, C. H. & DONALD, C. M., 1957. Changes in organic Matter and pH in a podzolic soil as influenced by subterranean clover and superphosphate. *Aust. J. Agric. Res.* **8,** *179—89.*

XX
RECENT STUDIES ON MARSUPIAL ECOLOGY

by

A. R. MAIN, J. W. SHIELD and H. WARING

(Zoology Department, University of Western Australia)

General Introduction

Marsupials, formerly more widely distributed, are at present restricted to the Australian, Nearctic and Neotropical faunal regions. The palaeontological evidence and zoogeographic affinities of these faunas have been most recently discussed by DARLINGTON (1957).

DARLINGTON says (p. 335) "Australian marsupials are the best existing example of evolutionary or adaptive radiation in an isolated place." This draws attention to one aspect of the uniqueness of the Australian marsupials which has been noted repeatedly by other zoologists and ecologists who have commented on the ecological parallelism between the fauna of Australia and other continental regions (e.g. ALLEE, EMERSON, PARK, PARK & SCHMIDT p. 470). Nevertheless the recognition of the various ecological analogues has not been the starting point for marsupial studies in Australia, rather these can be epitomised as follows:

(i) There is good overall cover of taxonomy of living forms with some areas covered in detail (TROUGHTON, this volume).

(ii) There has been detailed work on some aspects of cytology and reproduction, including embryology (SHARMAN, this volume).

(iii) Large numbers of fossil marsupials have been described from Australia but most of these are Pleistocene and give little clue to a better understanding of recent forms. STIRTON (1954 and 1957) has been responsible for great increases in our knowledge of Tertiary marsupials particularly from South Australia. However this work has not been interpreted and it is not possible to derive any ecological conclusions from it.

Any formal definition of animal ecology is so broad that it permits many shades of interpretation. It is probably true, however, that most people would share HARDY's view that ecology is scientific natural history which as usually interpreted means study of whole animals, and population ecology is devoted to relating fluctuation of numbers of animals to observed, or inferred, environmental changes. This may be a convenient arbitrary boundary to circumscribe ecology, but we believe that data on the individual physiology and biochemistry of animals is also called for if we are to understand the tolerance, or otherwise, of species to environmental changes; and this latter may be very significant to understanding present day distributions compared with the past. Clearly also the previous history of the animal, from the fossil record, and knowledge of climates of previously occupied areas is desirable to round off a total picture.

There has been no comprehensive ecology of Australian marsupials. Looking at what has been done one can discern roughly two approaches: (a) circumscribed self contained projects with limited objectives, characterised by the absence of initial long range assumptions and (b) investigations based on questions posed by speculations about past distributions, as evidenced by historical and fossil evidence, and in particular why some species persisted and have even become pests, while others died out.

Circumscribed self contained projects with limited objectives

Three motivations can here be distinguished:
(1) economic pressure to rid agricultural or pastoral areas of vermin,
(2) conservation of well known animals in danger of extinction and
(3) academic ends by people interested primarily in the refinement of trapping and marking techniques and the information this can yield for a picture of how population numbers behave.

Economic

TOMLINSON, GOODING & HARRISON in a series of papers (1953—55) record marked increase in kangaroo and wallaby populations in the north of Western Australia so that "the sandy wallaby, *Macropus agilis*, is now a menace in the Kimberleys." According to these authors, prior to the advent of white man's flocks the presence of aboriginal hunters and dingoes kept wallabies in check. As control from these sources diminished, the pest increased in numbers. With the sinking of bores in country away from river frontages, the wallaby population spread inland. "Information about feeding and drinking is limited (but) these animals ate almost the same food as sheep and cattle and were capable of substantially reducing stock carrying capacity." There is no internal evidence in the papers that the above statements are other than expressions of opinion. On the other hand their photographs witness how denuded of vegetation the Kimberleys can become and their statistics from poisoning and trapping are convincing evidence that there are large numbers of animals in the area.

STEWART (1936) recorded damage to pine plantations in Western Australia by quokkas *(Setonix brachyurus)* particularly in dry summer months towards the end of the season. Control was established by fencing and poisoning.

MCNALLY (1955) reported damage to exotic pine plantations in Victoria by *Trichosurus caninus*, *T. vulpecula* and *Wallabia bicolor*. The damage was bark stripping and eating phloem which is rich in starch and sugars. Again the control was by means of poison baits.

Conservation

MARLOWE (1958) carried out a survey of the status of marsupials in New South Wales, to form the basis for an effective conservation programme. His methods were trapping of small animals and counts of spot-lighted kangaroos; records were made on a gridded map. His conclusions were: (a) there has been a serious decline in all families except phalangers, so that of a total of 52 species, 44% are extinct (defined as not recorded since 1910) or rare and (b) animals of forested regions showed a relatively low extinction of 7%, the extinction figures for open grassland being 38% and scrub 50%.

MCNALLY (1957) has described a survey of a population of koalas on French Island off the coast of Victoria. Animals are not endemic to this island but were placed there in sanctuary when their total extinction appeared imminent on the mainland. During August 1955 it was decided to round-up the population for transport to the mainland, because it was judged that they were eating out available food supplies. Mass mortality had previously been reported on the island and it is thought that this was due to starvation; on that occasion the population had been so decimated that the trees regenerated. This animal is dependent on gum leaves *(Eucalyptus* sp.) for food and from zoo records it appears they mate in spring and summer, have a pouch life of 6 months, mature at 3—4 years, have a likely life span of 10 to 12 years, and the female produces one young a year. It is estimated that 80% of the total island population was captured and the sample consisted of 245 males, 278 females and 171 dependent juveniles.

Trapping and marking studies and analysis of commercial shooting returns

GUILER (1957b, 1958 and unpub.) has made a field study of *Potorous tridactylus* on Mt. Nelson in Tasmania by trapping methods with no animals removed from the area, supplemented by study of a domesticated colony. The chief results obtained are: Animals live at least 5 years and breed throughout the year but with a peak of mating between December and February. Pouch life is about 3 months and the reproductive potential of a female is 17 at least. It is widespread in low scrub and absent from plains and mountain tops. Density in the low scrub is patchy, ranging from one to ten per acre. Some individuals move freely, up to ½ mile in two nights; males tend to roam in the breeding season and females remain in a small home range.

GUILER (unpub.) working with the Tasmanian devil has identified the remains of 14 mammal species in the scats; only 2 of these species have been observed living.

GUILER (1957a) has mapped the relative abundance of *Thylogale* and *Wallabia* in Tasmania from fur trappers returns for the years 1923—55. He concludes that the data show no well defined rhythm, but suggests there is a possibility of a 10—11 year cycle which may be co-incident with the sunspot cycle, as has been shown to exist for some other mammals by ELTON (1925). Although predators are present, there is no evidence of a predator-prey balance similar to that seen in the arctic fox fur industries.

GUILER (1953), again from trappers' returns, has prepared a map showing the distribution of the black and grey forms of *Trichosurus* in Tasmania. In regions of high rainfall (north and west) there is a high percentage of blacks; in regions of low rainfall a high percentage of greys. PEARSON (1938) and HUXLEY (1942) had interpreted this to mean *Trichosurus* invaded the island originally from the N.E. corner and black became more common the further away from the point of invasion, but GUILER thinks it more likely that the black mutant arose where rainfall and humidity were high, it being according to him "an accepted principle that in regions of high humidity dark races are more frequent."

TYNDALE-BISCOE (1955) worked on *Trichosurus* in New Zealand, to which the species has been introduced, with box traps. By collation of trapping data with information from shot animals he elicited the following. *Trichosurus* is dioestrus and monovular, pregnancy is 15—24 days and is sometimes followed by post partum oestrus. As in Australia, there is anoestrus in February followed by reproductive activity in March. By May 90% of the females are pregnant, offspring stay in the pouch until August—September; females may produce a second offspring in October. The sex ratio is 50—50. Trapping and marking was not exhaustive enough to delimit the extent of movement but did show a tendency of animals to return to the same restricted area of forest at night to feed.

DUNNET (CSIRO unpub.) box trapped and marked *Trichosurus* in 6 areas within 2 miles of Canberra; he is fairly confident the whole population was marked. The object was to measure the role of dispersion of immature animals in population regulation. He confirmed the suspected territoriality in males which involves the outward movement of young males from the area. About 80% of the pouched young move out while still immature and only the young females remain to breed in their native area. DUNNET has information, not yet worked up, on changes of terrotories resulting from movement and disappearance of the males, individual ranges, breeding season, mortality in the pouch and population structure.

DUNNET has worked also on the population of quokkas on Rottnest Island *(vide infra)* off Perth, Western Australia. His aim was to measure and understand changes in population size and he was also concerned with behaviour and dispersion of quokka populations and their structure and organisation. His animals were caught with hand nets, and long fish nets that could be raised mechanically to enclose groups of animals at watering sites; box traps proved unsatisfactory (DUNNET, 1956). Animals were marked with metal ear tags with "scotchlite" tape to make them visible at night (EALEY & DUNNET, 1956). DUNNET tagged altogether 3,000 animals and 1,200 of these have been recaptured. He has calculated the dates of birth of 405 pouched young and shown that births occur from the third week in January to early November, with the majority of births in March and April. He has data which permit age determination of pouch young up to 150 days on the basis of hind foot measurement, and weight data for 1 and 2 year olds. His data on distribution have not yet been fully analysed, but his interim conclusion is that there are several more or less discrete populations on an island of 4,700 acres. In the hot dry summer season animals come long distances to the few fresh water soaks.

Investigations based on speculations about past distribution

Introduction

This approach is the basis of the work from the University of Western Australia on marsupials, frogs, spiders and snails. We think of ecology as embracing all aspects which conduce to the successful persistence in time of a population. It is implicit in this that the ecology of present day forms has an historical component reflected in the genetic selection imposed by the changing environments of the past. If an animal is closely adapted to an environment conditioned by a climate which changes, the animals' persistence, elevation to peak status or extinction, depends on its morphological and physiological adaptations and in particular to the range of conditions its physiological make up can tolerate. In this regard individual physiology is important in so far as it documents the range of variability in the population. One would expect, and it is in fact borne out *(vide infra)*, that there will be individual variability between members of the one population in almost any physiological character measured. Thus the important element say in regard to nutrition, or water balance, is how the structure of the population (e.g. number of frequent or infrequent drinkers) gives the population the requisite plasticity so that, through an indefinite period of time, the population can meet and survive the whole array of environmental, especially climatic, stresses which it is likely to meet. The work, largely uncompleted and unpublished, reported below from the Western Australian school is oriented to the demonstration of such differences between adjacent populations in situations differing climatically, but also in other environmental ways. In this context some comment is demanded on two matters: (a) the generally held view that marsupials are "second rate" animals which gave way to the eutherians introduced by Europeans. It is probably true that with the environmental changes imposed by the white man in Australia, the eutherians in many places ousted the marsupials but three things need to be remembered before accepting this as a satisfactory overall picture: (i) fauna now extant only in Eastern Australia was present in Western Australia in recent geological time and many marsupials became extinct shortly after European settlement and before either pastoral development, rabbits or foxes were established, (ii) many marsupials are still gazetted vermin because of their success and (iii) at present it is only an academic exercise to try to determine the "superiority" of the eutherian over its marsupial analogue. (b) If further work sustains the tentative view that many marsupials became extinct because of long range climatic changes, the present highly desirable move to establish fauna reserves will not be the complete answer to maintaining unique fauna threatened with extinction. Also needed will be data on their tolerances to water, food, temperature etc., so that these can be artificially provided.

Two ecological projects on marsupials, along the lines indicated, are in progress in Western Australia involving CSIRO and University personnel: quokka and euro.

Quokka *(Setonix brachyurus)*

The first of these, on the quokka, had purely academic beginnings. Broadly speaking its form was directed by the aforementioned speculations about previous climates etc., and by a desire to break new ground in the actual procedure of assessing and understanding population trends. Thus animal ecologists have generally confined themselves to countable attributes, total numbers of animals, the breakdown into age classes, sex ratios, reproductive rates and so on. Such information represents the static state of the population and can of itself, without similar studies suitably spaced in time, give no information on the vicissitudes of the population. The incidence of limiting factors such as starvation, disease and other stressing factors are usually inferred from the changes in population parameters and from a concurrent or subsequent analysis of the physical and biotic environment. Even the most complex ecological documents available, the Annual Report of the Registrar General for England and Wales and the Yearly Reports of Vital Statistics from the Department of the Census for the United States, contain little more than those countable attributes which specify the present composition of the population.

It has been felt that a step forward in population ecology would result if a debility, or morbidity evaluation, of a population were undertaken concurrently with a programme in which the conventional parameters were also estimated. Such an evaluation would allow of a more dynamic interpretation whereby a prognosis of the population's fate could be made. Such a prognostic evaluation has obvious advantages over a mere parametric description when the institution of control or conservation measures are dependent upon the knowledge of the state of the population. A population morbidity evaluation after the institution of conservation or control measures likewise allows of a much more accurate estimate of the success of such measures.

For this study a ready-made study area stocked with animals was available. Rottnest Island, about 4,700 acres in area, contains a fairly dense population of the quokka, (*Setonix brachyurus*). This herbivorous marsupial is the only mammal (except for house mice and an occasional rat and feral cat) upon the island. The population

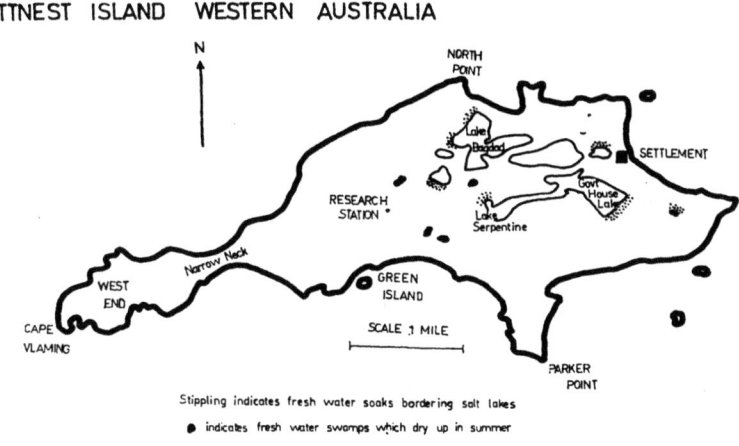

Fig. 1.

is thus almost ideally simplified for study. The population is definitely circumscribed with no immigration or emigration. No carnivorous predators complicate biotic factors and as the animals have enjoyed full protection under State conservation laws since 1914, the population is virtually unmolested by man. The animals themselves are small (about 3 Kg.), easily caught, docile when handled and take well to domestication.

Rottnest Island is composed of sandy limestones of Pleistocene age and lies on the continental shelf 11 miles west of the coastline adjacent to Perth. Formerly densely wooded with trees of *Acacia* sp., it is now primarily an open heath formation of low shrubs with occasional *Acacia* thickets, which are relics of the former *Acacia* woodland. The island lies in a region of winter rainfall with an annual precipitation of 29 inches. However, during the months of December through March there is an average rainfall of 1.50 inches (Commonwealth Met. Bureau Book of Normals). The porous limestone allows rain to soak away readily and fresh water is not available except in restricted localities where recent or Pleistocene dunes overly impervious marine beds. Under these circumstances the dune acts as a reservoir and the water is brought to the surface as a series of soaks wherever the marine beds outcrop. Commonly these soaks occur around the salt lakes which are a series of marine embayments at present cut-off from the sea.

Environmentally the situation on Rottnest Island is completely unlike the densely vegetated swamps now occupied by only a remnant of a formerly widespread population on the mainland. In the absence of a substantial mainland population of quokkas, interpopulation comparisons were possible because on Rottnest a narrow isthmus divides the island into two unequal parts. One of these, West End, lacks fresh water, while the other larger portion contains the fresh-water soaks and is the better vegetated part of the island. Animals marked by numbered ear tags indicate that there is no interchange between West End and the rest of the Island and, also, that within this latter area there are a number of virtually discrete populations which may then be compared for population structure. A convenient division of the populations into two study areas has been possible on the basis of access to fresh-water for drinking viz. West-End, (lacking water), and the areas adjacent to the fresh-water soaks where populations have fresh water at all times. In addition, for purposes of experimental control, a population having an adequate diet and water *ad lib* has been maintained in departamental compounds.

Field work has proceeded over a number of years but the most intensive phase, during which all clinical measures have been taken, extended from August 1955 to August 1957 during which time 28 visits were made to the Island totalling 600 man days.

From the data obtained it is possible to summarise the ecology of the quokka on Rottnest as follows:

Activity

Daily:

The quokka is essentially a nocturnal animal. During daylight animals lie-up beneath vegetation. At sunset animals leave cover and commence to feed; usually animals are abroad all night and return to cover shortly after dawn. Rain and strong, especially cold, winds may prevent the emergence of animals during the evening. Rain in long dry periods during the summer may cause the quokka to emerge during daylight and drink from roadside pools, and other temporary waters.

Annual:

During winter and spring when water is available either from free surface, or from growing plants, the quokkas appear to be widely though not uniformly dispersed throughout the island. As summer approaches and water is restricted to soaks, or succulent plants such as *Carpobrotus* sp., animals aggregate or move nightly to areas where water can be obtained. On West End water appears to be obtained from *Carpobrotus* on Cape Vlamingh.

Reproduction

SHARMAN (this volume) reports details of reproduction the ecologically significant part of which is an anoestrus period that, for the general population, ends late January i.e. before midsummer. After the birth from the pregnancy immediately following the termination of anoestrus, a post partum copulation leaves females with a pouched foetus and a blastocyst whose development is halted at about the 100 cell stage. The blastocyst remains quiescent unless the pouched foetus is lost. The maximum time during which development can be delayed is not known precisely, but recent work has shown that that "delayed" birth can take place up to when the original (suckling) joey has reached at least 300 grams, at which stage homeothermism has been established; blastocysts probably do not persist over the seasonal anoestrus. Experiments based on removal of suckling young from animals recently captured in the field and subsequently kept in the mainland yards indicate that delayed birth with establishment of a viable suckling pouch embryo can take place in only about 20% of cases.

Notwithstanding the fact that the adult females have survived the nutritional stress of late summer *(vide infra)* 90% have pouched young in July-August. These first leave the pouch during September—October and many are too large to return by October, however at this time they may still be suckling from outside the pouch. Until the re-commencement of reproduction of the following year the juveniles remain at-heel and accompany the mother. From one year of age, until sexually mature at two years, the juveniles are independent of their parents.

Sex ratios vary. At birth males are more frequent than females, as juveniles, males and females are approximately equal. The frequency among adults is difficult to assess since it is probable that females and juveniles are not associated with adult males throughout the year, and in particular when areas around soaks are sampled males may appear less frequent because they apparently do not have to drink as regularly as young animals which are accompanied by their mothers.

Control of population size

Before any systematic effort to assess clinical status of live animals relevant to population control was attempted, an estimate of population death pattern was made. Fortunately skeletal remains were plentiful and initially 600 crania were collected. After analysis of biometrical data and dental stage eruption, it was apparent that death of juveniles occurred in late summer. There is no way of knowing at what season adults die, but we accepted as a working proposition (subsequently confirmed by morbidity data *vide infra*) that environmental stresses which brought juveniles to the point of death would operate similarly on adults. Consequently the operational problem was twofold: (a) to substantiate the probable death season by a concurrent general population debility, and (b) to indicate which, of several probable limiting factors, was the actual cause of death. Answers were sought in measures which could be repetitively applied to the population throughout the year and which had little of the subjective judgement intrinsic in a conventional clinical evaluation. The measures used were a comprehensive haematological picture, together with several physiological and physical measurements. In all some 600 animals were handled and from each animal 6 ml. of blood was taken by cardiac puncture. From these samples the haemoglobin, haematocrit, red blood cell count (RBC), white blood cell count (WBC), erythrocyte sedimentation rate, total plasma protein, vitamin B_{12}, albumin and globulin were estimated. In addition the rectal temperatures, total body weights, various body length measures (including girth as a measure or rumen repletion), dental ages, sex and reproductive state of females were noted. The routine of handling was to capture the animals in the early morning and immediately deal with them on return to the field laboratory. Several hours were allowed to elapse to ensure that they suffered no ill effects from the cardiac puncture before the animals, which were now tagged, were returned to the places where they were captured.

Concurrently with this haematological sampling of the population experiments were conducted to determine the food preferences by means of exclusion quadrates on areas cleared by fire, and by STORR of this department who has worked out a method for identifying the remains of all the common island plants in the faecal pellets. These show that the quokka is a highly selective feeder for high protein vegetation. The reason for this is apparent from CALABY's (1958) laboratory experiments which showed the minimal nitrogen requirements and that these are satisfied only by a few plants. Plants forming the diet of the quokka have been sampled at monthly intervals and analysed for protein content. All plants showed a seasonal cycling in protein content and were highest in winter and spring and lowest in summer. The summer figures were about half those of the spring high and many plants nutritionally adequate in winter were inadequate in summer.

In particular it was demonstrated that the growth of *Acacia*, a main forage plant of the animals, was stopped because of animals eating out all the suckers and seedlings of this species. This demonstration of extreme control of the island's vegetation strongly implied that starvation was the chief limiting factor to population increase.

A second attractive possibility of a simple limiting factor lays in the known cobalt deficiency of the island soil. Cobalt deficiency, serious only to ruminants, leads to "coast disease" or "bush sickness". This condition is invariably fatal to sheep and early attempts to raise sheep on the island had failed for this reason. Since the quokka has a ruminant type digestion (MOIR et al., 1956), it was plausible to speculate that the cobalt deficiency was limiting the population.

The selection of the two study areas mentioned above permitted test of a third possibility, that the availability of free surface water was critical.

There were thus three obvious possible limiting factors and it is convenient to consider these in the following sequence:

Trace element deficiency.

Cobalt has been found to be the essential mineral component of the biologically active compound vitamin B_{12}. The deficiency of this compound brings about a typical syndrome of neurological and haematological symptoms. The easiest way to confirm cobalt deficiency is by assay of the circulating vitamin B_{12} in plasma. B_{12} estimations of the plasma in addition to the other haematological estimations were done on all animals in the field samples. There was no seasonal cycle of B_{12} level corresponding to the putative season of death for any age class of quokka. The average values for the population were at all times of the year well above the values of B_{12} which produced coast disease symptoms in sheep. It was concluded therefore that although the cobalt deficiency was sufficient to wholly eradicate a eutherian ruminant population from the area,

Starvation

All the available evidence, and particularly that from haematology, pointed to starvation during the late summer period. Haemoglobin, haematocrit and R.B.C. estimations demonstrated a severe seasonal anemia both for juveniles and adults. The average values of haemoglobin for the late winter were near 16 grams per cent which is the same as the well fed compound animals kept as controls. During the summer the average fell to near 12 grams per cent, a drop of 25%. Comparable falls for haematocrit and R.B.C. were also demonstrated. With these three established measures the mean corpuscular volume and the mean corpuscular haemoglobin values could be estimated. The summer anemia, by comparison with the normal values for the compound controls, was found to be of the slightly hyperchromic macrocytic type which is the type associated with starvation in human beings (KEYS et al, 1950). For one study area (West End) the average weight loss, corrected for age, is of the order of 25% during the late summer season. Again the rectal temperatures of both juveniles and adults show a marked decrease during the season of high mortality. An average drop of some 2° C was established for the West End population. This large decrease of temperature, at a time when environmental temperatures are high, is taken as confirmation that semi-starvation is the recurrent summer fate of the island animals. It has been established (BRODY, 1945) that rumen fermentation provides an appreciable amount of heat in eutherian ruminants and so the drop in rectal temperature during the summer for the quokka means a recession of the process of digestive fermentation. Belly circumference averages, which have been found to be a measure of the fullness of the rumen, confirm the deduction that the rumen contents are diminished in the study animals during this time of the year.

The foregoing measures are all consistent with a profound semi-starvation seasonally at the late summer period, but whether the deaths are due to starvation or whether disease is the ultimate factor accounting for the summer mortality remains as a further question. Two of the blood measures afford in general terms a measure of the incidence of disease within the population — the white blood cell count and the erythrocyte sedimentation rate. Both of these measures are non-specific in their reactions to disease. The sedimentation rate is in particular a good indicator of acute general infection. This measure is used in hospital practice to discriminate between debilitated states due to disease or organic disorder. From the fact that the sedimentation rates are on the average lower in the summer period, it seems that the attendant anemia is not due to any infection but that it is due to organic malfunction consequent upon semi-starvation. Likewise with leucocyte concentration, generally an increase in W.B.C. occurs with acute infection, localised infections such as pneumonia and certain general infections. Again the general average white cell count is at a minimum in the late summer period which does not support a thesis of a general increase in the level of disease. However mention must be made that haematopoetic disorders, certain infections of bacterial and virus origin such as tuberculosis and influenza, as well as general debilitated states, may have an attendant leucopenia. But in general haematopoetic disorders, and infections, exhibit increased sedimentation rates and can thus be distinguished from the general debilitated state consequent upon semi-starvation. Hence the low average values of these two indicators of disease strongly suggest that the summer starvation is not a mere preliminary to disease as the prime limiting factor. Consistent with this we have found that although a large number of the animals from the soaks are heavily infested with the intestinal parasite *Austrostrongylus thylogale* the number of parasites does not correlate with the haematological picture. No systematic attempt has been made to detect bacterial or virus infection by microbiological methods.

Summer dehydration.

It is generally known that animals suffering chronic restriction of drinking water limit their food intake and under prolonged conditions of drought may die with the typical symptoms of starvation. It was necessary to distinguish between these two conditions of true and secondary starvation. In order to establish a blood picture of animals which were chronically deprived of drinking water an experiment was set up in which a group of animals were gradually deprived of free drinking water but had food kept in excess at all times. The experiment lasted some 230 days, and blood values were estimated each month. These animals, suffering chronic water restriction, showed lowered rectal temperatures and all the attendant blood values of true starvation, except that total plasma protein values were higher than those from average well fed and watered animals. The field animals at the height of summer had a total plasma protein value lower than normal. On this basis we concluded that the field animals were exhibiting primary starvation due to lack of adequate food, consequent on its dying off, and in particular to the non-availability of *Acacia* which had been eaten out over the years. This conclusion is supported by the observation that the blood picture was essentially the same for the West End study area (no free water) as for the Bagdad "soak" areas.

The ecology sketched above indicates that animals experience semi-starvation each summer; the skull data shows that this is fatal in a large proportion of cases.

In the absence of predators and parasites and with complete protection it seems that this is the only regular population control on the island, exacerbated no doubt in the occasional harsh season, or succession of bad seasons. The historically documented reduction in *Acacia* (which is nutritionally adequate at all seasons and is a preferred item of diet) no doubt means (a) that population control by semi-starvation is now more acute than hitherto, through the observed grazing pressure on the *Acacia* seedlings and (b) that the present vegetation is approaching a grazing climax.

Our more limited data from the small mainland populations exhibits neither the seasonal death (skull data) nor the correlated seasonal debility (blood data). Animals on the mainland live in areas which are well watered throughout the year so that the vegetation does not suffer the severe seasonal fall in protein. The Rottnest study offers no suggestion as to why the species became almost extinct on the mainland, but the mainland zone of occurrence was subject to interference by man and his domestic animals and cultivation and these probably contributed an element towards extinction which was not present on Rottnest.

Euro *(Macropus robustus)*

The euro project was initiated in a quite different fashion. CSIRO started work in the Marble Bar area of Western Australia in response to requests from pastoralists for formulation of control measures for the euro kangaroo which had become a pest. Two derelict properties, formerly carrying sheep but now only euros, with little more than spinifex in place of the previous herbage, had been acquired previously by the Western Australian State Department of Agriculture and E. H. M. EALEY of CSIRO Wildlife set up his headquarters there and pursued a programme initially oriented to population control by poisoning at the water holes. After two years' work, including marking of individual animals (EALEY & DUNNET, 1956), it became apparent (EALEY, unpublished) (a) that while, as expected, during drought animals tended to converge on water holes there were animals that drank daily while others could go for very long periods between drinks, and (b) that euros thrived and reproduced on spinifex *(Triodia)* which experience showed was at the best minimal survival feed for sheep. The finding that some animals could endure long periods of dehydration destroyed any hope that poisoning water could quickly decimate the population, except possibly during times of extended drought, and further raised doubts as to the accepted proposition of the pastoralists that the euro had increased both in numbers and range to pest proportions (i.e. as successful competitors with the sheep for feed) as a direct result of opening up new water holes.

At this stage of the investigation one of us (A.R.M.) was invited to review and assess the situation. After examining both EALEY's records on euros and his analysis of the sheep populations and climate, in the light of the background of the work on *Setonix*, the following proposition was advanced for test.

Before white settlement the euro (sometimes referred to as the hills kangaroo or biggada) occupied poorer country, usually hilly areas vegetated principally by spinifex, while the red kangaroo occupied the flats which carried the nutritionally more adequate grasses. It was in these latter situations that the pastoralists established their flocks. With the customary cycle of droughts the sheep stocking was such that all the high protein feed was eaten-out. Under grazing pressure the vegetation was down graded to a grazing climax dominated by spinifex (*Triodia* sp.) which was unpalatable to sheep and in addition had a protein content insufficient to support their reproduction. In effect the foregoing has meant the expansion of the original habitat of the euro. If this interpretation is substantially true the nature of the euro as a pest needs revising; presumably under a changed pasture management programme the "good" grasses would return to the flats and with the contraction of the favourable habitat the euro would again be restricted to the hills. The fundamental problem in the foregoing proposition is of course, how can the euro exist and reproduce on a diet which is nutritionally inadequate for sheep. It is probable that all animals of equivalent weight have approximately the same protein requirement. However, in ruminants protein is devoted to feeding a bacterial flora and this, and its products of fermentation, are digested by the animal. Furthermore it is known that in sheep the bacterial flora can synthesise proteins from urea and this may be recycled through the saliva. Against the background of this information work on behaviour observation, seasonal analysis of plant and stomach contents for protein, urine analysis and haematology have been carried on jointly by CSIRO and the University with assistance from the Western Australian State botanists and the State Chemical laboratories. Much remains to be done but everything emerging so far is consistent with the substantial correctness of the original proposition, and in particular the case for believing that the population has frequent and infrequent drinkers has been strengthened. Whether an animal drinks freely or not, besides its gross obvious survival value, has implications with regard to its protein intakes; high protein demands water for the excretion of by-products. To see whether within this one species, the proportion of frequent drinkers to infrequent drinkers is different in different geographical areas, a parallel investigation is being done in the outer wheat belt.

At the present stage in the investigation the euro promises to be an ideal subject for the investigation of desert adaptations at the level of both the individual and the population.

Acknowledgement

We are grateful to the authors cited who have supplied unpublished information. Finance for the unpublished observations from our department has been from University Research grants and from CSIRO. The work on the quokka was done from the Rottnest Biological Station. B_{12} estimations have been done at the Royal Perth Hospital by arrangement with Dr. PITNEY.

REFERENCES

ALLEE, W. C., EMERSON, A. E., PARK, O., PARK, T. & SCHMIDT, K. P. 1949. Principles of Animal Ecology, Philadelphia. W.B. Saunders.

BRODY, S., Bioenergetics & Growth. Reinhold Publishing Co. New York (1945.)

CALABY, J. H., 1958. Studies in Marsupial Nutrition II. The rate of passage of food Residues and Digistibility of crude fibre and protein by the Quokka. *Setonix Brachyurus* (Quoy and Gaimard). *Aust. J. biol. Sci.* 11 (4), 571—580.

CHIEF STATISTICIAN, 1937-1951. Vital Statistics of the United States Yearly Reports 1937-1951. U.S. Govt. Printing Office.

DARLINGTON, P. J., 1957. Zoogeography with geographical distribution of animals. New York, John Wiley & Sons.

DUNNET, G. M., 1956. A Population Study of the Quokka, *Setonix brachyurus* Quoy and Gaimard (Marsupialia) I. Techniques for Trapping and Marking. *C.S.I.R.O. Wildlife Res.* I, (2), 73.

EALEY, E. H. M., & DUNNET, G. M., 1956. Plastic Collars with Patterns of Reflective tape for marking nocturnal mammals. *C.S.I.R.O. Wildl. Res.* I, (1), 59.

ELTON, C., 1925. Plague and the regulation in numbers in Wild Animals. *J. Hyg.* 24, 138.

GOODING, C. D. & HARRISON, L. A., 1953. The Wallaby Menace in the Kimberleys. *J. Agric. W. Aust.* 2, (3rd series), 333.

GOODING, C. D. & HARRISON, L. A., 1955. Trapping yards for kangaroos. *J. Agric. W. Aust.*, 4, (3rd series) 6, 3.

GUILER, E. R., 1953. Distribution of the Brush Possum in Tasmania. *Nature, Lond.* 172, 1091.

GUILER, E. R., 1957a. The present status of some Tasmanian mammals in relation to the fur industry of Tasmania. *Pap. Proc. Roy. Soc. Tasmania*, 91, 117.

GUILER, E. R., 1957b. Longevity in the wild Potoroo, *Potorous tridactylus* (Kerr). *Aust. J. Sci.* 20, (I), 26.

GUILER, E. R., 1958. Observations on a population of small marsupials in Tasmania. *J. Mammal*, 39, (1), 44.

HUXLEY, J. S., Evolution, the Modern Synthesis. Allen and Unwin, London, 1942.

KEYS, A., BROZEK, J., HENSCHEL, A., MICHELSEN, O. & TAYLOR, H., The Biology of Human Starvation. Univ. of Minnesota Press, Minneapolis, (1950).

MARLOWE, J. B., 1958. A survey of the Marsupials of New South Wales. *CSIRO Wildlife Res.* 3 (2), 71—114.

MOIR, R. J., M. SOMERS & H. WARING, 1956. Studies on Marsupial Nutrition. 1. Ruminant-like digestion in a herbivorous marsupial *Setonix brachyurus* (Quoy and Gaimard). *Aust. J. biol. Sci.* 9 (2), 293.

McNALLY, J., 1955. Damage to Victorian Exotic Pine Plantations by Native Animals *Aust. For.* XIX, (2), 87.

McNALLY, J., 1957a. A Field Survey of a Koala Population. *Proc. Roy. zool. Soc. N.S.W.* May 8th 1957.

McNALLY, J., 1957b. Koala Management in Victoria. Wild Life Circular No. 4, Fisheries and Game Department Victoria.

PEARSON, J., 1938. The Tasmanian Brush Opossum: Its Distribution and Colour Variation. *Pap. Proc. Roy. Soc. Tasmania*, 1937, **21.**

REGISTRAR GENERAL FOR ENGLAND AND WALES, 1921-1950. Statistical Review of England and Wales for the years 1921-1950. HMSO London.

STEWART, D. W. R., 1936. Notes on marsupial damage in pine plantations. *Aust. For.* **1** (2), *41*.

STIRTON, R. A., 1954. Late Tertiary marsupials from South Australia. *Bull. Geol. Soc. Amer.* **65** (12 : 2), *1351*.

STIRTON, R. A., 1957. Tertiary marsupials from Victoria, Australia. *Mem. Nat. Mus. Vict.*, **21,** *121*.

TYNDALE-BISCOE, C. H., 1955. Observations on the Reproduction and Ecology of the Brush-tailed possum (*Trichosurus vulpecula*) Kerr. *Aust. J. Zool.* **3,** (2), *162*.

TOMLINSON, A. R., GOODING, C. D. & HARRISON, L. A., 1954. The Walla by Menace in the Kimberleys. *J. Agric. W. Aust.* **3** (3rd Series) (5), *609*.

XXI
MARSUPIAL REPRODUCTION

by

G. B. SHARMAN

(Department of Zoology, University of Adelaide, Australia)

Introduction

The earliest recorded observations on reproduction in Australian marsupials deal mainly with the anatomy of the female reproductive system. JOHN HUNTER dissected and examined the reproductive systems of numerous marsupials but HOME (1795), who examined some of HUNTER's specimens, is credited with discovering that the marsupial foetus reached the exterior by way of a direct route between vaginal culs-de-sac and urogenital sinus — the pseudovaginal canal. Because of the absence of the pseudovaginal canal in all non-parous, and in many parous marsupials, HOME's conclusions were not generally accepted (see OWEN, 1834) until FLETCHER (1883) and STIRLING (1889) published their observations. FLETCHER showed the pseudovaginal canal to be absent in immature but present in parous kangaroos and STIRLING dissected a female kangaroo in which the young was passing down the pseudovaginal canal. HILL (1899) showed that the pseudovaginal canal in *Perameles* closed rapidly after parturition and entrapped within its tissues remnants of the foetal membranes. HILL's important observations ended a controversy, about the route taken by the marsupial foetus during parturition, which had continued for over 100 years. PEARSON's (1945 and later) papers have contributed greatly to knowledge of the comparative anatomy of the marsupial reproductive system.

Among pioneer observations on marsupial reproduction those of the late Professor J. P. HILL and his students and collaborators stand alone. HILL (1895) first recorded the occurrence of true allantoic placentation in *Perameles;* the whole marsupial group was previously regarded as "aplacental". The observations of HILL & O'DONOGHUE (1913) on *Dasyurus quoll*, although made incidentally during controlled breeding experiments for the collection of embryonic material, remain today the only work on the oestrous cycle of this species. These authors first recognised the functional significance of the corpus luteum in the non-pregnant marsupial and showed it to be responsible for a degree of uterine proliferation not less than that occurring in the uteri of pregnant animals. HILL and his students also contributed a series of papers on the early development of marsupials (HILL, 1910; FRASER & HILL, 1915; BUCHANAN & FRASER, 1918; TRIBE, 1918; etc.). Following HILL's initial studies and before the work of SHARMAN (1954a) the only significant paper dealing with reproduction in Australian marsupials is that of FLYNN (1930) whose observations on the rat-kangaroo *(Bettongia)* represent the first study of reproduction in a diprotodont marsupial. In contrast to the lack of knowledge about reproduction in Australian species the reproductive physiology of the North American polyprotodont marsupial *Didelphis*, is well known (REYNOLDS, 1952).

Recent studies have shown that the kangaroo-like marsupials, which are regarded on morphological grounds as being amongst the most advanced of the group, have achieved significant evolutionary advances in reproduction. *Dasyurus* and *Didelphis* ovulate large numbers of eggs which are progressively reduced by failure of fertilization, intra-uterine death and resorption, and immediate post-natal death (HILL & O'DONOGHUE, 1913; HARTMAN, 1923b) but the macropod marsupial *Setonix* is exclusively monovular and greater maternal care is taken of the single offspring. In *Dasyurus* and in *Didelphis* pregnancy does not outlast the uterine luteal phase and birth in *Didelphis* is coincident with the onset of the earliest degenerative changes in the corpus luteum. In *Setonix*, on the other hand, the luteal phase is of longer duration and the embryo is retained in the uterus during the initial period of regression of the corpus luteum of pregnancy (SHARMAN, 1955b).

It is well known that adaptive radiation has produced a range of marsupials more or less parallel to morphological counterparts amongst eutherian mammals. It is not so well known that a series of equally spectacular, but no less interesting, parallel physiological processes have been developed in marsupials. A ruminant type of digestion independently developed in macropod marsupials and eutherian mammals is a case in point (MOIR, SOMERS, SHARMAN & WARING, 1954). Examples of parallelism in the sphere of reproductive physiology include the occurrence of a type of delayed implantation in macropod marsupials analogous to that occurring in rodents (SHARMAN, 1955b, c). In this chapter it is suggested that allantoic placentation, which occurs in two widely separated marsupial groups, is also to be regarded as a parallel development.

The following marsupials, arranged according to the classification of SIMPSON (1945) are mentioned in the text. Alternative names, if used by earlier workers, are added in brackets.

DIDELPHOIDEA

Didelphidae
Didelphis azarae, D. virginiana – American opossums
Chironectes panamensis – water opossum

DASYUROIDEA
Dasyuridae

Phascogale flavipes – yellow-footed pouched mouse
Dasycercus cristicauda (= *Chaetocercus cristicauda*) – mulgara
Sminthopsis crassicaudata – fat-tailed pouched mouse
Dasyurus quoll (= *D. viverrinus*) – native cat
Sarcophilus harrisii – Tasmanian devil
Thylacinus cynocephalus – Tasmanian wolf
Myrmecobius fasciatus – banded anteater

Notoryctidae

Notoryctes typhlops – marsupial mole

PERAMELOIDEA
Peramelidae

Perameles gunnii, P. nasuta – long-nosed bandicoots
Thylacis obesulus (= *Perameles obesula* = *Isoodon obesulus*) – short-nosed bandicoot

PHALANGEROIDEA
Phalangeridae

Trichosurus vulpecula – brush possum
Cercaërtus concinnus (= *Dromicia concinna*) – pygmy possum
Phascolarctos cinereus – koala
Pseudocheirus convolutor (= *P. cooki*), *P. peregrinus* – ringtail possums

Phascolomidae

Phascolomis sp. – common wombat
Lasiorhinus latifrons – hairy-nosed wombat

Macropodidae

Protemnodon rufogrisea (= *Macropus* or *Wallabia rufogrisea* = *Halmaturus ruficollis*), *P. parma* (= *Macropus parma*), *P. dorsalis* (= *Macropus dorsalis*), *P. eugenii* – wallabies
Macropus giganteus (various subspecies are sometimes given specific rank – *M. major, M. fuliginosus*), *M. rufus, M. robustus* – kangaroos
Setonix brachyurus – quokka
Dendrolagus matschei – tree kangaroo
Bettongia cuniculus (= *Hypsiprymnus cuniculus*) – bettong
Aepyprymnus rufescens – rufous rat-kangaroo
Potorous tridactylus – potoroo

The Reproductive System

Anatomy of the female reproductive system

A general account of the anatomy of the reproductive system of female marsupials is given by ECKSTEIN & ZUCKERMAN (1956). Recent papers, not included in ECKSTEIN & ZUCKERMAN's account, are those of PEARSON (1946, 1950a, b), PEARSON & DE BAVAY (1951, 1953), DE BAVAY (1951) and SHARMAN (1954b). The reproductive system of the quokka *(Setonix brachyurus)* has been described by WARING, SHARMAN, LOVAT & KAHAN (1955).

Ovaries

The micro-anatomy of the ovaries and ova of marsupials has been described by HILL (1910) O'DONOGHUE (1912, 1916) and by HILL & O'DONOGHUE (1913). These papers deal mainly with the polyprotodont marsupial, *Dasyurus quoll*. BRAMBELL (1956) includes marsupials in his exhaustive treatment of ovarian histology.

The eggs of marsupials are intermediate in size between those of eutherian mammals and monotremes. Fatty yolk spheres are abundant in the ova of *Dasyurus* and the ova of other marsupials also have abundant yolk. The oocyte in the mature follicle is surrounded by a thin zona pellucida. A well developed membrana granulosa and cumulus are present but a corona radiata has never been recorded in marsupials and the eggs of *Setonix* are naked at ovulation. Theca interna and theca externa are present and well developed at all stages of the maturation of the Graafian follicles in *Setonix* but in *Dasyurus* the theca interna is progressively reduced as maturation proceeds. Ovulation, in all marsupials adequately examined, is spontaneous but the findings of MARTINEZ—ESTEVE (1937) on *Didelphis azarae* and of MATTHEWS (1947) on *Dendrolagus matschei* suggest that the stimulus of copulation may be necessary to induce ovulation in these species. Both granulosa cells and theca interna appear to contribute to the glandular elements of the corpus luteum in *Setonix* but SANDES (1903) considered that luteal glandular elements were derived only from the granulosa cells in *Dasyurus*. Blood vessels and connective tissue strands from the theca interna contribute to the tissues of the corpus luteum. O'DONOGHUE (1916) found that ovarian interstitial tissue was present in ten species of diprotodont marsupials and absent in six polyprotodont marsupials. Interstitial tissue was however reported by VAN DEN BROEK (1910) in the ovaries of the polyprotodont *Sminthopsis crassicaudata*.

Uteri

Two uteri are found in all marsupials and although some degree of fusion of superficial tissue may occur, the lumina are completely separate and open on separate papillae into an anterior vaginal chamber. The relationships of mesometrium, myometrium and endometrium are typical of mammals in general.

Vaginal complex

The structure of the marsupial vaginal complex differs significantly from that of the eutherian mammal because the ureters occupy different positions. The ureteric buds of mammals originate on the dorso-medial sides of the embryonic Wolffian ducts but during development the marsupial ureter shifts to occupy a medial position (BUCHANAN & FRASER, 1918) whereas in eutherians the shifting is such that the ureters finally arise from the lateral aspect of the Wolffian ducts. On further development the marsupial ureters open first into the urogenital sinus and later into the base of the bladder primordium, retaining their medial position in relation to the Wolffian ducts. The Mullerian duct, when it appears, grows backwards in close contact with the Wolffian duct so that, in the female marsupial, the ureters finally occupy a medial position between the Mullerian ducts. It is therefore impossible for the Mullerian ducts to form a single chamber by fusion in the vaginal region, as occurs in eutherian mammals, without including the ureters. The characteristic vaginal complex of marsupials consists of paired vaginae which pass backwards from the os uteri to form more or less elongated structures, the culs-de-sac, and then loop around the ureters to fuse into a common (posterior) vaginal sinus or to open independently into the urogenital sinus. A median vaginal structure, through which the young are born, is formed at the first parturition in most marsupials. This structure, the pseudovaginal canal, forms a direct route from the posterior end of the culs-de-sac to the urogenital sinus and passes between the ureters (Fig. 7). In primitive marsupials birth probably occurred by way of the lateral vaginal canals as it does in the recent marsupial *Potorous tridactylus* (FLYNN, 1923a). PEARSON (1945) states that this mode of birth in *Potorous* "is to be regarded not as a primitive characteristic but as a secondary reversion to a prototypal arrangement". The pseudovaginal canal remains open after the first parturition in the quokka (WARING, et al., 1955) and in most macropod marsupials (FLETCHER, 1883). A short pseudovaginal canal which closes after each parturition is found in the phalanger *Trichosurus vulpecula* (SHARMAN, 1954b) and a long pseudovaginal canal which closes rapidly after each parturition is found in *Perameles* and in *Dasyurus* (HILL, 1899, 1900).

The embryological paired condition of the vaginal culs-de-sac persists in the adult marsupial and is exemplified by the separation of the culs-de-sac into two chambers divided by a median septum. Rupture of the septum during parturition occurs in some species. The embryonic Wolffian duct may persist in female marsupials as a short blind diverticulum of the lateral vaginal canals (PEARSON & DE BAVAY, 1953; SHARMAN, 1954b).

In the quokka the lateral vaginal canals, the greater portion of the culs-de-sac, and the posterior vaginal sinus are lined with stratified squamous epithelium.

Urogenital sinus

The urethra and the posterior ends of the lateral vaginal canals unite to form the urogenital sinus which is lined by stratified squamous epithelium.

Pouch

The pouch is rudimentary, or absent, in the marsupial banded anteater *(Myrmecobius fasciatus)* and in some marsupial "mice" (Dasyuridae). The pouch opens backwards in Dasyuridae and Peramelidae and forwards in phalangers and macropod marsupials. In a few forms, such as wombats, the opening is placed so that some of the pouch is anterior to the opening and some (usually the greater part) posterior to the opening. Four pairs of teats occur in the dasyurid *Phascogale flavipes*, three to four pairs in *Dasyurus quoll* and two pairs in some other dasyurids, some phalangers and in all macropods. In some phalangers and in all wombats the teats are reduced to a single pair.

Anatomy of the male reproductive system

The testes of adult male marsupials are usually scrotal in position. The marsupial false mole *(Notoryctes)* has undescended testes (STIRLING, 1891) and those of wombats are inguinal except in the breeding season when they descend into a scrotum. The scrotum of marsupials is prepenial in position — the reverse of the usual mammalian relationship. For a full account of the reproductive system of male marsupials see ECKSTEIN & ZUCKERMAN (1956).

Seasonal Reproductive Periodicity

Reproductive periodicity in the female

Many wild animals exhibit an annual anoestrous period during which the maturation of female germ cells does not occur. The ripening of the germ cells begins at a more or less restricted season which may be so correlated that consequent mating will ensure the birth of the offspring at a time best suited to their nutritional and other requirements. Wild female quokkas were found in anoestrus between August and late January but domestic females showed a greatly shortened anoestrous period in their first year of captivity (SHARMAN, 1955a). Subsequent anoestrous periods were further shortened or failed to occur and the female of three years or longer domestication was potentially capable of breeding at any time of the year. Since the captive animals were maintained in open pens under the same conditions of temperature, light and humidity as their wild counterparts and differed only in receiving a constant uniform diet with adequate water, nutritional factors were thought to be of importance in determining the length of the breeding season. This was supported by observations on wild animals which indicated that undernourished females remained in anoestrus longer than well fed females. The way in which nutrition influences the onset of the breeding season is however not fully understood. In the wild quokka breeding begins in late January during a period of low rainfall, high temperatures and, often, of intense drought. The young emerge from the pouch in late August, September and later months of the year when green feed is relatively abundant. The inference from these observations is that the young become independant when grasses and herbage are undergoing spring and early summer growth.

The breeding season of the macropod *Bettongia* is longer than that of *Setonix* and lasts at least from March to December. FLYNN (1930) considered that there was an anoestrous period in January and February in this species. The breeding season is further extended to occupy the entire year in some marsupials. New born young of the red kangaroo *(Macropus rufus)* are found at all months of the year. In this species the absence of a seasonal anoestrous period may be connected with migratory activity as red kangaroos are said to undertake long migrations to areas where food is plentiful. This would presumably allow the young to emerge in areas of suitable feed regardless of the season at which they are born.

BOLLIGER (1940) gives the breeding season of *Trichosurus vulpecula* in the Sydney area as March to April with a second minor breeding season in August to September. DUNNET (1956) studied the same species throughout the year in the Canberra area and showed a bimodal distribution of births. Thirteen births were recorded in the period March to May and five in the period September to October. In the Adelaide area births have been recorded at all times of the year except January and February, most of the young being born in March and April (SHARMAN, unpublished). Observations on females from the same area have shown *Trichosurus* to be polyoestrous in captivity although the time of the year at which females enter anoestrus is variable. One female gave birth to a second young while an earlier offspring, less than five months old, suckled from her. These observations indicate that some females at least produce two young per year. TYNDALE—BISCOE (1955) studied *Trichosurus* in New Zealand where it has been introduced. His data indicates no substantial differences in the breeding season in New Zealand where, as in Australia, there is an anoestrous period in the early (summer) months of the year.

The hairy-nosed wombat *(Lasiorhinus latifrons)* appears to breed early in the summer in the Murray River area of South Australia and later in the summer in the more arid Nullarbor Plains area. According to JONES (1924) this species reproduces only once in the year. *Phascolarctos* breeds in October and November and *Protemnodon rufogrisea* in December in the New England district of New South Wales and *Macropus canguru* in December in the Gwydir district of New South Wales (CALDWELL, 1887). No information about the extent of the breeding season is given for these species. On Kangaroo Island, South Australia, *Macropus giganteus* begins breeding in December and the smaller *Protemnodon eugenii*, which carries the pouch young for a shorter period, in January and February. *Dasyurus quoll* has one breeding season each year extending from the end of May to the first fortnight in August (HILL & O'DONOGHUE, 1913). The young of *Sarcophilus* are born in late May or early June and emerge from the pouch 15—18 weeks later (FLEAY, 1952). The breeding season of *Dasycercus cristicauda* extends from June to September (JONES, 1923). *Phascogale flavipes* breeds once per year in August or September (B. J. MARLOW, pers. comm.).

It is concluded, from the data available, that most marsupial species are in anoestrus during the late spring and early summer. Breeding begins in the summer in those species which suckle their young for a long time (*Setonix, Bettongia* and other macropods) and later in the year in species which suckle for a shorter time (*Dasyurus, Phascogale* and other dasyurids). The young of most species leave the pouch in the spring and early summer.

Reproductive periodicity in the male.

The male quokka and males of other macropod marsupials continue active spermatogenesis at all seasons of the year including the time at which the female is in anoestrus. The male *Trichosurus* does not exhibit any seasonal variation in testis size (DUNNET, 1956) nor loss of fecundity once sexual maturity is reached (TYNDALE—BISCOE, 1955). BOLLIGER & CARRODUS (1938) have reported persistent spermatorrhoea in the same species. There is no seasonal fluctuation in reproductive activity of the male *Didelphis* (CHASE, 1939).

Some male marsupials do appear to have a seasonal cycle of spermatogenesis although only fragmentary evidence is at present available. In *Pseudocheirus convolutor*, observed in Tasmania in January, the testis was shrunken and sperms, spermatocytes and cells in division were absent. In *Sarcophilus harrisii* observed in June in Tasmania and *Myrmecobius fasciatus* observed in September in Western Australia spermatogonial mitoses were plentiful but later stages of spermatogenesis, and mature sperms, were absent. As previously mentioned the testes of wombats are inguinal during the non-breeding season and at this time spermatogenesis does not occur.

Reproductive Cycles

The oestrous cycle

Introduction

Generally speaking mammals may be grouped according to the way in which ovulation occurs: those in which ovulation is spontaneous and those in which copulation or other stimuli are necessary to induce ovulation. Most marsupials so far studied belong to the former class, but there are two possible exceptions (see page 335). Mammals which ovulate spontaneously may be further grouped into those in which there is a functional corpus luteum following each cyclic ovulation and those forms (rat, mouse, etc.) in which the corpus luteum does not initiate a luteal phase unless pregnancy, or pseudopregnancy, occur. In marsupials ovulation, whether fertilization occurs, or not, is followed by the growth of a functional corpus luteum and a luteal phase in the uterus (Fig. 1). A similar luteal phase may be induced in the uterus of the castrate marsupial by the injection of progesterone thus indicating that this phase is mediated by ovarian produced progesterone. The term pseudopregnancy was introduced by HILL & O'DONOGHUE (1913) to describe the post-ovulatory changes (including the luteal phase) in the uterus of the non-pregnant marsupial. LONG & EVANS (1922) used the same term to describe changes in the uterus of the rat following sterile copulation and, probably because of the wide use of the rat as a laboratory animal, this usage of the term is the one commonly accepted. In the rat (and also in the mouse and other rodents) pseudopregnancy is accompanied by changes mediated by the functional corpus luteum and is of such duration that several ovulations are missed. HILL & O'DONOGHUE's original usage of "pseudopregnancy" was to describe the changes, following ovulation, induced by the cyclic corpus luteum; it had no reference to the occurrence of sterile copulation. The changes induced by a functional corpus luteum, initiated by sterile copulation, in such forms as the rat and mouse are today properly called pseudopregnancy. To avoid confusion with this modern usage SHARMAN (1955a) described the post-ovulatory changes in marsupials in terms of their duration without using the term pseudopregnancy. This system is followed here.

Some mammalian species enter oestrus and ovulate but once per breeding season (monoestrous condition) whilst others have two or more recurring oestrous cycles per breeding season (polyoestrous condition). The quokka is polyoestrous and some observations on other Australian macropod and phalangerine marsupials indicate that this condition is general within these groups. Thus *Bettongia cuniculus* is polyoestrous (FLYNN, 1930) and O'DONOGHUE's (1916) observation that corpora lutea of different ages are found in the ovaries of the same individual of *Phascolarctos cinereus* and *Protemnodon dorsalis* indicate that these species are also polyoestrous. *Protemnodon eugenii* (SHARMAN, 1955c), *Pseudocheirus peregrinus* and *Trichosurus vulpecula* (SHARMAN unpublished) are also polyoestrous. TYNDALE—BISCOE (1955) states that *Trichosurus vulpecula* is "dioestrous" apparently because he frequently observed that females became pregnant at the second oestrus of the season. The use of "dioestrous" in this context is unfortunate as this term is usually used to describe the interval between two successive periods of oestrus. According to ECKSTEIN & ZUCKERMAN (1955) an animal is polyoestrous whether it has "two or twenty" cycles per breeding season. All available evidence indicates that *Trichosurus* is polyoestrous and that, in the non-fertilized female, recurring oestrous cycles continue through most of the year. HILL & O'DONOGHUE

341

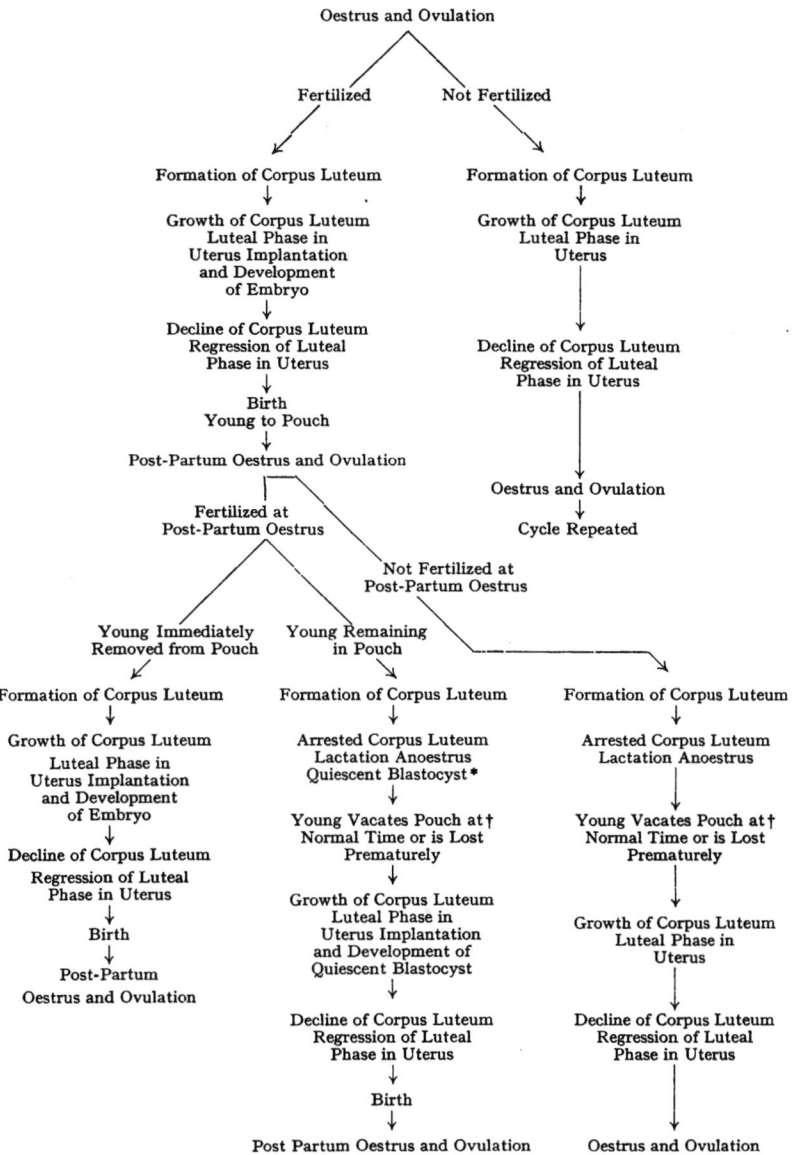

Fig. 1. Diagrammatic representation of the changes during the oestrous cycle, pregnancy and lactation in *Setonix*. Compare with Fig. 2 which gives the time relationships of the various reproductive phases. Reproduced from SHARMAN (1955b) by permission of the Editor, Australian Journal of Zoology.

* The blastocyst degenerates during anoestrus in wild females.

† Growth of the corpus luteum does not occur after the young vacate the pouches of wild females as these have entered anoestrus.

(1913) considered *Dasyurus* to be monoestrous but this view was contested by HARTMAN (1923b) after he demonstrated a polyoestrous condition in the American polyprotodont marsupial *Didelphis*. HILL & O'DONOGHUE's conclusions however receive support from studies on a related polyprotodont marsupial *Phascogale flavipes* which appears to be monoestrous (B. J. MARLOW, pers. comm.). Nothing is known of the nature of the oestrous cycle in the remaining polyprotodont marsupials.

The length of the oestrous cycle (the average time between successive ovulations) is 28 days in *Setonix* (Fig. 2). A cycle of 28 days has also been demonstrated in *Pseudocheirus*, *Trichosurus* and in the American marsupial *Didelphis*. The cycle of *Didelphis azarae* however lasts only about 7 days (MARTINEZ—ESTEVE, 1937).

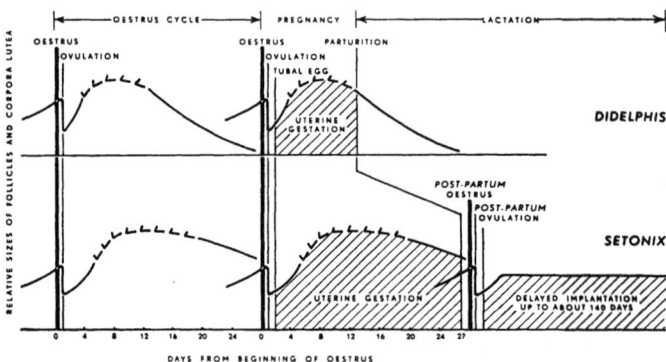

Fig. 2. The size relationships of ovarian follicles and corpora lutea during the oestrous cycle, pregnancy and lactation in *Didelphis* and in *Setonix*. The letter L repeated on the curve indicates the period at which the corpus luteum is active and a luteal phase is present in the uterus. Hatching indicates stages at which a uterine embryo is present. The gestation period of *Setonix* is characterised by a longer period of activity of the corpus luteum than in *Didelphis* and by the retention of the embryo in the uterus after the cessation of the luteal phase. Note that the uterine luteal phase is of the same duration in pregnant and non-pregnant females of both species and that the onset of lactation in *Didelphis* occurs before the maturation of the ovarian follicles of the next expected ovulation.

Anatomical and histological changes during the cycle

Pro-oestrus

This period, which is characterised by rapid growth of the follicle of the ensuing ovulation, has been arbitrarily defined as occupying the 5 days preceding oestrus in the quokka. In the first pro-oestrus following anoestrus only degenerate remains of old corpora lutea from previous breeding seasons are found in the ovaries, but in succeeding cycles the corpus luteum of the preceding cycle is prominent during pro-oestrus and oestrus. This is also true of *Trichosurus*

(Fig. 7) but contrasts with the condition in *Didelphis* where old corpora lutea involute and disappear rapidly (HARTMAN, 1923b). Coincident with the growth of the ovarian follicle the uterus of the pro-oestrus marsupial undergoes a period of hypertrophy and increased vascularity during which mitoses are common in the epithelia of both glands and lumen. This condition is general for all marsupials studied and can be duplicated in the castrate quokka by the injection of oestrogen. The vaginal canals of the pro-oestrus marsupial usually exhibit some cornification and cornified cells may appear in the vaginal smear.

Oestrus

In the oestrous *Setonix* the largest Graafian follicle measures about 2.8 mm diameter, the first maturation spindle is formed and mitoses are common in the granulosa cells. O'DONOGHUE (1916) reported, however, that mitoses were absent from the granulosa cells of nearly mature follicles in several other marsupials. At the second and subsequent ovulations of the breeding season there may be one or more large corpora lutea, formed after the earlier ovulation, in the ovaries. In the quokka and other monovular marsupials the degenerating corpus luteum of the earlier cycle is usually found in the alternate ovary to the one containing the ovulating follicle. The mitotic activity characteristic of gland and lumen epithelia during pro-oestrus continues during oestrus (Fig. 3). Vaginal smears taken at this time in the quokka show an increase in the number of cornified elements.

Changes following oestrus

Oestrus (as defined by the time at which the female will mate) is followed by ovulation. In *Setonix* (SHARMAN, 1955b) and *Didelphis* (HARTMAN, 1923b) ovulation occurs about one day after oestrus. HILL & O'DONOGHUE (1913) report the average interval between copulation and ovulation in *Dasyurus* as 5 or 6 days. Further observations with regard to the interval between oestrus and ovulation in *Dasyurus* and other Australian polyprotodonts are required especially as it appears that two or more matings, separated by an interval of some days, were necessary to ensure fertilization and embryonic development in *Phascogale flavipes* (B. J. MARLOW, pers. comm.). Ovulation is followed by the collapse of the Graafian follicle, the granulosa cells being retained so that they contribute to the tissues of the corpus luteum. O'DONOGHUE (1916) reported that the follicle collapsed after ovulation, before growth of the corpus luteum in *Trichosurus vulpecula*, but that follicular collapse did not occur in several other marsupial species. This observation must be interpreted with caution as it is probable that O'DONOGHUE had insufficient material available to draw valid conclusions about this matter. In general the microscopic structure of the corpus luteum appears similar in all marsupials although opinions differ as to whether the luteal cells are derived entirely from granulosa cells or whether the theca interna also contributes (see page 335).

The corpora lutea of *Dasyurus* (SANDES, 1903) and *Didelphis* (HARTMAN, 1923b) reach full size 3 days after ovulation; that of the quokka 5 days after ovulation. Degenerative changes in the corpora lutea of *Didelphis* are apparent 10 days after ovulation and at the succeeding oestrus (28 days after ovulation) the corpora lutea consist only of shrunken luteal cells and fat globules (MARTINEZ— ESTEVE, 1942). According to HILL & O'DONOGHUE (1913) the corpora lutea of the non-pregnant *Dasyurus* persist "some weeks at least and even then show no sign of degeneration". O'DONOGHUE (1912) states that he found perfectly preserved corpora lutea with "no signs of beginning to disappear" in an animal 20 days after ovulation. These observations are of little value for they are not related to possible degenerative changes in the luteal cells which may occur even while the corpus luteum is of large size. The first degenerative changes in the luteal cells of the quokka corpus luteum are apparent 20 days after ovulation. Its long duration after the onset of degenerative changes has already been referred to.

For the first two days after oestrus there is little change in the uterus. Mitoses are common in the epithelia during this period in the quokka. In *Didelphis* mitoses may persist, in greatly reduced numbers, 6 days after oestrus. Coincident with the cessation of active mitosis and correlated with the growth of the corpus luteum distinct changes become evident in the post-oestrous uterus of the non-pregnant marsupial (Fig. 4). These changes, subsequently designated luteal, may be induced in the ovariectomised quokka by the injection of progesterone after preliminary injections of oestrogen. They are most evident in the peripheral portions of the uterine glands where the epithelial cells undergo considerable increase in height. At the same time the cell nuclei become spherical (in contrast to their ovoid shape at oestrous and pro-oestrous stages) assume a basal or near basal position and display prominent basophilic nucleoli. The cell walls of the effected gland ephithelia readily take basic stains but the cytoplasm shows little affinity for either acid or basic stains. Staining with eosin-methylene blue at this time reveals that the cytoplasm shows no reaction for basophilia in contrast to its positive reaction at pro-oestrus, oestrus and post-luteal phases.

The luteal phase persists until about 12 days after oestrus in *Didelphis* and the onset of degenerative changes in the uterus is coincident with the first appearance of degenerative changes in the corpus luteum. In *Dasyurus* degenerative changes in the uterus occur 14 days after ovulation. The uterine luteal phase is further prolonged in *Setonix* and it only begins to lapse when the earliest degenerative changes occur in the corpus luteum; that is 20 days after oestrus (Fig. 2). Regression of the luteal phase is evident by decreased cell height of gland epithelia, the reappearance of a basophilic reaction in the gland cells of *Setonix* and by the appearance of cellular elements in the gland lumina. Epithelial cells are found in the glands of *Dasyurus* and *Didelphis* and leucocytes appear in the glands of *Setonix*.

The cornification of the lateral vaginae may increase during the first two days after oestrus in *Setonix* and extensive desquamation of lateral vaginal and urogenital sinus epithelia occur about 4 days after oestrus. This gives rise to one of the most characteristic changes in the vaginal smears seen during the oestrous cycle (SHARMAN, 1955a). TYNDALE—BISCOE (1955) described desquamation of the epithelium of the urogenital sinus of *Trichosurus* at "oestrus". As the animal in which desquamation occurred had a corpus luteum in one ovary this is actually a post-oestrous phase, perhaps corresponding to that at which desquamation occurs in the quokka.

Pregnancy

Gestation periods

The gestation period of the American opossum is 13 days ± 6 hours (REYNOLDS, 1952) and that of the quokka 27 days (SHARMAN, 1955b); few reliable observations have been made on other species. In *Dasyurus* (HILL & O'DONOGHUE, 1913) the variable period between insemination and ovulation made estimates uncertain and only two observations were made. One female was found with a single offspring in the pouch 8 days after mating but it was not known whether earlier copulation had taken place; a second produced young 16 days after observed copulation. The gestation period of *Trichosurus vulpecula* has been variously given as 21 days (JENNISON, 1927); 16 days (BOLLIGER & CARRODUS, 1940a); and 15—24 days (TYNDALE—BISCOE, 1955). LYNE & VERHAGEN (1957) record one instance of a gestation period being "not more than 21 days" in this species[1]. FLYNN (1930) recognised the occurrence of post partum oestrus and ovulation in *Bettongia* but mistook unimplanted quiescent blastocysts for stages in normal pregnancy, delayed implantation being then unknown in marsupials. He concluded that *Bettongia* had repeated pregnancies of "about 6 weeks duration" during the breeding season. This is almost certainly an erroneous conclusion. Because delayed implantation is

---

[1] The gestation period of *Trichosurus vulpecula* is now known to be 17^1/$_2$ days (LYNE PILTON & SHARMAN, *Nature, Lond.*, in the press).

now known to be a feature of reproduction in macropod marsupials (see page 354), earlier observations relating to the gestation period in this group are suspect. The pregnancy following post-partum oestrus may vary from the normal length to some five times this period depending on whether the first offspring is lost at some time during development or emerges from the pouch at the normal time (SHARMAN, 1955b).

Development

Fertilization and tubal journey

The marsupial egg is fertilized in the upper part of the Fallopian tube and fusion of the pronuclei occurs just before the egg reaches the uterus. Segmentation does not begin until the egg leaves the Fallopian tube. Fertilized (and unfertilized) eggs reach the uterus one day after ovulation in *Setonix* and also in *Didelphis* (HARTMAN, 1923b). These represent the shortest times in the Fallopian tubes known for the eggs of mammals.

Rate of embryonic development

The American marsupial *Didelphis* takes more than half the gestation period to arrive at differentiation of the cell layers (HARTMAN, 1952). In the last $5\frac{1}{2}$ days of the 13 day period elapsing between copulation and birth the embryo develops from a "delicate vescicle smaller than a pinhead to newborn opossum". Amongst eutherian mammals which have a short gestation period (e.g. mouse) mesoderm appears about one-third of the way through gestation, or earlier.

The quokka embryo 17 days after insemination measures approximately 5 mm in greatest length (greatest length at birth, 10 days later, is 16 mm but the foetus is considerably curved — see Fig. 6). At 17 days the gut is open to the yolk sac for a considerable portion of its length, the nerve cord is only incompletely closed off, the amnion folds are not completely fused and the allantois is represented only by a slight swelling of the posterior region of the gut. By contrast the allantois appears at $7\frac{1}{4}$ days in the mouse (SNELL, 1941). A number of small excretory tubules are present in the 17 day quokka which, following BUCHANAN & FRASER (1918), can probably be designated pronephric.

These observations indicate that the rate of embryonic development in marsupials is at first slow but is strikingly increased during the latter stages of pregnancy.

Foetal membranes

On the basis of placental structures marsupials may be grouped under four headings.

1. Marsupials with yolk-sac (chorio-vitelline) placentation in

which the allantois is buried in folds of the splanchnopleure and does not make contact with the chorion.

To this group belong the American marsupial *Didelphis* (SELENKA, 1887) the Australian phalangers *Trichosurus vulpecula* and *Pseudocheirus convolutor* (HILL, 1901, FLYNN, 1923b) and the Australian macropods *Macropus giganteus, M. robustus, Bettongia cuniculus, Aepyprymnus rufescens, Protemnodon rufogrisea, P. parma, Potorous tridactylus* and *Setonix brachyurus* (OWEN, 1837; SELENKA, 1892; SEMON, 1894; HILL, 1895, 1901; FLYNN, 1923b, 1930; SHARMAN, 1955b).

2. Those with yolk-sac placentation in which the allantois comes into contact, briefly, with the chorion but in which the allantoic vessels are degenerate and do not vascularise a placenta.

Amongst marsupials so far investigated only *Dasyurus quoll* (HILL, 1900) conforms to this pattern.

3. Those with yolk-sac placentation in which the allantois lies in contact with the chorion.

Phascolarctos cinereus is included here because, according to CALDWELL (1884) and SEMON (1894) the allantois does not vascularise a placenta. AMOROSO (1952) states however that true allantoic placentation is found in *Phascolarctos*. Because of the lack of clarity of CALDWELL's (1884) description *Protemnodon rufogrisea* has been considered to belong in this group but as shown by HILL (1895) it is properly placed in group (1) with other macropod marsupials.

4. Marsupials exhibiting both yolk-sac and allantoic placentation.

To this group belong *Perameles gunnii, P. nasuta* and *Thylacis obesulus* (HILL, 1895, 1898, 1901; FLYNN 1923b). AMOROSO (1952) states that allantoic placentation also occurs in the wombat *(Phascolomis)* and in the koala *(Phascolarctos)*.

These determinations of placental relationships rest only on superficial dissections in the majority of the species listed above. Detailed histological examinations have been made of the placentae of *Perameles* (HILL, 1898, 1901; FLYNN, 1923b), *Dasyurus* (HILL, 1900), *Bettongia* (FLYNN, 1930) and *Setonix* (SHARMAN, unpublished).

In *Bettongia* (FLYNN, 1930) fusion of placental and maternal tissues is most intimate in the region of the vascular trilaminar omphalopleure. Foetal and maternal blood streams are separated by uterine and foetal membrane epithelia, by the two endothelia and by maternal connective tissue. Cell debris, leucocytes and other material form "uterine milk" which is absorbed and ingested by foetal ectodermal cells in certain regions of the bilaminar omphalopleure.

The extra-embryonic membranes of *Setonix* 15 days after insemination consist virtually of bilaminar omphalopheure; only a little mesoderm is extended beyond the embryonal area and this, together with ectoderm and endoderm, constitutes the beginning of the trilaminar omphalopleure. The mesoderm of later embryos never extends more than about half-way beyond the adembryonic pole of the blastocyst and in the abembryonic hemisphere the bilaminar omphalopleure exists throughout gestation. In the 15 day embryo the membranes are prevented from making close apposition with the uterine wall because of an eosinophilic membrane which surrounds the entire blastocyst. This apparently represents the attenuated shell membrane which CALDWELL (1887) and HILL (1910) recognised in marsupial eggs. CALDWELL and HILL both consider the shell membrane to be homologous with the shell of the monotreme egg.

In the 17-day embryo the extension of mesoderm has converted a large portion of the bilaminar omphalopleure into a trilaminar vascularized structure. The shell membrane is still present in embryos of this age but two days later this has almost entirely disappeared.

The embryo *Setonix*, 19 days after insemination of the mother, is orientated with the head towards the anterior end of the uterus and an extensive pro-ammion is present in the head region. The yolk-sac placenta and a small area of true chorion are closely applied to the uterine epithelium. The allantois is small and does not reach the chorion. Two days later (Fig. 5) the embryo has turned so as to become orientated head down in the uterus and the allantois has increased in size. At 23 days after insemination the allantois is large but buried in folds of yolk-sac splanchnopleure which appear to effectively preclude contact with the chorion. The membranes investing the maternal epithelium consist of bilaminar omphalopleure in the upper half of the uterus, vascular trilaminar omphalopleure forming a broad band around the lower half and so extended as to leave only a small discoidal area of true chorion near the opening of the uterine neck. The maternal uterine epithelium is complete but flattened where contact is made with the foetal membranes.

Material giving a strong positive periodic acid-Schiff (P.A.S.) reaction occurred in the gland cavities, in the tissues of the vascular omphalopleure and chorion and to a lesser extent in the bilaminar omphalopleure of a 23 day quokka embryo. The material in the gland cavities (which in places exuded into the lumen from the gland mouths) was not digested by saliva nor did it give a positive reaction for glycogen, whereas much of the P.A.S. positive material associated with the membranes was undoubtedly glycogen. Droplets of glycogen were also found amongst the red blood corpuscles in the major vessels of the vascular omphalopleure — particularly those in the lower region of the uterus. These observations indicate that the vascular omphalopleure and chorion (in which the greater amounts of P.A.S. positive material occurred) are of importance in supplying nutritive materials to the embryo, a finding which should be compared with HILL's (1900) conclusion that the poorly vascularised trilaminar omphalopleure of *Dasyurus* served mainly for gaseous exchange. The P.A.S. positive material in the gland cavities

PLATE I

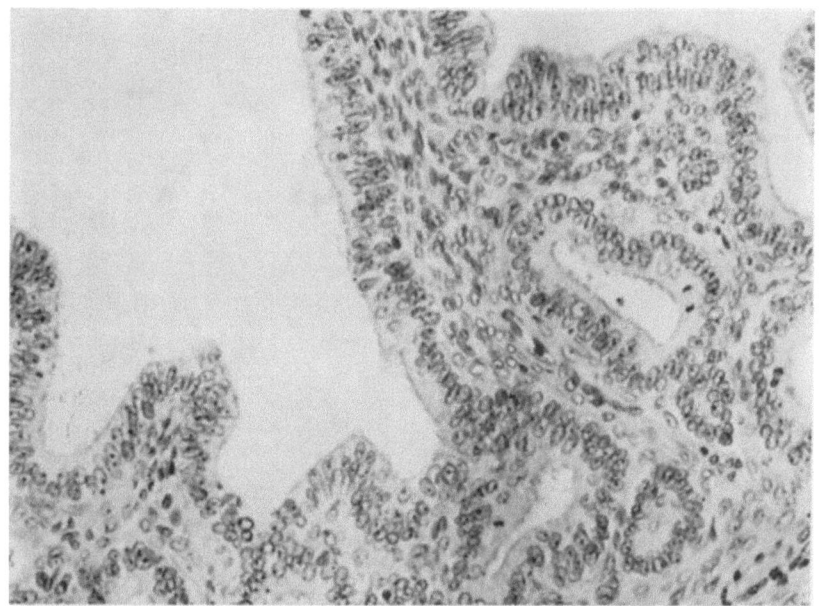

Fig. 3. Uterine glands of *Setonix* at oestrus. X 250.

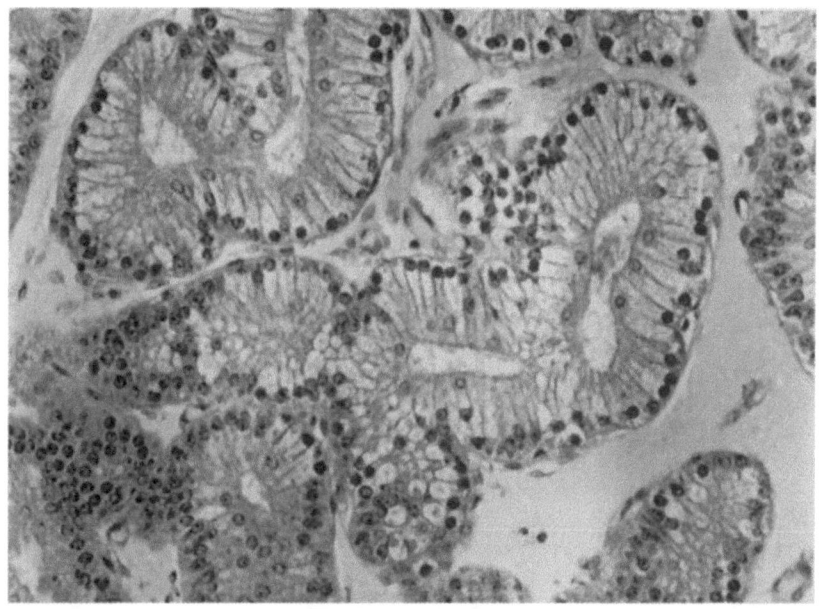

Fig. 4. Uterine glands of *Setonix* in luteal phase (10th day of oestrous cycle). X 250.

G. B. SHARMAN

PLATE II

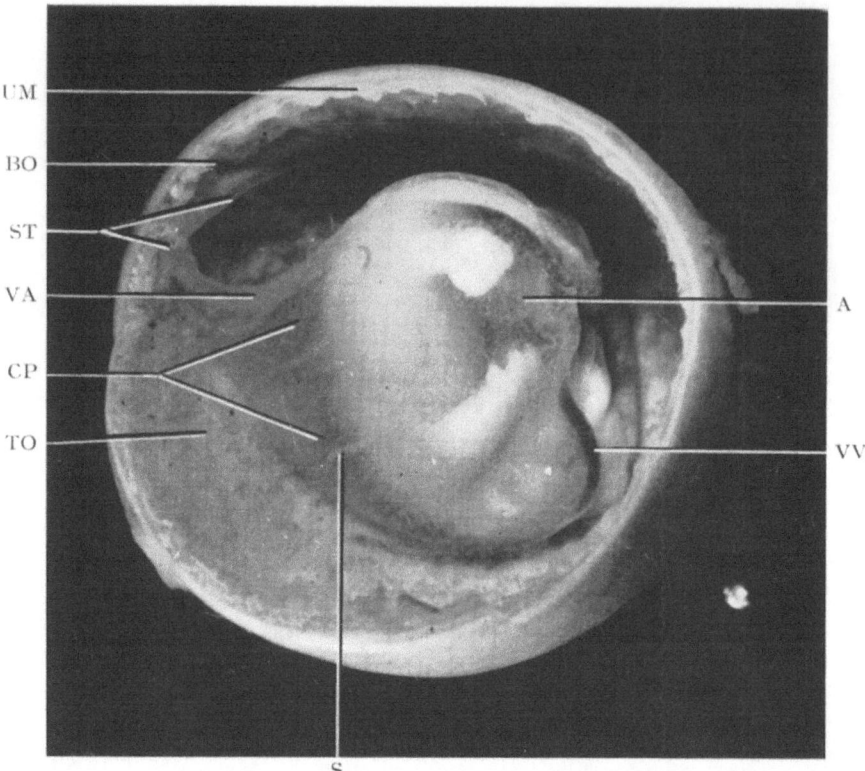

Fig. 5. *Setonix*. Dissected uterus from pregnant female 21 days after insemination. The top half of the uterus and nearly all the associated bilaminar omphalopleure and part of the trilaminar omphalopleure have been cut away. Photo taken looking into the uterine lumen from the anterior end. X 5. A, position of allantois; BO, bilaminar omphalopleure; CP, position of chorion; S, blood vessel in yolk-sac splanchnopleure; ST, sinus terminalis (cut near junction with vitelline artery on one side); TO, vascular trilaminar omphalopleure; UM, uterine mucosa; VA, vitelline artery; VV, vitelline vein.

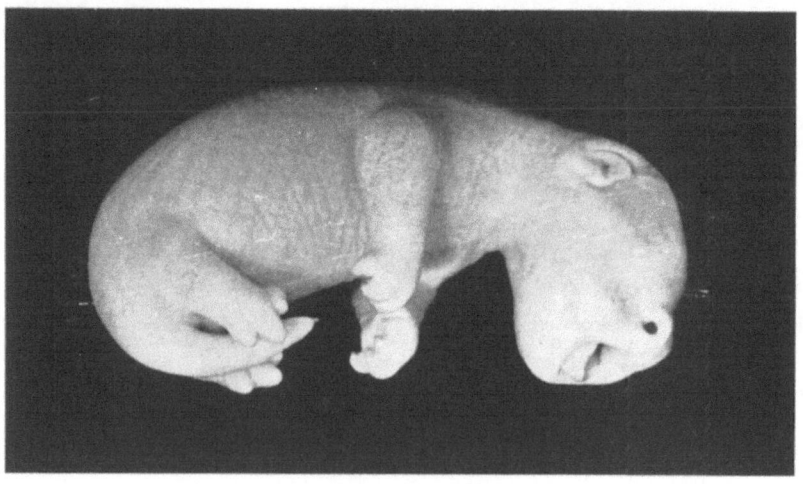

PLATE III

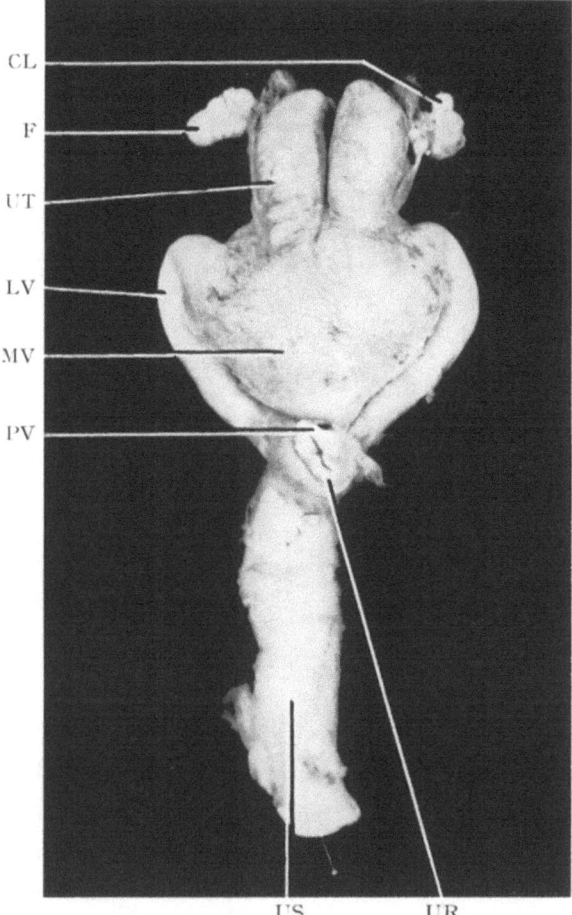

Fig. 7. *Trichosurus vulpecula*. Reproductive system of oestrous female, ventral view, X 1¼. CL, corpus luteum of preceding ovulation; F, mature Graafian follicle; LV, lateral vagina; MV, median vaginal cul-de-sac; PV, position where pseudovaginal canal is formed connecting culs-de-sac and urogenital sinus at parturition; UR, urethra; US, urogenital sinus; UT, uterus.

PLATE IV

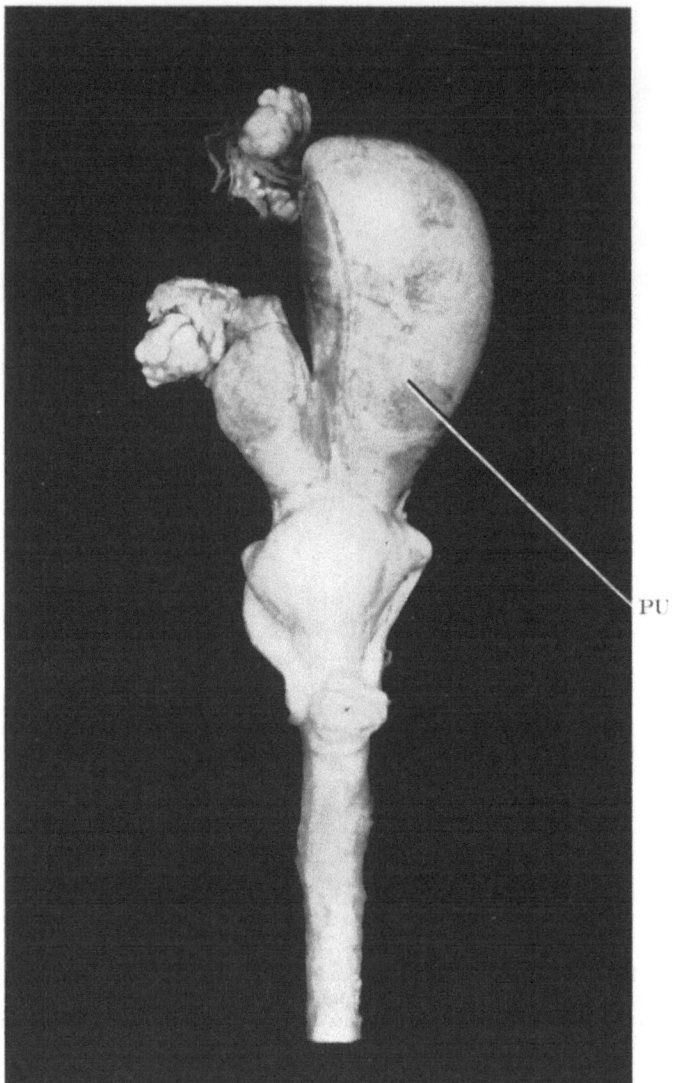

Fig. 8. *Trichosurus vulpecula*. Reproductive system of pregnant female, ventral view. X 1¼. Note that the vaginal complex is smaller than at oestrus (cf. Fig. 7) and that the pregnant uterus (PU) greatly exceeds the non-pregnant uterus in size

is undoubtedly histotrophic but the source of the material associated with the vascular omphalopleure is uncertain. Where the membranes were separated from the maternal epithelium material consisting of leucocytes and other cellular detritus was observed. An embryotrophic function was attributed by FLYNN (1930) to such material in *Bettongia*.

The yolk-sac placentae of *Protemnodon rufogrisea* and *Potorous tridactylus* closely resemble that of the quokka (SHARMAN, unpublished).

In *Dasyurus quoll* (HILL, 1900) the pro-amnion persists throughout gestation and invests the embryo as far back as the forelimbs. The unattached omphalopleural walls of adjacent embryos fuse where they come in contact and form common partitions separating the yolk-sac cavities. The allantois is long and band-like and there is no differentiated allantoic stalk. The vascular supply to the allantois was degenerate in the earliest stage studied and further reduced in later stages. In the third stage studied by HILL the allantois was spread beneath, and in places in close apposition with the chorion but devoid of blood vessels over its greater part. Later the degeneration of the allantoic blood vessels was complete and the allantois was not in contact with the chorion.

The vascular omphalopleure accurately follows the contours of the maternal uterine epithelium and is in intimate contact with this layer but both foetal and maternal capillary systems are poorly developed. In the earliest stage studied the bilaminar omphalopleure consisted of large cubical cells at the equatorial region and a layer of thin, mostly flattened cells at the lower pole. Later the equatorial bilaminar omphalopleure becomes attached to the uterine epithelium by pseudopodia like processes from the trophoblastic ectoderm cells which pass through the cells of the uterine epithelium engulfing the parts of cells occurring in their course. The trophoblastic ectoderm cells increase in size, their processes surround and enclose segments of uterine epithelium and maternal capillaries and the trophoblast, together with underlying endoderm, is converted into a syncytium by breakdown of cell walls. The separated portions of the maternal uterine epithelium degenerate and form nests of cells which, according to HILL, superficially resemble the nuclear nests in the syncytium of *Perameles*. In this equatorial region of the bilaminar omphalopleure the yolk-sac placenta therefore consists of a mixture of foetal and maternal elements. The foetal part consists of trophoblastic syncytium, with large deeply staining nuclei derived from ectoderm and endoderm, and the maternal part is of capillaries and isolated groups of degenerating uterine epithelial cells. HILL suggested that the vascular omphalopleure of *Dasyurus* serves as an organ for gaseous exchange and that the equatorial region of the bilaminar omphalopleure is concerned with nutrition; the engulfed material passing to the yolk-sac and hence to the gut of the embryo. The lower part of the bilaminar omphalopleure was not considered to be of functional importance in the nutrition of the embryo.

In *Phascolarctos* CALDWELL (1884) and SEMON (1894) have shown that a typical yolk-sac placenta, consisting of vascular and non-vascular regions, is present. Both authors describe the allantois as reaching the chorion but not forming a vascular placenta but AMOROSO (1952) asserted that *Phascolarctos* has allantoic placentation. PEARSON (1947) considered the allantois of *Phascolarctos* to have only a respiratory function. As CALDWELL's (1884) description is fragmentary and SEMON (1894) does not appear to have studied the histology in detail more work is needed on the foetal membranes of this species.

In the bandicoots *(Perameles* and *Thylacis)* the blastocyst becomes attached by enlarged chorionic ectodermal cells to a thickened discoidal area of uterine vascular syncytium (FLYNN, 1923b). AMOROSO (1952) considered this to be analogous to the symplasma of higher forms. It is formed of uterine epithelial cells which lose their boundaries and in which the nuclei clump in nests. Blood vessels form a network just beneath and on the surface of the symplasma. The chorionic ectodermal cells differentiate into a basal layer, the cytotrophoblast, and a peripheral layer, the syncytiotrophoblast, which invades the symplasma. The chorion is vascularised by allantoic vessels and as the cytotrophoblast is progressively converted to syncytiotrophoblast the foetal allantoic capillaries approach closer and closer to the maternal capillaries. Finally the foetal and maternal blood streams are apparently separated by only the closely opposed capillary walls — an arrangement referred to by AMOROSO (1952) as endothelio-endothelialis. AMOROSO (1955a) has compared the complex chorio-allantoic placenta of *Perameles* with that of eutherian mammals. A well developed yolk-sac placenta is present in addition to the allantoic placenta. In the bandicoots, as in *Dasyurus, Bettongia,* and *Setonix,* the foetal membranes are retained in the uterus at parturition where they undergo resorption.

HILL (1898) and FLYNN (1923b) interpreted the allantoic placenta of *Perameles* as evidence that marsupials were derived from ancestors with allantoic placentation. These authors however regarded *Perameles* as a primitive marsupial which is, as PEARSON (1947) has stated, unlikely. Furthermore the primitive marsupials, *Didelphis* and *Dasyurus,* have yolk-sac placentation whilst allantoic placentation is found in two specialised genera, *Phascolomis* and *Phascolarctos.* In this context FLYNN's (1923b) observation that a maternal "vascular syncytium" is found in the phalanger *Pseudocheirus,* which is closely related to *Phascolarctos,* is of interest. Syncytia are not found in the pregnant uteri of other phalangers (e.g. *Trichosurus),* which have purely yolk-sac placentation, but do occur in

Perameles and presumably may also occur in *Phascolarctos*. If so there remain no grounds for FLYNN's contention that the vascular syncytium of *Pseudocheirus* represents "the remains of an ancestral trophospongial proliferation": it may equally well represent a stage in the evolution of allantoic placentation. Such a view is acceptable on the analogy of reptilian allantoic placentation which has arisen independently in several lines (WEEKES, 1935). Finally HILL (1932, 1949), although maintaining his view that marsupials were derived from a protoplacental stock, concluded that the allantoic placenta of *Perameles* had been independently evolved within the limits of the marsupial order, if not within the limits of the family Peramelidae, and did not therefore possess the primitive significance he originally attached to it. This is in direct contradiction to the conclusions reached by FLYNN (1923b) who thought that the resemblances of the placenta of *Perameles* to that of certain eutherians ruled out the possibility of independent evolution. Until the placentae of marsupials are further studied no firm decision can be made but the hypothesis of independent derivation, from a primitive yolk-sac placenta, of allantoic placentation in at least two lines of marsupials is at present tenable. This view is supported by PEARSON (1947, 1949) who concluded that "the absence of allantoic placentation in most marsupials is unlikely to be the result of a secondary abbreviation of foetal life" and that the prototypal marsupial possessed only yolk-sac placentation.

Parturition

Pseudovaginal parturition in marsupials has already been discussed on page 336.

Parturition in *Didelphis* has been carefully described by REYNOLDS (1952) who witnessed the birth of 4 litters and observed 2 further females during parturition. In the hours preceding parturition the pouch was thoroughly cleaned by licking. Only one female licked a path in the fur between vulva and pouch which behaviour MCGRADY (1938) believed to be normal. According to REYNOLDS the females did nothing to assist the young in the journey to the pouch except to assume a sitting position with the tail extended between the hind legs and to relax the neck of the pouch. The young were born free of membranes and emerged swinging the forelimbs. New born young grasped the first object in reach: usually the mother's fur. The journey to the pouch was accomplished by continuing the swinging motion of the forelimbs and the young passed rapidly through the fur to the pouch as though swimming. In one case the journey from vulva to pouch was timed and occupied only $16\frac{1}{2}$ seconds. Only 60% of the 57 young borne by 4 observed females reached the pouch.

MATTHEWS (1943) has reviewed pertinent information about

parturition in kangaroos. The female kangaroo assumes a reclining posture, sitting on its haunches, with the tail extended along the ground between the hind legs. The maternal cloaca protrudes during parturition thus reducing the distance between it and the pouch. The female licks a track to the pouch up which the newborn kangaroo climbs but no further assistance is rendered by the mother. The time taken for young of various macropods to reach the pouch varies from 5 to about 30 minutes according to MATTHEWS.

Anatomical and histological changes during pregnancy
Fertilization and tubal life

Fertilization in marsupials (in common with other mammals — see AUSTIN & BISHOP, 1957) occurs in the upper part of the Fallopian tube. In most mammals the ova take 3—4 days to traverse the Fallopian tube but the time in *Didelphis* is one day and it is about 7 days in dogs and cats (see BOYD & HAMILTON, 1952). *Setonix* is similar to *Didelphis* in that fertilized (and unfertilized) eggs reach the uterus about one day after ovulation. Two days after copulation (one day after ovulation) the fertilized egg of *Setonix* is found in the lower part of the tube near the opening into the uterus. Male and female pronuclei are not fused at this time (SHARMAN, 1955b). The segmenting 4 celled egg of *Trichosurus* is found in the lower half of the uterus but in eutherian mammals (e.g. rat, mouse) 4 cell, 8 cell and even later stages are normally found in the tubes.

Post-oestrous changes in pregnant females

In *Dasyurus* (HILL & O'DONOGHUE, 1913) the corpora lutea of pregnant females are identical with those of non-pregnant females and pouch and mammary gland changes, as well as changes in the uterine mucosa, are similar in pregnant and non-pregnant animals. The secretions of the corpus luteum of pregnancy induce changes of the same degree and duration in the uterus of pregnant females as they do in the uterus of non-pregnant females. In monovular marsupials it is possible to accurately compare the changes in the non-pregnant uterus of a pregnant female with those occurring in females that have not been mated (Figs. 7, 8). In the quokka an identical sequence is found in every respect. Such changes as distinguish the pregnant from the non-pregnant uterus (Fig. 8), or from the uterus of a non-fertilized female, may be attributed to distension by the embryo and its membranes. In *Didelphis* and *Dasyurus* parturition is coincident with the beginning of degenerative changes in the uterine mucosa which may themselves be attributed to cessation of secretory activity by the corpora lutea. In the pregnant quokka degenerative changes in the corpus luteum begin 20 days after oestrus, as in non-pregnant females, but the quokka differs from *Didelphis* and *Dasyurus* in that pregnancy outlasts the uterine luteal phase. The latter stages of pregnancy in the quokka are therefore accompanied by maturation of an ovarian follicle and ovum and pro-oestrus changes in the uterine mucosa (Figs. 1, 2). Parturition occurs 27 days after insemination and is followed by a post-partum oestrus and ovulation which occur at the time expected had pregnancy not taken place, i.e. 28 days after the preceding oestrus and ovulation. In *Didelphis* the stimulus exerted by the young in the pouch appears to preclude ovulation during the nursing period since HARTMAN (1923a) has shown that oestrus recurs several days after the young are removed from the pouch of lactating females. The post-partum ovulation of the quokka is however imminent at the time of parturition because maturation of the ovum and follicle occur during the latter part of pregnancy before the beginning of suckling (Fig. 2).

SHARMAN (1956) concluded that pregnancy in *Setonix* is the equivalent of a specialised oestrous cycle during which all the hormone induced changes, which normally follow cyclic ovulation, are exploited by the developing embryo. In less advanced marsupials *(Dasyurus, Didelphis)* the embryo is retained in the uterus only until the corpus luteum ceases its active function and the young are less developed at birth than are those or *Setonix*. C. J. HILL (1933, 1941) has described phases in the uteri of monotremes, which essentially resemble the phases induced in the uteri of marsupials by ovarian hormones. The evolution of viviparity in mammals may well have been a process of adaptation of the developing embryo to those changes, mediated by follicular and luteal hormones, which were already present in oviparous ancestors.

Various explanations have been advanced to explain the reason for the short gestation period of marsupials. The vaginal anatomy does not appear, as had been suggested, to preclude the birth of larger young for pseudovaginal parturition makes it unnecessary for the foetus to reach the exterior through the tortuous lateral vaginal canals. The median birth passage, although small in some species, can readily be dilated under the influence of appropriate hormones. In *Trichosurus*, for example, the culs-de-sac become enormously distended at oestrus (SHARMAN, 1954b) and in *Setonix* distension prior to post-partum oestrus coincides with parturition (SHARMAN, 1955b). No marsupial is known in which the gestation period is longer than the time occupied by one oestrous cycle whereas in all polyoestrous eutherians pregnancy extends over the time normally occupied by several cycles. AMOROSO (1955b) suggested that the secretion of a luteotrophic principle, by the placenta, at the time of implantation was the initial step which made it possible for eutherian mammals to prolong the life of the corpus luteum and thus extend the gestation period beyond the limits of the oestrous cycle. The same author (AMOROSO, 1949) suggested that the relatively simple placentae of marsupials have not evolved those endocrine functions which enable the young to be retained in the uterus for prolonged periods. Some knowledge of possible endocrine synthesis by the complex allantoic placenta of *Perameles* is highly desirable for this species produces young which are much larger and structurally more advanced than those species of comparable size which have only yolk-sac placentation (e.g. *Dasyurus* and *Didelphis*). The inference is that the gestation period of *Perameles* is much longer than that of the other species and that this may have been made possible by the evolution of secretory activity in the placenta. Only one marsupial, the primitive polyprotodont *Didelphis*, has been studied from the point of view of possible hormone secretion by the placenta. HARTMAN (1925) showed that in this species castration performed at any stage of pregnancy caused death of the embryos. It is therefore assumed that the maintenance of pregnancy in this species does not involve hormonal mechanisms other than those that participate in the oestrous cycle (AMOROSO, 1955b).

Lactation controlled delayed implantation

In the quokka mating and fertilization may occur at post-partum oestrus and ovulation. If the foetus of the preceding pregnancy is immediately removed from the pouch a normal 27-day pregnancy ensues (Fig. 1). If, on the other hand, the foetus enters the pouch and becomes attached to a teat segmentation of the fertilized ovum of post-partum ovulation proceeds only as for as an early (0.3 mm diam.) blastocyst stage (Fig. 2). While the pouch foetus undergoes development this blastocyst remains in a quiescent state, a situation analogous to the lactation-controlled delayed implantation which occurs in rats and other rodents suckling large litters (reviewed by HAMLETT, 1935). The stimulus exerted by the sucking pouch foetus inhibits development of the corpus luteum of lactation formed following post-partum ovulation. This corpus luteum does not exceed 2.5 mm diameter during the period of delayed implantation whereas cyclic corpora lutea or corpora lutea of pregnancy exceed 4 mm diameter (Fig. 2). While it remains of small size the corpus luteum of lactation does not initiate luteal changes in the uterus. If, as an experimental procedure, the foetus be removed from the pouch the corpus luteum increases in size, a luteal phase occurs in the uterus and implantation and normal development of the hitherto quiescent blastocyst occur. Removal of the foetus before it has completed pouch life causes cessation of lactation but in some circumstances implantation and development of the quiescent blastocyst may occur in the presence of lactation. Thus the second pregnancy may proceed while the offspring of the earlier pregnancy is emerged from the pouch and suckling from the exterior. In one case the uterine blastocyst resumed development, after a long period of delayed implantation, while the foetus of an earlier pregnancy occupied the pouch. The amount of milk secreted or the time spent by the pouch foetus actually sucking are therefore probably important in controlling the delayed period. In this context it is important to note that the foetus may supplement its milk diet with herbage, cropped by extending its head from the pouch, in the latter stages of pouch life. The unimplanted blastocyst of the quokka may remain quiescent for periods of up to about 5 months (SHARMAN, 1955b).

Delayed implantation is known to occur in the quokka and in the macropod *Protemnodon eugenii* (SHARMAN, 1955c). It is suspected to occur in *Macropus rufus* in which JONES (1923) reports the case of a captive female producing young many months after death of its mate. In this instance it is significant that the female was suckling an earlier offspring when the male died. FLYNN's (1930) observation that early blastocysts are found in the uteri of lactating *Bettongia cuniculus* and *Potorous tridactylus*, and that post-partum oestrus occurred, may be interpreted as indicating that delayed implantation is also a feature of reproduction in these species. All these observations relate to macropod marsupials in which group delayed implantation is probably general. BOWLEY's (1939) observation that "delayed fertilization" occurs in the phalanger *Cercaërtus concinnus* and some other observations on this species (SHARMAN, unpublished) indicate that delayed implantation also occurs in some phalangers.

To date no experimental studies have been made on delayed implantation in marsupials. Available evidence suggests that the factor regulating progesterone secretion is not itself secreted so that the corpora lutea remain inactive and do not induce changes (which the embryo presumably requires for implantation and development) in the uterus (SHARMAN, 1955b). However COCHRANE & MEYER (1957) have shown that the injection of progesterone into rats ovariectomised on the 3rd day after mating produces a quiescent phase in the embryos which continues until oestrogen is given in addition to progesterone. COCHRANE & MEYER's findings support the conclusions reached by WEICHERT (1942a, b) who showed that oestrogen induced implantation and development in hitherto delayed blastocysts.

The precise adaptive significance of delayed implantation is conjectural. The post-partum oestrus is important in species producing a single offspring for if the foetus fails to complete the journey to the pouch the mother is immediately ready to undergo a second pregnancy. Delayed implantation may be important in those macropod marsupials which congregate at areas where feed and water are available during the summer mating season. Pregnancy and post-partum fertilization occur in a short space of time and females thereafter carry a foetus in the pouch and a quiescent blastocyst in the uterus. During later periods, when feed and water are relatively abundant, the parents may be widely dispersed so that following emergence of the first young, or after its premature loss, the female is able to produce a second offspring without a further mating. This is presumably of some significance if the animals are thinly dispersed over a wide area.

Post-partum oestrus occasionally occurs in *Trichosurus* in which the gestation period is shorter than that of *Setonix* (TYNDALE—BISCOE, 1955). If fertilization occurs however there is no period of delayed implantation so that two foetuses, differing in age by the length of the gestation period, eventually occupy the pouch. *Trichosurus* is thus an interesting example of a marsupial in which the gestation period is sufficiently long to sometimes allow the maturation of an ovarian ovum before the onset of lactation anoestrus. This species has not however evolved the mechanism by which lactation causes arrested development of the corpus luteum and subsequent delayed implantation.

Development of Young in the Pouch

Anatomy of the new-born young

HILL & HILL (1955) have described the precocious development of certain structures used by the young in its journey to the pouch and attachment to the nipple. Precocious features of the new-born marsupial (based on HILL & HILL's description of the new-born *Dasyurus* unless otherwise stated) follow.

(1) The forelimb musculature is greatly developed compared with that of the hindlimbs. Deciduous claws on the forelimbs, which are used to grasp the mother's fur and are shed after the young reach the pouch, are found in the new-born *Dasyurus*, *Didelphis* (HARTMAN, 1928; MCGRADY, 1938) and in bandicoots (LYNE, 1952). Strongly developed recurved claws, which HILL & HILL (1955) consider to be deciduous are found in *Macropus*, *Trichosurus*, *Phascolarctos* and *Phascolomis*. DUNNET (1956) states however that those of *Trichosurus* are the precociously developed permanent claws. Claws which also appear to be permanently retained are present on the forelimbs of the newborn quokka (SHARMAN, 1957).

(2) The voluntary musculature in the forepart of the body is highly developed.

(3) The new-born *Dasyurus* has an oral shield surrounding the mouth which enables the young to suck the nipple into the oral cavity.

(4) The tongue is large and has fully developed musculature innervated by hypoglossal nerves. The tongue of the new-born *Setonix* can be seen in Fig. 6.

(5) The epiglottis and associated structures are arranged so as to allow breathing and the swallowing of milk to go on simultaneously.

(6) The new-born marsupials has an "impermeable" keratinised external covering to prevent dessication while crawling from external genital opening to pouch.

(7) The olfactory apparatus is in an advanced condition in contrast to the embryonic condition of the eye and ear at birth.

(8) The medulla oblongata and cervical and thoracic regions of the spinal cord show a high degree of differentiation and the cranial, cervical, and thoracic spinal ganglia and nerves are early established. The parts of the nervous system necessary

for controlling the movements of the forelimb and sucking and respiratory movements are thus well developed at birth.

(9) The stomach and duodenum are in an advanced condition compared to the rest of the gut. The main absorptive region of the gut in the new-born marsupial is presumably the duodenum for villous folds are greatly developed in this region in *Dasyurus* and also in *Didelphis* (HEUSER, 1921).

The following are embryonic features of the new-born marsupial, not already mentioned.

(1) New-born marsupials are greatly exceeded in size by new-born eutherians. The greatest length (G.L.) of *Dasyurus* at birth is 7 mm (HILL & HILL, 1955) of *Didelphis* 11 mm (MCGRADY, 1938), *Perameles* 14 mm (FLYNN, 1923b), *Thylacis* 14.25 mm (HILL & HILL, 1955), *Trichosurus* 14.5 mm (BUCHANAN & FRASER, 1918), *Phascolarctos* 16.5 mm (O'DONOGHUE, 1916) and *Setonix* 16 mm. Birth weights are *Dasyurus* 12.5 mg, *Didelphis* 162 mg (HARTMAN, 1952), the quokka 450 mg and *Macropus* about 1.25 g. All of these weights, except that of *Macropus*, are exceeded by that of a new-born mouse (about 1 g is an average figure for this species).

(2) The lungs of the new-born *Dasyurus* are devoid of branchioles, alveolar ducts and alveoli and take the form of thin-walled sacs. The red blood corpuscles of the new-born *Dasyurus* are nucleated and those of the quokka are nucleated until at least the 23rd day of pregnancy.

(3) The mesonephros of the new-born marsupial is functionally active before and for some time after birth (BUCHANAN & FRASER, 1918). The gonads of *Dasyurus* are undifferentiated at birth and neither pouch nor scrotum may be distinguished in the new-born of this or other species of marsupials.

Composition of marsupial milk

BOLLIGER & PASCOE (1953) found that total sugar content of milk decreased and fat content increased with the age of the pouch young in *Macropus robustus*. The principle carbohydrate was thought to be other than lactose. BOLLIGER & PASCOE suggest that pentose and other carbohydrate, proteins and nitrogenous compounds, usually not associated with other types of mammalian milk, are present in the milk of *Macropus*. In different samples taken the fat content was equal to or much higher than that of most other mammals. A single sample of quokka milk analysed showed a high (19.6%) fat content (P. J. BENTLEY, unpublished) which compares closely with the amount of fat found in a single sample of monotreme milk (MARSTON, 1926).

O'DONOGHUE (1911) stated that no sign of milk was found in the alveoli or ducts of the mammary gland of *Dasyurus* with new-born young in the pouch. True milk appeared in the alveoli about 30 hours after birth and in the ducts 6 hours later. Colostrum-like secretion was present earlier. HILL & HILL (1955) state that only a clear serous fluid may be expressed from the nipple of *Dasyurus* at the time of parturition. These observations are of importance in any consideration of the hormone control of lactation in marsupials.

Growth rates of pouch young

HILL & O'DONOGHUE (1913) distinguish two phases in the nursing period of marsupials.

(1) The period of fixation. The lips are fused along the sides and the free end of the nipple becomes enlarged inside the mouth so that the foetus cannot become detached. This period in *Dasyurus* lasts some 7-8 weeks.

(2) The free period. The young are still dependent on the mother's milk but may release the nipple for short periods. This period lasts until the young are able to feed themselves.

It is generally accepted that the marsupial foetus cannot reattach if removed from the nipple during the first few weeks of pouch life. The young of *Pseudocheirus peregrinus* will however reattach when they weigh 0.87 g or less, measure 22 mm G.L. and are less than 3 weeks old (SHARMAN, unpublished). At this time the lips are fused along the sides and the mouth opening is circular and terminal. HARTMAN (1952) described several experiments which appear to indicate that the young marsupial, carefully removed from the mother's nipple, is able to reattach in the early stages of pouch life.

The following observations on the growth rates of pouch young are those of HILL & HILL (1955), REYNOLDS (1952), LYNE & VERHAGEN (1957) and some unpublished personal observations.

The sexes of pouch young of *Didelphis* become distinguishable at 13—14 days and of *Dasyurus* at 19 days. The pouch of the young female quokka appears at 14—15 days, the scrotum of the male two days later.

Papillae of facial vibrissae appear between the 11th and 15th days in *Didelphis* and at 19 days in *Dasyurus*. Vibrissae erupt from the papillae and short fine hairs appear on the head at 16—20 days in *Didelphis* and at 25 days in *Dasyurus*. Vibrissae and the earliest hair develop much later in *Trichosurus* and *Setonix*.

The young of *Didelphis* begin releasing the nipple for short periods at 48 days while the lips are still fused along the sides. The young of *Dasyurus* are able to reattach at 57—58 days. The young *Trichosurus* is able to reattach at 46 days and the young *Setonix* at 61 days. As previously mentioned some marsupials are able to reattach at much earlier ages.

The eyes open in *Didelphis* at 58—72 days, in *Dasyurus* at about 75 days, in *Trichosurus* at 87—122 days and in the quokka at about 105 days.

Dasyurus is completely furred at a little over 60 days, *Trichosurus* at 100 days and the quokka at 130 days. The young of *Dasyurus* emerge completely at about 3 months. *Didelphis* at 75 days, *Trichosurus* at $3\frac{1}{2}$—4 months and the quokka at 5—$5\frac{1}{2}$ months. Effective temperature control is established in *Didelphis* at 94 days. In quokka pouch young a degree of homoiothermy is established at 158 g weight (about 4 months of age) (BARTHOLOMEW, 1956).

HILL & HILL (1955) conclude that the nursing period of *Dasyurus* is longer than that of *Didelphis* because the young of the former are less advanced at birth than those of the latter. There are no valid reasons for this conclusion since the young of macropod marsupials are much more advanced at birth than those of either *Dasyurus* or *Didelphis* yet the nursing period is also much longer.

REYNOLDS (1952) studied the pouch environment in *Didelphis*. He found the CO_2 content of the pouch to be 2% in females with early pouch young and as high as 6% in females with later pouch young. The lowest pouch temperatures and the greatest range of variation were found in anoestrus females and in cyclic females without pouch young. The temperature of the pouch was consistently higher than that of the rectum and showed a lower range of variation (1.7° C vs. 2.3° C) in females with pouch young.

MINCHIN (1937) observed a young koala energetically eating a soft substance which exuded from its mother's rectum. The exuded substance differed in texture from faeces and consisted of "peptonised gum leaves from the upper bowel". MINCHIN subsequently observed that all young koalas, which had recently emerged from the pouch and were being weaned, ate this maternally predigested material at regular intervals over a period of several weeks. One of these doubled its body weight in two weeks on the diet. Apart from the nutritive value of this predigested food it is likely that it is the vehicle by which symbiotic micro-organisms are transferred from the mother's alimentary canal to that of the young. In the quokka and other marsupials symbiotic micro-organisms have considerable digestive importance (MOIR, et al., 1954).

Experimental Studies on Hormone Control

Effects of hormones on adult and foetal gonads

The anoestrous *Didelphis* injected on 4 successive days with follicle stimulating hormone (F.S.H.) and on the three following days with F.S.H. plus luteinising hormone (L.H.) ovulates on the 8th day (NELSEN & WHITE, 1941). The F.S.H. was obtained as a "purified powder' and the L.H. factor was Antuitrin S (Parke—Davis). Total doses were 8 mg F.S.H. and 300 rat units L.H. NELSEN & MAXWELL (1941) induced oestrus in 24 of 30 female *Didelphis* by injecting F.S.H. and L.H. during dioestrus. Three of these mated and produced young, 4 mated but produced no young and 17 failed to mate. Large doses F.S.H. (serum gonadotrophin) produce cystic follicles with absence of luteal cells in *Didelphis* whereas L.H. (chorionic gonadotrophin), or F.S.H. plus L.H., result in luteinisation (MORGAN, 1946). FARRIS (1941) used the technique of NELSEN & WHITE, cited above, in an attempt to induce oestrus in 6 tropical opossums *(Metachirops)*. All had an oestrous smear on the 4th day of injections and one mated on the 5th day.

MORGAN (1943) failed to modify the ovary of foetal *Didelphis* by the administration of oestrogens or androgens during the period of gonadal differentiation. Similarly the foetal testis was unmodified by either treatment (MOORE & MORGAN, 1942). BURNS (1956b) has however shown that if oestrogen administration is begun at "stage 34" (MCGRADY, 1938) the foetal presumptive testis may be transformed to an ovotestis or "ovary". The normally uterine stage 34 is occasionally found in the pouch.

Gonadotrophins did not appear to initiate secretory activity in the foetal *Didelphis* testis before day 70, or in the foetal ovary prior to day 100 (MOORE & MORGAN, 1943). MOORE (1947) interpreted this as evidence that reproductive hormones were not secreted by the foetal gonads. Precocious spermatorrhoea was induced in the adolescent *Trichosurus* by the injection of gonadotrophic hormones (BOLLIGER, 1942a).

Effects of hormones on accessory reproductive organs

Hormone control of sex differentiation

In male *Didelphis*, castrated before the 20th day of pouch life, the prostate developed normally in the absence of testes until day 50 (MOORE, 1941). In the gonadectomised pouch young the differentiation of Cowper's gland in the male, and of Bartholin's gland in the female, was normal until day 100 (RUBIN, 1943). Attempts to modify the development of accessory reproductive organs of this species by hormone treatment did not give results consistant with the theory of sex hormone control of sexual differentiation (MOORE, 1947). Androgen treatment before and during the sexually indifferent period stimulated the male duct system but the Mullerian (female) ducts were also greatly stimulated in both sexes. Conversely oestrogens, as well as stimulating the Mullerian duct system, caused pronounced increase in size of the Wolffian duct system of both male and female foetuses. MOORE (1947), after a consideration of this and other evidence, concluded that the control of mammalian sex differentiation was by genetic sex differentiating factors unconnected with sex hormone actions. BURNS (1949) however concluded that hormones of the embryonic gonads were identical with those of the adult and that these hormones behaved as specific differentiating agents. The same author (BURNS, 1945) showed that, as far as androgens were concerned, low dose levels which did not effect the Mullerian ducts were effective in stimulating the male duct system. This does much to remove MOORE's (1947) contention that the paradoxical response of the female duct system to male hormones can be used as evidence against the hormone control of sex differentiation. Male sex primordia however consistently attained a greater size in male than in female foetuses of *Didelphis* when both were treated with identical quantities of the same hormone. BURNS (1956a) did not regard this effect as due to secretion of extra hormone by the foetal testis, but to a capacity for growth inherent in the male sex primordia by virtue of its sex constitution.

Hormone induced organ transformations in adolescent marsupials

BOLLIGER (1946) concluded that the pouch and scrotum of *Trichosurus* were homologous structures developed from the same primary anlagen under the influence of appropriate male and female reproductive hormones. This conclusion was based on the reactions of pouch and scrotum to gonadotrophic and gonadal hormones. The following evidence was presented to support BOLLIGER's conclusion.

(1) Androgen administration resulted in precocious development of both pouch and scrotum (BOLLIGER & CARRODUS, 1940b) which was followed, in both sexes, by severe contraction of the organ if injections were continued (BOLLIGER, 1946).

(2) Progesterone given to non-castrated animals caused relaxation and enlargement of the pouch and also relaxation of the neck of the scrotum (BOLLIGER & CARRODUS, 1939). The testes withdrew into the inguinal canals following relaxation of the neck of the scrotum ("testicular ascent").

(3) Gonadotrophins caused an enlargement of the pouch after preliminary, short-lived contraction (BOLLIGER, 1942b) and a four-fold increase in size of the scrotum in both castrated and intact 4-6 month old males (BOLLIGER, 1943).

(4) "Testicular ascent" and collapse of the scrotum with formation of a pouch like structure in its place occurred in young males injected with oestrogens (BOLLIGER 1944).

(5) Injection of oestrogens following castration resulted in the formation of a "permanent" pouch reported to be present 6 months after the last injection in one case (BOLLIGER, 1943) and three months in another case (BOLLIGER, 1944). BOLLIGER (1946) considered that under certain circumstances testosterone administration converted the pouch of the castrated adolescent female into a scrotum.

Attempts to reproduce BOLLIGER's results, using the quokka, were unsuccessful (SHARMAN, unpublished). Oestrogen given to immature castrated males produced a slight depression round the scrotum which did not correspond to the pouch either in position or in size. Larger doses of oestrogen resulted in death of the animals due to excessive cornification, and hence occlusion, of the urethra. Even less conclusive results were obtained by injecting spayed females with androgens: in these the pouch increased in size faster than that of spayed or intact controls. A naturally occurring cryptorchid male was normal except for a well developed pouch. This animal, by its sexual libido and well developed interstitial cells, gave ample evidence of androgen secretion. This suggests that, if the pouch is produced by the action of gonadal hormones, it was in this case produced by androgen as no ovarian tissue was present. These results on the quokka of course do no more than indicate what may be the general case but several other factors seem to mitigate against the possibility of pouch and scrotum being homologous structures (which develop in one or other direction depending on the hormone secreted) in other marsupials. In *Didelphis virginiana*, in which species McGRADY (1938) also considered pouch and scrotum to be homologous, the two organs were reported to exist together in an essentially female "intersex" (HARTMAN, 1920). POCOCK (1926) also records that pouch and scrotum existed together in a male *Didelphis*. Furthermore BURNS (1956b) did not succeed in modifying either the pouch or scrotum of *Didelphis* by hormone treatments which extensively modified the other sex accessories. After treatment the genetic sex could only be recognised by the presence of pouch or scrotum. Finally in the marsupials *Notorcyctes* (STIRLING, 1891), *Thylacinus* (POCOCK, 1926) and *Chironectes panamensis* (ENDERS, 1937) the pouch is a normal development in the male. In the last two species a scrotum is present in addition to the pouch.

Effects of hormones on the female reproductive system

Injections of F.S.H. caused hypertrophy of the Fallopian tubes, uteri and lateral vaginal canals of the anoestrous *Didelphis*. Further development of the uteri was produced by the administration of progesterone but this hormone tended to suppress development of the lateral vaginal canals and Fallopian tubes (NELSEN, 1944). Castration of the anoestrous *Didelphis* resulted in a decrease in the number of ciliated cells in the epithelia of uterine lumen, uterine gland and Fallopian tube indicating that some hormones are secreted during the anoestrous phase. General atrophy of the reproductive organs occurred following castration of the cyclic female. F.S.H. administered to anoestrous females produced a uterine phase resembling the proliferative period of the oestrous cycle whereas L.H. alone or in combination with F.S.H. produced responses similar to those of the progravid phase. Oestrogens or androgens given to castrate females gave results similar to those obtained with F.S.H. and oestrogens and androgens given simultaneously were additive in effect (MORGAN, 1946). Six daily doses, each of 0.1 cm^3 oestradiol dipropionate, produced an oestrous-like response in the lateral vaginal canals of *Didelphis* (RISMAN, 1946). The vaginal culs-de-sac of *Trichosurus* are hypertrophied at oestrus (SHARMAN, 1954b); a condition which CARRODUS (cited by BOLLIGER, 1946) induced by the injection of oestrogen. An oestrus-like phase (Fig. 3) may be induced in the uterus of the castrate quokka by the injection of oestrogen and a luteal phase (Fig. 4) may be induced by the injection of progesterone after earlier priming with oestrogen (SHARMAN, LEAK & WARING, unpublished).

These results do not conflict with the assumption that the cyclic reproductive processes of marsupials are controlled by the same hormones as are produced by eutherian mammals.

Acknowledgements

This chapter is based largely on the results obtained during a 3 year research fellowship in the Zoology Department, University of Western Australia. The section on the foetal membranes of *Setonix* is based on a study carried out at the Department of Physiology, Royal Veterinary College, London under the guidance of Professor E. C. AMOROSO, F.R.S. A full account is in preparation. Figures 3 and 4 are from photographs taken by Mr. A. R. GOFFIN of the Department of Physiology, Royal Veterinary College. I wish to thank Professor A. A. ABBIE of the Anatomy Department, University of Adelaide, who read and criticised the manuscript.

REFERENCES

AMOROSO, E. C., 1949. Discussion of paper on "Placentation of the Marsupialia" by Dr. J. Pearson. *Proc. Linn. Soc. Lond.* **161**, *1—9*.

AMOROSO, E. C., 1952. Placentation. In "Marshall's Physiology of Reproduction" 3rd Edn. Vol. 2. Ed. A.S. Parkes, Longmans Green & Co., London.

AMOROSO, E. C., 1955a. The comparative anatomy and histology of the placental barrier. Josiah Macy, Jr. Foundation, 1st Conference on Gestation, New York.

AMOROSO, E. C., 1955b. Endocrinology of pregnancy. *Brit. med. Bull.* **11**, *117—25*.

AUSTIN, C. E. & BISHOP, M., 1957. Fertilization in mammals. *Biol. Rev.* **32**, *296—349*.

BARTHOLOMEW, G. A., 1956. Temperature regulation in the macropod marsupial *Setonix brachyurus*. *Physiol. Zool.* **29**, *26—40*.

DE BAVAY, J. M., 1951. Notes on the female urogenital system of *Tarsipes spenserae* (Marsupialia). *Pap. roy. Soc. Tas.* **1950**, *143—50*.

BOLLIGER, A., 1940. *Trichosurus* as an experimental animal. *Aust. J. Sci.* **3**, *59—61*.

BOLLIGER, A., 1942a. Spermatorrhoea in marsupials, with special reference to the action of sex hormones on spermatogenesis of *Trichosurus vulpecula*. *J. roy. Soc. N.S.W.* **76**, *86—92*.

BOLLIGER, A. 1942b. The effect of gonadotrophin obtained from human pregnancy urine on the pouch of *Trichosurus vulpecula*. *J. roy. Soc. N.S.W.* **76**, *137—41*.

BOLLIGER, A. 1943. Functional relations between scrotum and pouch and the experimental production of a pouch-like structure in the male of *Trichosurus vulpecula*. *J. roy. Soc. N.S.W.* **76**, *283—93*.

BOLLIGER, A. 1944. An experiment on the complete transformation of the scrotum into a marsupial pouch in *Trichosurus vulpecula*. *Med. J. Aust.* **1944** (2), *56—58*.

BOLLIGER, A. 1946. Some aspects of marsupial reproduction. *J. roy. Soc. N.S.W.* **80**, *2—13*.

BOLLIGER, A. & CARRODUS, A. L. 1938. Spermatorrhoea in *Trichosurus vulpecula* and other marsupials. *Med. J. Aust.* **1938** (2), *1118—9*.

BOLLIGER, A. & CARRODUS, A. L., 1939. Ambisexual action of progesterone as observed in the common Australian opossum. *Nature, Lond.* **144**, *671*.

BOLLIGER, A. & CARRODUS, A. L., 1940a. The effects of oestrogens on the pouch of the marsupial *Trichosurus vulpecula*. *J. roy. Soc. N.S.W.* **73**, *218—27*.

BOLLIGER, A. & CARRODUS, A. L., 1940b. Effect of testosterone propionate on pouch, scrotum, clitoris and penis of *Trichosurus vulpecula*. *Med. J. Aust.* **1940** (2), *368—73*.

BOLLIGER, A. & PASCOE, J. V., 1953. The composition of kangaroo milk (wallaroo, *Macropus robustus*). *Aust. J. Sci.* **15**, *215—7*.

BOWLEY, E. A., 1939. Delayed fertilization in *Dromicia*. *J. Mammal.* **20**, *499*.
BOYD, D. & HAMILTON, W. J., 1952. Cleavage, early development and implantation of the egg. *In* "Marshall's Physiology of Reproduction" 3rd Ed. Vol. 2. Ed. A.S. Parkes. Longmans Green & Co., London.
BRAMBELL, F. W. R., 1956. Ovarian changes. *In* "Marshall's Physiology of Reproduction" 3rd. Ed. Vol. 1. Pt. 1. Ed. A. S. Parkes. Longmans Green & Co., London.
VAN DEN BROEK, A. J. P., 1910. Entwicklung und Bau des Urogenital-Apparates der Beutler und dessen Verhältnis zu diesen Organen andrer Säuger und niederer Wirbeltiere. *Morphol. Jb.* **41**, *437—68*.
BUCHANAN, G. & FRASER, E. A., 1918. The development of the urogenital system in the Marsupialia, with special reference to *Trichosurus vulpecula*. Part 1. *J. Anat.* **53**, *35—95*.
BURNS, R. K., 1945. Bisexual differentiation of the sex ducts in opossums as a result of treatment with androgen. *J. exp. Zool.* **100**, *119—40*.
BURNS, R. K., 1949. Hormones and the differentiation of sex. *In* "Survey of Biological Progress" Vol. 1. Ed. George S. Avery Jr. Academic Press, New York.
BURNS, R. K., 1956a. Hormones versus constitutional factors in the growth of embryonic sex primordia in the opossum. *Amer. J. Anat.* **98**, *35—68*.
BURNS, R. K., 1956b. Transformation du testicule embryonaire de l'opossum en ovotestis ou en "ovaire" sous l'action de l'hormone femelle, le dipropionate d'oestradiol. *Arch. Anat. micrs. Morph.* **45**, *173—202*.
CALDWELL, H. W., 1884. On the arrangement of embryonic membrances in marsupial animals. *Quart. J. micr. Sci.* **24**, *655—58*.
CALDWELL, H. W., 1887. The embryology of Monotremata and Marsupialia. Part 1. *Philos. Trans.* (B) **178**, *463—86*.
CHASE, E. B., 1939. The reproductive system of the male opossum, *Didelphis virginiana* Kerr and its experimental modification. *J. Morph.* **65**, *215—39*.
COCHRANE, R. L. & MEYER, R. K., 1957. Delayed nidation in the rat induced by progesterone. *Proc. Soc. exp. Biol. N.Y.* **96**, *155—9*.
DUNNET, G. M., 1956. A live trapping study of the brush-tailed possum *Trichosurus vulpecula* Kerr (Marsupialia). *C.S.I.R.O. Wildl. Res.* **1**, *1—18*.
ECKSTEIN, P. & ZUCKERMAN, S., 1955. Reproduction in mammals. *In* "Comparative Physiology of Reproduction". *Mem. Soc. Endocrinology* No. 4. Ed. I. Chester Jones and P. Eckstein. University Press, Cambridge.
ECKSTEIN, P. & ZUCKERMAN, S., 1956. Morphology of the reproductive tract. *In* "Marshall's Physiology of Reproduction" 3 Edn. Vol. 1. Pt. 1. Ed. A.S. Parkes. Longmans Green & Co., London.
ENDERS, R. K., 1937. Panniculus carnosus and formation of the pouch in Didelphids. *J. Morph.* **61**, *1—26*.
FARRIS, E. J., 1941. Behaviour responses of tropical opossum *(Metachirops)* to gonadotrophic hormones. *Anat. Rec.* **81** (suppl.), *105* (abst.).
FLEAY, D. B., 1952. The Tasmanian or marsupial devil. Its habits and family life. *Aus. Mus. Mag.* **10**, *275—80*.
FLETCHER, J. J., 1883. On some points in the anatomy of urogenital organs in females of certain species of kangaroos. Part 2. *Proc. Linn. Soc. N.S.W.* **8**, *6—11*.
FLYNN, T. T., 1923a. Photograph illustrating method of parturition in *Potorous tridactylus* exhibited at meeting Linn. Soc. N.S.W., 27th Sep. 1922. *Proc. Linn. Soc. N.S.W.* **47**, xxviii.
FLYNN, T. T., 1923b. The yolk-sac and allantoic placenta in *Perameles. Quart. J. micr. Sci.* **67**, *123—83*.
FLYNN, T. T., 1930. The uterine cycle of pregnancy and pseudopregnancy as it is in the diprotodont marsupial *Bettongia cuniculus* with notes on other reproductive phenomena in this marsupial. *Proc. Linn. Soc. N.S.W.* **55**, *506—31*.
FRASER, E. A. & HILL, J. P., 1915. The development of the thymus, epithelial bodies and thyroid in the Marsupialia. 1. *Trichosurus vulpecula. Philos. Trans.* (B) **207**, *1—85*.

HAMLETT, G. W. D., 1935. Delayed implantation and discontinous development in mammals. *Quart. Rev. Biol.* **10**, *432—47*.
HARTMAN, C. G., 1920. The freemartin and its reciprocal: opossum, man, dog. *Science* **52**, *469—71*.
HARTMAN, C. G., 1923a. Breeding habits, development and birth of the opossum. *Rept. Smithson. Instn.* year ending June 30, 1921, *347—64*.
HARTMAN, C. G., 1923b. The oestrous cycle of the opossum. *Amer. J. Anat.* **32**. *353—421*.
HARTMAN, C. G., 1925. The interruption of pregnancy by ovariectomy in the aplacental opossum: a study in the physiology of implantation. *Amer. J. Physiol.* **71**, *436—54*.
HARTMAN, C. G., 1928. The breeding season of the opossum *(Didelphis virginiana)* and the rate of intra-uterine and post natal development. *J. Morph.* **46**, *142—215*.
HARTMAN, C. G., 1952. "Possums". Univ. Texas Press, Austin.
HEUSER, C. H., 1921. The early establishment of intestinal nutrition in the opossum - the digestive system just before and soon after birth. *Amer. J. Anat.* **28**, *341—56*.
HILL, C. J., 1933. The development of the Monotremata. Part. 1. The histology of the oviduct during gestation. *Trans. zool. Soc. Lond.* **21**, *413—43*.
HILL, C. J., 1941. The development of the Monotremata. Part. 5. Further observations on the histology of the oviduct prior to and during gestation. *Trans. zool. Soc. Lond.* **25**, *1—13*.
HILL, J. P., 1895. Preliminary note on the occurrence of a placental connection in *Perameles obesula*, and on the foetal membranes of certain Macropods. *Proc. Linn. Soc. N.S.W.* **10**, *578—81*.
HILL, J. P., 1898. The placentation of *Perameles*. *Quart. J. micr. Sci.* **40**, *385—446*.
HILL, J. P., 1899. Contributions to the morphology and development of the female urogenital organs in the Marsupialia. 1. On the female urogenital organs of *Perameles* with an account of the phenomenon of parturition. *Proc. Linn. Soc. N.S.W.* **14**, *42—82*.
HILL, J. P., 1900. On the foetal membranes, placentation and parturition of the native cat *(Dasyurus viverrinus)*. *Anat. Anz.* **18**, *364—73*.
HILL, J. P., 1901. Contributions to the embryology of the Marsupialia. 2. On a further stage of placentation of Perameles. 3. On the foetal membranes of *Macropus parma*. *Quart. J. micr. Sci.* **43**, *1—22*.
HILL, J. P., 1910. Contributions to the embryology of the Marsupialia. 4. The early development of the Marsupialia with special reference to the native cat *(Dasyurus viverrinus)*. *Quart. J. micr. Sci.* **56**, *1—134*.
HILL, J. P., 1932. Croonian Lecture. - The developmental history of the Primates. *Philos. Trans.* (B) **221**, *45—178*.
HILL, J. P., 1949. Discussion of paper on "Placentation of the Marsupialia" by Dr. J. Pearson. *Proc. Linn. Soc. Lond.* **161**, *1—9*.
HILL, J. P. & HILL, W. C. O., 1955. The growth stages of the pouch young of the native cat *(Dasyurus viverrinus)* together with observations on the anatomy of the new-born young. *Trans. zool. Soc. Lond.* **28**, *349—452*.
HILL, J. P. & O'DONOGHUE, C. H., 1913. The reproductive cycle in the marsupial *Dasyurus viverrinus*. *Quart. J. micr. Sci.* **59**, *133—174*.
HOME, E., 1795. Some observations on the mode of generation of the Kangaroo, with a particular description of the organs themselves. *Philos. Trans.* **1795**, *222—30*.
JENNISON, G., 1927. "Table of Gestation Periods and Number of Young". A. & C. Black Ltd., London.
JONES, F. WOOD, 1923. "The Mammals of South Australia". Part. 1. The Monotremes and Carnivorous Marsupials. Government Printer, Adelaide.
JONES, F. WOOD, 1924. "The Mammals of South Australia." Part. 2. The Bandicoots and Herbivorous Marsupials. Government Printer, Adelaide.

LONG, J. A. & EVANS, H. M., 1922. The oestrous cycle of the rat and its associated phenomena. *Mem. Univ. Calif.* **6**, *1—148*.
LYNE, A. G., 1952. Notes on external characters of the pouch young of four species of bandicoot. *Proc. zool. Soc. Lond.* **122**, *625—49*.
LYNE, A. G. & VERHAGEN, A. M. W., 1957. Growth of the marsupial *Trichosurus vulpecula* and a comparison with some higher mammals. *Growth* **21**, *167—95*.
McGRADY, E., 1938. The embryology of the opossum. *Mem. Wistar Inst.* No. 16.
MARSTON, H. R., 1926. The milk of the monotreme - *Echidna aculeata multi-aculeata*. *Aust. J. exp. Biol. med. Sci.* **3**, *217—20*.
MARTINEZ-ESTEVE, P., 1937. Le cycle sexual vaginal chez le marsupial *Didelphis azarae*. *C.R. Soc. Biol. Paris* **124**, *502—4*.
MARTINEZ-ESTEVE, P., 1942. Observations on the histology of the opossum ovary. *Contr. Embryol. Carneg. Instn.* **30**, *17—26*.
MATTHEWS, L. H., 1943. Parturition in the kangaroo. *Proc. zool. Soc. Lond.* (A) **113**, *117—20*.
MATTHEWS, L. H., 1947. A note on the female reproductive tract in the tree kangaroos *(Dendrolagus)*. *Proc. zool. Soc. Lond.* **117**, *313—333*.
MINCHIN, A. K., 1937. Notes on the weaning of a young koala *(Phascolarctos cinereus)*. *Rec. S. Aust. Mus.* **6**, *1—3*.
MOIR, R. J., SOMERS, M., SHARMAN, G. B. & WARING, H., 1954. Ruminant like digestion in a marsupial. *Nature, Lond.* **173**, *269*.
MOORE, C. R., 1941. Embryonic differentiation of the opossum prostate following castration and responses of the juvenile gland to hormones. *Anat. Rec.* **80**, *315—27*.
MOORE, C. R., 1947. "Embryonic Sex Hormones and Sexual Differentiation". Charles C. Thomas, Springfield.
MOORE, C. R. & MORGAN, C. F., 1942. Responses of the testis to androgen treatment. *Endocrinology* **30**, *990—9*.
MOORE, C. R. & MORGAN, C. F., 1943. First response of developing opossum gonads to gonadotrophic treatment. *Endocrinology* **32**, *17—26*.
MORGAN, C. F., 1943. The normal development of the ovary of the opossum from birth to maturity and its reactions to sex hormones. *J. Morph.* **72**, *27—85*.
MORGAN, C. F., 1946. Sexual rhythms in the reproductive tract of the adult female opossum and effects of hormonal treatment. *Amer. J. Anat.* **78**, *411—63*.
NELSEN, O. E., 1944. Possible control of the lateral vaginal canal in the opossum during reproduction. *Anat. Rec.* **89** (suppl.), *563—4* (abst.).
NELSEN, O. E. & MAXWELL, N., 1941. Induced oestrus and mating in the opossum *(Didelphis virginiana)*. *Anat. Rec.* **81** (suppl.), *105* (abst.).
NELSEN, O. E. & WHITE, E. L., 1941. A method for inducing ovulation in the anoestrous opossum *(Didelphis virginiana)*. *Anat. Rec.* **81**, *529—35*.
O'DONOGHUE, C. H., 1911. The growth changes in the mammary apparatus of *Dasyurus* and the relation of the corpora lutea thereto. *Quart. J. micr. Sci.* **57**, *187—234*.
O'DONOGHUE. C. H., 1912. The corpus luteum in the non pregnant *Dasyurus* and polyovular follicles in *Dasyurus*. *Anat. Anz.* **41**, *353—68*.
O'DONOGHUE, C. H., 1916. On the corpora lurea and interstitial tissue of the ovary in the Marsupialia. *Quart. J. micr. Sci.* **61**, *433—73*.
OWEN, R., 1834. On the generation of the marsupial animals, with a description of the impregnated uterus of a kangaroo. *Philos. Trans.* **1834**, *333—64*.
OWEN, R., 1837. Exhibition of a foetal kangaroo, proving the existence of an allantois. *Proc. zool. Soc. Lond.* **1837**, *82—3*.
PEARSON, J., 1945. The female urogenital system of the Marsupialia with special reference to the vaginal complex. *Pap. roy. Soc. Tas.* **1944**, *71—98*.
PEARSON, J., 1946. The affinities of the rat-kangaroos (Marsupialia) as revealed by a comparative study of the female urogenital system. *Pap. roy. Soc. Tas.* **1945**, *13—25*.

PEARSON, J., 1947. Some problems of marsupial phylogeny. *Rep. Aust. N.Z. Ass. Advanc. Sci.* **25**, *71—102*.
PEARSON, J., 1949. Placentation of the Marsupialia. *Proc. Linn. Soc. Lond.* **161**, *1—9*.
PEARSON, J., 1950a. A further note on the female urogenital system of *Hypsiprymnodon moschatus* (Marsupialia). *Pap. roy. Soc. Tas.* **1949**, *203—10*.
PEARSON, J., 1950b. The relationships of the Potoroidae to the Macropodidae (Marsupialia). *Pap. roy. Soc. Tas.* **1949**, *211—29*.
PEARSON, J. & DE BAVAY, J. M., 1951. The female urogenital system of *Antechinus* (Marsupialia). *Pap. roy. Soc. Tas.* **1950**, *137—42*.
PEARSON, J. & DE BAVAY, J. M., 1953. The urogenital system of the Dasyurinae and Thylacininae (Marsupialia, Dasyuridae). *Pap. roy. Soc. Tas.* **87**, *175—99*.
POCOCK, R. J., 1926. The external characters of *Thylacinus, Sarcophilus* and some related marsupials. *Proc. zool. Soc. Lond.* **1926**, *1037—84*.
REYNOLDS, H. C., 1952. Studies on reproduction in the opossum *(Didelphis virginiana)*. *Univ. Calif. Publ. Zool.* **52**, *223—83*.
RISMAN, G. C., 1946. Experimental stimulation of the female reproductive tract in the opossum *(Didelphis virginiana)*. *Anat. Rec.* **94** (suppl.), *408—9* (abst.).
RUBIN, D., 1943. Embryonic differentiation of Cowper's and Bartholin's glands of the opossum following castration and ovariectomy. *J. exp. Zool.* **94**, *463—73*.
SANDES, F. P., 1903. The corpus luteum of *Dasyurus viverrinus* with observations on the growth and atrophy of the Graafian follicle. *Proc. Linn. Soc. N.S.W.* **28**, *364—405*.
SELENKA, E., 1887. "Studien über Entwickelungsgeschichte der Thiere". 4. Das Opossum. C. W. Kreidel, Wiesbaden.
SELENKA, E., 1892. "Studien über Entwickelungsgeschichte der Thiere". 5. Beutelfuchs and Kanguruhratte. C. W. Kreidel, Wiesbaden.
SEMON, R., 1894. Zoologische Forschungsreisen in Australien und dem malayischen Archipel. 2. Die Embryonalhüllen der Monotremen und Marsupialier. *Denkschr. med.-naturw. Ges. Jena* **5**, *19—58*.
SHARMAN, G. B., 1954a. Reproduction in marsupials. *Nature, Lond.* **173**, *302—3*.
SHARMAN, G. B., 1954b. The relationships of the quokka *(Setonix brachyurus)*. *W.A. Nat.* **4**, *159—68*.
SHARMAN, G. B., 1955a. Studies on marsupial reproduction. 2. The oestrous cycle of *Setonix brachyurus*. *Aust. J. Zool.* **3**, *44—55*.
SHARMAN, G. B., 1955b. Studies on marsupial reproduction. 3. Normal and delayed pregnancy in *Setonix brachyurus*. *Aust. J. Zool.* **3**, *56—70*.
SHARMAN, G. B., 1955c. Studies on marsupial reproduction. 4. Delayed birth in *Protemnodon eugenii*. *Aust. J. Zool.* **3**, *156—61*.
SHARMAN, G. B., 1956. Some aspects of marsupial reproduction. *Proc. zool. Soc. Lond.* **127**, *141—3* (abst.).
SHARMAN, G. B., 1957. The quokka. *In* "The Ufaw Handbook on the Care and Management of Laboratory Animals". 2nd. Ed. Alistair N. Worden and W. Lane-Petter. Ufaw, London.
SIMPSON, G. G., 1945. Principles of classification and a classification of mammals. *Bull. Amer. Mus. Nat. Hist.* **85**, *1—350*.
SNELL, G. D., 1941. "Biology of the Laboratory Mouse". Blakiston, Philadelphia.
STIRLING, E. C., 1889. On some points in the anatomy of the female organs of generation of the Kangaroo especially in relation to the acts of impregnation and parturition. *Proc. zool. Soc. Lond.* **1889**, *433—40*.
STIRLING, E. C., 1891. Description of a new genus and species of Marsupialia, *Notoryctes typhlops*. *Trans. roy. Soc. S. Aust.* **14**, *154—87*.
TRIBE, M., 1918. The development of the pancreas, the pancreatic and hepatic ducts in *Trichosurus vulpecula*. *Philos. Trans.* (B) **212**, *147—207*.
TYNDALE-BISCOE, C. H., 1955. Observations on the reproduction and ecology of the brush tailed 'possum *Trichosurus vulpecula* Kerr (Marsupialia), in New Zealand. *Aust. J. Zool* **3**, *162—84*.

WARING, H., SHARMAN, G. B., LOVAT, D. & KAHAN, M., 1955. Studies on marsupial reproduction 1. General features and techniques. *Aust. J. Zool.* **3**, *34—43*.
WEEKES, H. C., 1935. A review of placentation among reptiles with particular regard to the function and evolution of the placenta. *Proc. zool. Soc. Lond.* **1935**, *625—45*.
WEICHERT, C. K., 1942a. The experimental control of prolonged pregnancy in the lactating rat by means of oestrogen. *Anat. Rec.* **83**, *1—18*.
WEICHERT, C. K., 1942b. A case of parallel embryonic development in the rat and its bearing on the question of superfoetation. *Anat. Rec.* **83**, *511—20*.

XXII
THE CONTRIBUTION OF BANDING TO AUSTRALIAN BIRD ECOLOGY

by

ROBERT CARRICK

(Wildlife Section, C.S.I.R.O., Canberra)

Introduction

The systematic marking of wild birds with numbered metal legbands was initiated by the Danish schoolmaster MORTENSEN in 1899, and many countries have had national bird-banding schemes in operation for several decades. These are designed primarily to document bird movements and migrations, and they also provide an essential tool for more intensive and detailed studies based upon repeated records, by trapping and observation, of birds identifiable as individuals. The Australian Bird-banding Scheme, inaugurated in 1953 by the Wildlife Section, C.S.I.R.O. (CARRICK, 1956a) is now serving both of these purposes, and its first four years have added significantly to our knowledge of the ecology and behaviour of Australian birds. Seasonal movements, hitherto unknown or at best only surmised, are becoming clear, especially those which take place within the geographical range permanently occupied by the species, and the age, sex and regional elements involved in them are being defined. Prior to banding, information was general and inconclusive, but now the widely-scattered origins of birds which occur in the Australian region, and the unsuspected patterns of dispersal of some species, are becoming known. Data of fundamental importance in population studies — such as breeding ecology; dispersal, survival and mortality rates; age structure of populations; duration of immaturity and longevity; and so on — are accumulating in terms of banded individuals. Apart from its intrinsic scientific interest, the information which bird-banding in Australia is beginning to document is essential for the proper understanding of conservation, economic status and the sporting value of birds. Effective legislation on these subjects awaits the accumulation of knowledge, much of which can be obtained only by extensive, nation-wide banding and by intensive ecological studies of marked birds.

The C.S.I.R.O. Scheme was preceded by several State and private bird-banding activities (CARRICK, 1956a; 1956b). The earliest efforts were in 1912, when the Bird Observers' Club, Melbourne, and the Royal Australasian Ornithologists' Union began to band short-tailed shearwaters, *Puffinus tenuirostris* (TEMMINCK), and white-faced storm-petrels, *Pelagodroma marina* (LATHAM) near Melbourne. In the 1930s, some 4000 wedge-tailed shearwaters, *Puffinus pacificus* (GMELIN), are said to have been banded by the Embury Barrier Reef Expedition at North-west Island, Queensland. Between March 1947 and March 1957, SERVENTY (1957a) banded 18,121 *P. tenuirostris* in the Furneaux group of islands, Bass Strait, as part of the joint C.S.I.R.O.—Tasmanian Fauna Board investigation into the biology of this commercial species. The need for a national bird-banding scheme in Australia has been widely discussed since the late 1940s, but in 1948 D'OMBRAIN used bands with the Victorian Fisheries and Game Department stamp on the white-winged (= GOULD's) petrel, *Pterodroma leucoptera* GOULD, at Cabbage Tree Island, New South Wales, and in 1950 the Bird Observers' Club used similar bands for their study of the silver gull, *Larus novae-hollandiae* STEPHENS, at Altona near Melbourne (WHEELER, 1952). The Australian National Antarctic Research Expeditions to Heard and Macquarie Islands began banding seabirds in 1949 (HOWARD, 1954), using a variety of bands including those of the Tasmanian Fauna Board, the Dominion Museum, New Zealand, and several stamped ANARE. SERVENTY also used New Zealand bands on the Australian gannet, *Sula serrator* GRAY, at Cat Island, Tasmania. Large-scale banding of ducks was initiated in 1951 by the Fisheries and Game Department, Victoria (MCNALLY & FALCONER, 1953; DOWNES, 1954), and June 1958 30,758 birds have received their bands. The Fisheries Department, Western Australia, started a duck-banding scheme in 1952 and has banded 5826 birds, and the Department of Agriculture and Stock, Queensland, followed suit in 1957. Since 1953, all bird-banding in Australia, except the Tasmanian short-tailed shearwater study and the three State duck-banding schemes mentioned, has been done under the Australian Bird-banding Scheme with headquarters at Canberra. The duck-banding studies organised by FRITH et al. (1959) throughout New South Wales, Northern Territory, South Australia and Tasmania account for 27,500 of the 66,741 birds, comprising 182 species, banded by this Scheme up to June 1958. The 7059 recoveries give an average rate of 10.1 % of individual birds recovered at least once, but this figure varies widely according to the characteristics of each species and the human (scientific and sporting) interest in it. Those species in which more than a thousand birds have been banded have yielded the following results (HITCHCOCK & CARRICK, 1958b):

Species	Banded	Recovered	Per cent
White-faced storm-petrel *Pelagodroma marina*	1357	18	1.3
Short-tailed shearwater *Puffinus tenuirostris*	1307	2	0.2

Species	Banded	Recovered	Per cent
Giant petrel *Macronectes giganteus*	4476	45	1.0
Crested tern *Sterna bergii*	5160	153	3.0
Silver gull *Larus novae-hollandiae*	12,255	432	3.5
Southern skua *Catharacta skua lonnbergi*	1331	281	21.1
Straw-necked ibis *Threskiornis spinicollis*	3070	12	0.4
Pied goose *Anseranas semipalmata*	2108	314	14.9
Black duck *Anas superciliosa*	10,495	1591	15.2
Grey teal *Anas gibberifrons*	13,267	2035	15.3
Black-backed magpie *Gymnorhina tibicen*	1345	674	50.1

The officers responsible for fauna protection in each State cooperate and issue permits to banders, of whom there are seventy actively operating. This number and the scale of operations are likely to increase from now on, following the recent appointment of an experienced Australian ornithologist, Mr. W. B. HITCHCOCK, as full-time Secretary. In addition to widespread, opportunist banding of many species, the Scheme supports seven economic studies and twelve others, while nineteen species are being colour-banded, ten of them individually. Reports are published annually in C.S.I.R.O. Wildlife Research (HITCHCOCK & CARRICK, 1958b, and earlier reports). Intermittent reports on recoveries of foreign bands in Australia appear in the same journal (HITCHCOCK & CARRICK, 1958a).

The standard aluminium split-ring type of band is used for most species, and the particular problems presented by shearwaters and penguins have led to the development of special bands for them. SERVENTY (1956) has described the overlapping monel metal band designed to overcome abrasion of softer metals in the burrowing *P. tenuirostris*. The short, soft tarsus of penguins cannot carry a band with safety, and trials on the royal penguin, *Eudyptes chrysolophus schlegeli* FINSCH, at Macquarie Island have led to the elaboration of an aluminium flipper-band shaped to eliminate abrasion of feathers or skin (CARRICK et al., 1959), as shown in Plate 1. The difficulties of permanent colour-banding under the wide range of climatic conditions from bright sunshine and heat to Antarctic cold have been overcome by the use of overlapping, celluloid bands cemented with acetone and covered with reflective scotchlite, and strips of the latter on royal penguin flipper-bands have even withstood a winter's immersion in the Southern Ocean.

Ecological Significance of Territory

Territorial behaviour in birds is widespread, was known ninety years ago, and has been much discussed in ornithological literature during this century; yet, in a recent comprehensive review, HINDE (1956) had to conclude: "The functions of territorial behaviour are extremely diverse, and the quality of the evidence available for assessing them is little different from that available to HOWARD." The latter's classics on this subject were published during 1907—20. HINDE also states: "Territorial behaviour may reduce disease, but this is unlikely to be a significant consequence except in some colonial species," and "There is no direct evidence that territory limits the total breeding population in all habitats." Some proof on these two points has been obtained in a population study of the Australian black-backed magpie, *Gymnorhina tibicen* (LATHAM). This bold, pied bird, about the size and build of the European jackdaw, *Corvus monedula* VIEILLOT, is a member of the Australo-Papuan family Cracticidae.

During the past three years 1345 magpies have been banded in the study area of five square miles at Canberra; 367, including 91 of known age, have been individually colour-banded. This has shown that the species, in an open savannah woodland habitat, consists of two discrete but inter-acting elements. Wooded areas are subdivided into territories, each strongly held by small groups of 2—8 birds, which fight as a team and never leave their home under normal circumstances; open treeless country is populated by loose flocks, which may contain several hundred birds, some of which remain sedentary while others undergo limited movements of, at most, a few miles. Breeding is confined to the territories, in which the immature birds may remain for one or two years before passing out into an adjacent flock. Group-formation takes place in the flocks, and new groups continually attempt to force their way into the more favoured and adequate habitat. Death of the dominant male may cause a group to disintegrate, and loss of a female may be followed by acquisition of one, or more, new ones. A group of two is always an adult pair, larger groups contain up to three adults or immature birds of each sex, and polygamy is common.

Since the breeding season of 1955, practically every bird has been colour-banded in the forty-odd groups which inhabit the area of woodland most intensively studied. In June—September 1956, an obvious and heavy mortality from disease occurred in flock birds adjacent to this area, and elsewhere; the causative organism was identified as *Pasteurella pseudotuberculosis* (MYKYTOWYCZ & DAVIES 1959), and the cold wet weather of that period doubtless contributed to the epizootic by lowering the resistance of the birds. This contact-spread disease decimated the flocks, yet banded birds in territories nearby, carolling their challenge to keep intruders at bay, suffered no losses from it. A year later, during frosty weather, the flock magpies tended to concentrate on haystacks for feeding, and further mortality resulted from inhalation of the spores of *Aspergillus*

Royal penguin, *Eudyptes chrysolophus schlegeli* FINCH, at Macquarie Island, wearing aluminium numbered flipper-band. (Photograph by K. KEITH).

R. CARRICK

fumigatus. Once again the group birds escaped, for the softer ground in their territories provided an adequate available supply of food. Other diseases, such as benign insect-borne fowl pox, have affected both group and flock birds. These results indicate the survival value of the territorial habit, which buffers that element of the population against diseases which are capable of causing significant mortality in this species.

The flocks contain a high proportion of immature individuals, some of which have been banded as nestlings during the past four years. The duration of immaturity is not yet known, but two females in territories, presumed from plumage to be first-year birds when banded, nested the following year. Also, the flocks contain older birds, including some colour-banded individuals of each sex, which are known to have bred when they were members of successful territory-holding groups. So the question arises whether territorial behaviour limits the breeding potential of the species. It does, for these older birds have been observed in the flock during the breeding season, and one of each sex, dissected before egg-laying ceased, had gonads less developed than in territorial individuals. The female was a member of a well-knit group which attempted, throughout the season, to gain a foothold in the woodland habitat, but failed. These findings also throw light on the complex nature of proximate stimuli for gonad development; the physical environment is not enough, and behavioural stimuli such as possession of an adequate nesting-place, building the nest, and so on are also essential. A completely adequate territory, and freedom from undue conflict along its frontiers, are necessary for successful breeding; one group, recently established in a tiny territory which lacked feeding-ground, built the foundation of a nest but no eggs were laid, and another similar group achieved two clutches which became addled due to neglect because the females constantly assisted the single male to repel the neighbours.

Observation of colour-banded magpies has revealed aspects of social behaviour, such as communal feeding of young by members of the group other than the parents. A similar study of the superb blue wren, *Malurus cyaneus* LATHAM, (ROWLEY, 1957) has made an even more interesting contribution on this point. In addition to assistance from other wrens in the group, the nestlings and fledglings of later broods are frequently fed by their recently-fledged brothers and sisters. Thus the parents can telescope more broods into each breeding season, and the dependent young birds are afforded the benefits of "parental" care and experience for a much longer period than is usual in small passerines. Hence the territorial group habit increases the number of young produced each season and would enable the species to recover more rapidly from the serious setback in numbers which small birds of sedentary habit are apt to suffer under adverse weather and food conditions.

Trans-Equatorial Migration

To date, the most dramatic Australian recoveries of banded birds are the common tern, *Sterna h. hirundo* L., and the arctic tern, *Sterna macrura* NAUMANN, found in Western Australia (DUNNET, 1956a; 1956b). They are the only banded members of these species to reach Australia, and each was banded as a nestling in July 1955, the former in Sweden and the latter in north-west Russia near the White Sea; recoveries were made in January 1956 and May 1956 respectively. The common tern record is a particularly remarkable one, for the eleven previous specimens of *S. hirundo* collected in Australia have been assigned to the eastern Asiatic race *S. h. longipennis* NORDMANN, and but for the fact that DUNNET salvaged the carcass, the doubt would have remained that the specific identity of the chick had been mistaken, easily done in mixed colonies (AUSTIN, 1958). Despite the volume of banding evidence for constancy of return of individuals to their birth-place, there would appear to be greater latitudinal mixing of population stocks than has been imagined.

There are now five recoveries of *P. tenuirostris* banded as nestlings in Tasmania or South Australia and reported in Japanese waters, the far north Pacific, and the Bering Sea, the last only seven weeks after banding (SERVENTY, 1957b). These, and recoveries on the coast of New South Wales, confirm the known migration route of this species round the north Pacific Ocean.

Migrations and Breeding of Subantarctic Sea-Birds

The giant petrel, *Macronectes giganteus* (GMELIN), has been the most extensively banded species and has provided many recoveries of young birds far distant from their birthplace. During the past decade about 10,000 giant petrels have been banded at Australian, British, and French stations; 7333 of these, comprising 5623 nestlings and 1710 adults, were done at Heard and Macquarie Islands. The 100 recoveries, analysed by HITCHCOCK & CARRICK (1958a), show that young giant petrels disperse from their breeding places and, influenced by westerly winds, make extensive circum-polar journeys in their first autumn and winter (Fig. 1). They move into the lower latitudes of southern Australia, South America and South Africa, and some approach the Equator, e.g. Tahiti. They are also recovered there in their second winter, and one Heard Island bird was recovered at Kerguelen after five years. Adults return annually to the same colony.

The larger albatrosses have produced fewer migration records but their constancy of return to the same nest-site and often the same mate has been established by banding. The wandering albatross,

Diomedea exulans exulans L., the black-browed albatross, *Diomedea melanophris* TEMMINCK, and the light-mantled sooty albatross, *Phoebetria palpebrata* (FORSTER), have been banded, and the second of these has twice been recovered in New South Wales, an adult

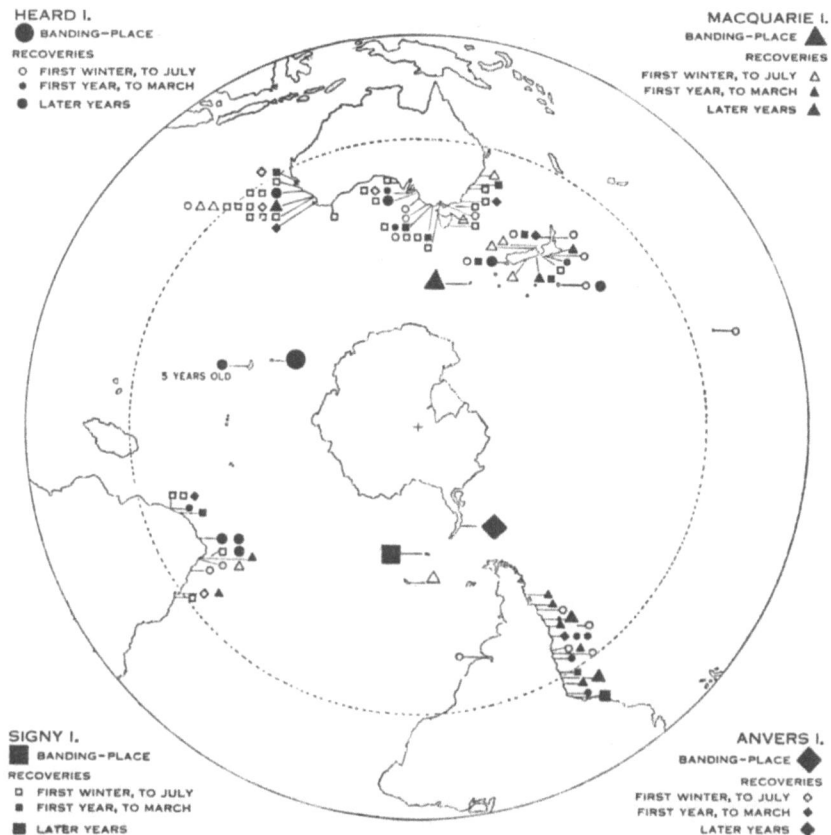

Fig.1. Map showing places of banding and distant recovery, up to end of March 1958, of giant petrel, *Macronectes giganteus* GMELIN. (From HITCHCOCK & CARRICK, 1958a.)

from Macquarie Island and a nestling from Heard Island (Fig. 2). The southern skua, *Catharacta skua lonnbergi* (MATHEWS), has provided similar records (Fig. 2) and the Macquarie Island cormorant has been shown to be very sedentary indeed and to breed in the colony of origin at two years of age. A detailed study of royal penguins marked with flipper-bands has commenced at Macquarie Island, and there is evidence of constancy of pairs in successive

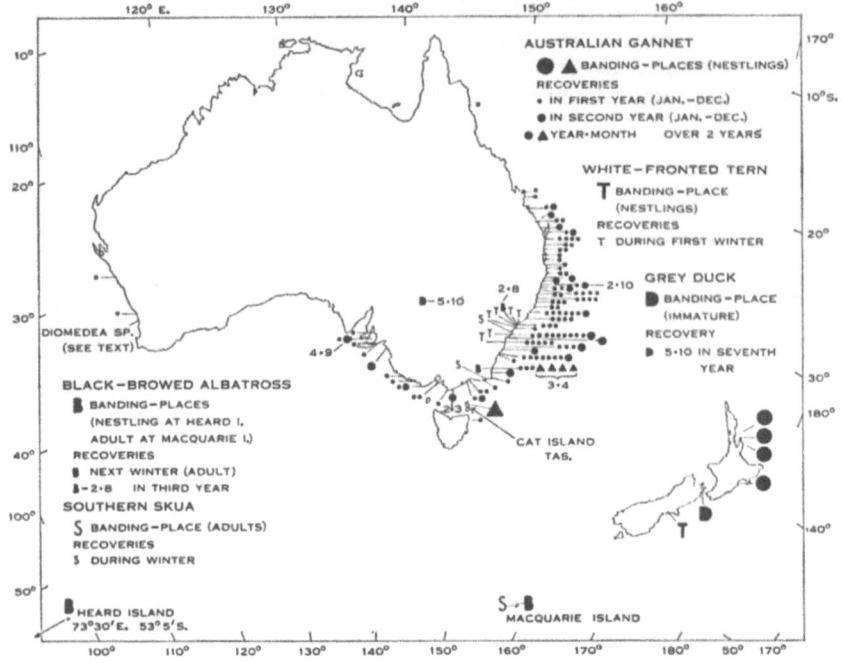

Fig. 2. Map showing places of recovery in Australia, up to end of March 1958, of birds (except giant petrel) banded in New Zealand, Macquarie Island, and Heard Island. The species are:- Australian gannet, *Sula serrator* GRAY; white-fronted tern, *Sterna striata* GMELIN; southern skua, *Catharacta skua lonnbergi* (MATHEWS); black-browed albatross, *Diomedea melanophris* TEMMINCK; black duck, *Anas superciliosa* GMELIN. (From HITCHCOCK & CARRICK, 1958a.)

seasons and of return to the same location even in vast colonies containing hundreds of thousands of birds.

The ultimate reasons for the differing patterns of seasonal migration and dispersal shown by subantarctic sea-birds remain obscure in the absence of better knowledge of the needs of each species, especially their food requirements, and of the important causes of mortality. The available evidence suggests that life in off-shore temperate waters during winter may be easier than a pelagic existence at higher latitudes. Many albatrosses and larger petrels, among which immature birds predominate, rely upon a man-made food supply, such as the effluent from meat factories, and young giant petrels have been recovered ashore in poor condition, often during storms at sea. Banding is establishing the facts about direction and length of the journeys, and the categories of each species which perform them; on the basis of this information, studies to determine the survival value of such behaviour can be planned.

Trans-Tasman Migration

As yet, no birds banded in Australia have been reported from New Zealand, although some sea-birds regularly make the crossing and water-birds do so in drier years. The New Zealand banding scheme, in operation since 1950, has had 119 Australian gannets, *Sula serrator* GRAY, 6 white-fronted terns, *Sterna striata* GMELIN and one black duck, *Anas superciliosa* GMELIN, reported from Australia (Fig. 2). The last of these was a six-year-old bird, and the east-west movement was less expected than one in the opposite direction.

Movements Within Australia

The patterns of seasonal movement of some species of birds in Australia are beginning to emerge from the results of several years' banding, and for details the reader is referred to the annual reports of the Australian Bird-banding Scheme (HITCHCOCK & CARRICK, 1958b, and earlier reports), to the first report on the short-tailed shearwater (SERVENTY, 1957a), and to papers on the silver gull and crested tern, *Sterna bergii* LICHTENSTEIN (CARRICK et al., 1957) and on ducks (FRITH et al., 1959). Partial migration by particular age, sex or regional elements of the species, especially when it is confined within the breeding range, is often difficult to define, and these band records are documenting with some accuracy facts which have so far been realised only vaguely, if at all.

In eastern Australia, young silver gulls (Fig. 3) disperse mainly northward about 250 miles, extending to 800 miles, during their first winter; the longest record is a Victorian bird which reached Queensland, 1050 miles north. During the second year, most are within 200 miles of their birth-place. All gulls banded as nestlings and subsequently found in breeding colonies have been at their own place of birth when two years old or more. This suggests a measure of inbreeding consistent with the development of the regional forms described in this species. However, recent observations have queried the actual breeding distribution of these forms, for some adults in New South Wales colonies are the Tasmanian race *gunni*. It is evident that further widespread and large-scale banding, in conjunction with study of plumage types, is required. Young and adult crested terns dispersed some 600 miles both north and south from colonies in New South Wales (Fig. 4), and some colour-banded breeding birds nested 130 miles south of their previous location. The different dispersal patterns and constancy of breeding-place may be related to the generalised habits, especially feeding, of the gull compared with the more specialised tern; the natural scavenging habits of the silver gull, which frequents urban refuse tips in large numbers, may explain why young birds from Five Islands and

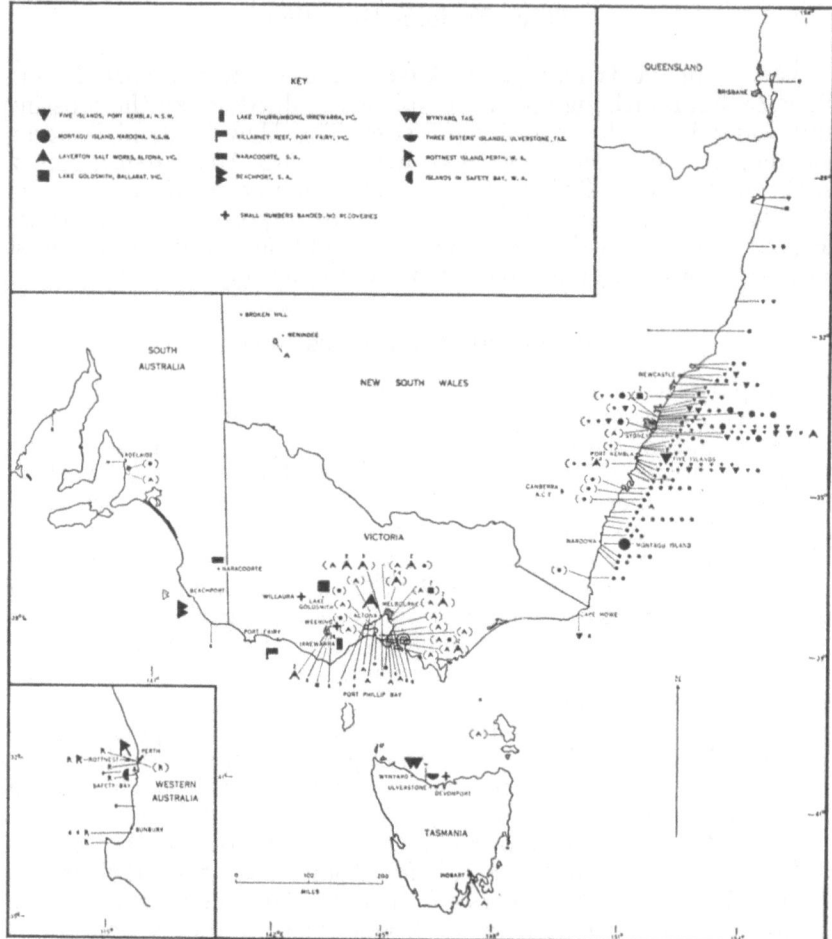

Fig. 3. Map showing the places of banding and recovery of silver gulls *Larus novae-hollandiae* STEPHENS in Australia during 1950-57.
Each symbol without brackets indicates a reported band; those in brackets represent one or more banded birds observed with binoculars, either together or on different dates at the same place. The largest symbol is the banding-place. Solid symbols are birds banded as young and recovered before the next breeding season (smaller size) or later (larger size). The figure above a larger symbol is the year since banding (each year commences at the beginning of the breeding season).
The open symbol containing dots is a bird banded as an adult. (From CARRICK, WHEELER & MURRAY, 1957.)

379

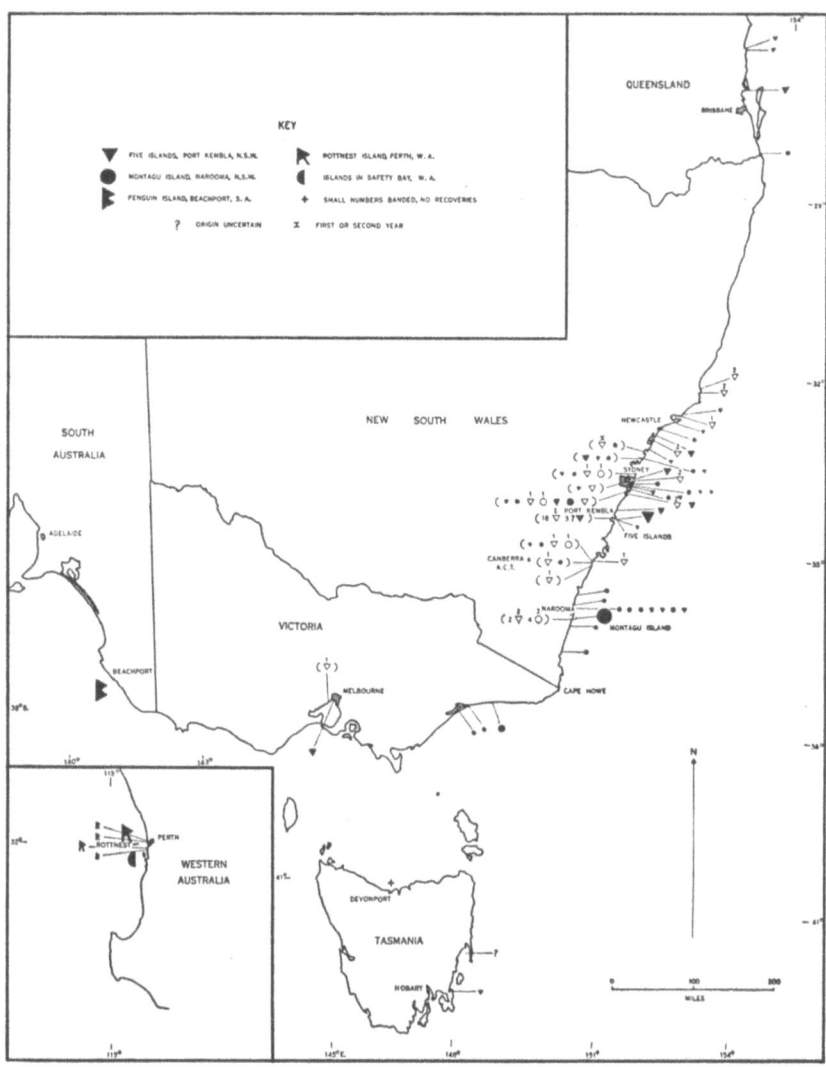

Fig. 4. Map showing the places of banding and recovery of crested terns *Sterna bergii* LICHTENSTEIN in Australia during 1955-57.

Each symbol without brackets indicates a reported band; those in brackets represent one or more banded birds observed with binoculars, either together or on different dates at the same place.

The largest symbol is the banding-place. Solid symbols are birds banded as young and recovered before the next breeding season (smaller size) or later (larger size). Open symbols are birds banded as breeding adults. The figure in front of a symbol gives the number of individuals observed, and the figure above it is the year since banding (each year commences at the beginning of the breeding season); x above a symbol is an observation which may refer to an adult banded in 1955 and 1956. (From CARRICK, WHEELER & MURRAY, 1957.)

Montagu Island on the south coast of New South Wales have evolved a northward movement into the region of denser human population.

Water-birds are the other species which, banded in large numbers, have yielded significant results. A roving habit and opportunist breeding where floodwaters are available are shown by the Australian white ibis, *Threskiornis molucca* (CUVIER), and straw-necked ibis, *T. spinicollis* (JAMESON), which, banded as young in the Macquarie Marshes in the centre of New South Wales, have been recovered near Cape York, Queensland and the Gulf of Carpentaria, Northern Territory, after two years.

The far-flung and contrasting patterns of nomadism in ducks have been revealed in the banding studies by FRITH et al. (1959). During the drought year 1957, 10,026 grey teal, *Anas gibberifrons* MÜLLER, 5284 black duck, *Anas superciliosa* GMELIN, and 796 wood duck, *Chenonetta jubata* (LATHAM) were banded, in addition to 307, 251, and 77 ducks of these species which had been banded in earlier years. Most of this work was done in south-western New South Wales and at Darwin, also around Sydney and in South Australia and Tasmania. The recovery rates of each species from shooting and re-trapping, up to the end of March 1958, were 7, 14 and 19% respectively. The distribution of recoveries in time and space, under conditions of drought and consequent food shortage, shows the extraordinary mobility of the grey teal, the lesser tendency of the black duck to disperse, and the more local habit of the wood duck. As inland breeding floodwaters dried up during the first half of 1957, ducks redistributed to the more permanent inland swamps, and there was a partial exodus, mainly of grey teal but also black duck, which led to coastal concentrations of these species. As the drought continued and coastal waters also dried, the band recoveries revealed that grey teal were crossing and re-crossing the continent in all directions, often over short periods of time; trapping and shooting at permanent waters frequently resulted in the simultaneous recovery of birds banded at any of the stations on coastal or inland Australia. Such extensive and apparently random nomadism by a bird has not been demonstrated before, and could only have been proved by means of marked birds. It merits illustration in some detail:

Three grey teal, banded and released together on January 18, 1957, at Gum Creek, an area of residual floodwater in south-western New South Wales, were recovered separately, one on April 18 on a permanent lagoon 8 miles distant, another on April 30 at Mandurah, Western Australia, 1745 miles west, and the third on July 24 at Humpty Doo, near Darwin, Northern Territory, 1760 miles north-west.

Grey teal banded near Darwin during July-August 1957 were recovered from July 1957 to March 1958 all round Australia, in every State except Tasmania, and even at Alice Springs, Northern Territory, in the centre of the continent.

During the summer of 1957-58, grey teal recovered at Hume Reservoir on the River Murray near Albury represented all regions where the species had been banded, namely inland and coastal New South Wales, Northern Territory (Darwin), South Australia and Tasmania.

The ecological and behavioural explanations of these varying degrees and patterns of mobility which duck-banding has demonstrated are still somewhat speculative. FRITH et al. (1959) suggest that the more conservative feeding habits of the black duck may restrict its movements compared with those of the adaptable grey teal. He also raises the intriguing possibility that the rapid colonization of waters by large numbers of ducks as soon as they form appears to necessitate some positive response which directs the birds towards them, and is not due merely to chance encounter by ducks dispersing aimlessly from unfavourable areas.

REFERENCES

AUSTIN, O. L., 1958. Verification of Australian common tern recovery. *Bird-banding* **29**, 41—2.
CARRICK, R., 1956a. The Australian Bird-banding Scheme. *C.S.I.R.O. Wildl. Res.* **1**, 26—30.
CARRICK, R., 1956b. Early Australian bird-banding. *C.S.I.R.O. Wildl. Res.* **1**, 135.
CARRICK, R., GWYNN, A. G., KEITH, K. & HINES, M. P., 1959. Flipper-bands for penguins. *A.N.A.R.E. Rep.* Ser. B. **1**. Zoology (in press).
CARRICK, R., WHEELER, W. R. & MURRAY, M. D., 1957. Seasonal dispersal and mortality in the silver gull, *Larus novae-hollandiae* Stephens, and crested tern, *Sterna bergii* Lichtenstein. *C.S.I.R.O. Wildl. Res.* **2**, 116—44.
DOWNES, M. C., 1954. Waterfowl conservation in Victoria. *Emu* **54**, 169—80.
DUNNET, G. M., 1956a. Common tern, *Sterna hirundo hirundo* L. banded in Sweden and recovered in Australia. *C.S.I.R.O. Wildl. Res.* **1**, 68.
DUNNET, G. M., 1956b. Arctic tern, *Sterna macrura* Naum. banded in Russia and recovered in Western Australia. *C.S.I.R.O. Wildl. Res.* **1**, 134.
FRITH, H. J., BROWN, B. K., & BOOTH, G. F. 1959. The ecology of wild ducks in inland Australia. II. Movements. *C.S.I.R.O. Wildl. Res.* (in press).
HINDE, R. A., 1956. The biological significance of the territories of birds. *Ibis* **98**, 340—69.
HITCHCOCK, W. B. & CARRICK, R., 1958a. First report of banded birds migrating between Australia and other parts of the world. *C.S.I.R.O. Wildl. Res.* **3**, (1) 54—70.
HITCHCOCK, W. B. & CARRICK, R., 1958b. Fourth Annual report of the Australian Bird-banding Scheme, July 1957 - June 1958. *C.S.I.R.O. Wildl. Res.* **3**, 115—141.
HOWARD, PATRICIA, 1954. A.N.A.R.E. bird banding and seal marking. *Vic. Nat.* **71**, 73—82.
MCNALLY, J. & FALCONER, D. D., 1953. Trapping and banding operations Lara Lake, 1952. *Emu* **53**, 51—70.

MYKYTOWYCZ, R. & DAVIES, D. W., 1959. *Pasteurella pseudotuberculosis* in the Australian black-backed magpie *Gymnorhina tibicen* (Latham). *Aust. Vet. J. C.S.I.R.O. Wildl. Res.*

ROWLEY, IAN, 1957. Co-operative feeding of young by superb blue wrens. *Emu* 57, 356—7.

SERVENTY, D. L., 1956. A banding technique for burrowing petrels. *Emu* 56, 215—18.

SERVENTY, D. L., 1957a. The banding programme on *Puffinus tenuirostris* (Temminck) I. First report. *C.S.I.R.O. Wildl. Res.* 2, 51—59.

SERVENTY, D. L., 1957b. Recovery of a South Australian ringed *Puffinus tenuirostris* in the Bering Sea. *S. Aust. Orn.* 22, 56.

WHEELER, W. R., 1952. A Review of the Altona Survey Group. *Emu* 52, 206—8.

XXIII
ECOLOGY OF WILD DUCKS IN INLAND AUSTRALIA

by

H. J. FRITH

(Wildlife Survey Section, C.S.I.R.O., Canberra)

Introduction

In many countries and in the coastal regions of Australia, where the seasons are reasonably regular, the movements, breeding and feeding patterns of waterbirds are well established. In the Northern Hemisphere it has been possible to accurately chart migration routes and breeding seasons and to forecast with some expectation of accuracy the numbers of birds liable to reach the extremities of their migration routes and the dates of their movements.

In inland Australia, however, the climate is arid and erratic. The depth and extent of water areas fluctuate widely from year to year. In some years, following flood, there may be hundreds of square miles of swamp in a region, but this may dry up and the land remain dry and parched for many years. Under conditions such as these it is apparent that waterbirds must develop adaptations to enable them to deal with the rapidly and erratically changing situations. This paper summarizes the results of studies on various aspects of the ecology of wild ducks in inland New South Wales and seeks to explain how wild ducks can exist in very large numbers in this semi-arid environment.

The Environment

The climate of inland Australia has been discussed in detail by many authors (e.g. LAWRENCE, 1937). The boundaries of the State of New South Wales enclose a typical cross section of all environmental conditions that are found on the continent ranging from the well watered coast to the semideserts in the far western parts of the State. The majority of the work, on which this paper is based, was carried out in mid-western New South Wales between the 20-inch isohyet in the east and the 8-inch in the west; the rainfall is erratic and seldom approximates the annual average. In the north most of the rain falls in the summer and in the south most of it falls in the winter. Summer temperatures are high and the annual evaporation varies between 50 inches in the east and 100 inches in the west. The region is semi-arid with unreliable rainfall.

The land is flat and forms part of the flood plain of the Murray-

Darling river system. The principial rivers, the Murray, Murrumbidgee, Lachlan, Macquarie, and Darling are sluggish and characterised by extensive systems of meanders, "billabongs" (ox-bows) and effluent streams. The levels of the rivers vary greatly; in times of flood the waters of the Murray, Murrumbidgee, and Lachlan may actually join across the plains but at other times the rivers may cease to flow and even dry up. The level of these rivers and flooding of the plains are mainly determined by conditions on the catchments hundreds of miles to the east and local rainfall is seldom sufficient to affect their level. They need have no relation to the local climate at the time and may occur at any time of the year.

Waterfowl Habitat

Streams

None of the rivers is entirely permanent but although they often cease to flow they seldom dry up entirely. Flooding is rather frequent and each 4 or 5 years on an average much of the region is inundated. The rivers in general have steep bare banks and under these conditions few aquatic plants grow. Herbs, however, periodically flourish on exposed mud banks and beaches.

The effluent streams are normally dry but when the river reaches a sufficient level water flows into them and is carried far into the plains. The extent of the development of effluents and their type varies greatly between rivers and accordingly minor flooding gives different effects. In some cases a minor increase in water level merely leads to an increased rate of flow through the numerous streams, but in others quite a small rise in river level forces water to flow through the numerous effluents and sends shallow, temporary water across very large areas of swamps and reedbeds.

Irrigation areas, and Domestic Stock and Water Supply districts have been developed in the region. The channels carry water only periodically but, in some places, their banks support dense growths of *Eleocharis*, *Juncus*, *Polygonum*, and *Carex*, and provide thin strips of green herbage through otherwise dry plains.

Billabongs

For the present purpose three distinct types are recognised. The juvenile billabong is one which has apparently only recently been separated from the river. Its banks are tree lined, steep, deep, and bare of herbage. It has virtually no shallows. The inlet and outlet creeks are always present and almost as deep as the river itself. This means that a rise of a few feet in river level will cause water to flow through the billabong, but also a fall of a few feet will cause it to drain out again. Juvenile billabongs are usually dry or nearly so.

In more mature billabongs the banks are eroded to a gentle slope and the water course itself has partly filled with silt. The water is relatively shallow and supports perennial aquatic plants including *Potomogeton*. They usually have quite extensive shallow edges well colonised by *Carex* and *Eleocharis*. As the billabong silts up further if often loses its characteristic shapes and the normal water level recedes further from the tree line which remains to mark flood level. Due to the relative shallowness of the water the whole area is usually colonised by aquatic plants. A lagoon usually has relatively wide shallow edges carrying *Juncus, Scirpus,* and *Polygonum,* and in the deeper centre may be *Azolla, Marsilea drummondii,* and other aquatic plants.

Swamps

Swamps with dense emergent vegetation are formed in the more permanent water areas. Three distinct types are developed depending on the permanence of the water. Where the water is deep and permanent the dominant vegetation is cumbungi, *Typha angustifolia,* which forms dense pure stands. The deeper parts are occupied by *Azolla* and *Myriophyllum,* and the shallow edges usually carry stands of *Eleocharis, Carex, Marsilea drummondii, Agrostis avenaceae,* and *Paspalum distichum*.

In those depressions in which the water is less permanent and shallower, cane grass, *Glyceria ramigera,* is dominant forming a cane grass swamp. Normally *Marsilea drummondii* and *Azolla* do not occur and the more open parts of the swamp support heavy growth of *Eleocharis, Carex, Scirpus,* and other aquatic plants.

Lignum, *Muehlenbeckia cunninghamii,* is the dominant vegetation in areas subject to less frequent inundation, the more frequent the flooding the denser the lignum; few other aquatic plants occur in lignum swamps.

Temporary Water

On flooding of the rivers very large areas of the plain are covered by water to a shallow depth. These areas of residual floodwaters make available to the ducks an abundance of submerged dry-land plants and become colonised by large numbers of aquatic animals. They are sometimes sufficiently permanent for aquatic plants to develop.

Temporary water is also held in claypans, naturally occurring, circular depressed areas on the treeless plains. The bottoms are flat and composed of heavy clay, and support no plant growth. In times of heavy rain these collect water and form shallow temporary lakes but this water quickly evaporates and although claypans are often the source of some animal, particulary insect, life, they never support any plant growth.

Habitat Utilization

Twelve species of waterfowl, listed below regularly occur and breed in the region.

Black swan	*Cygnus atratus* (LATHAM)
Mountain duck	*Casarca tadornoides* (JARDINE & SELBY)
Wood duck	*Chenonetta jubata* (LATHAM)
Black duck	*Anas superciliosa* GMELIN
Grey teal	*Anas gibberifrons* MULLER
Chestnut teal	*Anas castanea* (EYTON)
Blue-winged shoveler	*Anas rhynchotis* LATHAM
Pink-eared duck	*Malacorynchus membranaceus* (LATHAM)
Freckled duck	*Stictonetta naevosa* (GOULD)
White-eyed duck	*Aythya australia* (EYTON)
Musk duck	*Biziura lobata* (SHAW)
Blue-billed duck	*Oxyura australis* GOULD

The plumed tree duck, *Dendrocygna eytoni,* (EYTON) is an irregular visitor and may sometimes breed. Among these species some distinct habitat preferences exist.

Streams

The main stream of the rivers forms the principal wood duck and mountain duck habitat in the region. Both these species are almost exclusively grazing animals and feed on the banks. The rivers are used to a very limited extent by black duck and grey teal, but are completely avoided by the other species. Even in times of drought, when the main streams retain some water, few ducks are found on them. Presumably river water does not provide sufficient food for the wild ducks.

The effluent streams, which are not so deep nor permanent as the rivers, provide important feeding habitat for all species except the swan and diving ducks. These are apparently excluded by lack of breeding sites and of suitable food. The species utilizing effluent streams vary according to the vegetation. Where the stream flows through living timber wood ducks, black ducks, and grey teal are abundant; where the timber fringe is not continuous wood ducks and black ducks are less numerous and grey teal the commonest species. Where the stream crosses treeless plains grey teal and pink-eared ducks may be found in very large numbers but rarely are other species seen. Where the stream is deep and flows through lignum or cane grass the white yed duck and freckled duck are the commonest species.

The larger storage dams are deep and usually contain few aquatic plants. They are not generally suitable for the breeding or feeding of waterfowl and few birds are found on them. In times of drought, however, they serve as temporary refuges for congregations of all species, when all other water has dried up. The irrigation channels provide some water habitat through large areas of otherwise dry country. They serve as focal points for wood duck flocks and frequently support small numbers of black duck and grey teal. In general, however, they are not an important waterfowl habitat.

Billabongs

Billabongs are the principal relatively permanent habitat for wild ducks in the region and are used for breeding, feeding, or refuge by all species. Immature billabongs seldom contain wild ducks apart from small numbers of black duck and grey teal, but the mature billabongs and lagoons form very important and extensive semi-premanent breeding places for many species and, after breeding, support very large numbers of all species except the diving ducks.

Swamps

The cumbungi swamps are the breeding and feeding habitat for musk and blue-billed ducks and the usual habitat for black swans. A few black duck and grey teal may breed in them if some trees are available, but they are usually avoided by wood ducks. During the summer very large concentrations of black duck, grey teal, white-eyed and freckled ducks congregate in them but usually feed elsewhere (FRITH, 1957a).

Lignum and cane grass swamps form almost the sole breeding and feeding habitat for freckled ducks and white-eyed ducks, and sometimes support large numbers of non-breeding grey teal. Over the western part of the region studied lignum and cane grass swamps are almost the sole waterfowl habitat.

Temporary Water

When the rivers flood all vegetation associations are submerged and the waterfowl habitat in the region is increased many hundredfold. The utilization of the floodwater by waterfowl varies according to the type of country flooded and the birds' breeding and feeding requirements. On flooding, especially in the lightly timbered and treeless plains, vast areas of new waterfowl habitat are created and these areas are invaded by very great numbers of grey teal, pink-eared ducks, and shovelers, and smaller numbers of black duck, and sometimes white-eyed duck. Normally floodwaters are ignored by other species. The extensive breeding that may occur has been described (FRITH, 1957b and unpublished).

Large and relatively deep claypans are an important site for opportunist breeding by grey teal and small numbers of blue-winged shovelers. They are also important opportunist feeding places for grey teal and pink-eared duck. On the whole, however, their filling is infrequent and permanence slight.

Movements

It has frequently been observed that the numbers of wild ducks in different localities usually vary greatly from year to year (e.g. MORGAN, 1954; DOWNES, 1955; HOBBS, 1957) and that, in the southern and coastal parts of the continent, a tendency exists for the numbers to increase during the summer. Banding returns and field observations, however, show that these movements cannot be explained as a simple migration. It is apparent that both migration and nomadic movements, dictated by climatic conditions, must be considered in the interpretation of waterfowl movements. Among the different species and, in some cases, within the one species, every movement pattern from completely sedentary, through erratically nomadic to regularly migratory can be found. The movement patterns of the different species seem to be determined largely by their habitat requirements and alterations in the extent of this habitat.

Those species which inhabit the permanent swamps, exclusively, are quite regular in their movements. Thus the musk duck, which is confined to the heavily vegetated permanent swamps, is sedentary and is practically never seen beyond these swamps. The blue-billed duck which inhabits the same swamps has very regular seasonal movements. Both species can apparently afford to be regular in habit, in the one case sedentary and in the other migratory, because their habitats are permanent and "safe".

The white-eyed duck, whilst mainly confined to the deep permanent swamps is also able to utilize deep semi-permanent floodwater. In accordance with these habitat requirements a regular north-south movement occurs in the permanent swamps, the species being more numerous in southern regions during the summer. In times of drought, when the permanent swamps decrease in area, this movement, however, decreases in volume or may not occur. In addition to this regular movement the birds are sufficiently adaptable to utilize deep floodwaters at any time of the year wherever they occur. Severe flooding is usually followed by an influx of white-eyed ducks, but this influx when compared to the species discussed below is slight except into those areas where lignum and cane grass are flooded.

The black duck prefers deep heavily vegetated swamps but is more elastic in its habitat requirements than the white-eyed duck and is able to utilize, to some extent, most other habitats for breeding and feeding. Each permanent swamp contains some black ducks always, but annually the number of birds in southern and coastal regions increases each summer and decreases each winter. The movements are, however, strongly affected by climatic conditions and, although dry seasons are characterized by an increased volume of movement to the coast, the majority of birds remain in the inland where great concentrations occur. In times of flood a movement to the flooded area occurs in any direction and at any time of the year. This movement however, in volume, is only a fraction of that of the grey teal.

The grey teal is a true nomad and may move over the whole continent in all directions and at all seasons in search of suitable living conditions. In the inland each creek, swamp and billabong supports small numbers of birds and as in the black duck, these numbers increase annually in the better watered areas of the south and on the coast; this movement superficially resembles true migration. However it has been shown by banding (FRITH, unpublished data) that these movements are erratic, birds from one breeding place may disperse over the whole continent in all directions. In time of flood immense numbers of grey teal arrive in that area within a few days. In time of drought virtually the whole population vacates the interior and concentrates on the coast or wherever rain has fallen. This is in contrast to the black duck which at these times has a tendency to concentrate in increasing numbers on the permanent waters inland.

The pink-eared duck is even more strongly nomadic than the grey teal. Whereas, in the grey teal, there are birds permanently in most areas and some semblance of regular movements, at times, in the pink-eared duck there is none. The whole population is nomadic and may appear in a district in very great numbers in one year but then disappear and not be seen again for many years. The birds require for their habitat expanses of shallow water, dirty, stagnant and dense with plankton. This type of water in the inland is only provided by flooding and under the conditions of high evaporation that exist is very temporary, and as it dries up the pink-eared ducks vacate the district completely. Extreme mobility is apparently essential for their survival.

Food Habits

In the period 1952—56 about 4,000 gizzards of the six common species of ducks in the inland were examined. From this study it was apparent that most of the movements of wild ducks could be explained by fluctuations in the food supply due to flooding or rainfall.

Among the nomadic species, the grey teal had a very variable diet. On the average the food consisted of 27% of plants usually growing on dry land, 40% aquatic plants and 33% aquatic animals, predominantly insects. The composition of the diet at any time, however, rarely approximated the average and varied greatly from place to place and from time to time according to the weather and the flooding of the rivers. There was no regular annual cycle of food but a cycle existed that depended on the stage of flooding in the area. Thus as the streams increased in level, or flooded, the birds fed almost entirely on dry-land plants and dry-land insects but as these were exhausted and the swamp plants (e.g. *Carex, Polygonum, Eleocharis*) were established the seeds of these swamp plants became the most important source of food. As the waters fell in level or evaporated the aquatic insects became more concentrated and these in turn became the most important source of food until ultimately the diet consisted entirely of them.

The blue-winged shoveler ate mainly aquatic animals, insects being most important, and was less dependent on vegetable food than the teal. Of the vegetable food the majority was collected from the bottom of shallow water and negligible amounts only were derived from growing swamp plants. In conformity with this food preference the shoveler occupies principally floodwaters at a rather later stage in their development arriving as the waters are receding and these foods are abundant.

The pink-eared duck fed almost exclusively on aquatic animals, including large quantities of microscopic forms collected by filtration of water and not from the bottom or edge as in other species. Accordingly pink-eared ducks do not compete with other species for food and at the same time the type of water in which they can live is limited. They are extremely mobile and arrive in an area as the floodwaters are receding and evaporating and can remain longer than most other species.

The diet of the less mobile black duck was similar to that of the grey teal but included greater quantities of swamp plants and aquatic animals. The black duck fed predominantly on the animals and seeds of larger size and characteristic of more permanent water. They were not adapted to collecting small submerged grass seeds and so were not able to exploit freshly flooded areas as efficiently as the grey teal; accordingly floodwaters were only used to a limited extent. Similarly the white-eyed duck, musk duck, and blue-billed duck fed predominantly on the larger animals characteristic of deep permanent water. These species were unable to utilize temporary floodwater and their movements correspondingly restricted.

In inland Australia where the water areas fluctuate rapidly both in extent and depth, it is apparent that for a water bird to exist permanently in very great numbers it must be sufficiently adaptable, in its food habits, to deal with a food supply that may alter rapidly in both composition and abundance. The seeding swamp plants being used as food one day may be covered by several feet depth of clear water the next, whereas the dry grass seed far from the river may equally suddenly become available due to flooding.

The grey teal has evolved great adaptability in both the food eaten, the methods of collection and feeding habitat. This adaptability enables it to exploit most types of water and food as soon as they occur. There is no doubt that this adaptability to the food supply accounts, at least in part, for the grey teal being the commonest and most widespread and mobile species of wild duck in Australia.

The movements of the other highly nomadic species, the pink-eared duck, may also be explained on its food requirements. The birds are completely dependent on plankton and insects which are only common in drying waters. In order to secure this food regularly the birds must be prepared to move very widely and rapidly as water conditions alter.

The black duck and white-eyed duck are adapted to utilize the foods produced by more permanent water than the other species. Their distribution and movements are controlled by the availability of this water, and as changes in its extent cannot be widespread nor rapid these species are accordingly comparatively local in distribution and relatively low in numbers in the inland.

Breeding

Breeding Seasons

The breeding of many species of wild ducks are strongly affected by rainfall and flooding. Among the common species every stage exists between those having regular annual breeding seasons and those that may breed at any time of the year when conditions are suitable; they may breed at a different time and in a different place in successive years. In the period 1950—1957 observations were made on the breeding of the common species of wild ducks in the Murrumbidgee and Lachlan regions, New South Wales. In this period conditions varied from extensive unprecedented flooding in 1951 and 1956 to droughts in 1954 and 1957. The volume of breeding also of many species varied directly with the seasonal conditions in each year.

Those species which inhabit the permanent stable water areas have regular breeding seasons. Thus the musk duck which is sedentary and confined to the cumbungi swamps where the water is deep and permanent with dense emergent growth of bulrushes has a regular breeding season beginning in late August and continuing until mid-October. Neither the extensive flooding that occurred in some years nor the droughts in others caused any significant departure from these dates or differences in the volume of breeding in different years. Similarly the blue-billed duck and black swan which occupy the same habitat are regular spring breeders and are not noticeably affected by flood or rain.

In the nomadic species the position is quite different and no regular breeding season exists. The birds breed wherever and whenever suitable conditions occur. Small numbers of grey teal are widely distributed throughout but they do not breed unless ex-

Fig. 1. In 1956 there was widespread flooding and breeding of wild ducks in the inland, but in 1957 a serious drought developed and little rain fell.

The upper figures refer to black duck and the lower to grey teal; in each case the figures on the left show the initial dispersal from the breeding areas at Gum Creek, N.S.W., and Joanna, S.A., those on the right show the subsequent dispersal from the drought concentration areas.

It can be seen that during 1957 the grey teal dispersal occurred from the breeding grounds to the extremities of the continent; movement by black duck was less – they concentrated on permanent swamps in the inland.

As the coastal refuges dried with continued drought there was a widespread scatter of grey teal over the whole continent, including the driest and most drought stricken parts. There was however little movement of black duck from the few permanent swamps.

ceptionally heavy rainfall or, more usually, a fresh in the river causes an increase in water level. Such an increase in level is followed immediately by sexual display and, within a few days, by ovulation. In addition to the resident grey teal, however, very large nomadic flocks move about the inland from place to place as water areas change in extent. In 1955 for instance extensive breeding of these nomadic birds occurred in widely separated parts of the country throughout the whole year as floods or heavy rain occurred at different times. This breeding season began in February in south-western Queensland, in March in northern New South Wales, in April in central New South Wales, in May on the Lachlan River (south-western N.S.W.), in July on the Murrumbidgee River, in August on the Murray River, and again in October on the Lachlan River. The movements and breeding of the pink-eared duck are completely dominated by the climate and there is no regular breeding season, they move and breed wherever and whenever the correct habitat is provided. When breeding is finished and the water dries up the whole population moves elsewhere and no residue remains.

The black duck is intermediate in habitat requirement; it is commonest in permanent swamps but is less regularly found in the temporary waters. Similarly the black duck is intermediate in its degree of nomadism and regularity of breeding season. The species tends to have a regular breeding season but apparently individuals differ greatly in the sensitivity of their response to the proximate factors initiating breeding. In unfavourable seasons, those with low rainfall and shrinking water levels, very few birds breed. In seasons of normal rainfall all the local birds breed and when floods occur the local birds are reinforced by newcomers who occupy and breed in the extra habitat created. In the grey teal unusually good conditions are utilized by a great influx of nomads and an extension of the length of the breeding season, but in the black duck there is no such extension of the breeding period but the good conditions are exploited by a greater proportion of the local birds breeding.

Effect of Water Level on Breeding

There have been several observations that, in inland Australia, breeding seasons of birds are closely associated with rainfall (SERVENTY & WHITTELL, 1948; KEAST & MARSHALL, 1955). SERVENTY & MARSHALL (1957) concluded that in W.A. photoperiodicity was of little importance as a regulator of most birds, and that the critical stimuli to breeding were environmental conditions arising after rainfall in relatively high temperatures. In wild ducks where some species have completely erratic breeding seasons ranging from midwinter to midsummer, clearly fixed annual factors such as daylength or air temperature could have little effect in determining the onset of breeding.

In studies of the sexual cycle of wild ducks (FRITH, unpublished) it has been shown that in the grey teal every outburst of sexual activity followed an increase in water level and every increase in water level was followed by sexual activity whether rain had fallen or not. The response to a rise in water level is very rapid; sexual activity begins immediately and eggs may be laid 7—10 days later. There is little doubt that in the grey teal the breeding season is initiated by an increase in water level.

The pink-eared duck carries the adaptation towards a breeding season initiated by a rise in water level one stage further. In the grey teal an increase in level sufficient to fill the lagoons is followed by ovulation. The pink-eared duck, even if already present in the district is not affected sexually by such a change in level. The sexual cycle, leading to ovulation does not begin until actual flooding of low-lying land occurs — the species only breeds when these conditions occur.

In both the preceding species it has been shown that the rising water level which initiates the breeding season also initiates an increase in the amount of food available for the ducklings. Grey teal ducklings feed entirely on animal food, principally Corixidae and Dytiscidae (Insecta). It has been shown that each increase in water level is followed by the sudden appearance of large numbers of juvenile forms and ultimately an increase in the number of insects available. Apparently the breeding season of the insects is also initiated by a rise in water level. The result is that a greatly increased food supply is available for the ducklings.

Similarly the ducklings of the pink-eared duck require large quantities of plankton. These organisms are only abundant in drying floodwater; the flooding of low-lying country, whilst initiating the sexual cycle of the ducks, at the same time provides conditions suitable for an increase in the duckling food supply. Synchrony of the ducklings and their food supply, which increases slowly, is achieved by a relatively slow development of the pink-eared duck gonad following its initial stimulation.

Summary

In inland Australia waterfowl habitats fluctuate greatly in extent from year to year. Small areas of permanent habitats occur but the most extensive habitats are those that are formed periodically and erratically by flooding of the rivers.

The movement patterns of the various species of wild ducks that inhabit the region vary according to the habitat occupied by the species and vary in regularity according to the permanence of the habitat. Species which are confined to the permanent swamps are very regular in movement, being either sedentary or regularly migratory, but those that utilize the more temporary habitats have developed nomadic habits to a very high degree.

The degree of mobility of the different species is related to their food requirements. The species having regular movements have regular food cycles. Some nomadic species have very adaptable food habits and can utilize a very wide variety of foods, thus being able to exploit all types of water as they occur. One nomadic species, however, is a food specialist and accordingly has developed an extreme type of nomadic wandering.

The species characteristic of permanent swamps have very regular breeding seasons but the nomadic species are able to breed at any time of the year whenever suitable conditions occur. The sexual cycle culminating in breeding is initiated by increasing water level in rivers. In this manner whenever flooding occurs the birds breed in the newly developed floodwater. The same factor, increasing water level, that initiates breeding in the birds initiates breeding in the animals forming the food of the ducklings so that abundant food is available.

REFERENCES

DOWNES, M.C., 1955. Where are the ducks? The Bird Observer. Leaflet No. 286.
FRITH, H. J., 1957a. Wild ducks and the rice industry in New South Wales. *C.S.I.R.O. Wildl. Res.* **2**, *32—50*.
FRITH, H. J., 1957b. Breeding and movements of wild ducks in inland New South Wales. *C.S.I.R.O. Wildl. Res.* **2**, *19—31*.
HOBBS, J., 1957. Notes on the pink-eared duck. *Emu* **57**, *265—8*.
KEAST, J. A. & MARSHALL, A. J., 1955. The influence of drought and rainfall on reproduction in Australian desert birds. *Proc. zool. Soc. Lond.* **124**, *493—9*.
LAWRENCE, E. F., 1937. A climatic analysis of New South Wales. *Austr. Geographer* **3**, (3), *3—24*.
MORGAN, D. C., 1954. Seasonal changes in populations of Anatidae at the Laverton Saltworks, Victoria 1950-1953. *Emu* **54**, *263—78*.
SERVENTY, D. L. & WHITTELL, H. .M., 1948. "A HANDBOOK of the Birds of Western Australia." (Patterson Brokensha: Perth.)
SERVENTY, D. L. & MARSHALL, A. J., 1957. Breeding periodicity of Western Australian birds with an account of unseasonal nesting in 1953 and 1955. *Emu* **57**, *99—126*.

XXIV
ECOLOGY OF AUSTRALIAN FROGS

by

A. R. MAIN, M. J. LITTLEJOHN and A. K. LEE

(Zoology Department, University of Western Australia)

Introduction

Four families of Anura are represented in Australia, these, with the contained genera and the number of species in each genus are listed in Table I. Only the Leptodactylidae and Hylidae occur throughout the continent, the Ranidae and Microhylidae are apparently recent migrants at present restricted to Cape York Peninsular. The Leptodactylids (occurring also in South and Central America), an ancient and relatively successful group so far as Australia is concerned, occupy a great variety of ecological situations; some are purely aquatic frogs while others are highly modified and specialised as burrowers.

The physiography and climate of Australia has been discussed at length in earlier papers in the present volume. Generally speaking

Table I

Families and Genera of frogs recorded from Australia with the number of species at present recognised in each genus.

Family	Genus	No. of species
Ranidae	*Rana*	1
Microhylidae	*Spenophryne*	2
	Cophixalus	1
Hylidae	*Hyla*	44
Leptodactylidae	*Lechriodes*	1
	Mixophys	1
	Cyclorana	8
	Adelotus	1
	Heleioporus	5
	Neobatrachus	5
	Limnodynastes	8
	Philoria	2
	Notaden	2
	Glauertia	3
	Uperoleia	2
	Crinia	15
	Myobatrachus	1
	Metacrinia	1
	Pseudophryne	9

the climate of more than half the continent suggests that the environment would be unfavourable for all except specially arid adapted forms of life. Frogs in particular might be expected to be limited by arid conditions because:
(a) Water is usually required for the completion of the larval stages of the life history
(b) Adults cannot control water loss through the skin and are thus, in arid environments, liable to death by desiccation.

Well developed morphological adaptations i.e. feet and body shape for burrowing as well as physiological adaptations and a capacity to aestivate are common to the frogs of the central desert and the seasonally arid areas e.g. with a Mediterranean type climate. The peculiarities of the Australian physiography and climate control the ecology of frogs as is shown by the number of genera which are represented in the fauna. Figure 1 compiled from the literature and unpublished data shows areas containing equal numbers of genera. No doubt this generalised figure is not completely accurate, nevertheless it serves to give perspective when considering the paucity of the desert fauna relative to areas having a more diverse frog fauna.

Fig 1. Generalised map with areas of equal number of genera.

Several points stand out:
  i. the frog fauna is probably as diverse as can be expected for such a featureless land-mass.
  ii. although nineteen genera are found on the continent the greatest number of genera found in any one geographical area is only ten which are found in the uniform rainfall area of the central east coast.
  iii. the only area which appears to be without frogs is the waterless Nullarbor region although *Neobatrachus centralis* (PARKER) might yet be collected there.

Desert Adaptations

Species of *Hyla, Limnodynastes, Cyclorana, Neobatrachus* and *Pseudophryne* occupy the central desert, however this does not imply a high degree of adaptation in all these frogs.

Desert adaptations in frogs may be viewed in two complementary ways viz:
  (a) From the view point of Regional Climate; one may define desert and non-desert in climatic terms e.g. annual rainfall, smaller regions can be delimited on the basis of season of rainfall. One can then consider the frogs in each of the regions delimited, or
  (b) From the viewpoint of the Species Biology; one may take well defined stages in the life history and from a knowledge of the biology of familiar species it should be possible to classify the different life history patterns so that what might be conceived as adaptation or pre-adaptations to deserts are revealed. With a knowledge of the biology one can then plot the geographical distribution of species and in this way emphasis is thrown on those species which transgress the boundaries of climatically defined regions.

In the present state of knowledge of the Australian frog fauna one has to generalise about desert adaptations from the limited knowledge of the biology of a few species and so the approach is necessarily as outlined under (b) above.

Biology

Different combinations of survival responses are illustrated by most genera, but the major innovations have been in relation to one of the following:
  (a) Breeding Biology
       (i) breeding season
       (ii) site of oviposition
       (iii) larval life
  (b) adult behaviour

(c) adult physiology
(d) habitat selection
(e) feeding
(f) adult size

The frogs of the southern part of Australia below about 25° S. latitude, are best known to the authors and the frogs of Western Australia will be used as examples to illustrate the aspects of the biology listed above. The picture of arid land adaptation which emerges after consideration of the western frogs will probably be valid for the rest of the continent because the western part of the continent includes the great desert areas of Australia and the fauna of this region should reveal the most complete range of desert adaptations likely to be displayed in Australian frogs.

Breeding biology

Breeding season

The breeding habits of Australian frogs have previously been discussed by FLETCHER (1889) and HARRISON (1922). HARRISON classified frogs into three categories:
  (a) Small frogs, the same individuals breed more than once through the year after rain.
  (b) Frogs in which each individual breeds only once in the year, but there is no definite breeding season, and individuals breed when they are ready to do so.
  (c) Frogs which have a definite breeding season.

While this classification is adequate, a more serviceable grouping for emphasising arid-land adaptations can be made if the preferred adult habitat is taken into account as a primary factor, thus:
 (1) Adults require permanent water or permanently wet situations. All frogs of this group have extended breeding seasons and egg laying occurs at intervals throughout this time as local conditions of rainfall and temperature are suitable.
 (2) Adults able to tolerate arid conditions either of a general or seasonal nature. In Western Australia the frogs of this group fall into two sections, only the first being well defined:
    (a) Frogs which are obligate winter breeders (May to August). These are the successful occupants of seasonally arid regions having a Mediterranean-type climate.
    (b) Frogs which have a very restricted but non seasonal breeding activity (usually one or two nights in the year).
       Time of breeding may vary widely from season to season, i.e. breeding may occur in any month. These are the successful desert frogs which are adapted to take immediate advantage of locally favourable conditions. Typically oviposition occurs after summer rains, usually thunderstorms, when there are high night temperatures.

All breeding activity appears to be related to specific weather conditions of which the most important appear to be (a) the quantity and type of rainfall (thunderstorms, or general cyclonic in which case it is important whether rainfall is derived from activity of a warm or cold front) and (b) the air and water temperature during the evenings following rainfall. Mating and oviposition occur, over a very restricted period of time following very specific environmental conditions which may last no longer than an hour or so in some sensitive species. In no species does mating occur over the whole range of environmental conditions during which calling males can be collected. Data on oviposition are difficult to obtain and at present are very incomplete, however, time of male call is easily recorded and despite the fact that male call period is more extended in time than oviposition it can be used when plotted against season and temperature to define the maximum length of the breeding season. This had been done for the frogs of South-western Australia (Fig. 2 and 3). The ecological importance of much of this is apparent and discussed when presenting data on eggs and larval life.

Site of oviposition

The site of egg-laying has been briefly discussed by LUCAS & LE SOUEF (1909) who classified frogs, according to the deposition and development of the ova, into two groups:
(a) small ova laid in water
(b) large yolk and part or whole of development inside the egg without water.

Those genera in which eggs are laid in water include *Hyla*, *Crinia* (except as discussed below), *Limnodynastes*, *Neobatrachus* and *Cyclorana*. Some species of the genera *Limnodynastes*, *Neobatrachus* and *Cyclorana* are successful occupants of large tracts of desert.

The genera which fall into category (b) above with large-yolked eggs laid out of water include *Heleioporus*, *Pseudophryne* and *Crinia (rosea)*. In Western Australia all the *Heleioporus* species are obligate winter breeders (Fig. 2), unpigmented eggs being laid in a frothy mass in a burrow specially excavated by the male in a site which will later be flooded by heavy winter rains. *P. guentheri* BOULENGER is a winter breeder while *P. occidentalis* PARKER breeds during summer and early autumn. Eggs of both species are pigmented and laid in wet clay in shallow horizontal burrows, usually beneath dead grass and other vegetable debris in situations which will be flooded by subsequent rains. When *P. occidentalis* breeds during the summer eggs may be laid in the moist clay at the edge of temporary pools. Upon hatching the larvae escape into the water.

The one Western Australian species of *Crinia*, (*C. rosea* HARRISON), which regularly lays eggs away from water is restricted to karri forest (*Eucalyptus diversicolor*).

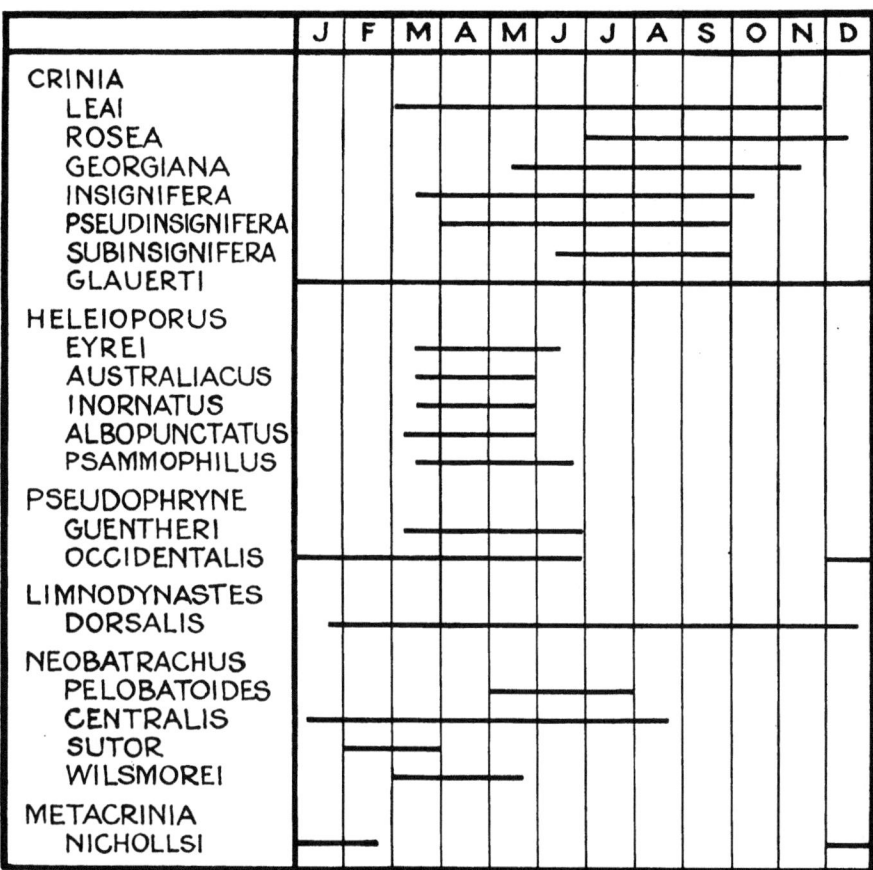

Fig. 2. Months of the year during which males of species listed have been heard calling.

Larval Life

The most outstanding fact about larval development is that all the species of *Heleioporus* and *Pseudophryne* which lay eggs out of water must complete their larval life in water. In the five species of *Heleioporus* found in Western Australia embryonic development to hatching is completed before water is required. At the hatching stage further growth and development is halted until the egg-burrows are flooded and the larvae released into the stream or swamp where normal larval development proceeds. *H. inornatus* LEE & MAIN appears to be an exception in not requiring large amounts of free water for larval development. Eggs are laid in burrows in sandy-peat, when regular winter rains raise the water-table the

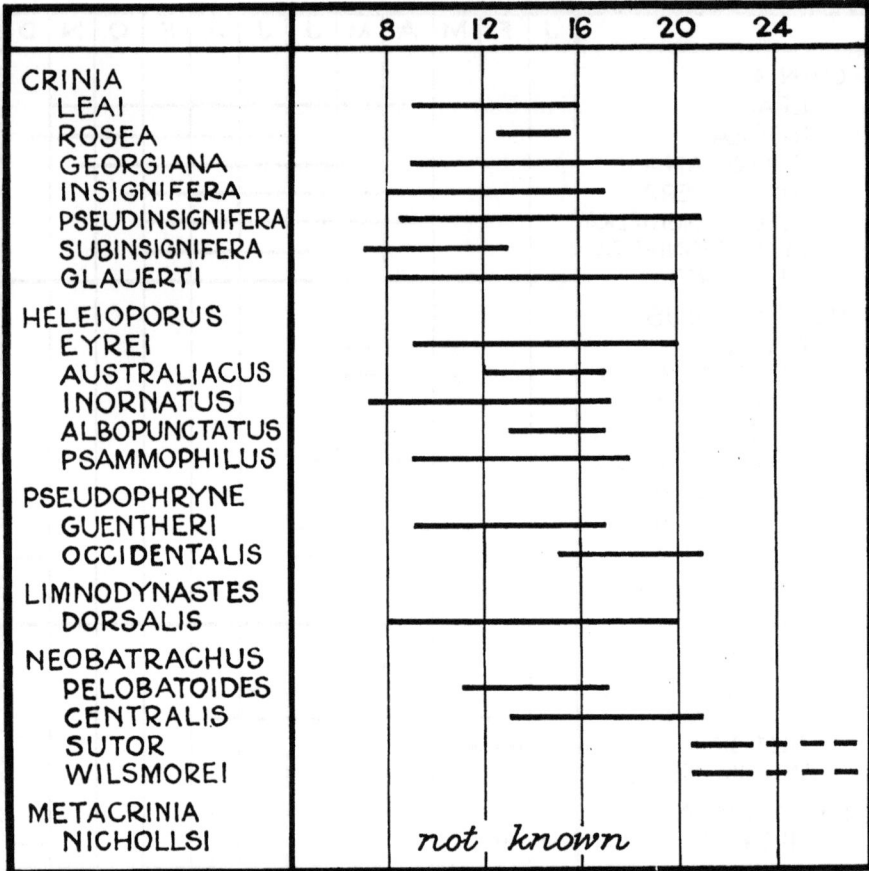

Fig. 3. Ranges in degrees centigrade over which males of the species listed have been heard calling.

burrows are flooded thus releasing the larvae from the frothy egg-mass. Larvae develop to metamorphosis within the water-filled burrow containing not more than 500 cc of fluid.

Pseudophryne eggs can also complete embryonic development before free water is required and if free water is not present hatching may be delayed for several weeks (HARRISON, 1922). In the two Western species of *Pseudophryne* aquatic larval development, usually longer than 40 days, takes place in the most ephemeral type of pond which rarely lasts longer than 50 days. With such narrow margins of safety for completing larval life, embryonic development, in this case usually taking 6—8 days, without free water, is a decided advantage.

Three species of Western frogs lack, or are suspected of lacking, an aquatic larval life namely *Crinia rosea, Metacrinia nichollsi* (HARRISON), *Myobatrachus gouldii* (GRAY). None of these is an inhabitant of arid land. *Crinia rosea* is restricted to the wet karri *(Eucalyptus diversicolor)* forests around Pemberton (annual rainfall 40 inches). The male of this species excavates a shallow depression in very damp soil or within the rotten centre of fallen logs. The clasping, oviposition, and larval development occurs within this depression. Larvae are extremely inactive inside the broken down egg capsules which fill the depression but show no obvious morphological adaptations for this type of development.

In the case of *Metacrinia* and *Myobatrachus* the evidence for terrestrial development is circumstantial. The larvae of all other frogs are known and no unidentifiable larvae have yet been found. This in conjunction with the adult biology discussed below suggests that development lacks an aquatic stage. *Metacrinia nichollsi* has a more extensive geographical range than *C. rosea* but is still restricted to wet forests. Ovarian eggs are largest in females in spring. Males call during summer and especially after warm summer rains from beneath forest litter in damp situations which, however, are never liable to flood. Furthermore, calling males have never been heard or collected in association with free water. In all other species known oviposition takes place at the site where the male calls and if this is true for *Metacrinia* it seems likely that embryonic and larval development will proceed without free water probably in shallow burrows in moist soil or rotting vegetation. Egg masses are reported by GLAUERT (1945) but have not been seen by the present authors. Mature eggs from the ovaries never exceed 11 in number and are 2.0 mm in diameter which would support non aqueous development.

Myobatrachus is always found associated with termites in sandy situations, frequently great distances from the water. Males are not known to call so that no clue to the time of breeding has been given. The smallest juveniles (14—17 mm long) have been collected during June and July and this suggests that earlier development took place in late summer and early autumn. Both adults and juveniles have been found associated with termite galleries in sand beneath rotting timber. In none of these situations is normal aquatic larval development possible. It may be that eggs are deposited in deep burrows where the soil is saturated with water, but if this is so it is unlikely that the suppositions regarding larval life will be easily confirmed.

The breeding biology of the genera which occur in Western Australia south of the 20th degree S. latitude is summarised in Table II. From this table and previous discussion on breeding biology the following conclusions can be drawn:

(i) *P. occidentalis* is the only species laying eggs out of water which occupies a desert environment.
(ii) No species of frog with a terrestrial larval development occurs in the desert.
(iii) No winter breeding species occurs in the desert.

Table II

Frogs of southern western Australia summary of breeding biology.

	Eggs laid		Larvae develop		Type of water		Season	
	In water	w/out water	In water	w/out water	still	running	winter	summer
Pseudophryne		2	2		2		1	1+
Heleioporus		5	5		4	1	5	
Neobatrachus	4		4		4		1	3+
Limnodynastes	2		2		2		1	1+
Cyclorana	2		2		2			2+
Hyla	4		4		4		3	1+
Crinia	6	1	6	1	6		7	1(part)
Glauertia	1		1		1			1+
Uperoleia	1		1		1			1
Metacrinia		1		1				1
Myobatrachus		1		1				1
Total	20	10	27	3	25	1	18	13

* + desert or arid land species.

Adult behaviour

The frogs of Australia fall naturally into two broad groups:
(a) those which are burrowers having morphological modifications such as a globose body, thick fleshy webbing between the toes and large metatarsal tubercles which act as "digging shovels" when burrowing. Burrows are usually excavated by digging backwards solely by using the hind feet except for *Myobatrachus gouldii* which can burrow head forward by using the fore and hind limbs. All species in the genera *Heleioporus*, *Neobatrachus* and *Notaden* are burrowers. In some genera, e.g. *Limnodynastes* and *Cyclorana*, the species can be classified into two categories, (i) burrowers and (ii) slender non-burrowing forms similar to those discussed in (b) below.

(b) Those non-burrowers which can be classified as aquatic species or "grass-frogs" (PARKER, 1940: 16). If the toes are webbed the webbing is delicate. In some genera webbing is absent, e.g. *Crinia*.

These frogs are not efficient burrowers. When ponds dry-up and conditions become temporarily unfavourable frogs of this group become cryptozoic and shelter beneath loose rocks, logs and plant debris where the soil remains damp for a long period after the pond has dried.

From a consideration of the two groups of frogs discussed above it is suspected that all successful arid-land species need to be efficient burrowers. While this is so for the species as *Neobatrachus wilsmorei* PARKER, *N. sutor* MAIN, *N. centralis* PARKER, *Cyclorana platycephalus* (GÜNTHER) and *Limnodynastes spenceri* PARKER it does not follow that only burrowers can survive in the desert regions. *Hyla rubella* GRAY is found widely spread through the arid parts of the continent where a summer rainfall predominates but this species is restricted to reasonably permanent water-holes. Should these dry-up as occasionally happens, during drought, *H. rubella* may survive for a few months as discussed in the previous paragraph. *Pseudophryne occidentalis* also appears to survive in arid areas by burrowing in the still plastic mud of the pond bottom before this dries completely.

Burrowing frogs are most successful (as measured by species diversity) in seasonally arid regions which enjoy a Mediterranean-type climate and animals in such situations may be considered as pre-adapted for desert existence insofar as burrowing allows occupation of such an environment by adults. That all burrowers do not occupy desert environments suggests that factors in other than burrowing ability limit the penetration into dry habitats.

The burrowing species of *Limnodynastes* and all species of *Heleioporus* except *H. australiacus* (SHAW) excavate their burrows in friable sandy soil which permits the frog to penetrate deep into the cool moist sub-soil. Specimens of *H. albopunctatus* GRAY have been dug from burrows in which they were aestivating 26 to 33 inches vertically from the surface. On the other hand, all species of *Neobatrachus* generally frequent habitats where the soil is hard, compact and has a high clay content. These species together with *Cyclorana platycephalus* (referred to as *Chiroleptes platycephalus* by SPENCER, 1896) are apparently unable to burrow deeply in impervious clay soils, consequently burrows are usually less than 12 inches deep and may be expected to be hotter than the deeper burrows of *Heleioporus*. Nevertheless the frogs in these shallow burrows are apparently protected from the heating effects of solar insolation by surface litter and the shade cast by shrubs and other vegetation beneath which the burrow is dug.

Physiology

Frogs are unable to control the loss of water through the skin. However, it is possible that some of the species of frogs occupying arid or desert environments in Australia are capable of tolerating a greater water loss than species from moister environments. On the other hand it is likely that some of the burrowers discussed in the previous section avoid dessication. It is therefore of interest to examine the ways in which the frogs of arid Australia have adapted to the stress of life without free water throughout the greater part of their adult life.

It has been known since SPENCER (1896) that *Cyclorana platycephalus* enters the period of aestivation with the bladder and coelom filled with water. In nature this store of water obviously allows a longer period of time to elapse before dehydration of tissues becomes critical. There appears to be no published work relating to the degree of dessication which can be endured experimentally by *C. platycephalus*.

Species of *Neobatrachus* range over large areas of desert and seasonally arid regions but ranges of some of the desert species *(N. sutor* and *N. centralis)* extend into the higher more regular rainfall area of southwestern Western Australia to which region all the western species of *Heleioporus* are confined. Nevertheless, while a considerable geographical region is occupied by species of both genera it seems that generally there is a habitat separation, the salient points have been noted earlier, e.g. *Heleioporus* friable soil *Neobatrachus* clay.

BENTLEY, LEE & MAIN (1958) have, in a series of laboratory experiments, compared all western species of the two genera *Neobatrachus* and *Heleioporus* with regard to (a) ability to tolerate dessication and (b) speed of re-hydration after dehydration to 75% of the initial body weight.

With regard to toleration of dessication there is no significant difference between the species within each genus; also there is no difference between the species of the two genera. All species die when about 40—50% of the body weight is lost.

When species are compared with regard to their capacity to re-hydrate after dehydration to 75% of body weight the results differ markedly from the foregoing. The species can be arranged according to speed of re-hydration (measured as mg/cm^2/hour).

In the genus *Heleioporus*; *H. psammophilus* LEE & MAIN 44; *H. eyrei* GRAY 52.5; *H. inornatus* 55.0; *H. albopunctatus* 57.7; *H. australaicus* 60.0. In the genus *Neobatrachus*; *N. pelobatoides* (WERNER) 33.3; *N. centralis* 55.7; *N. sutor* 84.8; *N. wilsmorei* 99.4.

An analysis of variance showed that:
(a) there was no significant difference between the species of *Heleioporus*.
(b) the differences between the species of *Neobatrachus* were significant at the following levels; *N. pelobatoides* v *N. centralis* $P < 0.01$; *N. centralis* v *N. sutor* $P < 0.01$; *N. sutor*

v *N. wilsmorei* not significant. Desert areas are characterised by irregular rains which fall on only a few days per year. Under such conditions speed of re-hydration may be an advantage and, as perhaps would be anticipated, *N. wilsmorei* and *N. sutor* from the extreme northern regions with the fewest rainy-days per year show the greatest capacity to re-hydrate rapidly.

Habitat selection

In the climatically defined Australian desert regions there is a considerable heterogeneity in the environment which with regard to frogs can be grouped as:
  (i) springs or permanent and semipermanent water holes where a minor wet-land assemblage of plants such as reeds and sedges and the trees of *Melaleuca* sp. persist.
  (ii) sandy water courses commonly distinguished by the presence of trees of *Eucalyptus camaldulensis*.
  (iii) the sandy or rock strewn wastes usually covered with spinifex (*Triodia* spp.) and the clayey heavier soils usually carrying mulga (*Acacia* spp.) scrub.

Whether frogs which are restricted to environments of category (i) in the above classification are to be regarded as true desert frogs is a matter of definition. From the point of view of the biology of species so restricted it can be said that they occupy desert regions because of the presence of minor favourable non-desert habitats. Such species as *Hyla rubella* and *Glauertia russelli* LOVERIDGE fall into this group of frogs.

Frogs which occur in group (ii) habitats are *Limnodynastes spenceri* PARKER and *Cyclorana cultripes* PARKER. *Cyclorana platycephalus* aestivates in the clay banks of such streams as well as in clay associated with less well defined waterways.

Group (iii) type of environments may be expected to contain the distinctive desert frogs. As has already been suggested *C. platycephalus* occupies environments of category (ii) and category (iii). In the southern arid and desert areas the three species of *Neobatrachus* viz. *N. centralis, N. wilsmorei* and *N. sutor* occur over vast areas especially those with clay soils and poorly defined drainage channels.

Feeding

The necessary data from which the detailed picture of the diet of Australian frogs can be drawn are not yet recorded. It is not possible to state whether sympatric species of the same genus have dietary preferences or even whether the diet of a species changes over the feeding period. Nevertheless the few data available indicate that frogs are not specialised feeders. MAIN & CALABY (1957) after analysing the gut contents of a number of specimens of various species collected by E. H. M. EALEY at Woodstock (Near Port Hedland) concluded that these desert species are opportunistic feeders.

CALABY (1956) analysed the gut contents of 46 specimens of *Myobatrachus gouldii* and concluded that this frog is predominantly a termite feeder although an occasional ant may be taken.

The difficulty of classifying the food habits of frogs on a few observations is well illustrated by *Glauertia russelli*. LOVERIDGE (1933: 89) describes this species as being myrmecophagous. MAIN & CALABY (1957) record termites, ants, beetles, Hemiptera, Arachnida as well as presumably accidentally ingested material such as vegetable matter and sand grains.

The data on *Myobatrachus* along with other fragmentary information suggest that specialised feeding is associated only with reliable climatic conditions and is unlikely to be found in desert frogs.

Adult size

Among species of the same genus it might be expected that those occupying desert environments would be larger because water loss has a surface area/body volume relationship. The species already discussed afford two examples on which to test these expectations.

Neobatrachus wilsmorei and *N. sutor* occur sympatrically over a large area stretching from the Murchison River to Kalgoorlie. Population samples from breeding congresses within the sympatric zone have the following range in length: *N. wilsmorei* 49 to 66 mm; *N. sutor* 35 to 43 mm.

Neobatrachus sutor indicate that small size per se is not a barrier to successful penetration of the desert. In the genus *Pseudophryne* the smaller *P. occidentalis* occupies the southern drier interior of Western Australia while the larger *P. guentheri* occupies the more coastal regions enjoying a typical Mediterranean climate. In terms of total rainfall *occidentalis* occupies the arid region. However, in terms of summer rainfall *guentheri* has a decidedly dryer climate. In the region occupied by *P. occidentalis* summer rainfall is apparently regular enough for breeding to be almost exclusively dependent on it. Thus it may be that in the interior, summer rains (e.g. thunderstorms) are frequent enough to relieve the drought and save *occidentalis* from death by dessication.

It would be surprising if, in desert frogs, size and danger of death by dessication, were the only factors apart from breeding biology which were of selective advantage to the species population. A hint that other factors are involved is indicated by the apparent differences in reproductive potential between *N. wilsmorei* which probably matures two years after metamorphosis and lays about 2,000 eggs and *N. sutor* which probably matures one year after metamorphosis and lays only about 1,000 eggs. Early sexual maturity would then give *N. sutor* an advantage (COLE, 1954) and other things being equal, populations of this species might be expected to recover in numbers more rapidly after a decline than the slower maturing species.

Discussion

In his discussion of the amphibia collected by the Horn Expedition SPENCER (1896) was the first to consider the desert adaptations of Australian frogs. SPENCER generalised from field observations on the following desert species; *Cyclorana platycephalus* (referred to by SPENCER as *Chiroleptes platycephalus*); *Neobatrachus centralis* (referred to by SPENCER as *Heleioporus pictus*); *Limnodynastes spenceri* (referred to by SPENCER as *Limnodynastes ornatus*); and *Hyla rubella*; nevertheless his conclusions remain valid today and have been confirmed by the study of other Australian frogs chosen because their ecology might suggest ways in which frogs adapt and survive in arid situations.

The review already given emphasises the fact that our knowledge of desert frogs shows only slight advance on that of SPENCER's time. However, what is known suggests innumerable future lines of enquiry which become almost impossible because of the geographical remoteness of many of the areas where further studies ought to be pursued.

The limitations and advantages of various aspects of the biology in relation to desert adaptation can be illustrated by comparing the species of *Neobatrachus* and *Heleioporus* all of which burrow. The significant aspects of the biology suggest that the principal reason for the failure of any species of *Heleioporus* to enter the desert is undoubtedly the absolute dependence of all species of this genus on early winter rains which must not be so heavy or frequent that the burrows are flooded before the larvae are ready to hatch. When the larvae are ready to hatch rain must be heavy and frequent enough to flood the burrows and maintain temporary streams and ponds for from 100 to 120 days. Furthermore the genus *Heleioporus* is further limited because during larval life all species require low water temperatures, *H. albopunctatus* being the only species in which normal development can proceed with water temperatures above 22° C. On the other hand the three arid-land species of *Neobatrachus* viz. *N. centralis*, *N. sutor* and *N. wilsmorei* are opportunistic breeders whose larvae can tolerate water temperatures higher than 25° C. Useful comparative work can be continued on the two genera *Neobatrachus* and *Heleioporus* but the interpretation of the findings will only be possible when a satisfactory comprehensive zoogeographical and evolutionary study of the frog fauna is completed. A preliminary analysis along the lines envisaged for the whole fauna has already been made (MAIN, et al. 1958), for the genera *Crinia*, *Neobatrachus* and *Heleioporus* in South-western Australia.

Summary

The areas of equal generic frequency of frogs in Australia shown in Fig. 1 has a concentric arrangement very similar to that of the average annual rainfall distribution. However, it appears that frogs occupying the desert regions of Australia are not adapted solely to conditions of low rainfall, in particular it seems that:

(a) Frogs with the following life history characteristics are not found in Australian deserts;
  (i) Terrestrial development during larval life.
  (ii) Terrestrial development during embryonic life.
  (iii) Oviposition only in the cooler season of the year i.e. obligatory winter breeders.
  (iv) Long larval life and an inability of larvae to tolerate high water temperatures.

(b) Frogs with the following life history characteristics appear to be successful in Australian deserts.
  (i) Aquatic embryonic and larval life.
  (ii) Non seasonal oviposition usually when temperatures are high and as soon as sufficient rain has fallen at any time of year to create temporary breeding sites in ponds and pools i.e. opportunistic breeders.
  (iii) Short larval life and larvae able to tolerate high water temperature.
  (iv) Efficient adult burrowing ability and a capacity to aestivate.

Acknowledgments

The authors wish to acknowledge the receipt of University of Western Australia research grants during the course of the work reviewed. They also desire to thank Mr. G. M. STORR who drew figures 1—3.

REFERENCES

BENTLEY, P. J., LEE, A. K. & MAIN, A. R. 1958. Comparison of Dehydration and Hydration of two Genera of Frogs *(Helioporus* and *Neobatrachus)* that live in areas of varying Aridity. *J. exper. Biol.* **35**, *677—684.*

CALABY, J. H., 1956. The Food Habits of the Frog, *Myobatrachus gouldii* (Gray). *W. Aust. Nat.* **5**, *93—96.*

COLE, LA MONTE C., 1954. The population consequences of life history phenomena. *Quart. Rev. Biol.* **29**, *103—137.*

FLETCHER, J. J., 1889. Observations on the Oviposition and habits of certain Australian Batrachians. *Proc. Linn. Soc. N.S.W.* (ii) **4**, *357—387*.

GLAUERT, L., 1945. Some Australian Frogs. *Aust. Mus. Mag. March April 379—382*.

HARRISON, L., 1922. On the Breeding Habits of Some Australian Frogs. *Aust. zool.* **3**, *17—34*.

LOVERIDGE, A., 1935. Australian Amphibia in the Museum Comparative Zoology, Cambridge. *Bull. Mus. Comp. Zool.* **78**, *1—60*.

LUCAS, A.H.S. & LE SOUEF, W. H.D., 1909. The Animals of Australia. Whitcombe & Tombs, Melbourne 327 pp.

MAIN, A. R. & CALABY, J. H., 1957. New Records & Notes on the biology of frogs from North-Western Australia. *W. Aust. Nat.* **5**, *216—228*.

MAIN, A. R., LEE, A. K. & LITTLEJOHN, M. J., 1958. Evolution in three Genera of Australian Frogs. *Evolution* **12**, *224—233*.

PARKER, H. W., 1940. The Australian Frogs of the family Leptodactylidae. *Nov. Zool.* **42**, *1—106*.

SPENCER, B., 1896. Report on the work of the Horn Expedition to Central Australia. Pt. 2. Zoology pp. *1—431* (Melbourne).

XXV
ORNITHOSIS RESEARCH* IN AUSTRALIA

by

J. A. R. MILES

(Department of Microbiology, University of Otago)

The ornithosis group of viruses infect a wide variety of birds and those infecting psittacine birds are often called psittacosis viruses.

Because the infection is often transmitted to man they are of considerable public health interest. The disease is transmitted to man mainly in dust from aviaries and feathers via the respiratory tract and causes a pneumonia with severe toxic manifestations and the mortality may be as high as 20%. Even with modern antibiotic treatments the mortality of severe outbreaks has been 6%. The disease was first described in 1874 and it became the subject of special field investigation in Australia in 1934 when a shipment of recently captured budgerigars was found to be infected on arrival in California. These studies remain as one of the few investigations of the incidence and transmission of a virus disease in wild animals. As such they are of special ecological interest.

In 1929—30 there was widespread psittacosis in Europe and America due to the dissemination of a large collection of infected Amazonian parrots, and the interest aroused by the occurrence of more than 700 human cases of what had previously been regarded as an exceedingly rare disease, led to careful aetiological investigations which resulted in the isolation of the virus responsible. (BEDSON et al., 1930). In the next few years investigations (MEYER & EDDIE, 1934a) showed that the majority of the sporadic human cases and small outbreaks affected people in contact with the budgerigar *(Melopsittacus undulatus)*, a native of Australia, rather than with South American parrots. MEYER & EDDIE showed that 52% of aviaries breeding budgerigars in California were infected and that, in the infected aviaries, the proportion of birds from which virus could be isolated varied between 10 to 90%. THEY (1934b) attempted to obtain a healthy stock of budgerigars by importing 200 birds freshly caught in South Australia, but deaths occurred among these birds in California while they were strictly isolated from other birds. These deaths were shown to be due to psittacosis.

In 1934 also, MERRILEES reported a fatal epizootic in budgerigars and other cage birds which had occurred in an Australian town in 1929. The agent responsible was transmissible in mice and on retrospect MERRILEES was of the opinion that it was probably

---

* The term psittacosis is usually reserved for the infection of psittacine birds and ornithosis for other birds, but these terms have historic rather than biologic significance and ornithosis is used here to cover the whole group.

psittacosis virus. It seemed probable that one human case was associated with this epizootic.

Early Work in Australia

BURNET, therefore, commenced an investigation to find out the extent of enzootic infection among psittacine birds in Australia (BURNET, 1935). He obtained birds from all states except Tasmania. Both wild birds and birds bought from dealers were examined and psittacosis virus was isolated from 50 of 274 birds tested. Wild birds were obtained from Victoria, New South Wales, Queensland and Western Australia and birds were obtained from dealers in Victoria and South Australia. The results are summarised in tables I and II.

Table I

Summary of isolations of virus from psittacine Birds
(from BURNET, 1935)

State	Wild Birds	
	No. tested	No. infected
Queensland	37	8
New South Wales	28	0
Victoria	130	19
Western Aust.	11	0
Totals	206	27 (13%)
	Bought from Dealers	
Victoria	24	8
South Australia	44	15
Totals	68	23 (34%)

Virus was isolated from wild birds only from Queensland and Victoria, but the samples from other states were inadequate since there were only eleven birds from Western Australia and all 28 tested from New South Wales were eastern rosellas *(Platycercus eximius)*. The percentage of infected birds among those bought from dealers was much higher than that of the wild birds (34 as against 13%), clearly indicating the extent to which birds become infected in captivity and showing how unwise it is to attempt to draw conclusions about the position in wild birds from information obtained from birds which have been in captivity even for a brief time. Among the wild birds

Table II
Isolations of virus from wild psittacine birds (from BURNET, 1935)

Species	No. tested	No. infected
Scaly Breasted Lorikeet (*Trichoglossus chlorolepidotus*)	6	4
Blue Mountain Lorikeet (*Trichoglossus moluccanus*)	6	3
White Cockatoo (*Kakatöe galerita*)	12	0
Galah (*Kakatöe roseicapilla*)	33	1
Cockatiel (*Leptolophus hollandicus*)	6	2
Crimson Rosella (*Platycercus elegans*)	31	0
Pale Headed Rosella (*Platycercus adscitus*)	7	0
Eastern Rosella (*Platycercus eximius*)	34	0
Western Rosella (*Platycercus icterotis*)	3	0
Twenty-eight Parrot (*Barnardius semitorquatus*)	5	0
Red-Rump Parrot (*Psephotus haematonatus*)	45	17
Budgerigar (*Melopsittacus undulatus*)	15	0
Elegant Parrot (*Neophema elegans*)	3	0
Totals	206	27

the two species of lorikeet *(Trichoglossus)* tested and the red-rumped parrot *(Psephotus haematonatus)* had notably high infection rates; while none of the substantial number of rosellas *(Platycercus)* belonging to four species yielded virus, only one of 44 cockatoos *(Kakatöe)* was found to be infected and none of the fifteen budgerigars *(Melopsittacus undulatus)* gave any evidence of infection.

BURNET's investigation clearly established the existence of enzootic psittacosis among wild psittacine birds in Eastern Australia and showed that there were significant differences in infection rate in different species in the areas in which his investigations were carried out. There was, at that time, no evidence that there was an unusual or excessive death rate in the heavily infected species.

In 1938 TREMAIN reported the isolation of virus from several species of psittacine birds which had been trapped in New South Wales. Several species which had not been investigated previously were included in this series, but since these birds had been in captivity for some time it is uncertain which of them had been infected in the wild state. He also referred to infection of finches trapped in the Northern Territory near Darwin and brought to Sydney and also of an exotic finch. These were the first records in Australia of such a virus being isolated from a bird other than one of the *Psittaciformes* and the first justification for the use of the wider term "Ornithosis" rather than the restrictive term "Psittacosis" for this disease in Australia. However it still remains uncertain whether finches are infected with viruses of this group in the wild state. They are extremely susceptible in captivity and, in this, contrast with most of the species which have been proved to be infected in the wild (MEYER, 1942).

In the intervening years there had been several reports of human cases of psittacosis in Australia which are irrelevant for the consideration of the ecology of the virus and which have recently been reviewed by FRENCH et al. (1954).

The 1938—39 Ornithosis Epizootic

In 1938 there were fatal epizootics in wild parrots in Australia which may have continued into the early part of 1939.

These outbreaks were briefly reported by BURNET (1939 a and b). In May 1938 two recently captured king parrots *(Aprosmictus scapularis)* were found to have died of ornithosis. Mr. DAVID FLEAY at that time stated that in the same area (Healesville, Victoria) he had seen an occasional dead king parrot in the bush. These deaths continued in July and August and two dead crimson rosellas *(Platycercus elegans)* were also found. A king parrot which fell dead from a tree on July 18th was found to show the typical clinical picture of ornithosis and virus was isolated. Mr. J. A. DUMARESQ reported deaths in eastern rosellas *(Platycercus eximius)* near Launceston, Tasmania in October. A dead bird showed the typical clinical picture and BURNET was able to isolate virus. DUMARESQ reported that on this property "dead parrots have been found for many years, but never in large numbers and most commonly they have been rosellas in the moulting stage". BURNET suggested that this might mean that there was an unusually virulent strain of virus enzootic in that area of Tasmania. The outbreak was apparently only local and 30 specimens of the green rosella *(Platycercus caledonicus)* and nine eastern rosellas *(P. eximius)* sent from Tasmania in August 1938 were all uninfected. It is, therefore, quite likely that the Tasmanian outbreak was separate from that on the mainland.

In October and November there was an intense epizootic in parrots in the South East of South Australia. A correspondent from Tarpeena wrote to WESTON HURST: "The matter concerns the large number of wild parrots of this district which have been dying during the past month (October). Literally hundreds have dropped, and are lying about in paddocks, found caught up in trees, etc. Other birds are not affected, so it seems there is some disease, possibly the Parrot disease, which is affecting them." HURST isolated ornithosis virus from a dead parrot sent to him from Tarpeena. This bird was described as a "grass parrot" a term used for the parrots of both the genera *Psephotus* and *Neophema*, but, in view of the area from which it came, it is highly probable that it was a specimen of *Psephotus haematonatus*.

In January 1939 BELLCHAMBERS noticed numbers of Adelaide rosellas *(P. elegans adelaidae)* dropping dead from trees in Humbug Scrub near Adelaide. He attributed this to the exceptionally hot weather at the time and no investigations were made, but in view of the widespread epizootics of ornithosis it would seem more likely that this was another epizootic focus of the disease.

In his paper (1939b) BURNET also gave presumptive evidence for the occurrence of ornithosis in Western Australia. He stated that in one of the Australian zoological gardens, a high mortality amongst several species followed the addition of a few *Barnardius* parrots from Western Australia to the aviary concerned.

Since the 1938—39 die-off of parrots, there has been no report of a similar widespread epizootic and none of the few localised outbreaks of disease which are reputed to have occurred have been investigated aetiologically.

"STANLEY (i.l.) has in 1957 isolated ornithosis virus from *Barnardius semitorquatus* taken in the south west of Western Australia, thereby confirming the indirect evidence that ornithosis is present in that state as well as in the rest of Australia."

Enzootic Ornithosis in South Australia

After the 1938—39 epizootic work was not carried out on ornithosis in wild birds in Australia until the author's unit in Adelaide commenced work early in 1951. The objects of this investigation were to follow the infection rate in a number of common species in suitable areas in the settled areas of South Australia over a series of years, to obtain evidence on whether species existed which were not affected, at least locally, with a virus of this group and to compare the infection rate in wild or feral with that in aviary or domestic birds.

The area in which the collecting was done is shown in Fig. 1, but a few birds collected in other areas for different purposes were also used for this survey.

Six species were chosen for the survey because they were common in the area conveniently near to Adelaide and limited surveys were

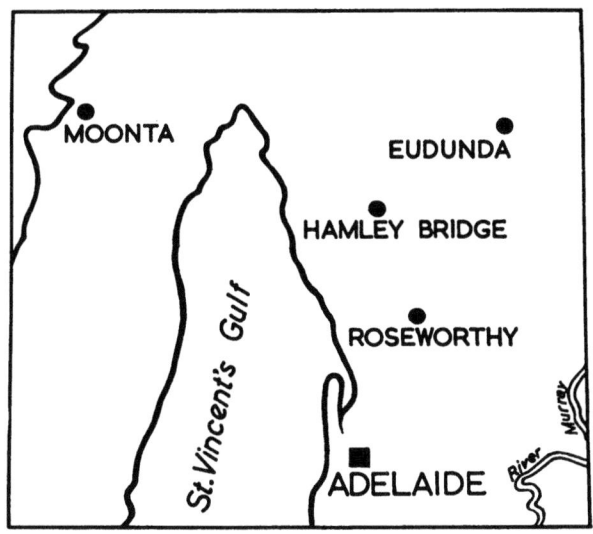

Fig. 1.

made of other species when birds which had been brought into the laboratory for other purposes were available. A small survey was also made of starlings and the survey of budgerigars was regrettably limited owing to their irregular appearances in small numbers in the survey area. Table III shows the results of all positive investigations and of all investigations on groups of more than 10 birds. Negative results on very small numbers of birds have not been included.

The first point of note was that over the four years of the survey no infected birds were found in the following three of the six species. The crested pigeons *(Ocyphaps lophotes)* all came from the Hamley Bridge area and the Indian doves *(Streptopelia suratensis)*, which had originated from aviary escapes, all came from the Adelaide area. Most of the galahs *(Kakatöe roseicapilla)* came from the Eudunda area. In each case a reasonable number of birds was tested and it is fair to conclude that infection in these species is very rare if it occurs at all in the area investigated. Burnet examined 27 wild galahs from Victoria all of which were uninfected, but he isolated virus from one of six sent from Queensland. It is reasonable to conclude that infection of galahs is rare or absent in the wild in Southern Australia, but does apparently occur in Queensland. By contrast, severe or fatal infection is not uncommon in tame birds and the two tame galahs in Burnet's series were both received dead from ornithosis.

The other two uninfected species have not been tested previously, but the extremely heavy infection rate in domestic pigeons and the

Table III
Ornithosis infection in certain common species in South Australia

Species	1951 Tested	1951 Infected	1952 Tested	1952 Infected	1953 Tested	1953 Infected	1954 Tested	1954 Infected	Totals Tested	Totals Infected
Adelaide Rosella *(Platycercus elegans adelaidae)*	24	11	36	7	45	1	41	2	146	21
Red Rump Parrot *(Psephotus haematonatus)*	13	1	19	3	20	1	14	2	66	7
Musk Lorikeet *(Glossopsitta concinna)*	—		—		2	2	—		2	2
Galak *(Kakatöe roseicapilla)*	20	0	23	0	16	0	8	0	67	0
Budgerigar *(Melopsittacus undulatus)*	—		—		11	0	2	0	13	0
House Sparrow *(Passer domesticus)*	30	1	30	0	56	11	73	1	189	13
Feral Pigeons *(Columba livia*)*	—		—		—		12	2	12	2
Crested Pigeon *(Ocyphaps lophotes)*	8	0	11	0	17	0	9	0	45	0
Indian Dove *(Streptopelia suratensis)*	8	0	9	0	18	0	37	0	72	0
Starling *(Sturnus vulgaris)*	—		16	0	15	0	1	0	32	0
White Backed Magpie *(Gymnorhina hypoleuca)*	12	0	2	0	—		—		14	0
Moorhens *(Gallinula tenebrosa)*	—		—		13	0	—		13	0
Totals	115	13	146	10	213	15	197	7	671	45

* Tested serologically only.
Where pools of birds were tested, the number of positives given in this table is the probable number assuming that infected birds are distributed according to Poisson's Distribution.
Data from BEECH & MILES, 1953; BEECH 1951-'52, 1952-53, DANE & BEECH, 1953-'54.

occurrence of infection in feral pigeons *(Columba livia)* both in South Australia (DANE & BEECH, 1955) and elsewhere made the position of other wild *Columbiformes* of special interest. All these three species are very common and gregarious birds in the settled areas of South Australia.

None of the thirteen wild budgerigars *(Melopsittacus undulatus)* taken near Moonta were infected, similarly the fifteen from Victoria tested by BURNET were all uninfected. This raises the possibility that despite the very widespread infection of this species in captivity it is not infected in the wild. Another very gregarious species in which we found no infection was the starling *(Sturnus vulgaris)*.

A varying number of isolations was made in the different years from the other three species. The results are shown in Table III and expressed as percentages in fig. 2. Usually spleens and livers from more than one bird were pooled and tested together (usually three in the case of the Adelaide rosella and five in the sparrows) and the number of infected birds was estimated, assuming a Poisson distri-

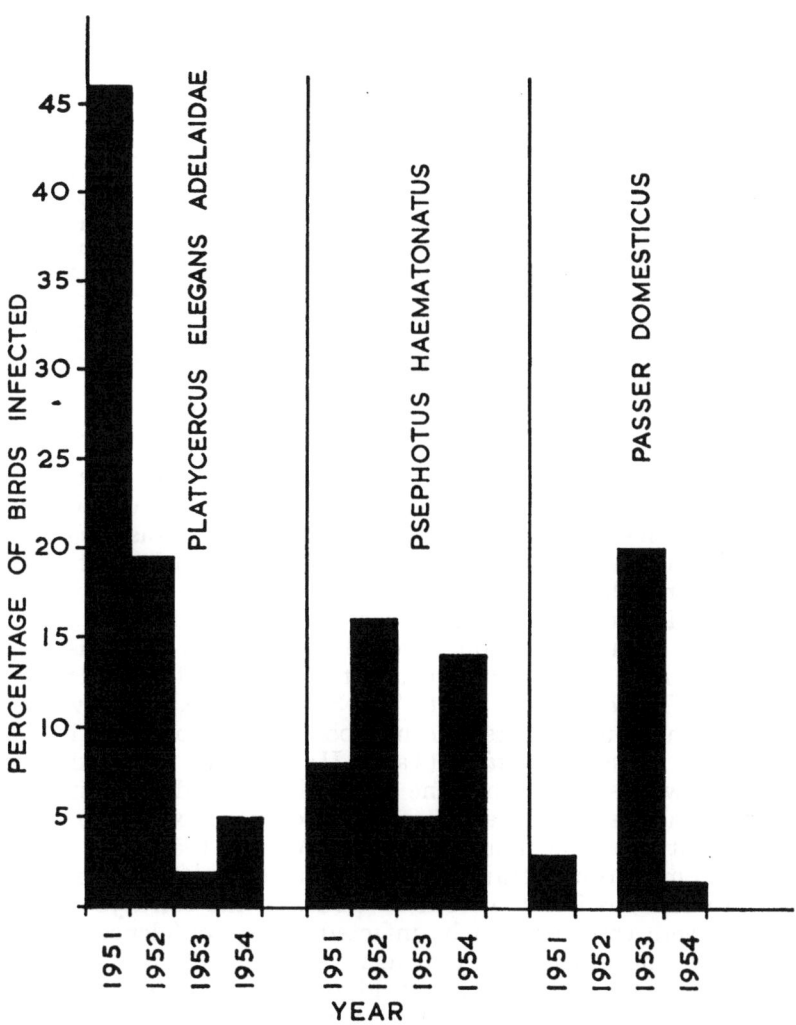

Fig. 2.

bution of infected birds, by Mr. G. F. S. SPEARS the statistician to the University of Otago, Medical School. His results for the Adelaide rosellas are shown in detail in Table IV.

The vast majority of the red rump parrots *(Psephotus haematonatus)* and of the Adelaide rosellas *(Platycercus elegans adelaidae)* were collected in the Eudunda area. All the sparrows *(Passer domesticus)* save for a small number, none of which were infected, were collected at Roseworthy Agricultural College.

During the four years of the survey the variations in the infection rate of *Ps. haematonatus* were not significant and it is interesting that

Table IV
Estimation of the number of infected *Platycercus elegans adelaidae* assuming Poisson distribution of infected birds

Year	No. birds tested	No. pools tested	No. pools positive	Probable No. birds infected	95% Confidence interval of positive total
1951	24	8	6	11.2	± 3.8
1952	36	12	5	6.6	± 2.9
1953	45	15	1	1.1	± 1.2
1954	41	14	2	2.1	± 1.6

in all years the rate was lower than that reported by BURNET for the same species at Kerang, Victoria in 1934. He obtained 17 isolations from 45 birds tested (38% infected) which was the highest infection rate in wild birds of any species of which he tested a substantial number.

The position in the two other species was quite different. In the first year of the investigation, a very high infection rate was found in *P.e. adelaidae*, when virus was isolated from 11 of 24 birds tested (Fig. 2 and Table IV) this fell to 7 of 36 in the next year, while in the third and fourth years, the number of infections was low. In the case of *Passer domesticus* (Table III and Fig. 2), in the first year virus was isolated from one of 30 birds tested. This strain of virus was interesting in being, on first isolation, extremely low in pathogenicity for the mouse (BEECH & MILES, 1953) and in this differed from all other strains examined in this series. In the second year no virus was isolated from 30 birds. In the third year, 11 of 56 birds were estimated to be infected (95% confidence interval 11.0 ± 2.9) and in the final year again only one of 73.

In neither case was there any field evidence of a die-off in these species and the numbers did not seem to be abnormally low in the area investigated in the year with the high infection rate, although the number of *P.e. adelaidae* collected in a given time in the same area in 1953 and 1954 was larger than in the first two years. This may suggest an increase in numbers coincident with the fall in the infection rate. The difference in numbers infected in the different years was clearly significant and the most likely interpretation of these findings is that in 1951—52 in *P.e. adelaidae* and in 1953 in *Passer domesticus* there was an epizootic of ornithosis with a low mortality. The high percentage of *Ps. haematonatus* which BURNET found to be infected in 1934—35 suggests the possibility that they were undergoing a similar non-fatal epizootic at that time.

The results on wild birds may be compared with those on domestic and aviary birds shown in Table V. Particularly interesting are the

Table V

Ornithosis in domestic or aviary birds in South Australia

Species	No. tested	No. infected
Fowls *(Gallus gallus)*	60*	31
Pigeons *(Columba livia)*	100*	84
Budgerigars *(Melopsittacus undulatus)*	9	3
Pileated Parrot *(Purpureicephalus spurius)*	1	1
Hybrid Rosellas	3	1
Various Aviary Finches	11	1
Muscovy Ducks *(Cairnina moschata)*	8	0
Total	192	121

* Tested serologically.
Sources of data as Table III plus DANE & BEECH, 1955.

pigeons *(Columba livia)*, of which 84% were shown by serological testing, using both the complement fixation and complement fixation inhibition tests (MILES, 1954), to be infected. Virus could only be isolated from half the birds which gave serological evidence of infection (DANE & BEECH, 1955). By contrast serological evidence of infection was only found in two of 12 feral pigeons trapped near the centre of Adelaide.

The domestic fowls tested were from general farm flocks which were free ranging. The only specialist poultry farm run on the deep litter system tested was found to be free from infection. The results on aviary birds showed infection to occur commonly and they differ in no way from those obtained elsewhere.

Mutton Bird (Puffinus tenuirostris) Ornithosis

Early in 1953 Dr. D. L. SERVENTY sent us twelve mutton birds from the Bass Strait Islands for another investigation. One of these birds died shortly after reaching Adelaide. The post mortem showed signs suggestive of ornithosis and virus was isolated in eggs. Dr. SERVENTY and his colleagues working on the *Puffinus tenuirostris* had noted a considerable number of birds affected with a disease known by the mutton-birders as "limey bird" disease. Dr. MYKYTOWYCZ, therefore, sent material from these diseased birds and blood from a larger number of apparently normal birds for investigation. It became clear that the "limey bird" disease was not ornithosis (MYKYTOWYCZ, et al., 1955), but it was shown serologically that approximately 20% of both juvenile and adult *Puffinus tenuirostris* had evidence of ornithosis infection (Table VI). Since virus had

Table VI

Complement fixation inhibition tests on mutton bird *(Puffinus tenuirostris)* sera

	Titre					No. tested	No. positive	% positive
	<5	5	10	20	≧40			
Juveniles	131	16	9	3	6	165	34	21
Adults	57	12*	0	1	0	70	13	19

* 8 of these sera were positive at 1 : 8

(Modified from MYKYTOWYCZ et al. 1955)

already been isolated from this species, an extensive series of virus isolations was not attempted. The results do not suggest that the endemic ornithosis in this species was responsible for any appreciable number of deaths in the year in which the investigations were made, and no human cases of ornithosis have been recognised among the mutton-birders. There is no evidence as to whether this virus is ever responsible for fatal epizootics in *Puffinus tenuirostris*.

Cross-infection between species

Evidence has been repeatedly obtained that under aviary and pet shop conditions an epizootic of ornithosis affects most if not all the species exposed, although there may be differences in susceptibility between different species. The evidence for such cross-infection is not easy to obtain in the wild and the rarity or absence of infection in certain species of common and gregarious birds, indicates that it cannot occur commonly. However BURNET (1939) reported both *Aprosmictus scapularis* and *Platycercus elegans* to be affected in the fatal epizootic in 1938 at Healesville, Victoria, and later in the year a "grass Parrot" *(?Psephotus haematonatus)* died in large numbers at Tarpeena in South Australia also from ornithosis. It must be highly probable that both species were affected by the same highly virulent strain of virus at Healesville and it remains possible that the same strain had spread sufficiently widely to affect a different species in the South East of South Australia later in the year.

The Adelaide work gives no good evidence of cross-infection. In the main the strains of virus isolated from the same species in different years did not vary significantly in their pathogenicity for mice, but strains from non-psittacine birds were usually of lower virulence for mice by the intraperitoneal route than those from psittacine. Such evidence is against, rather than for cross-infection. The only notable variation in virulence was between the first sparrow strain isolated, which was of unusually low virulence for mice, whereas those obtained in the epizootic period and later were similar to other strains from non-psittacine birds. This can be interpreted in two ways. Firstly exaltation of virulence may have occurred leading to the epizootic or, secondly the virulent virus may have been acquired from some other bird (possibly fowl or domestic pigeon) and then have become epizootic in the new host. We have no way of distinguishing between these possibilities.

Discussion

Our knowledge of the epidemiology of ornithosis in birds is mainly due to the work of MEYER & EDDIE (MEYER, 1942) on aviary bred budgerigars *(Melopsittacus undulatus)*. They found that many birds in their first season were carrying the virus, but apparently healthy. During the physiological stress of nesting, the balance between host and parasite was upset and they again began to shed virus. The young were thus infected in the nest and became ill. A small percentage (up to about 10%) died, but the remainder recovered slowly and in due course appeared quite healthy until either the stress of travelling, unsuitable diet or their own breeding upset the host-parasite balance and they began to shed virus again, often appearing ill with ruffled feathers, nasal discharge and soiled vent. Thus the next generation would be infected in the nest and sometimes other adult birds might be infected. In this species most birds seemed to clear themselves of virus during their second season, but some would carry active virus for two to three years at least.

All work suggests that this is the normal method for the perpetuation of the virus and that the usual time for infection to take place is in the nest. For instance the figures for *Puffinus tenuirostris* show as high a proportion of juvenile as adult birds with evidence of present or past infection (Table VI).

Apparently this method of spread leads to a very high rate of enzootic infection in domestic birds, but only a low rate in the wild.

It is well known that there is a very high mortality in wild birds during their first year of life which decreases in succeeding years and it may well be that infected juveniles have an even higher mortality than normals because they are smaller and weaker at the time of leaving the nest.

Under domestic conditions the vast majority of birds survive and, particularly where breeders are seeking to establish new variants, breed in their first year when they are likely to be carrying active virus. The average age of the breeding population of wild birds is much higher and, therefore, the proportion of infected breeding birds will be likely to be lower. Further if there is a rather higher mortality in the nests of infected parents than in those of normals and a rather lower survival rate of juvenile birds, then, if the virus were only transmitted in the nest, the proportion of infected birds would steadily fall and in due course the disease would die out.

This disease cannot perpetuate itself in the bird purely by nest infections as Herpes simplex does in man by parent to child infections (BURNET, 1945) because, whereas the Herpes virus survives with the host throughout his life causing only minimal inconvenience, ornithosis virus only survives for a relatively limited part of the bird's breeding life. Occasional epizootics, in which the virus spreads among fledged birds, are necessary for the survival of ornithosis virus under natural wild conditions.

There is good evidence both for the occurrence of fatal and non-fatal epizootics of ornithosis in Australia which, since the pre-epidemic infection rate has been shown at least in the case of *Passer domesticus* to have been very low, must have been due to infection derived from some source other than the parent bird in the nest. We have no evidence on whether the epizootic virus was a variant virus from the same host with enhanced virulence or transmissibility, or whether it was obtained from another species of bird; but there is evidence that the highly virulent strain responsible for the 1938—39 fatal epizootic was transmitted from one species to another. On the other hand the 1951 epizootic which was non-fatal did not appear to affect either *Kakatöe roseicapilla* or *Psephotus haematonatus* in the district in which the infection rate of *Platycercus e. adelaidae* was 46%, although the three species might commonly be seen in the same trees in quite close contact.

It would then appear that the amount of contact between species under natural wild conditions is inadequate to allow transmission from one to another save when strains of very unusual virulence and transmissibility are epizootic. If transmission between species under natural conditions occurred often, then the finding of common gregarious species with little or no infection would be quite inexplicable.

Finally it is interesting to speculate as to whether ornithosis is of any importance in controlling population levels in infected species. Clearly a major die-off like that described from the South East of South Australia in 1938 must have a substantial temporary effect on numbers, but such a die-off has not been recorded for the last nineteen years and, therefore, cannot be regarded as one of the major factors in controlling the population. On the other hand milder epizootics are more common and, although there is no field evidence of a die-off, this does not exclude the possibility of a high mortality in the nest which might explain the apparently smaller numbers of *Platycercus e. adelaidae* in 1951 than in 1953 and 1954. However, in view of the low enzootic rate and the relative infrequency of epizootics, it seems improbable that this disease is a major factor in controlling population size under natural conditions.

Acknowledgment

It is a pleasure to acknowledge the assistance of Mr. G. F. S. SPEARS in the statistical analysis of certain of the findings.

REFERENCES

BEDSON, S. P., WESTERN, G. T. & SIMPSON, S. L., 1930. 'Observations on the Aetiology of Psittacosis', *Lancet* **1**, *235—236*.

BEECH, M. D., 1951-'52. 'Psittacosis', *Inst. Med. Vet. Sci. Adelaide, Ann. Rep.*, **14**, *26—28*.

BEECH, M. D., 1952-'53. 'Psittacosis', *Inst. Med. Vet. Sci. Adelaide, Ann. Rep.*, **15**, *30—34*.

BEECH, M. D. & MILES, J. A. R., 1953. 'Psittacosis Among Birds in South Australia', *Aust. J. exp. Biol.*, **31**, *473—480*.

BELLCHAMBERS, R. F., 1953. Personal communication.

BURNET, F. M., 1935. 'Enzootic Psittacosis amongst Wild Australian Parrots', *J. Hyg.* **35**, *412—420*.

BURNET, F. M., 1939a. 'A Note on the Occurrence of Fatal Psittacosis in Parrots Living in the Wild State'. *Med. J. Aust.* **2**, *545—546*.

BURNET, F. M., 1939b. 'Psittacosis in Australia', *Proc. Sixth Pacific Sci. Congr.* **5**, *349—351*.

BURNET, F. M., 1945. 'Virus as Organism', Harvard University Press, Chs. 4 & 6.

DANE, D. S. & BEECH, M. D., 1953-'54. 'Psittacosis' *Inst. Med. Vet. Sci. Adelaide, Ann. Rep.*, **16**, *26—29*.

DANE, D. S. & BEECH, M. D., 1955. 'Psittacosis among Birds in Contact with Man', *Med. J. Aust.* **1**, *428—429*.

FRENCH, E. L., JOSKE, R.A., BODYCOMB, D. H., MACKENZIE, E. F. & RIGG, W. R., 1954. 'Psittacosis in Australia', *Med. J. Aust.*, **1**, *392—397*.

MERRILEES, C. R., 1934. 'Psittacosis in Australia', *Med. J. Aust.*, **2**, *320—321*.

MEYER, K. F., 1942. 'The Ecology of Psittacosis and Ornithosis', *Medicine*, **21**, *175—206*.

MEYER, K. F. & EDDIE, B., 1934a. 'Über Papageienpest. Auf Grund von epidemiologischen und experimentellen Studien in Kalifornien', *Klin. Wschr.*, **13**, *865—870*.

MEYER, K. F. & EDDIE, B., 1934b. 'Psittacosis in the Native Australian Budgerigars', *Proc. Soc. exp. Biol. Med.*, **31**, *917—920*.

MILES, J. A. R., 1954. 'Observations on Complement Fixation and Complement Fixation Inhibition using Certain Avian Sera', *Aust. J. exp. Biol.*, **32**, *57—68*.

MYKYTOWYCZ, R., DANE, D.S. & BEECH, M. D., 1955. 'Ornithosis in the petrel, *Puffinus tenuirostris* (Temminck)', *Aust. J. exp. Biol.*, **33**, *629—636*.

TREMAIN, A. R., 1938. 'Some Aspects of Psittacosis and the Isolation of the Virus', *Med. J. Aust.*, **2**, *417—421*.

XXVI
VEGETATION OF HIGH MOUNTAINS IN AUSTRALIA IN RELATION TO LAND USE

by

A. B. COSTIN

(C.S.I.R.O. Division of Plant Industry, Alpine Ecology Section, Canberra)

Introduction

Australia, although predominantly a country of low relief, low rainfall, and arid and semi-arid vegetation and soils, also contains considerable areas of elevated land which regularly receive heavy snowfalls and where the vegetation and soils are distinctly "high mountain" in character. The lower limit of this high mountain country is defined naturally by the winter snow line; this varies from about 5,500 to 5,000 feet in the Australian Capital Territory and New South Wales to about 4,500 feet in southern Victoria and 3,000 feet in Tasmania. The total area above the winter snow-line on the Australian mainland is about 2,000 square miles of which approximately 1,000 square miles are in New South Wales, 870 square miles in Victoria, and 140 square miles in the Australian Capital Territory. (Fig. 1). In Tasmania, the area of snow country aggregates about 2,500 square miles. Most of this discussion deals with mainland conditions as information on the Tasmanian high mountain vegetation is too incomplete to enable more than passing reference to be made at this stage.

In addition to the intrinsic interest of the Australian high mountains and their vegetation they are of great importance to the national economy on account of the large and reliable rivers which have their headwaters there. As seen from Fig. 1 these include the Cotter River in the Australian Capital Territory; the Goodradigbee, Upper Murrumbidgee, Upper Murray and Snowy-Eucumbene systems in N.S.W.; and the Upper Murray, Goulburn, Tambo, Mitchell, Macalister-Thomson-Latrobe, and Yarra River systems in Victoria. The great potential of these areas to supply water for irrigation and hydro-electric power and for domestic and industrial use is now being developed on an increasing scale.

The High Mountain Environment

The high mountain environment is naturally defined as that situated above the winter snowline where the ground is usually snow covered continuously for not less than one month of the year.

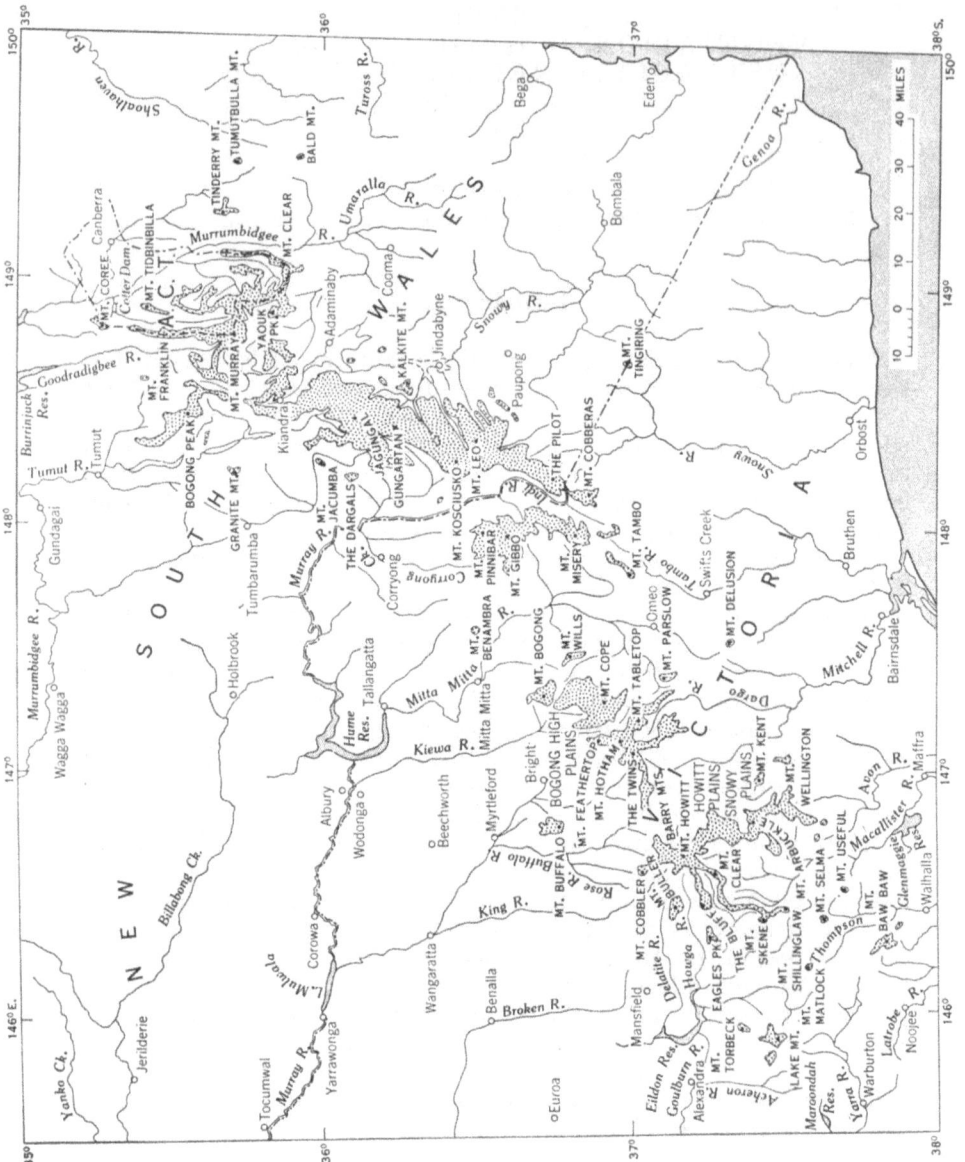

This environment includes two well-defined tracts, subalpine and alpine. The subalpine tract, extending from the lower level of the winter snow line to the treeline, is normally snow-covered continuously for about 1—4 months. The alpine tract, extending from the tree-line upwards, is snow-covered for longer periods, and locally in protected snowpatch situations for most of the year. Large areas of the Australian Alps were glaciated in Pleistocene times but today permanent ice- and snow-fields no longer exist. (Plates 1 and 2).

Often blowing with gale to hurricane force, the prevailing north westerly to south westerly winds, associated with cyclones originating in the Antarctic, bring most of the precipitation which falls mainly as snow with a relative winter incidence. In summer, heavy rainstorms are also common. In the sub-alpine tract the average annual precipitation varies from about 30 to 80 inches and the number of rain days from about 120 to 140. In the alpine tract precipitations range from about 70 to 120 inches per annum, with about 130 to 150 rain days. It is of interest to record that precipitations do not decrease at the higher levels as they commonly do on higher mountains overseas.

Mean monthly maximum temperatures rise to about 60—75° F. in midsummer in the subalpine tract and probably about 50—60° F. in the alps. The corresponding mean monthly minima fall to about 25—30° F. and to less than 25° F. in mid-winter, and do not exceed freezing point for about 6 and 6 to 8 months of the year respectively. During the snow free months, however, when there is often a regular alternation of warm days and freezing nights, daily temperature fluctuations are considerably higher and lower than the monthly means would indicate, the diurnal range not infrequently approaching 50° F. or more.

Geologically, the Australian high mountains show considerable variation, all of the main rock types such as acid granites, slates, basalt and limestone being represented. In contrast to Europe, however, the high mountain soils and vegetation are not strongly differentiated according to rock type, similar plant communities and soils being developed on quite different rocks. Thus, of the eight well-defined soil groups (COSTIN, 1954), alpine humus soils, lithosols, grey podsols, acid marsh soils, snow patch meadow soils, bog peats, poor fen peats and humified peats, the alpine humus soils develop as the climatic climax on all kinds of parent material. These soils, although strongly acid and base unsaturated, are neither peaty nor podsolised, and in contrast to the mountain soils of Britain and Scandinavia which are considered to be undergoing progressive base depletion leading to moor formation and further restriction of high mountain species, they appear to be stabilised against further leaching by a small but significant return of bases and mineral matter by decomposing herbs and earthworm activity

(COSTIN, HALLSWORTH & WOOF, 1952). Compared with the high mountains of Europe, moreover, where the surfaces are predominantly peaty or rocky, the Australian Alps are in the nature of soil mountains with soil and decomposing parent material often many feet in depth. This stronger soil development in Australia is related to the milder glacial history, the gentler slopes, the favourable climate for soil weathering and the unusually vigorous biological conditions in the soil. The association of these deep organo-mineral soils with summer storms of high intensity and wide daily temperature fluctuations capable of causing frost heave constitutes a high natural erosion hazard, necessitating great care in land use, if stability is to be maintained.

Conditions in Tasmania differ in several main aspects from those on the mainland. Most of the Tasmanian snow country consists of a large central plateau of resistant dolerite, with the smaller Ben Lomond massif in the northeast. The climate is distinctly more oceanic, especially in summer, and the long dry spells broken by summer thunderstorm activity are not so marked. In drier sites, the soils are shallow and stoney, and in wetter situations, peaty with gleyed or podzolised mineral layers beneath. Thus, there is little or no development of the "soil mountains" found on the mainland. In general appearance, the Tasmanian mountains resemble parts of the British mountains more than the Australian Alps.

Flora and Fauna

The high mountain flora and fauna* differ greatly from those of the rest of the continent, not only in their ecology but also in their unique spectra of life-forms, geographical origin, and evolution.

Many of the high mountain plants show much the same adaptations to alpine conditions as are found in other countries, such as nanism (dwarfing), the rock-clinging habit, bright and conspicuous flowers, perenniality, the tussock habit, ericoid, divided and lanceolate leaves, spinescence, strong vascular development, pubescence, and physiological resistance to fluctuating high and freezing temperatures, intense insolation, fierce winds, periodic moisture stress, and burial by snow.

The geographical affinities of the flora are best considered in terms of the following main elements: Australian, South African, Zealandic, Andine, Palaeo-antarctic, Palaeotropic and Cosmopolitan (including the European element). The distribution of these elements in the high mountain tracts (alpine and subalpine) compared with their distribution in the lower montane and tableland areas is shown in the case of the Monaro Region of New South Wales, which

---

* Detailed lists are given in COSTIN (1954).

Table I
Geographical elements of the flora of the Monaro Region

Tract	Geographical Elements														Total No. of Species
	Australian		South African		Zealandic		Andine		Palaeo-antarctic		Palaeo-tropic		Cosmopo-litan		
	No. spp.	%	No. spp.	%	No. spp.	%	No. spp.	%	No. spp.	%	No. spp.	%	No. spp.	%	
Tableland (2,000–3,000 ft.)	193	50.2	9	2.3	4	1.0	7	1.8	2	0.5	34	8.8	136	35.4	385
Montane (3,000–5,000 ft.)	242	53.1	10	2.2	6	1.3	8	1.8	5	1.1	33	7.2	152	33.3	456
Subalpine (5,000–6,000 ft.)	93	35.6	11	4.2	10	3.8	5	1.9	11	4.2	6	2.3	125	48.0	261
Alpine (> 6,000 ft.)	58	30.6	5	2.6	15	7.9	7	3.7	15	7.9	2	1.1	88	46.2	190
Total Monaro	360	49.6	16	2.2	20	2.8	13	1.8	20	2.8	49	6.7	248	34.1	726

contains the largest and highest parts of high country on the mainland (Table I).

It will be seen that the total flora of the Monaro Region is composed largely of Australian and Cosmopolitan species but the relative proportions of these and the other elements change characteristically with increasing elevation. In the tableland and montane floras Palaeotropic species are prominent and species with southern affinities unimportant, in keeping with the relative mildness of the past and present climates of these lower tracts. In the subalpine and alpine floras, on the other hand, in which severe climatic conditions have operated, there is a marked increase in the proportions of the Zealandic, Andine, Palaeoantarctic and Cosmopolitan species at the expense of those with Australian and Palaeotropic affinities.

Analysis of the life-form composition of the flora (Table II) by the method of RAUNKIAER (1934) is also instructive. Developed in Northern Europe, this method assumes that the unfavourable winter season of the year limits plant distribution and that the extent of this limitation is expressed in the life-form composition of the flora, which can thus be used to define characteristic "plant climates". In the case of the Monaro Region, however, neither the tableland nor the montane floras have any counterpart among RAUNKIAER's characteristic plant climates of the Northern Hemisphere. This is not surprising in view of the milder climate, both past and present, which these areas have experienced. On the other

Table II

Biological spectra of the flora of the Monaro Region.

Flora	No. of Spp.	Percent Distribution of the Species among the Life-forms										
		MM	M	N	Ch	H	G	HH	Th	S	E	Pt.-Qt.
Tableland (2,000-3,000 ft.)	373	7	9	21	12	23	11	7	9	0	1	0.8
Montane (3,000-5,000 ft.)	439	6	8	24	14	20	12	10	5	0	1	1.0
Subalpine (5,000-6,000 ft.)	255	1	1	18	16	32	13	9	10	0	0	0.6
Alpine (> 6,000 ft.)	184	0	0	1	31	41	10	8	9	0	0	0.8
Total Monaro	703	5	7	21	16	23	11	8	8	0	1	0.8
Raunkiaer: Normal Spectrum	400	6	17	20	9	27	3	1	13	1	3	1.0

hand, the subalpine and alpine floras show spectra which can be matched with RAUNKIAER's, the subalpine spectrum approximating to his Hemicryptophyte and Chamaephyte Climate of the European Boreal Zone and of the European Alps at an elevation of about 6,000 feet, and the alpine spectrum to the Chamaephyte Climate of the Arctic Zone verging on the Arctic Nival Region. Thus, with increasing cold in the subalpine and alpine tracts, there is a characteristic increase in the proportions of cold-resistant hemicryptophyte and chamaephyte life-forms, as in Northern Europe.

In the absence of much palynological and radio-carbon data providing direct evidence bearing on the evolution of the flora and fauna, it is impossible to do more than suggest the most likely course of evolution, in the light of geological and bio-geographical evidence. During most of the Tertiary, south-eastern Australia experienced a tropical to subtropical climate which in the prolonged Early and Middle Miocene Stillstand produced virtually a peneplain surface with extensive laterisation. Apart from a minor uplift in the Late Miocene, these conditions persisted until the commencement of the Kosciusko Uplift in the Late Pliocene when extensive physiographic, climatic and biological differentiation began. Up to this time a high mountain flora and vegetation would not have been able to exist on the Australian mainland. During the Early Pleistocene an ice-cap formed on the Kosciusko Plateau and smaller glaciations affected considerable areas of the Victorian Alps. Under these conditions the winter snow-line is estimated to have descended as low as about 3,000 feet in the Kosciusko area (2,000 feet lower than at present); it would have been as low as 2,000 feet in parts of Victoria, and from 1,000 feet almost to sea level (near Port Davey) in Tasmania. As Tasmania was then connected to the mainland by recently-emerged land bridges which, being largely unvegetated, would provide suitable conditions for the invasion and spread of mountain plants, this large-scale lowering of the winter snowline would have provided virtually continuous "high mountain" conditions from the sub-antarctic extensions of Tasmania to Victoria and New South Wales. Under these conditions, the broad outlines of the plant communities as they occur in the high mountains today probably took shape. Glacial recessions and the re-development of smaller glaciations during the Pleistocene would have produced various range changes and discontinuities of vegetation, and the final glacial recession of Late Pleistocene-Earlier Recent Time would have produced the distribution patterns much as they occur today.

The impact of the native fauna on the development of the flora and vegetation appears to have been largely negative. The larger marsupials were only occasional summer influents, and grazing pressure from this source would have been very slight. With the advent of the white settler on the tablelands surrounding the mountains, however, new species and new practices of land use were superimposed. A considerable number of exotic plants and animals have now become naturalised, the latter, including the hare, rabbit, brumby (wild horse), fox and feral cat, now dominating the fauna and exerting a considerable influence on the vegetation. Apart from these "accidental" effects of white settlement, the major influence has been the practice of summer grazing of sheep, cattle and occasionally horses with the associated practice of "burning off" the high mountain vegetation to improve its palatability. The effects of these practices on the native vegetation and soils, and their compatibility with the now more important need to conserve water-supply, recreational and scientific values is discussed later.

Description of the Vegetation

As in other countries there is a distinct zonation of vegetation on the Australian Alps with increasing elevation. In most areas the zonation is masked at the lower levels by the fact that precipitations still remain high, but in certain localities, notably the Benambra-Omeo-Bogong area in Victoria and the Monaro Tableland-Snowy Mountains area in N.S.W., the lower tracts receive little precipitation and a full climatic sequence of soil and vegetation is developed. The best examples are found on the eastern slopes of the Snowy Mountains from the highest point at Mt. Kosciusko (7,328') to the dry rain-shadow areas about 30 miles eastwards around Berridale and Dalgety (less than 3,000'). This sequence is depicted in Fig. 2. It will be seen that between the extremes of Berridale with an average annual precipitation of about 20" and a mean annual temperature of about 54° F. and Mt. Kosciusko with an average annual precipitation of about 120" and a mean annual temperature of about 36° F., the vegetation and soils range from dry tussock grassland with arid-climate soils resembling brown soils of light texture, progressively through open savannah woodland with slightly podzolised grey-brown podzolic soils, dry sclerophyll forest with distinctly podzolised iron podzols, wet sclerophyll forest with brown podzolic soils and transitional alpine humus soils (resembling acid brown forest soils), into the high mountain subalpine woodland with unpodzolised alpine humus soils giving way above the tree line to tall alpine herbfield on similar soils.

Within the sub-alpine and alpine tracts shown in Fig. 2 as dominated by sub-alpine woodlands and tall alpine herbfields, topographic and geological variation give rise to a large number of other high mountain communities including sod tussock grassland, heaths, bogs and fens (swamps), fjaeldmark (fell field vegetation), chomophyte communities (vegetation of cliffs, rock ledges, etc.), and snow patch vegetation (Plates 1 & 2). These communities are briefly described below.

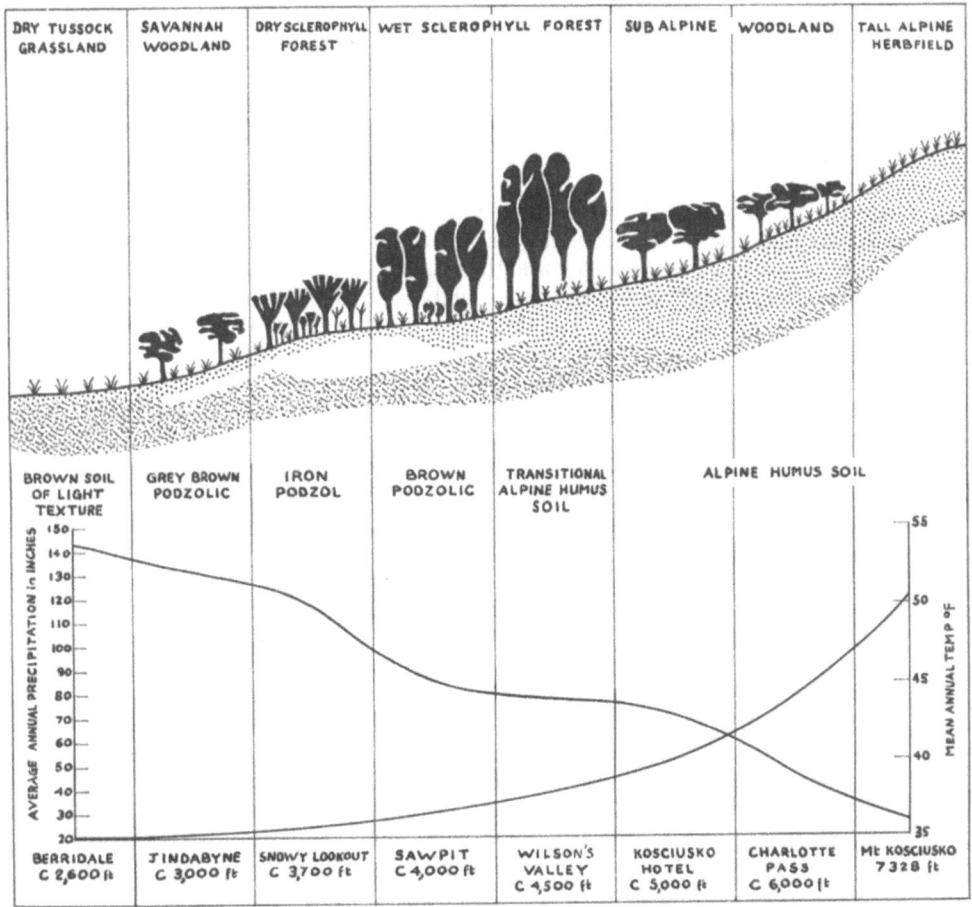

Fig. 2. Climatic sequence of vegetation and soils, Monaro Region of New South Wales.

Subalpine woodland: *Eucalyptus niphophila* alliance.

Subalpine woodlands of the *Eucalyptus niphophila* (snow gum) alliance are the climatic climax of the subalpine tract. They dominate the greater part of the high mountain country under discussion and form the tree-line vegetation in the highest areas. The upper limit of tree-growth varies between about 6,500 feet on Jagungal north of Mt. Kosciusko, to a little more than 6,000 feet on the Kosciusko Plateau, down to 5,750 and 5,500 feet on the more southern Victorian mountains of Bogong, Buller and Stirling. Exposed, isolated summits or their windward aspects are commonly bare of trees to considerably lower levels, as on Mt. Cobbler (5,342′) and Mt. Wellington (5,355′)

in Victoria. Another interesting latitudinal variation which appears in the southerly subalpine woodlands of the Baw Baw Plateau is the presence of myrtle beech (*Nothofagus cunninghamii* OERST.), a species which is entirely absent from New South Wales, and which elsewhere in southern Victoria occurs only under montane conditions. Its presence in the sub-alpine tract of the Baw Baws is related to the higher humidities and greater cloudiness of this area. It provides a strong link, together with an interesting group of Tasmanian species, with conditions in Tasmania, where the myrtle beech often ascends to the treeline as a scrubby rainforest in the moister subalpine environments, with scrubs of the endemic Tasmanian eucalypt *Eucalyptus coccifera* HOOK. f. in drier situations. In Tasmania endemic conifers (e.g. *Arthrotaxis cupressoides* DON.) may also be present as co-dominants and in pure stands giving the Tasmanian subalpine vegetation a quite different physiognomy from the subalpine woodlands of the mainland.

Mature communities of the *Eucalyptus niphophila* alliance are parklike in character with spreading single-boled snow gums underlain by a dense herbaceous stratum with relatively few shrubs except in locally more rocky situations where shrubs are more common. This original structure has been much modified by bush fires and intentional burning designed to improve the palatability of the native snowgrass. The addition of grazing influences complicates the picture further.

It so happens that sheeps (but not cattle) are fond of the new snow gum shoots and these are persistently nibbled back when a tree is attempting to make new growth after a fire or when a seedling is attempting to establish itself. If this defoliation is frequent enough the ligno-tuber fails to regenerate. In this way large areas of snow gum country have been made treeless and additional areas now regenerating with difficulty will soon become treeless if current grazing practice continues. Once snow gums in an area have been killed and the seed reserves exhausted, recovery takes place slowly and with difficulty. In initially difficult snowgum environments, furthermore, as near the treeline and next to frost hollows, it is problematical if natural regeneration can ever be achieved, as the removal of the woodland en masse creates more extreme climatic conditions as regards exposure to wind and frost which exceed the tolerance of the snowgum species.

Tall alpine herbfield: *Celmisia longifolia-Poa caespitosa* alliance.

At about 5,500 to 6,500 feet, depending on latitude, the subalpine woodlands give way to treeless communities of which the *Celmisia longifolia-Poa caespitosa* tall alpine herbfield alliance is the alpine climatic climax. Its most extensive development in Australia is

PLATE I

Plate 1. Alpine landscape, Kosciusko, showing typical distribution pattern of some of the vegetation. Climatic climax tall alpine herbfield has the widest occurrence: the community in the foreground has been seriously damaged following heavy grazing. Short alpine herbfield occurs below the snow patches, and fjaeldmark communities along the upper margins of the snowpatches and on the adjacent windswept cols. Chomophyte herbfield vegetation is abundant on the steep slopes of the rocky mountain shown on the right. Bogs, fens and grassland develop along the poorly drained valleys below the snow patches. Moraine and similar rocky environments (not shown) support heaths.

A. B. Costin

PLATE II

Plate 2. Typical subalpine landscape and vegetation pattern, Kosciusko. Subalpine woodland covers most of the slopes with heaths in rockier situations. Grassland, bogs, fen, and wet heath occupy the valley floor. A few of the surrounding peaks rise just above the tree-line into tall alpine herbfield.

in the Kosciusko area but it is also well developed in Victoria in the Bogong High Plains area, on Mts. Buller and Stirling, Pinnibar, Gibbo and the Cobberas, and on the leeward slopes of Mt. Speculation, the Crosscut Saw, Mt. Howitt, Mt. Magdala, Mt. Lovick and The Bluff. The occurrence of alpine herbfield as a climatic climax over a wide range of rocks from basalt to granite, sandstone and slate, means that the influence of parent material is relatively slight. This is in strong contrast to the corresponding herbfield vegetation of Europe which develops mostly in association with rocks or ground waters of adequate base status. Under the moist climatic conditions of the British and Scandinavian mountains, furthermore, progressive base depletion and mor formation are thought to be further restricting the already limited distribution of this type of vegetation. On the Australian mainland, on the other hand, the herbfield communities appear to be in stable equilibrium owing to the special soil stabilising processes of base and mineral circulation referred to previously.

Mature communities of this alliance usually consist of a continuous sward of herbs with associated minor chamaephytes; in summer time the massed flowering of *Celmisia longifolia* and similar flowering forbs (species of *Ranunculus, Gentiana, Euphrasia, Wahlenbergia, Brachycome, Craspedia, Erechtites, Podolepis, Helichrysum* and *Senecio*) presents pictures of striking colour, variety and beauty. Modification by grazing with occasional fires has virtually eliminated several palatable co-dominant herbs which old records show to have been important about the turn of the century, notably *Aciphylla glacialis* and *Danthonia frigida*. This stage of floristic modification has now proceeded in most areas towards structural change involving the opening up of the continuous sward with the appearance of minor herbs. Many of the most severely damaged and/or unfavourably situated areas have been eroded down to the underlying rock mantle as on parts of the Kosciusko Plateau, and the destruction is increasing each year. When such extensive depletion occurs, natural recovery may be impossible except over a long period of geological time.

Tall alpine herbfield: *Brachycome nivalis-Danthonia alpicola* alliance.

In crevices, rock ledges and similar steep rocky situations the closed herbfield communities of the *Celmisia longifolia-Poa caespitosa* alliance give way to chomophyte vegetation of the *Brachycome nivalis-Danthonia alpicola* alliance. These communities usually have a fragmentary distribution, often forming small colonies of a single species. They occur through the alpine areas of the Australian Alps but particularly on the steeper mountains such as Bogong and

Feathertop where the chomophyte habitat is well developed. The relative inaccessibility of these communities has protected many of them from grazing influences.

Short alpine herbfield: *Plantago muelleri-Montia australasica* alliance.

On leeward aspects on the highest mountains of the alpine tract where the snow drifts heavily and persists for at least eight months of the year other distinctively alpine communities develop: the lower snowpatch situations irrigated by the melt waters from the snow patch support a short carpet-like vegetation of the *Plantago muelleri-Montia australasica* short alpine herbfield alliance and the drier rocky snow patch margins an open fjaeldmark vegetation of the *Coprosma pumila-Colobanthus benthamianus* alliance.

The short alpine herbfields attain their best development in the Kosciusko area and are not uncommon in the Victorian Alps on Mt. Bogong, the Bogong High Plains, Mt. Hotham, Mt. Feathertop, Mt. Stirling and Mt. Buller. Mature communities usually consist of a single continuous herbaceous stratum closely appressed to the soil surface in a mat-like fashion. Physiognomically, they are very similar to the more familiar snowpatch vegetation of Europe but differ in the comparative absence of dwarf woody chamaephytes such as the dwarf willows, and in the greater importance of flowering plants rather than mosses and liverworts.

These snowpatch communities are subject to heavy pressure even under natural conditions by virtue of the fact that they act as stabilising outwash aprons for the meltwaters from the snow patches. Once the surface mat is broken by livestock-trampling, however, and the evenly distributed surface flow from the snowpatch is channelised, gully erosion and drying out of the snowpatch vegetation advance rapidly often down to the underlying rock.

Fjaeldmark: *Coprosma pumila-Colobanthus benthamianus* alliance.

The upper snowpatch margins which rapidly dry out on becoming free of snow and for the remainder of the short snow-free season are exposed to drought and considerable daily temperature changes, support a very open fjaeldmark vegetation in which the mat plants *Coprosma pumila* and *Colobanthus benthamianus* are the characteristic dominants. The Kosciusko area is the only place on the mainland where this vegetation occurs. In other countries homologous communities ascend higher than any others to the level of permanent ice and snow; their presence in Australia is thus of considerable historical value. An interesting feature of the Australian fjaeldmark vegetation is the absence of cushion plants of the type found in Tasmania, New Zealand and the sub-Antarctic Islands. It seems unlikely, in the light of other phytogeographical evidence, that such plants failed to reach the mainland; post-glacial climatic changes towards drier summers, leading to the elimination of the cushion plants, is a more likely cause.

Fjaeldmark: *Epacris petrophila-Veronica densifolia* alliance.

Another fjaeldmark alliance, in this case characterised by *Epacris petrophila* and *Veronica densifolia*, also develops in the Kosciusko area, on the most wind-exposed alpine cols and summits. Unlike the fjaeldmark of the previous alliance, this community often exhibits a characteristic pattern of coseral development in which the vegetation is being renewed and destroyed simultaneously on the leeward and windward sides. This continuous process of destruction by wind and wind blast, and renewal, results in a slow migration of the community in the direction of the prevailing winds. In this way a given area of ground becomes successively bare and covered by fjaeldmark vegetation in its various upgrade and downgrade stages. The *Rhacomitrium* "heaths" of the British mountain tops show a similar pattern.

The Australian fjaeldmark communities are not attractive to livestock and have not undergone much modification.

Sod tussock grassland: *Poa caespitosa-Danthonia nudiflora* alliance.

The sod tussock grasslands of the *Poa caespitosa-Danthonia nudiflora* alliance are naturally closed communities dominated by the snowgrass and other perennial herbs. They occur both in the sub-alpine and alpine tracts mainly in the level and undulating situations of broad valleys and plateaux. In some cases excessive soil wetness is the limiting factor preventing the invasion by sub-alpine woodland or alpine herbfield but in most cases cold air drainage into broad valleys producing frosts lethal to the establishment of trees appears to be the main cause. Where cold air drainage is pronounced, as in the headwater region of the Murrumbidgee River, the lower limit of this alliance is depressed by as much as 1,000 feet below the normal sub-alpine level. Widespread in the Snowy Mountains, particularly in the Murrumbidgee headwaters and the Kiandra area, these communities also occur extensively in Victoria as on the Bogong High Plains, Dargo High Plains, Howitt Plains, Snowy Plains and similar areas further south. An interesting outlier of this alliance, apparently the northernmost of any size, occurs on the Barrington Tops of New South Wales.

Under natural conditions bare ground is rarely exposed in sod tussock grassland and when it is the bare area is of a small enough size to be overhung and protected by the leaves of adjacent tussocks.

This protection seems to be essential for the early ececis of the snowgrass which, though hardy in the adult stage, is extremely tender and susceptible to disturbance as a seedling. As in the snow gum and alpine herbfield country, fires and grazing have produced a general opening up of the tussocks with the development of intertussock spaces and more extensive areas of minor herbs, and shrubs. It is the area of minor herbs which now provide most of the palatable grazing.

Heath: *Oxylobium ellipticum-Podocarpus alpinus* alliance.

Locally exposed areas in the subalpine tract and very rocky environments both in the subalpine and alpine tracts such as rock outcrops, stream banks and glacial moraine support heaths of the *Oxylobium ellipticum-Podocarpus alpinus* alliance. The dominants of this alliance usually grow as dense shrubberies which under more extreme conditions of exposure to wind and cold frequently assume a rock clinging habit or become prostrate. The massed flowering of these communities in summer is one of the most spectacular sights in the mountains. Like the associated subalpine woodlands, the heathlands have been burnt frequently to secure more palatable grazing, resulting is a greater proportion of fire-favoured species. Where the fires have been frequent and combined with grazing, disclimax herbaceous communities have been produced and accelerated soil erosion has frequently followed. Closely related heath vegetation is also widespread in the high country of Tasmania where the rockier soils provide an ideal environment.

Heath: *Epacris serpyllifolia-Kunzea muelleri* alliance.

The heath communities of the *Epacris serpyllifolia-Kunzea muelleri* alliance, unlike those described above, are found on poorly aerated but not necessarily wet soils of level to gently sloping situations both in the alpine and subalpine tracts. The dominants themselves usually grow less densely so that a subordinate grassy herbaceous stratum often develops. Unlike the *Oxylobium-Podocarpus* heaths the *Epacris-Kunzea* heaths are not particularly fire-resistant and consequently they may be largely eliminated by frequent fires. On the other hand, deterioration of climax grassland communities in the vicinity with resultant exposure of bare soil commonly leads to invasion by *Kunzea muelleri*.

Bogs: *Epacris paludosa-Sphagnum cristatum* alliance; *Carex gaudichaudiana-Sphagnum cristatum* alliance.

Two bog alliances have been studied in the Australian Alps: the

Epacris paludosa-Sphagnum cristatum raised bog alliance and the *Carex gaudichaudiana-Sphagnum cristatum* valley bog alliance. In their active condition the raised bogs consist of alternating moss hummocks and hollows associated respectively with acidopholus shrubs and with helophytes and hydrophytes. The valley bogs lack this well marked hollow-hummock pattern, consisting instead of a series of low moss hummock banks without shrubs, inclosing shallow pools containing helophytes. These communities have an interesting distribution in south-east Australia, the northernmost apparently being on the New England Plateau and on the Barrington Tops. It is also of interest to record the increasing relative importance of bog with increasing humidity and cloudiness, especially in summer, from north to south; for example there is an estimated 5% of bog above the winter snowline on the Bogong High Plains and about 10% above the corresponding level on the Baw Baw Plateau further south.

The bogs contain some very palatable herbs, especially *Carex gaudichaudiana*, and are selectively overgrazed, particularly by cattle. The resultant trampling, together with associated fires, leads to rapid deterioration of the bog with the breakdown of moss hummocks and banks and the resultant initiation of drainage lines through the bog. Entrenchment, gullying and drying out of the peat soon follow. Already an estimated 50% of the bogs of the Australian Alps as a whole have undergone fairly advanced deterioration; in many areas this deterioration is almost complete.

Fen: *Carex gaudichaudiana* alliance.

The fens of the *Carex gaudichaudiana* alliance also develop on wet soils, but under slightly higher (though still low) conditions of base status than the bogs. In the natural condition these communities have a rather sparse cover of *Carex* sometimes with associated helophytes. Typically, they occupy permanently wet basin situations, but they can also develop even on gentle slopes on account of the surface water retaining properties of the peats. The compactness of the peat enables the water table to be raised locally several feet above the general level of the water table, and there is abundant evidence in the form of deep uniform deposits of peat that this has been happening in many places for a long time. Once the surface drainage of the fen is destroyed, however, enabling water which was formerly pooled on the broadly concave fen surface to run off, this building up process is stopped and often reversed until, even within a few years after disturbance, the fen is eroded down to the underlying gravel or rock. Selective overgrazing of the fens, mainly on account of the palatable *Carex*, has produced widespread trampling damage, drying out and erosion.

Community interrelationships

The interrelationships of the high mountain communities can now be summarised as follows:

Under alpine conditions of climate, tall alpine herbfield of the *Celmisia longifolia-Poa caespitosa* alliance forms the climatic climax and under subalpine conditions, subalpine woodland of the *Eucalyptus niphophila* alliance.

Physiographically determined climaxes which replace the alpine climatic climax are the localised tall alpine herbfields of the *Brachycome nivalis-Danthonia alpicola* alliance in crevices, on rock ledges and other relatively protected rock faces; the short alpine herbfields of the *Plantago muelleri-Montia australasica* alliance in moist, snow-patch environments, on wet gravelly areas and along associated streams; fjaeldmark of the *Coprosma pumila-Colobanthus benthamianus* alliance in adjacent relatively dry and exposed snowpatch situations; and fjaeldmark of the *Epacris petrophila-Veronica densifolia* alliance on the most wind-exposed alpine plateaux and cols.

On imperfectly aerated though rarely waterlogged soils in the alpine and sub-alpine tracts and under conditions of cold air drainage in the subalpine tract the alpine and subalpine climatic climaxes are replaced by sod tussock grasslands of the *Poa caespitosa-Danthonia nudiflora* alliance. On somewhat drier soils the grasslands grade into heaths of the *Epacris serpyllifolia-Kunzea muelleri* alliance.

In contrast to the above conditions producing sod tussock grassland or *Epacris-Kunzea* heath, relatively rocky and freely drained situations support heaths of the *Oxylobium ellipticum-Podocarpus alpinus* alliance both in the alpine and sub-alpine tracts. In the alpine tract the shorter duration of snow cover as determined by local physiography also appears to be important.

In permanently wet situations, under conditions of increasing acidity and base deficiency the above communities are replaced respectively by fen (*Carex gaudichaudiana* alliance), valley bog (*Carex gaudichaudiana-Sphagnum cristatum* alliance) and raised bog (*Epacris paludosa-Sphagnum cristatum* alliance) developing as physiographic climaxes.

In most parts of the Australian alps the relationships indicated above and the relationships within the alliances themselves have been much modified by land use, largely the summer grazing of sheep and cattle associated with recurrent fires and more locally by engineering and tourist operations. The nature of this modification is now considered in greater detail.

Land Use and Economic Importance

The high mountains of Australia, all removed from the main centres of population, remained for many years largely unknown except to a handful of scientists, miners, graziers and timber cutters, and occasional hikers and skiers for whom these areas had some special attraction. This situation still remains substantially the same in much of the Victorian Alps and in Tasmania, although it has now altered in those parts of the Snowy Mountains and Bogong High Plains in which engineering works for the development of hydro-electric power and the storage and diversion of water for irrigation are in progress.

An inevitable result of this disinterestedness has been that the various forms of land use in the mountains have remained of a pioneering nature in which the individual with no other resources but his own has employed exploitative means to force quick returns without care or knowledge of the long-term consequences. The basic philosophy of conservation — that for each natural resource the methods of use must maintain and, if possible, increase the permanent productive capacity of that resource — has never been fully recognised. This philosophy of productive use without detriment to the resource being used implies that priorities according to land use capability should be established. Management for one purpose may be incompatible with permanent management for the type of production for which the resource is most suitable, even if this most suitable use still remains to be applied. This is the essence of the ecological approach to land use which insists that whilst the incidence of factors influencing the soil-vegetation system may change, the overall stability of the system must be maintained. A given form of land use should be capable of maintaining this stability if it is to have justification, even before economic criteria are applied.

Before the coming of the white settler to the lower country surrounding the mountains little more than 100 to 125 years ago, the impact of the native fauna including the Australian aboriginal on the high mountain soils and vegetation seems to have been very slight. The early settlers with their sheep and cattle found themselves surrounded by open grassy forests leading up to the high mountains in which palatable grazing was immediately available. From this it may be assumed that the palatability of the natural mountain vegetation was greater than that of supposedly natural vegetation of today, in which burning is considered by many to be necessary to encourage new and palatable growth. There is little doubt that this was so: the bogs and fens contain a number of very palatable plants, particularly the sedge *Carex gaudichaudiana*, and field surveys indicate that these palatable areas have been reduced by at least 50% over the last few decades. There is also good evidence that certain palatable native species (e.g. *Hemarthria uncinata, Danthonia frigida, Aciphylla glacialis*, etc.) recorded as common by earlier botanists have now become infrequent and in some cases have virtually disappeared leaving a sward more predominantly composed of one of the various forms of the snow grass *Poa caespitosa*. This presents a problem which high country grazing in most of Europe and North America does not have to face, namely, that the main dominant herbaceous plant available for grazing is not attractive, except at the seedling and seed head stages. In the U.S.A., for instance, high range management methods are based on maintaining the dominant herbs in as near climax condition as possible, but in the Australian Alps (and in New Zealand) present range management, if it can be said to exist, consists in damaging the dominant herb as this somehow produces more palatable grazing. The effect of this damage, as shown earlier, is to produce bare areas on which palatable secondary herbs develop, and it is these bare areas which provide most of the grazing. Thus it can be seen that at present a condition of widespread, moderate soil and vegetation deterioration is the one which, apart from an artificial improved pasture, most favours the grazier, and the paradoxial situation has arisen where the areas least able to stand grazing are being heavily grazed whilst the more extensive, sound areas better able to stand grazing are supporting very few stock. This is one of the key problems for which a solution would need to be found before the continuation of high country grazing would be placed on an ecologically sound, permanently productive basis.

This problem can be resolved into two main issues. If grazing were to become more efficient it would first need to be made more uniform so that most of the total area — not just a small fraction of it — were being utilized for stock feed. This introduces the second issue — that if more uniform grazing were to be achieved either the widespread unpalatable snow grass would need to be made palatable or replaced by an improved pasture sward. Experience in the Australian Alps now indicates that the establishment of pastures of recognised grazing value would offer greater possibilities and there is no doubt from several accidental and intentional trials that a considerable number of species could be established and persist. Costs of pasture establishment and management would, however, be very high: approximately 30/— per acre per annum for topdressing and another 10/— for the erection and maintenance of the fencing required. Opposed to this minimum estimate of £2 per acre per annum for typical high mountain country, is the low present and potential value of grazing from these areas: this varies from about 5/— to 20/— per acre per annum and represents little more thans 0.1 % of the value of production from sheep and cattle in Victoria and New South Wales (approximately £335,000,000 in 1954—55). Clearly, pasture improvement under these conditions would be uneconomic, except perhaps in some of the lower, more uniform areas where economies might be effected.

Land use, in the conservation sense, should have more than economic justification, however, before it can lay claim to a particular land resource. More important, it should not reduce the productive capacity of actually or potentially more important forms of land use by more than it itself is producing. Thus, the case for grazing involving either the maintenance of the vegetation in a deteriorated condition or its replacement by improved pastures should also be examined in relation to management requirements for the supply of water for domestic, irrigation and industrial needs, hydro-electric power, mountain recreation areas and nature reserves.

The use of water as such and in the production of hydro-electric power involves storage in dams and reticulation in race lines, pipes or tunnels. There are few other countries where the efficient use of water for these purpose is as important as in Australia; here water is recognised as one of the main factors limiting present and future production and maintaining and increasing the standard of living. Furthermore, suitable storage sites are comparatively few and mainly restricted to the mountain areas themselves. This means that storages should have their lives extended as much as possible towards an indefinite life, the purely economic criterion being insufficient since there may be no other suitable storage sites. There is little need to pursue the argument that the condition of high mountain vegetation should be managed in such a way that any works for storage and diversion of water should have maximum protection from sedimentation.

Considering the use of water as water, priorities of use would probably be given to domestic use, then primary industry and industrial purposes. In the first instance the water should be as pure as possible. This means that maximum infiltration of precipitation on the catchment should be achieved. The Federal Capital, Canberra, derives almost all of its domestic water supply from the high mountain section of the Cotter catchment, whilst in Victoria the rapidly growing towns of the Latrobe Valley and the city of Melbourne itself are largely or partly dependent on the high mountains for their domestic supplies. With its many uses for drinking, household, sanitary and garden purposes, and its relation to health and illness in the community, it is almost impossible to put a true monetary value on high quality domestic water beyond stating that it is indispensable.

Water is also an absolute necessity in industry, very large amounts being required for many industrial purposes, and for the removal of wastes at a rate sufficient to prevent obnoxious pollution. In the rapidly growing industrial areas of the Latrobe Valley in Victoria where brown coal is being mined on an increasing scale for gasification and electrification purposes, and where secondary industries are developing around the sources of power, not to mention the many factories processing primary products of this rich rural area, including paper pulp and dairy products, good quality water will be needed in greatly increased amounts and the southern areas of the Victorian Alps provide the only suitable sources. In these cases, therefore, management for domestic and industrial uses should take priority over management as high country grazing land and even for electric power for which there are several more suitable alternatives.

One of the greatest developments in Australian primary industries has been that of irrigation in the dry, formerly unproductive areas of inland New South Wales, Victoria and South Australia. The water now supplied by the Australian Alps for irrigation has an estimated production value of about £ 20,000,000 in N.S.W., and £ 30,000,000 in Victoria. This total* of £ 50,000,000 will be increased to about £ 80,000,000 by the diversion works of the Snowy Mountains Project. On the basis that the 1,875 square miles of high mountain country in N.S.W. and Victoria supply about half of this water, it can be seen that present value for water exceeds £ 20 per acre and the expected value is more than £ 33. The full utilization of the waters of the Victorian Alps would increase this value still further. (f. Australian Academy of Science, 1957).

Other values which should be added to these are taxation revenue (about 20% of the above production values), and the decentralization of population and industry. Finally, not only is the water of the high mountains of great value for irrigation and out of all proportion to its value for mountain grazing, but these mountains are the major source of irrigation water in southern Australia, for which there are no other alternatives.

In the case of hydro-electric power in New South Wales and Victoria, the Snowy Mountains, Kiewa, Eildon and Hume generating systems will produce power valued at more than £ 30,000,000 per annum, representing an annual return of more than £ 12 per acre of high country. This amount can be increased by further development in Victoria. In these instances, however, the contribution of hydro-electric power to total power will be considerably less than in the case of water for use as water, representing less than 20% of the total power output in these States. The situation in Tasmania is a little different. Here the value of hydro-electric power is expected to be about £ 20,000,000, representing a return exceeding £ 20 per acre per annum of high country. This power is indispensable to the Tasmanian economy as there are no suitable sources of coal, which means that the management of the Tasmanian mountains for power production is more necessary than on the mainland.

---

* Figures for South Australia are not available.

In the case of recreational values assignable to the high mountains, the economic criterion is almost impossible to apply. In this instance the role these areas do and will play in providing outdoor relaxation of a special kind, and the lack of alternative areas for obtaining such recreation, are probably the main criteria for attributing priorities.

Visiting scientists who have seen the unique flora and vegetation of the Australian Alps are among the strongest supporters of the move to preserve these unique features for their intrinsic value as a special kind of high mountain vegetation and for future study. This particularly applies to the smaller areas above the treeline where a diverse and unique alpine vegetation and flora are preserved. In this instance there are absolutely no alternatives, and the need for preservation as a primitive area can stand on its own merits.

Against this background information of actual and potential value of the high mountain areas for various purposes, the case is strongest for the use of water as such for domestic, irrigation and industrial needs in view both of its greater economic value for these purposes, and the virtual lack of alternative catchments from which water can be obtained. The use of water for hydro-electric power is also very important economically, but except in Tasmania where there are as yet no suitable alternative sources of power, the needs of hydro-electric power production should be second to those of water use as such. Recreational and scientific needs are almost impossible to estimate in economic terms, but most would agree in view of the lack of alternative areas that they should be given far greater emphasis than at present, probably third to water and power in most areas and second to water use as such at least in the unique alpine areas above the treeline. Grazing, though historically first in the field, can claim only lowest priority.

In ascertaining which types of vegetation best meet the requirements of these various needs in the priorities mentioned, it can be stated at the outset that only the natural vegetation is acceptable for scientific purposes and that this also ensures highest values for recreation. It is also clear that a deteriorated native cover or a highly improved pasture would be most desirable for grazing purposes but this would be incompatible with preservation for scientific and recreational needs. Optimum management for water supply purposes will of course be governed by the hydrological properties of the various cover types which either exist or might be produced in the mountains. The more important of these properties are soil stability (conversely, erosion), surface infiltration (conversely, run off), and evapo-transpiration, including the more special aspects of snow storage and cloud and fog drip. Final answers to all of these problems cannot be given until research has been carried further, but a good deal of preliminary information is now at hand to provide a reasonable understanding of present and longterm management needs.

Hydrological Research

The most important hydrological property of the vegetation as regards catchment values, namely the rate of soil loss with the associated property of surface run-off, is being studied by means of numerous small experimental plots on which the effects of natural and artificial rains and snowfalls are being regularly measured. The cover types include all of the important ones found naturally and under current grazing management and also include the type of improved pasture which intensive development for grazing might hope to produce. Table III, based on about 50 individual plots and 14 individual rains, gives comparative soil loss and surface runoff incides for the various cover types:

Table III.

Soil loss and surface runoff indices for high mountain cover types, Kosciusko

Cover Type	Soil Loss Index	Surface Runoff Index
Subalpine woodland		
– Mature: Parklike with dense snowgrass sward	0	2
– Regrowth, about 30 years after fire, with shrubs and grass.	0	1
– Shrubs, about 15 years after fire	1.5	13
– Bare Soil, 3 years after fire	1300	56
Sod tussock grassland		
– Mature: Continuous coarse sward	0	3
– Damaged: Bare spaces between tussocks	110	42
– Improved Pasture: Short-grazed continuous sward	1	22

These results show clearly the value of the native vegetation, especially a dense herbaceous layer of coarse snow grass, in reducing soil losses and surface runoffs: i.e., of producing water which is low in sediment and sustained in yield. Any deterioration in the cover type involving exposure of the surface soil produces accelerated soil losses and increased overland flows. A short grazed improved pasture sward also minimises soil loss but results in appreciably greater surface runoff which may cause erosion in valleys lower down; this also means less infiltration and hence less sustained stream flow.

The programme of soil moisture measurement is not yet sufficiently advanced to enable many comparisons to be made except that the tree and shrub types use more water than the purely herbaceous communities. The greater water use by the trees, however, appears to be more than offset by the greater amounts of snow which they accumulate and by the considerably slower melting of this snow under the protection of the trees. The presence of trees in snow country thus contributes to more sustained and probably greater stream flows. Snow-melt rates are also slower under the forest and in the small clearings sheltered by trees than in the open

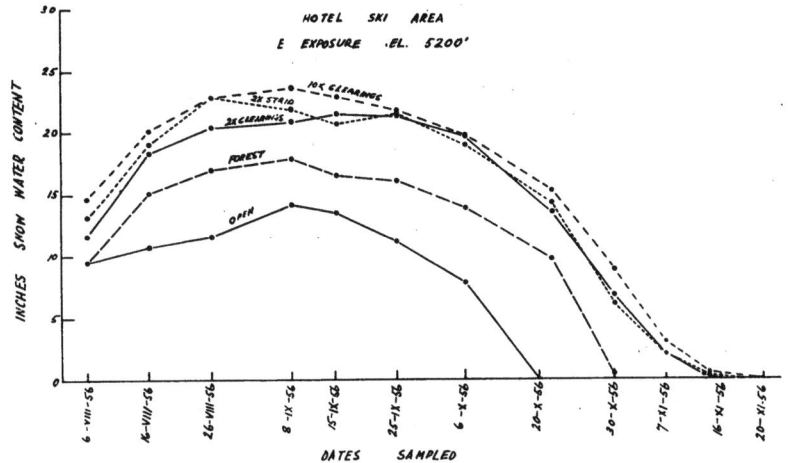

Fig. 3. Relation between size of clearing and snow-water content in dense subalpine woodland (sheltered aspect), Kosciusko.

where insolation and wind are greater*. (Fig. 3). Preliminary experiments also show that the trees strain out appreciable amounts of water from clouds when, as frequently happens in the mountains, the cloudlayer descends to ground level.

These hydrological measurements have also been supplemented by measurements of water flow and water use in *Sphagnum* bogs** through which a large part of the water flows which eventually enters the mountain streams. These measurements have been carried out by keeping a continual record of the input of water into selected small bogs from springs and precipitation, and the output of water from the bog. It is apparent from these measurements that the bog plays a part in regulating stream flow. During the snow-free months the yield from a given precipitation is spread over one or more days in the case of a single bog and probably over considerably longer periods when a whole downslope series of inter-connected bogs is considered. During the winter the bogs make a continuous con-

---

* Snow survey data obtained by Mr. L. GAY, American Fullbright Scholar working in Kosciusko area in 1956.
** Bog experiments conducted by Mr. G. LLOYD, American Fullbright Scholar working in Kosciusko area in 1957.

tribution to stream flow, by slowly melting the overlying snow, due to the thermal properties of moving ground waters near the bog surface. In non-ground-water areas, there is little snow-melt until spring. On the other hand, evapotranspiration losses are obviously considerable during the drier summer months, when potential evapotranspiration exceeds precipitation, so that the regulating function of the bogs is probably achieved at the price of total yield. However, as sediment content and regulated flow are given priority over gross yield, the bogs are desirable catchment assets.

With this framework of hydrological data and more coming to hand all the time, it is now possible to cross-check and supplement the large amount of interpretative field data accumulated over the years with more precise quantitative information. The results outlined above show that the existence of a continuous, coarse growing herbaceous vegetation to induce maximum infiltration with resultant sustained yield of pure water is far superior to bare or partly bare ground, shortgrazed improved pasture, or shrubs. The snow survey data also show that the presence or absence of trees has a considerable effect on the accumulation and subsequent maturing and persistence of snow to the extent that a high-mountain catchment with trees produces a more sustained and probably greater total flow of water than a deforested catchment. It would seem, however, that a fairly low density of trees or clumps of trees would produce more water than a densely forested catchment where interception and transpiration losses would be greater, the optimum diameter of the clearing between clumps of trees being approximately five times the height of the trees on exposed aspects and up to ten times their height on sheltered aspects. Quality and continuity of yield are further improved by the presence of bogs and swamps.

Management Requirements

These hydrological data show that the natural or near-natural vegetation is the most desirable of the several economically possible cover types which might be developed for the production of water, although in the more densely timbered subalpine woodland areas judicious light group selection felling might be expected to show some increase in water yield. By and large, the optimum vegetation requirements for water production are the same as those for recreation and scientific purposes, but they conflict with the present and possible future requirements of grazing.

The most important management procedure for the development of a near-natural vegetation over the bulk of the mountains would be the removal of domestic livestock and complete protection from fires. This policy already exists in the Australian Capital Territory, and from 1958 it will also operate in New South Wales.

These management methods, which are cheap and easy to

apply, could be expected to restore the vegetation over most of the high mountain catchments within the next 30—50 years. In more severely damaged or critically exposed situations, however, artificial conservation measures involving re-seeding under a protective mulch, control of runoff water by diversion banks and stone-pitched drains, and the construction of snow fences, have been initiated by the Soil Conservation Service of N.S.W. Experimental re-afforestation is also visualised for extensively deforested areas where natural regeneration has so far failed. The control of stream bank erosion by water spreading and gully control structures will also be necessary in many areas. These measures to restore the catchment as a whole are also being accompanied by increasing attention to the more localised but often severe erosion connected with engineering works, and to the compatability of hydro-electric and national park values which so far have received little consideration. It is also becoming clear that a much more precise definition of responsibilities with co-ordinated control is essential among the many government and semi-government organisations concerned in these areas, if the scientific and technical knowledge available is to be effectively applied.

From being remote, little known, of small interest, and divorced from the realities of the day, the high mountains of Australia and their vegetation are becoming recognised as of vital importance to the nation and its inhabitants. Every man, woman and child of the ten million population already contributes more than £2 per annum to national works in these areas and it is apparent that the security of this huge investment depends largely on preserving a vegetation able to stabilise the soil. Lovers of the outdoors are now discovering the uniqueness of the Australian Alps and their vegetation, in a continent otherwise lacking in these resources. Scientists also realise the intrinsic interest of the unique plant communities and of the fauna which depends on them. Thus, the case is extremely strong for the preservation of the high mountain vegetation largely in its natural condition, of readjusting land use so that this near-natural condition can redevelop, and in cases of more severe soil-vegetation deterioration of implementing active soil conservation measures.

REFERENCES

Australian Academy of Science, 1957. "On the Condition of the High Mountain Catchments of New South Wales and Victoria." Australian Academy of Science, Canberra.
COSTIN, A. B., 1954. "A Study of the Ecosystems of the Monaro Region of New South Wales." Govt. Printer, Sydney.
COSTIN, A. B., HALLSWORTH, E. G. & WOOF, M., 1952 Studies in Pedogenesis in New South Wales. III. The Alpine Humus Soils. *J. Soil Sci.*, **3,** *190.*
RAUNKIAER, C., 1934. "The Life-Forms of Plants and Statistical Plant Geography". Oxford.

XXVII
SOME ASPECTS OF ECOLOGICAL RESEARCH IN SEMI-ARID AUSTRALIA

by

N. C. W. BEADLE

(Professor of Botany, University of New England, Armidale, N.S.W.)

The Australian continent, with its vast area of semi-arid land has developed a xerophytic flora which, though unique in some respects, is similar to the floras in other areas of the world which experience a similar climate. Perhaps the most outstanding and exceptional superficial feature of the vegetation is the presence even in very dry areas (mean annual rainfall of 8 inches) of woody shrubs or even trees. Yet in many respects the flora is similar to that of the dry areas of other continents. Genera which occur elsewhere and which may be mentioned here because they feature in the discussion that follows are *Acacia, Atriplex* and *Kochia*.

Before the advent of white man the semi-arid communities were stable and underwent only minor changes with the changing of the seasons. Indeed, the changes did not affect the dominants, but rather the herbaceous strata which were relatively unimportant in maintaining the stability of the virgin stands.

Explorers first visited the dry interior during the middle of last century and the salient geographical features of the eastern half of the continent were mapped before the turn of the century. Settlers with flocks of sheep and herds of cattle followed closely in the steps of the explorers so that by 1890 the semi-arid portion of eastern Australia teemed with grazing animals, and later with rabbits. Since carrying capacity, rainfall, techniques for grazing, and rates of regeneration were unknown, or never thought of, it was inevitable that a few dry seasons, culminating in the great drought of 1900—1, should decimate the stock population and at the same time greatly change the vegetation, in some cases beyond recognition. This great drought and its attendant losses provoked a Royal Commission. Furthermore this drought and those that followed led to the gradual development of a land policy whose greatest achievement has been the fixing of carrying capacity of holdings, thereby limiting stock numbers.

Ecological work on the semi-arid vegetation was pioneered in South Australia. This work stimulated research in Australia in general, so that now a wealth of descriptive data on plant communities, soils, the effect of rabbits, cattle, and sheep are available.

The data which are to be included below deal chiefly with inter-

PLATE II

Fig. 3. General view of Saltbush in reasonably good condition north of Broken Hill, N.S.W. A few bushes of *Kochia* are present in the centre. Note also the quartz "gibbers" in patches.

PLATE I

Fig. 1. Mulga scrub in fair condition on low stable dunes near Wilcannia, N.S.W.

Fig. 2. Mallee *(Eucalyptus oleosa)* with the porcupine grass in the foreground. Near Balranald, N.S.W.

N. C. W. BEADLE

pretations and reactions within communities rather than with their description, but for the sake of clarity a brief description of the environment and of the communities on which research has been done must be given.

General Features of the Arid Environment

The area in which detailed research has been done is the western portion of New South Wales, though observations have been made westward and north into Northern Territory. The whole area discussed here receives less than 10 inches of rain per annum, with no pronounced seasonal peaks, though the northern part of the area north of latitude 32 can expect more rain in the summer than in the winter. The driest part of the area receives an average of about 6 inches. Annual rainfall registrations for any of the stations where rainfall is recorded show that the minimum annual rainfall may be as low as 2 inches whereas maxima may be as high as 25. It is unusual for one above-average wet year to be followed by another wet year. Individual falls of rain may exceed 2 inches, and it is not uncommon to record one such heavy fall in the year and no others of any significance for the remainder of the 12-month period. In general terms, rainfall is low and unreliable — a feature which greatly increases the task of the ecologist concerned with regeneration work. Temperatures during the summer months frequently exceed 100° F. and under these conditions evaporation rates are so high that summer rains, though highly beneficial to established plants are of little value in establishing seedlings: For example, from observation and experimentation it has been established that 3 inches of rain during mid-summer is inadequate for the germination and establishment of the seed of perennial saltbush. Minimum winter temperatures for a few weeks of the year approach or are just below freezing point, consequently growth rates during the winter are always conspicuously retarded. Annual evaporation rates are of the order of 100 inches. Wind velocities are not as high as in many coastal and tableland districts and it is only when the countryside becomes bare that winds become an important ecological factor. Light intensity throughout the whole area is always high, as would be expected from the latitude.

In general terms the semi-arid portion of Australia is flat. Most of the area lies at a level of a few hundred to a little over 1000 feet above the sea. Low rocky hills occur at widely separated intervals; in New South Wales these hills do not exceed 1000 feet above sea level, but those of the interior approach, and in a few cases exceed, 4000 feet. Excluding the hills, which cover a relatively small area, the country is either flat, or ribbed with sand-dunes, or undulating and strewn with stones. These topographically different areas bear different types of vegetation which are mentioned below.

The Plant Communities

In the area which has been studied extensively five major plant communities occur, viz: the mulga *(Acacia aneura)* scrub, the mallee (with dominants of *Eucalyptus oleosa* and *E. dumosa)*, the saltbush *(Atriplex)* and bluebush *(Kochia)* communities, and the woodlands of river red gum *(Eucalyptus camaldulensis)* which follow the watercourses and of which no more will be said. The first four communities are distributed as follows:

The mulga scrub (Fig. 1) occurs in the area which receives predominantly a summer rainfall. It is confined to soils of light texture either on the hills where the soils are gravelly and shallow (but the roots penetrate cracks in rocks) or on stable sand-dunes. Local dominants other than the mulga sometimes occur, for example the "pine" *(Callitris glauca)* and rosewood *(Heterodendron oleifolium —* Family Sapindaceae). A lower shrub stratum rarely occurs, though scattered shrubs, notably species of *Cassia* are sometimes common; the herbaceous stratum was well developed and consisted chiefly of grasses, of which the love grass *(Eragrostis eriopoda)* was common.

The mallee occurs southward of the mulga scrub and the dominants here are eucalypts which reach a height of 6 to 12 feet and which have large underground lignotubers producing many stems. Shrubs are infrequent and the only herbaceous plant of note is the hummock-forming grass *Triodia irritans* (Fig. 2). The soils are sandy.

. The saltbush and bluebush communities consist of dwarf shrubs about 3 feet high and spaced at intervals of 3 to 15 feet. In the spaces between the bushes annual plants appear in varying numbers according to the rainfall. Saltbushes and bluebushes sometimes occur intermixed (Fig. 3). Saltbush occurs for the most part on clayey soils, many of which are saline. Bluebushes on the other hand occur in a variety of habitats — hillocks containing much lime, watercourses and sandy flats.

The Effect of Man and Grazing Animals

In the virgin state the area was grazed by moving populations of indigenous animals, chiefly kangaroos and emus. Under these conditions the country was probably not eroded. The aborigines did not disturb the vegetation to any great extent because they did not practise agriculture and were not accustomed to felling timber. They did light fires however, and it is possible that for this and possibly for other reasons they disturbed or even destroyed small, but insignificant, areas of vegetation.

The white man on the other hand has wrought vast changes by pasturing sheep and cattle, and through the introduction of the rabbit which spread into the drier areas from the south. Also the cutting of timber, in the first instance for household supplies and fence posts and later to provide stock-feed in time of drought, had disastrous effects in some communities. For example, in the mulga scrub the removal of timber frequently leads to death of the stand. The reason is not certain: death may be due to the alteration of microclimate, but other factors may be concerned. The mallee, on the other hand, does not suffer similarly when it is disturbed; on the contrary, mallee plants regenerate rapidly from their lignotubers and it is only when the mallee plants are removed completely that soil erosion follows.

In all communities grazing by stock — including the husbanded animals, rabbits and the indigenous fauna — has led to degeneration of the pastures. Details of the changes are not necessary here; it is sufficient to say that in all areas the following trends are evident: (a) a decrease in density of the perennial fodder species, culminating in their complete removal; (b) an increase in the density of less palatable species culminating in the dominance of useless weed species. Even further degeneration can occur — and this is frequently the case — if the surface layers of soil are removed by erosion.

Soil Erosion and its Consequences

As long as a protective covering is provided to the soil, erosion does not occur. Protection can be afforded by vegetation of any kind, by debris, or by a surface layer of stones which occur naturally in certain areas known in Australia as the "gibber downs" (Fig. 3). The stones afford protection against wind erosion, but not against water. Contrary to general belief, damage caused by water erosion, even in these dry climates, in considerable.

It is not possible to state precisely the exact area of country which has suffered from erosion. It is safe to say however that the degree of erosion is correlated with the length of time that the country has been settled. For example Western New South Wales in particular, which has been extensively grazed for more than half a century, is most seriously eroded. On the other hand, areas of the Northern Territory especially where watering places for stock are few and far between are still in a condition which is similar to if not identical with the virgin.

The kind of erosion that occurs when the vegetation is removed or thinned depends largely on the degree of compaction of the soil profile and on the texture of the various horizons. When the profile is sandy throughout, as in the dune country supporting mulga scrub or mallee, sand-dunes result. These may be temporarily stabilised by annual plants for short periods during good seasons. On the other hand, when the subsurface layer is compact — it may be either sandy or clayey — erosion ceases at this layer and bare shiny surfaces, known as "scalds", result.

Natural regeneration on eroded surfaces does occur; the rate, which is governed by a host of factors, is always slow. At this point it is sufficient to say that recolonisation rarely produces a stable condition, chiefly because one or another of the inhibiting factors (defined below) wipes out the struggling vegetation either completely or partially. The stages in the recolonisation can be set out in general terms as follows: herbaceous annual plants are followed by herbaceous perennials and these are followed by woody perennials. It seems that algae may play a significant rôle in some cases in so far as some of the algae, which form scums on the bare surfaces, are capable of fixing atmospheric nitrogen.

Factors Retarding the Colonisation of Bare Areas

A bare area in semi-arid country, whether it is eroded or not, is subjected to far more rigorous climatic conditions than soil surfaces which are protected or partially protected by vegetation. The factors which must be considered when one is dealing with regeneration of these pastures on either non-eroded or eroded country are as follows:
(a) Rate of stocking.
(b) Rabbit populations.
(c) Native fauna.
(d) Climatic factors (including water, temperature, wind, evaporation).
(e) Seed supplies.
(f) Soil physical properties (particularly in so far as they determine the rate of infiltration of water).
(g) Soil chemical properties.

Some of these factors can be dismissed with a minimum of discussion. For example, stocking rates can be controlled and there is ample evidence to show that, if all kinds of grazing animals are excluded from an area, regeneration of a stable community will result. The time required naturally varies with the degree of erosion, the rainfall, and the kind of vegetation. Likewise the rabbit factor can be passed over cursorily on the grounds that rabbit populations can be controlled over small areas and, it is hoped, will be effectively controlled over large areas. This must not be taken to mean that the rabbit is an unimportant factor; on the contrary destruction by the rabbit includes not only the eating of pasture plants but also the ring-barking of shrubs and the removal of shrub-seedlings, thereby precluding the development of stable communities. The native animals, destructive though they are, play only a minor rôle in most cases. Apart from the mammals, birds may be mentioned here, in particular the enormous flocks of galahs which not only reduce seed supplies locally but also destroy perennial seedling by lopping off, but not eating, the aerial parts.

The discussion therefore turns to a consideration of the reaction of regenerating plants to the physical environment. One aspect of mineral nutrition which has hitherto been neglected will be considered.

Until recently it has been the general impression that low water supply is the limiting factor in establishment and growth. While water is in low supply for long periods and may be limiting during these periods, there is now ample evidence to support the contention that when water conditions are adequate, the supply of soil nitrogen retards growth and selects the annual herbage. The condition of inadequacy of nitrogen for the establishment and growth of the original virgin species has been brought about by the loss through erosion of the surface soil. These findings have diverted attention from the water-factor and have led to an intensive study of the nitrogen economy of certain plant communities. The preliminary observations made with regard to this study are presented below, together with a discussion of the possible channels along which future research is to be directed.

Levels of Soil Nitrogen and Soil Organic Matter

In virgin communities (as far as can now be judged) the level of soil organic matter in the surface three inches of soil, was probably of the order of at least 2% in the mulga scrub and 4.5% in the mallee. Levels in saltbush and bluebush communities were probably between those two figures. Since the carbon/nitrogen ratio is of the order of 10 (frequently less), soil nitrogen was fairly high and doubtless these high levels accounted for the then rapid growth of plants, particularly of annuals in the bare spaces between the perennials, when rainfall was good. Soils which approach these levels are still to be found and the occurrence on these soils of good stands of vegetation supports the contention that soil-nitrogen plays an important role in the production of vegetation during good seasons. Two points are of special interest here. Firstly, the highest nitrogen levels are found in the mallee — a plant community in which legumes are so rare that they are undoubtedly of little consequence in the nitrogen economy. This point is mentioned again below. Secondly, by quadrating to ascertain percentage ground cover on various soil types at various stages of erosion, it has been established that 100% cover can be produced (when soil moisture conditions are excellent) only when the percentage organic carbon in the soil is of the order of 1.5—1.8% — except in the case of legumes which may form mats over soils whose organic carbon content is as low as 0.3%. If this figure is correct (more data are required, particularly on the nitrogen-requirements of the various herbs concerned) then it follows that in virgin communities, except the mallee, nitrogen levels in the soil were not far in excess of the demands of the community.

Let us now consider present day values for soil organic matter. A series of samplings from good to bad vegetation (i.e. from little to extreme erosion) was done. The analyses show marked decrease in soil nitrogen with increasing erosion. Most, if not all, of the nitrogen has been lost by the washing or blowing away of the organic matter. The following figures to illustrate this point are quoted: Scalded areas which support no vegetation contain between 0.3 and 0.5% organic matter whereas loose sand dune may contain as little as 0.01%.

Sources of Nitrogen

A survey of the micro-organisms which are known to fix atmospheric nitrogen shows that in all communities both nonsymbiotic and symbiontic nitrogen fixing organisms do occur. Their significance has been assessed as follows:

Non-symbiotic Organisms

Azotobacter (but not *Clostridia*) and the Blue-green algae, *Anabaena* and *Nostoc*, have been investigated.

Azotobacter was found to occur in the soils of the ranges, especially where soil accumulates in pockets among rocks; but it was lacking from the sand dunes which support a similar vegetation (mulga scrub). It occurs in the sandy soils of the mallee and also in some, at least, of the clayey soils in the treeless country. In all cases where it occurred the frequency was less than 50 cells per gram of soil and its contribution therefore is but slight. It is estimated that under field conditions its maximum fixation is less than 0.1 pound of nitrogen per acre per annum.

The contribution made by the algae, *Anabaena* and *Nostoc*, is probably greater, especially when the algae are protected under translucent pebbles or occur in slight depressions where moisture conditions are more favourable for growth. However, their theoretical maximum contribution is 2 to 3 lbs. of nitrogen per acre per annum.

The Symbiotic organisms

A survey of these organisms has not yet been completed. However, sufficient data have been collected to draw some fairly decisive conclusions. Firstly, the distribution of legume species is of interest. The mulga scrub with its dominant of *Acacia aneura* and local societies of *Cassia* spp. is at least well supplied with legumes. However, no nitrogen-fixing nodules have yet been recovered from any of the three species of *Cassia* which have been thoroughly investigated. On the other hand nodules on only a few mulgas have been found. These occurred on plants which were in good condition; plants which are in poor condition do not appear to be nodulated.

Is it possible then that altered microclimatic conditions have wiped out *Rhizobia* in certain areas and that this has led to the death of the mulgas and the cutting off of the nitrogen supply to the community? Whether this is the case or not, the practical problem of replacing the nitrogen and maintaining a source of supply has to be faced and in order to replace it, a nitrogen-fixing legume must be established.

The mallee contains no legumes in sufficient quantity to lead one to presuppose that any significant quantity of nitrogen is derived from this source. The high level of nitrogen in these communities cannot at present be accounted for. However, the herbaceous legumes which are abundant in the saltbush and bluebush communities undoubtedly add large quantities of nitrogen to the soil. For example it is estimated that the purple pea *Swainsona swainsonioides*, which carpets the ground under certain conditions, may fix nitrogen at the rate of 250 lb. nitrogen per acre per annum, which is a highly significant figure. Such an addition, however, is likely to occur only about once every ten years. However, there is little doubt that these herbaceous legumes provide the treeless areas with most of their nitrogen supply.

Of special interest is the legume *Psoralea patens* which not only nodulates freely but also is tolerant of salt at the levels which obtain in some of the saline stony country. This plant is relatively uncommon except in good seasons. Work is in progress to ascertain its value for reclamation work.

Conclusion

The ecologist's position is now more clearly defined: How can he replace in a short period of time that vast amount of combined nitrogen which accumulated in the communities over decades or even centuries and was lost by erosion over a few years. It is clear now that the non-symbiotic organisms are too infrequent to be of any value and that the legume-*Rhizobium* system is to be encouraged, or established in those areas where it is lacking. Where herbaceous native legumes occur, a gradual increase in soil nitrogen can be expected and the problem in such areas will be solved by controlling the grazing animals. Elsewhere the herbaceous legumes with appropriate *Rhizobia* must be established. It is expected that sown legumes will be established only in restricted areas, for example, where water conditions are best. Elsewhere it is probable that soil-moisture conditions will have to be improved by loosening the surface (on eroded areas) or by the damming of water channels. Finally, it may be added that this theoretical scientific solution may not be applicable to large areas because of its high cost relative to the low value of the land.

REFERENCES

Additional references can be obtained from the bibliographies in the publications listed below.

BEADLE, N. C. W., 1948. The Vegetation and Pastures of Western New South Wales. Govt. Printer, Sydney.
CROCKER, R. L., 1946. The Simpson Desert Expedition, 1939. *Trans. Roy. Soc. S. Aust.* **70**, *235—258*.
TCHAN, Y. T. & BEADLE, N. C. W., 1955. Nitrogen Economy in Semi-Arid Plant Communities Part II. *Proc. Linn. Soc. N.S.W. l x x x, 97—104.*
WOOD, J. G., 1937. The Vegetation of South Australia. Govt. Printer, Adelaide.

XXVIII
SPECIES DISTRIBUTION AND ASSOCIATION IN EUCALYPTUS

by

L. D. PRYOR

(Department of Botany, Canberra University College, Canberra)

Eucalyptus is one of the most distinctive botanical features of Australia, and since most of the species are trees and occur mainly in the non-arid quarter of the Continent with a rainfall of more than about 15 inches, they are a very prominent feature of the natural landscape of most of the more closely settled areas and a major component of the plant communities of this zone.

According to the latest systematic treatise of Eucalyptus (BLAKELY, 1955), the number of species and varieties taken together approaches 700. Eucalyptus does not pass to the Asian side of WALLACE's Line*, and is almost entirely confined to Australia, but there are amongst the 700 species found in the country half a dozen which also extend outside Australia, but probably only two which occur outside Australia and not at the same time on the Continent. Thus it is clear that the genus is overwhelmingly an Australian endemic.

It is a matter for speculation as to why a genus of trees, which is the principal component of most of the Australian forests and woodlands should be so closely confined to Australia and be without very close relatives elsewhere, but one simple conclusion from this occurrence is that, in common with other systematic groups of both plants and animals, this results from a long period of isolation. At the same time it is clear that in this isolation evolutionary development in Eucalyptus has not been at a standstill, since although the genus does not provide species which occupy the really arid, major portion of the Continent, an extremely wide range of climate and soils has been successfully occupied by species of the genus.

Eucalyptus as a genus is relatively uniform in its morphology. For example, nearly all species are evergreen (two or three may be monsoon deciduous) with the leaves generally similar in shape; the flowers have the same general structure, the inflorescence is broadly similar and the same kind of fruit is found throughout. In short, the generic characters by which the group is distinguished are pretty clearly displayed through all the species. This, at first sight, might lead to the supposition that there is greater uniformity (in the plant community sense) in vegetation dominated by species of Eucalyptus

---

* Using the term in the HUXLEY sense, *Eucalyptus deglupta* extends to Mindanao.

than would be the case for the same range of environmental conditions in another continent. For example, in California in comparing altitudinal changes in the latitude of San Francisco with those of Australia in the vicinity of Mt. Kosciusko, the group of different plant communities each occupying a climatic zone represented by an interval of 1,000 to 2,000 feet in altitude frequently includes, at least so far as the tree component is concerned, species belonging to quite different systematic groups and at the same time species with widely different life forms. For example, at high levels on the Sierra Nevada there are conifers of different genera forming high-level forest, followed at lower elevations by still other conifers belonging to still different genera and, below this, a zone of evergreen oak mixed with different conifers followed by a zone of deciduous oak woodland, and finally giving place to broad-leaved evergreens of oak and even other families in some parts of the coastal region. Each separate community often involves, even in its tree components, species not only belonging to different genera but often to different families, and at times still more widely separated systematic groups.

In about the same range in physical environment taken from the tree-line on Mt. Kosciusko to the sea, in the vicinity of Eden, N.S.W., all the communities are either forest or woodland and they have species of Eucalyptus, and only Eucalyptus, as the tree component. This characteristic confers an apparent uniformity on the vegetation which on closer inspection is found to be somewhat misleading. It is true that this uniformity is partly real, since for example the species of Eucalyptus which grows at the highest elevations (Snow Gum, *E. niphophila*), which reaches the tree line on Mt. Kosciusko at about 6,500 feet, still retains the evergreen habit even though there is regularly a very cold winter with a snow fall of several feet and minimum temperatures often below 10° F. (COSTIN, 1954). While Eucalyptus has had (from the physiological point of view) the capacity to survive such extremes and has successfully provided species to form the trees in communities in such an environment, it has not the capacity which many genera (for example, *Quercus*) possess, to provide separate species of different life form such as evergreens in warmer climates, but deciduous in colder climates. In contrast with Eucalyptus, which cannot do this, the capacity is found in a genus which is thoroughly indigenous in Australia, that is *Nothofagus*. This is evergreen in two of the species, *N. Cunninghamii* and *N. Moorei*, which are characteristic of more equable habitats, but is represented by the only winter-deciduous species in the entire Australian flora by *N. Gunnii*, which occurs in very restricted areas in the central highlands of Tasmania.

On the other hand, if the wood of Eucalyptus species from sites of high altitude and cold winters is examined, it is found to be rather similar to the ring-porous species of the deciduous forests of the Northern Hemisphere, and annual rings can be distinguished and counted just as easily as in species belonging to genera such as Ash, Oak and Elm.

As an environment is approached which favours either temperate or tropical rain forest, the species of Eucalyptus thriving there display a capacity for very fast growth and a capacity to survive in some degree in competition with the rain forest communities by virtue of this fast growth, if certain other necessary environmental factors are present. At the same time the genus has not been able to produce a species which can regenerate under heavy rain forest cover as is the case with many species of genera belonging to this latter type of community. Therefore, where habitats are presented in which physical conditions have been reached which favour the persistence of rain forest species, Eucalyptus is ultimately eliminated unless apparently aided by recurring catastrophic events such as fire, excessive drought or storm damage, (GILBERT, 1958).

At the arid side of the range of sites in the environmental range occupied by Eucalyptus communities various specializations occur, and in particular the distinct life-form "Mallee", in which the plant is no longer a single-stemmed tree but a woody plant which produces many shoots of from ten to twenty feet tall from a common underground rootstock, so that each individual is many-stemmed. The Mallee form is displayed by all Eucalyptus species of these highly successful xerophytic Mallee communities. They are, however, with very few exceptions within the climatic zone in which they occur, confined to a special soil type which is highly calcareous.

As the rainfall drops below about 10 or 12 inches per annum in southern areas or below about 25 inches in northern summer rainfall areas, Eucalyptus has again not been able to provide species which can survive and be a component of the widely spread plant communities of such regions. By far the major portion of Australia, by area, has a rainfall of less than 10 inches and is occupied by communities which do not contain Eucalyptus. When Eucalyptus does occur in this zone it is usually confined to special habitats such as the sandy water-courses of Central Australia, where *E. camaldulensis* is an especially characteristic plant. Another situation where in favoured sites in areas in this zone some other species occur is where there are rocky cliffs and ranges. In these cases Eucalyptus species occur because they avoid the conditions of the general environment and escape the extreme physical conditions which otherwise prevent their survival.

So far as the provision of species to thrive in the more arid areas is concerned, the genus lacks the plasticity of one such as the rutaceous *Flindersia*, which in Australia is a common component of the Indo-Malayan type Rain Forest but which occurs in at least one species, *F. maculosa*, as a component of the Scrub, a community occurring widely in areas in western New South Wales and Queensland where the rainfall is generally less than 12 inches per annum.

However, within the habitat limits favourable to it, Eucalyptus present an extraordinarily large diversity of species. Even more particularly, this diversity is closely associated with the Eucalyptus communities occupying soils of very low fertility. The distribution of Eucalyptus species in Dry Sclerophyll Forest, which is the community characteristic of the poorer soils of the moderate rainfall areas (20"—40" per annum) of New South Wales, Victoria, South Australia, Tasmania and the south-western part of Western Australia, is of especial interest not only because of the diversity of species but because of the pattern of distribution. This displays a number of interesting features which may be unique. Certainly in the Australian environment they are peculiar to the genus.

Distribution of Eucalyptus Species

Within the zone which is most favourable for the development of Eucalyptus there is a very wide range of species, and this intense display of forms is associated with very precise correspondence of each species with a particular micro-habitat. Again, by contrast with a common situation in a forest such as Beech in Europe, it is a characteristic of communities dominated by Eucalyptus that in many areas a change of aspect will produce a change in tree species, and a series of species one following the other will be found associated with exposed, moderate or sheltered aspects in a graded sequence. Since most areas present habitats which are mosaic in this regard the sequence may be repeated numberless times as the appropriate microhabitats are presented in their recurring pattern. In microhabitat changes of similar magnitude in the Beech forests of Europe (TANSLEY, 1939), or Pine in Finland (CAJANDER, 1926), changes in specific composition of the community are found largely in the herb or shrub layer and the response of the trees is confined largely to dimensional changes such as total height growth. The intricate pattern of distribution of species in Eucalyptus in accordance with micro-habitat changes has been somewhat obscured by the rather unsatisfactory state of taxonomy in the genus. This arises firstly because taxonomic treatment has been inconsistent and at times species have been named from types which are taken either from hybrids or alternatively from individuals in populations varying in accordance with some clinal pattern (PRYOR, 1957). Secondly, there is an inherent difficulty in classifying Eucalyptus because it is, by any taxonomic standard, a large genus with at least some hundreds of species.

While some species are widespread in their total geographic

extent, the mosaic pattern characteristic of many areas of Eucalyptus is not affected by the occurrence in that area of the widespread species. In almost all cases such a species, though widespread, is also tied closely to a circumscribed habitat in the particular limited area. This implies that the habitat factors which finally limit a widespread species evoke a different physiological response in the species and are perhaps different from those which cause species to change from site to site in a restricted area.

Variability in Eucalyptus Populations

Clinal Variation

There are two main causes of variability in populations of Eucalyptus, and especially in the kind of variation which has led to taxonomic difficulty as mentioned above. The first of these is clinal variation, which is regular and is associated with some regularly changing factor of the environment. It always has a strong genetic component in its determination. This is, of course, particularly characteristic of widespread species, and though such species, in the concrete sense, are made up of innumerable stands each of which is tied to and found only on the particular micro-habitat which it favours, as these stands occur in countless repetition, so there are gradual environmental changes often associated either with latitude or altitude changes, or perhaps "continentality" as represented by distance from the coast. Where a species is sufficiently widespread, that is, where it has a geographic range of some hundreds of miles, there are always, at least as far as has been tested, differences in the genetic make-up of the populations from separated areas. This is probably most clearly displayed in physiological differences, as for example in frost resistance. In experiments recently carried out, seedlings of *E. viminalis* raised from seed taken from trees growing at nearly 4,000 feet elevation in a latitude of about 35° South are found to be fully frost resistant when planted at this elevation, whereas seedlings of the same species from stands closely similar in the taxonomic sense and of the same species, but taken from trees growing at an elevation of only 1,000 feet in southern Victoria and similarly in the vicinity of Adelaide, are killed to ground level by the normal winters which are experienced at 4,000 feet. The same characteristic appears in *E. coccifera* and *E. pauciflora*, and more casual observations make it clear that it is likely to be a general characteristic of species of the genus. On the other hand, when two closely related species, such as *E. globulus* and *E. bicostata* (which may be reduced by some to the rank of Subspecies), the former being characteristic of low levels of eastern Tasmania and the latter of the inland side of the Dividing Range of Victoria and New South Wales, are compared it is found that planted in an environment a little drier than that in which *E. bicostata* normally occurs, the *E.bicostata* seedlings endure both drought and low temperature better than *E. globulus*. Such differences are most clearly displayed by experimental trials which are made by subjecting populations to an environment which passes the threshold for survival of one species but not the other, but it is probable that species in general, if they are widespread, display such variations, although they may be rather subtle and not easily perceived unless put to a rather rigorous experimental test.

When populations from within one species which are known to behave in this way are examined it is found that there are often various minor morphological features which are characteristic of separate populations and which are an index of the behaviour of the various populations. It is reasonable to assume that such morphological difference may well have adaptive significance, since they are frequently graded and subject to quantitative variation in accordance with a trend in physiological behaviour displayed by populations taken from a graded series of habitats. The best studied of these features is glaucousness, which has been investigated by BARBER (1955). If an extrapolation is made from the limited critical study so far available it can be said that it is likely that differences of this kind are very widespread in many species. The stimulus given by the idea that gene controlled differences of this kind have a value in relation to speciation in terms of natural selection in the Darwinian sense is likely to lead to a valuable field of experimental study in the genus. At the same time it has extremely important practical consequences. Though this characteristic of clinal variation exists within many well-studied tree genera of the Northern Hemisphere, for example, *Pinus, Betula* and *Pseudotsuga*, and its importance in practice is well known through seed provenance studies, this possibly applies even more critically still in Eucalyptus. In short, provenance of Eucalyptus seed from within Australia is certainly as important as has ever been imagined for a tree crop for any of the species commonly used in the Northern Hemisphere.

The taxonomic approach to the naming of a cline has not yet been resolved properly, and in Eucalyptus in particular much difficulty has been made by names having been set up (to describe separate species) and applied to specimens taken from different parts of a cline in what is essentially one Linnean species. The study of the genus is aided by recognizing which species erected in the various taxonomic works on Eucalyptus are of this kind and which names should strictly be relegated to a status other than that of species.

Hybridization

The second source of variation is due to hybridization. The historical systematic study of the genus reflects very different attitudes to the recognition of hybrids within the genus. As early as 1810 CALEY decided that certain individuals occurring in the vicinity of Sydney were hybrid between two well-recognized (both then and now) species, whereas later WOOLLS (1867) considered that hybridization did not occur, and a botanist as eminent as FERDINAND MUELLER gave reasons, though they were inadequate to explain it, as to why he considered that such could not exist. On the other hand MAIDEN presented views in 1904 which, while not backed by experimental tests, conformed well with what has been later well established. In recent years the first evidence presented on a sound basis to indicate that hybrids occur between well-recognized species was that of BRETT (1937) and McAULAY (1937). Since then, considerable study has been carried out which establishes clearly that hybrids do occur quite commonly under natural conditions and also a good many have been synthesized by manipulation. In the debate which has been associated with the propounding of rules following the accumulation of evidence in this field there has been a tendency to forget the known frequency of occurrence of hybrids naturally within Eucalyptus. The major portion of the populations of Eucalyptus, in fact a very large majority of such populations, is made up of species which are well recognized, incorporating such clinal variation as has been described above. At the same time a small proportion of the individuals usually on the fringes of such populations is hybrid between the species concerned and a second one.

The number of hybrids varies considerably and no doubt for several reasons. Nevertheless, within these limits some of these are probably the direct result of settlement and the consequent breakdown of the operation of some of the environmental factors, especially stand stability, which is one factor which led to very precise correspondence of species with habitat. For example, where populations of Eucalyptus proceed to their normal physical maturity, which for a great many species is between 200 and 400 years, the boundaries between stands, each occupying its distinct microhabitat, are generally very sharp. This condition can be discerned in many instances still today in spite of the great vegetational changes following settlement in Australia, if regard is paid only to the mature trees which are generally over 150 years old. In such cases it is found that hybrid individuals of similar age exist, but that they are closely confined to the narrow zone which forms the boundary between the stands of two species; in short, a common means of maintenance of species of Eucalyptus as discrete populations of distinct entity is at least partly ecological. This is apparent in many cases where species which can interbreed are in contact. The capacity of the hybrid individuals to survive when in competition with the pure species is found to be inferior to that of the pure species on either side of the marginal zone where these two species areas meet. It is clear that many hybrids were, in the virgin condition in Eucalyptus, confined very strictly to an extremely limited zone. Where, however, such populations have been subject to burning and partial clearing, and perhaps other disturbance, the average age of the populations has been much reduced, and the precise correspondence of populations, each with its appropriate micro-habitat, has often been somewhat broken down. In such cases it is quite likely also that hybrid individuals, or hybrid swarms which undoubtedly result and spread in many cases at such junctions, are heterotic, at least so far as rate of growth is concerned up to 40 or 50 years of age, and if the rotation length is reduced to periods approximating this it is clear that in many cases hybrids would be favoured over the species from which they are derived. In a good many areas the narrow zone of hybrid individuals characteristic of virgin communities has been considerably broadened and a hybrid swarm is now thriving with the consequent blurring of the precise species/area boundary. It has been noticed in such cases that where there is an opportunity for hybrids to reach greater ages, if they are growing on sites considerably removed from the marginal zone of their presettlement occurrence, their length of life is less than that of the parents.

While shorter life and faster growth during that lifetime may be a feature of considerable importance in technological attempts to improve Eucalyptus for various cultural uses, from the biological point of view it suggests that a return to the presettlement conditions of long rotations would favour the re-establishment of the species distribution pattern which was characteristic of the Eucalyptus vegetation in the days of pre-European settlement. The frequency of occurrence of hybrids at the junction zones of two species areas is not uniform and is also apparently conditioned by certain characteristics of these zones. For example, survival is favoured when there is a graded habitat, such as that which results where there is a species change-over simply in accordance with the passing of a critical limit within a regular habitat gradient such as that tied to regular change of altitude. On the other hand, where two species populations are brought into contact by an abrupt change in habitat such as those due to edaphic features, as where there is an outlier of basalt on an area otherwise sandstone, in many cases stand margins may be searched almost in vain for hybrids, although interbreeding between the two species may be known to occur elsewhere.

On the other hand it is known that in some combinations, such as between *E. cinerea* × *E. Blakelyi* (PRYOR, 1956), the hybrids, and segregates from these which more or less resemble the hybrids, are at the best weakly viable and thus would present little challenge to the pure species and would not lead to a breakdown of precise species/area boundaries.

Hybridism in Relation to Interbreeding Groups

One limitation in the occurrence of hybrids is a much more deep-seated one, and this appears to result from basic breeding incompatibilities within the genus which correspond with Subgeneric groups. The genus may be divided into five Subgeneric groups, namely Eudesmia, Corymbosa, Adnata, Renanthera and Macranthera, of which only the latter four are at all widespread. In the south-eastern part of the Continent the species are largely confined to the latter three of these, and in Tasmania the species are still more closely restricted to only the final two. Within each of these Subgeneric groups a large majority of species has been found to hybridize one with the other, although with great variations in frequency and in relative viability. There are certain exceptions, which indicate the development of breeding barriers, to some degree, within a particular Subgenus, but the great majority of species within each such group can interbreed with one another in the field and thorough search will reveal hybrids between them, although the occurrences may be erratic. From studies of the natural occurrence of hybrids, reinforced by experimental crossing, it may be stated as a generality in Eucalyptus that interbreeding Eucalyptus species form separate stands each occupying distinctly different ecological situations.

Mixed Stands in Eucalyptus

The presentation above of the pattern of Eucalyptus distribution has been over-simplified, since in most areas of south-eastern Australia each available habitat is occupied by a stand made up of trees of Eucalyptus of at least two species. In such mixed stands the two species concerned do not interbreed, and always each is derived from a different Subgeneric group. A second rule may therefore be stated relating to the occurrence of Eucalyptus species which is that mixed stands of Eucalyptus are made up of species which are genetically isolated. It must be borne in mind, of course, that there are many minor exceptions to this near stand boundaries, but if looked at from the 1000 acre scale the position is clear. In accordance with topographic pattern, each species of any given pair in a particular stand gives way appropriately to other species with the hybrid boundary pattern which has already been described. However, the point of change-over from one species to another is not coincident so far as the two species are concerned which go to make a single stand in a particular area. In fact, it is almost always distinctly different. Since it is a characteristic of Eucalyptus that the trees are approximately equal in their demands upon the habitat (in forestry terms they are almost all "light demanders" and form equal co-dominants in a given stand), this pattern of distribution poses some interesting problems. It is clear from field evidence that two species which can interbreed do occur together as a mixed stand in the long-term ecological sense which was that seen in Australia prior to European settlement. This is reasonable, since it follows that if gene exchange were possible between them it would almost certainly lead to the emergence of a gene combination — a new species — better adapted to the particular habitat than either of the two parents. It seems that wherever this was appropriate such a process would have taken place a very long time ago, since it would be only in quite exceptional circumstances that two species would be brought together in recent times in such a way as to allow this process to take place again. It may well be that there are areas in which this occurs, but in the few cases where this is suspected there are quite unusual circumstances and good reasons for supposing recent placing together of the species.

There is a further interesting pre-requisite for the stable occurrence of mixed stands of long duration which follows from the fact the species which occur mixed together have not the same habitat limits, although the range may be more or less the same. Since their limits are almost always different, it may be supposed that this indicates also that their optimum requirements are different. If this is so, it follows that they can grow together in a mixed stand only if there is some positive biological benefit in such growing together, and it is postulated that the mixed stands of the virgin Eucalyptus populations are in combinations in which there is such positive biological benefit. If they were not, it must be presumed that pure stands of each species would occur in a graded sequence and not the mixed groups which are actually found. There has so far been no experimental verification of this hypothesis of positive benefit or "mutual aid" between different Eucalyptus species in mixed stands, but in view of the fact that low nutrient supply is characteristic of the soils of the Australian environment which supports Eucalyptus Sclerophyll Forest showing this distribution pattern, it seems likely that such a pattern could have a nutritional basis. It may be anticipated that studies in the future will lead to a critical examination of the association in Eucalyptus in this regard and it seems that the Subgeneric group Renanthera plays a critical part in this scheme. This is especially so since it has been demonstrated (PRYOR, 1956) that Renanthera, uniquely amongst the Subgenera of Eucalyptus, seems to have an obligate need for mycorrhiza for really effective growth. In the pattern of mixed stands described above species of the Subgenera Renanthera are always one of the pair, although the second member may be derived from at least three other Subgenera.

The Association in Eucalyptus

The regular occurrence on closely circumscribed micro-habitats of Eucalyptus populations made up of a pair, or even three or four, species as equal co-dominants in the stand leads to easy recognition, in practice, (with or without the theoretical views outlined above), of units for designation as associations (BEADLE & COSTIN, 1952, PIDGEON, 1942, PRYOR, 1954). It is considered very likely that such units which it has now, so far as Eucalyptus ecology is concerned, become the practice to describe as Eucalyptus associations, have a reality which goes beyond the simple coming together of plants without any integrated, positive interaction. It is argued that the Eucalyptus association, especially of those areas where rainfall is more than 20 inches per annum, and where nutritional stress results since the soils are of generally lower fertility, is also backed by a biotic component in which the Eucalyptus species which grow together to make the mixed stand and so characterize the association are also a more effective community than either would be should they ever grow separately in pure stands. The distribution of Eucalyptus species is a rather peculiar one which is consistent with such a hypothesis. Experimental proof or disproof of it awaits studies yet to be carried out.

REFERENCES

BARBER, H. N., 1955. Adaptive Gene Substitutions in Tasmanian Eucalypts. *Evolution*, **9** (1), *1—14*.

BEADLE, N. C. W. & COSTIN, A. B., 1952. Ecological Classification & Nomenclature. *Proc. Linn. Soc. N.S.W.*, **77**, 1-2, *61—82*.

BLAKELY, W. F., 1955. A Key to the Eucalypts. F. & T. B., Canberra.

BRETT, R. G., 1937. A Survey of Eucalypt Species in Tasmania, *Proc. Roy. Soc. Tas. 75—109*.

CAJANDER, A. K., 1926. Theory of Forest Types. Helsinki.

COSTIN, A. B., 1954. A study of the Ecosystems of the Monaro Region of N.S.W. Govt. Printer, Sydney.

GILBERT, J. M., 1958. Personal communication.

MAIDEN, J. H., 1904. Report A.N.Z.A.A.S., Dunedin.

MCAULAY, A. L., 1937. Evidence for the Existence of a Natural Hybrid between *Eucalyptus globulus* and *E. ovata. Pap. & Proc. Roy. Soc. Tas.*

PIDGEON, I. M., 1942. Ecological Studies in N.S.W. D. Sc. Thesis, Sydney University.

PRYOR, L. D., 1954. Vegetation of the A.C.T., A.N.Z.A.A.S. Handbook.

PRYOR, L. D., 1956. The Identity of *Eucalyptus subviridis*. *Proc. Linn. Soc. N.S.W.* **81**, *101—104*.

PRYOR, L. D., 1956. Chlorosis and Lack of Vigour in Seedlings of Renantherous Species of Eucalyptus caused by Lack of Mycorrhiza. *Proc. Linn. Soc. N.S.W.* **81**, 1, *91—96*.

PRYOR, L. D., 1957. Silvae Genetica. J. D. Sauerländer's Verlag, Frankfurt on Main, Germany.

TANSLEY, A. G., 1939. The British Islands and their Vegetation. Cambridge University Press.

XXIX
THE ECOLOGY AND PREVENTION OF SOIL EROSION

by

R. G. DOWNES

(Deputy Chairman, Soil Conservation Authority of Victoria)

Introduction

Accelerated soil erosion is a result of the operation of the physical forces of wind and water on soil which has become vulnerable because of man's interference with the natural environment. For this reason soil erosion can be viewed as an ecological catastrophy; an upset in the balance of an environment which can frequently lead to such significant changes that a new succession is required to re-establish an ecological equilibrium.

Before the advent of civilized man, ecological catastrophies probably occurred only at infrequent intervals, but for the past 3,000 years man has had a devastating influence in changing the face of the earth. It is mainly because of his activities in certain kinds of environments that soil erosion has occurred.

The most recent devastating changes in natural environments have occurred in countries settled by white men, where a new human culture has replaced an older one, and where forms of land use which may have been suitable for one kind of environment have been applied in others in which there are significantly different conditions.

Australia is one of the most recently settled countries, and the effects of such treatment are now reaching their peak of severity.

New people in a new environment

The natural environments of Australia were first subjected to the changes imposed by civilized man, when the British people established a penal colony in 1788 on the site where Sydney now stands. For about twenty years the new settlement had a precarious existence. There was a constant struggle to produce the necessities of life on a relatively small area of useful land between the mountains, which presented a seemingly impenetrable barrier to the inland, and the sea. The colony was frequently on the point of starvation which was only averted by the timely arrival of supply ships.

There were three reasons for these agricultural difficulties. The soils were poor, the strains of crop plants were not suited to the different conditions of moisture, light and temperature, and there was a shortage of skilled agriculturalists.

Since this inauspicious beginning, exploration and settlement along with the breeding and adaptation of crops and animals suitable for the new environments have advanced in stages.

At first there was a broad-scale use of the land. The holdings were large and open range grazing was practised. Small holdings close to towns provided the major food requirements. About a hundred years later, the large holdings in more favoured environments were subdivided into farms on which a broad-scale agriculture developed. By improvements in agricultural technology and machinery this type of farming has reached a peak of efficiency with respect to production per man.

During both of these stages of development there was an attitude of exploitation which still persists in some environments today. However, in others, the soil erosion resulting from the systems of land use and management made men realize that their problems were not at an end, and that they had not yet fitted into their environment.

In the meantime, the closer settlement in the better rainfall areas had virtually forced the broad-scale grazing into the more arid environments and again there was trouble with soil erosion.

These events have drawn attention to the need to move toward the final stage of the settlement and development of the continent, that of soil conservation. It is now becoming more generally accepted that by the use of all possible technological information, permanent systems of land use can and must be devised and introduced for each of the many types of environments. Only when this stage has been reached can civilized man be said to have reached a balance with his environment.

However, the achievement of this ambition is not without its problems which are plainly ecological in character. Their solution will only come from a more intimate knowledge of the different environments, and the reasons why the land use systems which have already been imposed have upset the balance, and what changes need to be made to re-establish an equilibrium. It will need too, a more general appreciation that proper land use is in fact applied ecology.

In spite of the problems which still remain to be solved, the effect of settlement and development has been dramatic. From a small isolated penal settlement unable to produce sufficient food for its needs, there has developed a country which now depends on its exports of primary produce for most of its overseas income. This achievement has been gained at a price, the price of soil erosion in many kinds of environment and even complete destruction in some. Such a result is not surprising when the habits of the white settlers and their domesticated animals are compared ecologically with those of the native people and fauna.

The effect of settlement

Continental Australia has been isolated from other land masses of the world since the early Tertiary period. In such isolation a characteristic flora and fauna have evolved and survived without competition from species which have subsequently been evolved in other parts of the world. In relation to area, the aboriginal and fauna populations were small. These species were able to survive because of their nomadic habit in seeking their food requirements over vast areas in accordance with seasonal conditions. The vegetation evolved under these conditions of light pressure.

With the influx of white man and his domesticated animals, the whole system was changed. Larger numbers of people and of hard footed, closer grazing animals were confined on specific areas in a settled existence. The resulting constant pressure on the environment irrespective of the variable climatic conditions has had significant effects. Clearing of vegetation, seasonal burning, cultivating and constant hard grazing were all radical changes which have upset the ecological equilibrium in various kinds of environments.

The various manifestations of wrong land use and the upsetting of the ecological equilibrium are merely reflections of how different environments have been able to react to the imposed conditions. In some places, there has been little actual loss of soil but merely a declining productivity due to deterioration of the physical condition, chemical fertility, or moisture status of the soil. In other places, there has been complete loss of vegetative cover and a hardening of the soil to an arid and inhospitable environment for any form of vegetation. In others, there have been tremendous losses of soil from the surface and from scoured gullies. In some wetter environments landslips have become frequent occurrences; while in drier environments, sand dune systems have become unstable and are being redistributed around the countryside.

These different effects of upsetting the ecological equilibrium are indicators of the inherent weaknesses of the different environments. In some places highly specialized vegetation could not stand up to the imposed conditions, in others, poorly structured soils collapsed completely under cultivation and the pounding action of rain, and in others, naturally unstable topographic conditions have become even more unstable.

Although soil erosion due to wind is an important problem in large areas of the dry inland parts of Australia, it is erosion by water which has caused the greatest amount of damage and economic loss. Water erosion is more common in the better rainfall areas where it affects the more productive and more valuable land. These are also the more populated and more highly developed areas and so erosion by water is more likely to cause damage to public utilities such as roads and railways. Furthermore, some of the most spectacular erosion, even in the dry lands of the interior, is the result of water action.

It would be wrong to imagine that soil erosion in Australia is as bad or so widespread as in the Middle East, or that it is comparable even with the erosion which has occurred in the United States of America. There are large areas of most hazardous country in Australia which have not yet been affected by erosion to any appreciable extent. In fact there is more erosion in some of the less hazardous environments in both the Middle East and the United States and even on the moors of Scotland than can be found in areas of comparable hazard in Australia. Maps of soil erosion do not provide a satisfactory comparison of the conservation needs of different countries. They merely provide a qualitative assessment of the damage without giving any indication of the nature of the environment or its inherent hazard.

The erosion which has taken place in Australia has occurred on areas with a high degree of erosion hazard which formerly was not readily recognised. This has created numerous problems of soil conservation, but fortunately it is not too late to make their solution worthwhile over most of the continent.

The problem of soil conservation

Soil conservation in any environment is fundamentally a problem of determining the correct form of land use and management. The correct form of land use and management is one which provides a higher level, or different form of productivity than that available in the natural state, but this new productivity must be capable of being maintained indefinitely. This means that the balance of the natural environment must be replaced by another balanced system under the changed form of land use. The determination of correct land use is therefore a problem of applied ecology.

Natural environments have evolved to a condition, where from the available constellation of plant and animal species, communities have developed in which there is a relative abundance of the various species best able to survive in association and competition with each other under the existing soil, climate and topographic conditions.

A natural environment represents a maximum productivity of plants and animals, at that stage of its succession, of the species available during the developmental and evolutionary processes. It represents a permanent natural productivity which will be maintained indefinitely or even increased by successional changes unless there is some catastrophic force imposed on it. These naturally infrequent catastrophies may result from geological upheaval, vulcanism, climatic change, or the occurrence of mutant species having overwhelming advantage in competition with others.

However, except for seasonal changes temporarily favouring one species against another, natural environments are in a state of equilibrium over long periods. Even a run of seasons favouring certain groups of species does not significantly upset the balance to any marked extent, because these same conditions inevitably favour predators or parasites which will tend to reduce the numbers of the favoured species. If not for these reasons, competition for food or water supply tends to restore the environment to its normal condition once more.

Man's objective in land use is to either raise the productivity of the environment or to produce in it other plants and animals which are of more value to him. This requires a change in the environment and the balance must be upset, but unless a new equilibrium is established under the changed land use, the interference will set off a chain of reactions comparable to those which could be expected only rarely under natural conditions.

The ecological problem of land use and soil conservation, is the provision of more desirable species to occupy artificially created niches from which a new equilibrium of maximum productivity will result.

Since Australia has been isolated from other land masses for a long period of geological time, there is every reason to believe that many plant and animal species which were unavailable during the evolution of its natural environments, might find suitable niches. In fact the history of some introduced species, their rapid colonization and adaptation to the environmental conditions confirms this. Some introduced species have become pests but this merely indicates the need for balance. The European rabbit and prickly pear (*Opuntia* sp.) are two such examples of introductions which were too successful and for which suitable opposition has now been introduced. Of the more useful species, Monterey Pine *(P. radiata)*, subterranean clover *(T. subterraneum)* and domestic sheep are good examples. These have not yet been incorporated into properly balanced environments, although sheep and subterranean clover together, are getting close to a desirable balance in some localities.

There is a wide range of environments and although there may be superficial similarity between some of them, each presents its own particular problems of land use and development to achieve a high level of permanent productivity and soil conservation.

Some environments and their problems

To outline the ecology of erosion and conservation in many of the Australian environments would be a formidable task. Only brief mention of a few will be made here and one environment will be treated in more detail later.

COSTIN (1954, 1957) has shown how man's activity has lead to instability and damage in the alpine and subalpine environments in S.E. Australia. Grazing and burning of these natural alpine tussock grasslands *(Poa caespitosa)* has caused a deterioration and vegetative change which has enabled both wind and water erosion to occur. In some places the damage has been severe. In addition moss beds and bogs which normally occur in the lower situations on the peneplain have been dried out and destroyed as a result of stock trampling. The reclamation and re-establishment of a balanced environment in these areas having an elevation of more than 4,500 feet above sea level is difficult because of the harsh climatic conditions, and the depletion of the plant species specialized for life in such an environment.

In the wet sclerophyll forests successive forest fires have obliterated certain valuable timber species in some areas, and in others the more fire sensitive and more valuable species have been replaced by more fire resistant forms. Clearing in the wet forest areas for use either for crops or pasture has not been entirely successful. In the tropical areas high intensity rains, even on the naturally well structured soils have caused considerable erosion. In the south, although good pastures have been established in some places, the hydrological balance has not been maintained. The replacement of deep rooting trees by perennial grasses apparently enables the soils to become excessively wet at depth and landslips become common.

In the drier areas BEADLE (1948) has outlined the consequences of use of many kinds of environments in western New South Wales. Here the problem is one of excessive grazing of vegetative types not evolved to cope with such treatment. In some areas the result has been large areas of "scalded" plains on which there is not a vestige of vegetation and the bare soil is in such a condition that it will not readily permit the entry of water.

In even drier environments of the arid interior, RATCLIFFE (1936) has described the ecological imbalance of certain kinds of environments as a result of man's occupation and use. His discussion of ecological differences between the natural and present use of the Saltbush *(Atriplex vesicarium)* environment is particularly revealing. His conclusion concerning the proper use of the dry country virtually means that it should be treated in the way it was accustomed to being treated under natural conditions. — "Inconstant stocking, the figures varying between wide limits so as to take full advantage of the flush feed in good seasons and to avoid damage to the perennial vegetation in bad." This is precisely the way in which this country was used by the native fauna.

A difficult but interesting environment

The ecological implications of erosion control and soil conservation are exemplified by one particular type of environment which occurs widely in Victoria, the south-eastern State of the continent. In this environment there has been considerable instability and consequent erosion of various kinds as a result of what had appeared to be a reasonable and relatively mild form of land use.

The environment has a rolling to hilly topography on which the original vegetation was a dry sclerophyll forest of *Eucalyptus* spp. with a sparse under-storey of shrubs and perennial grasses. The rainfall ranges from 20"—30" per annum most of which falls during the winter months which is the growing season. Rainfall during the summer consists mainly of isolated thunderstorms which are of little value for plant growth because of the hot conditions and high evaporation. The hydrology of the environment is such that rainfall and evapotranspiration approximately balance and consequently there are no permanent streams. Flood flows can occur when rainfall intensity exceeds the infiltration capacity of the soils or late in the winter when steady rain falls on already saturated soils. Under natural conditions there was a delicate hydrological balance. The soils are formed on fine sandstones and shales and they are relatively shallow having an average depth of from 3 to 4 feet.

The environment is subject to accession of oceanic or "cyclic" salts. ANDERSON (1941, 1945) and later LESLIE & HUTTON (1958) have shown that over these areas about 10 to 30 lbs. of salt per acre can be brought in by rain each year. These amounts of salt are in themselves relatively insignificant except in those areas where the relation between rainfall and evaporation precludes their complete leaching out of the soil season by season, thus enabling their accumulation. DOWNES (1954) has put forward the thesis that the soils in this environment are solodic, and have been formed as a result of the salinization and subsequent leaching of the pre-existing soils. In recent geological time there have been significant changes of climate which would have enabled considerable accumulation of salt during dry periods and its subsequent leaching during wet periods, within certain zones which can be correlated with present day climatic limits. The zone in which this environment occurs is one in which the most intensive solodization would have been possible.

It is because of such a genesis that the soils have certain properties which make them susceptible to the curious forms of erosion which have subsequently occurred. The soils have a relatively shallow, poorly structured, compact loam surface horizon over a bleached structureless subsurface horizon beneath which there is a sharp transition to a heavy clay. The heavy clay subsoil has a moderate medium sub-angular blocky structure when dry, but when it is wet it disperses so readily that local farmers talk of the sub-soils as being "sugary" because they "melt" so easily. In common with other solodic soils they are acid throughout, have a low content of soluble salts and low amounts of exchangeable calcium. In fact the sub-soils are hydrogen-magnesium clays.

Although the problems of this type of country are largely due to the character of the soils, the factors operating to produce such soils are themselves of importance in enabling the operation of processes which produce the problems. Man's activity in this environment has merely been a reduction of tree cover in favour of grass and subsequent overgrazing of that grassland, but the results of such treatment have been spectacularly bad.

The first significant disturbance of the environment occurred about eighty years ago when the land was more closely settled. Trees were thinned out and in some places completely cleared to encourage a better growth of the existing perennial grasses of the *Danthonia* spp. and *Stipa* spp. This grassland was grazed by sheep. However the grasses themselves had evolved under a condition of occasional browsing by relatively few marsupials and were unable to maintain density and vigour under the constant grazing pressure of sheep, irrespective of the seasonal conditions. In addition, the European rabbit had increased considerably in numbers and this added to the grazing pressure on the vegetation.

Under such treatment the pastures deteriorated in density and vigour and the exposed soil became hard and compacted, the surface became impermeable, productivity declined, and soil erosion became evident. The increased runoff scoured watercourses and they became eroding gullies. But these were only preliminary warnings.

About forty years ago it became evident that as well as the more obvious effects of instability, more insidious troubles had been developing. At that time the first signs of subsoil or tunnel erosion appeared and it has subsequently developed into a widespread and difficult problem.

Yet another problem emerged — the development of salinity in the soils along some of the creek lines and watercourses and on the lower parts of some slopes. At first it appeared to be a minor or transitory problem, but like tunnel erosion, it too increased in incidence and extent, more particularly during the past ten years.

Associated with both of these problems there was the difficulty of establishing improved pastures. Many attempts by landholders had met with outright failure, or at the best only poor germination and lack of persistence. These failures were for many years attributed to climatic conditions, although the climatic data offered no support for such contentions.

Within a space of eighty years, a logical and reasonable system of land use had resulted in a degree of degradation in many places which could never have been imagined by the early settlers and appeared to offer nothing to the present generation but further deterioration at an ever increasing rate.

With the information now available about this environment, it is easier to understand how these results were inevitable, and a closer examination of the problems will reveal this.

Tunnel erosion is a most insidious form of erosion because of the amount of deterioration which occurs before there is any visible sign of damage. The earliest stages are marked by small patches of yellow clay which have oozed through a small crack or ant hole to the surface. At a later stage there may be conspicuous "fans" of yellow clay material which has been washed down slope from small holes. At an even later stage there may be a line of holes upslope from the point where the clay is being washed out. These holes occur where parts of the surface soil have collapsed into the tunnel which has been eroded out of the subsoil below.

The mechanism of tunnel erosion and the reasons for its occurrence were put forward by DOWNES (1946). Basically it is due to deterioration of pasture cover which enables the naturally poorly structured surface soil to develop an impermeable surface condition. This enables increased runoff from a large part of the area and less moisture for plant growth. But in certain places there is an increase of infiltration by the concentration of runoff water. Small natural hollows and old stump holes have better growth of grass and an infiltration capacity of more than 50 times that of the surrounding bare areas. In these places, particularly after a dry summer, the water soaks in rapidly and as it passes into the cracked subsoil it disperses some of the clay and carries it downslope until wetting and swelling of the clay prevents any further such movement.

After many wet and dry seasons much clay from beneath the hollow has been removed, the hollow enlarges and downslope from it there is a partly formed tunnel in the subsoil but nothing visible above ground. At some critical time, possibly the first autumn rain after a drought or prolonged dry summer, the quick movement into the subsoil develops sufficient hydrostatic pressure downslope for some of the liquid clay to be forced through a crack or ant hole to the surface. After the next dry season, when the clay has dried and cracked, there is a complete channel into and out of the subsoil and rapid scouring takes place from this stage to produce characteristic clay "fans". Once tunnels have been formed, they provide a harbour for rabbits in country which previously had not been a desirable habitat for them because the soils were too hard and compact for easy burrowing. The invasion of rabbits adds to the grazing pressure and tends to accentuate the trouble.

Tunnels deepen and widen until the roof of the tunnel can no longer support its own weight and it collapses to form a gully.

Salting in this kind of country was first observed by HOLMES & LEEPER (1939) and later by DOWNES (1949) but it was so limited in extent that it was thought to be of academic interest only. However recent investigations by COPE (1957) have indicated that it is widespread and increasing; his work also confirms a hypothesis for salting which was accepted but not investigated in detail by the previous observers.

Salting results from an upset of the hydrologic balance of the environment. The removal of trees to grow pastures is probably in itself insufficient upset to cause salting, but when there is overgrazing in the catchments and insufficient water use, the excess water seeps to lower levels carrying with it the small amount of salt brought in by the rain. In this way slightly saline water, if not used by vegetation in the areas of seepage, enables a concentration of salt by evaporation and a level can be reached which will inhibit plant growth. Stock tend to concentrate on and eat out pastures growing on these seepage areas and this tends to intensify incipient trouble. There is a characteristic succession of vegetation on the seepage areas with the increasing accumulation of salt, the normal species giving way to more halophytic species and ultimately even these will die.

Early recognition of salting enables rapid reclamation merely by improving water use in the catchment and fencing out the incipient salt area to prevent overgrazing by stock. In advanced stages however reclamation is beset with all the problems normally associated with soils of high salinity.

It is now becoming clearer that adequate water use in catchments at least in some of the wetter areas, cannot be properly achieved by improved annual pastures alone, and to restore a hydrologic balance either perennial pastures or some trees will be needed if not on the catchment, at least as buffer areas just above the seepage areas.

For the correction of both tunnelling and salting the establishment of better pasture cover is essential and until recently this has not been easy. ANDERSON & MOYE (1952) showed that soils in New South Wales similar to these were deficient in molybdenum and that they were too acid for the quick multiplication of *Rhizobium* spp. to enable proper nodulation of subterranean clover. Later this was confirmed for these soils by NEWMAN (1955) and pastures having subterranean clover as the legume constituent can now be readily established using lime-pelleted seed and molybdated superphosphate as fertilizer.

It is not surprising that these soils should be deficient in molybdenum in view of their mode of formation outlined by DOWNES (1954). For the same reason it is not surprising that the character of the soils should enable tunnelling, and that salting should also occur.

In fact investigations show that the whole environment is potentially unstable and requires careful handling to avoid trouble. Possibly a little less intensive grazing could have avoided all of these problems.

Nevertheless the better understanding of the total environment has enabled the solution of the problems and there has been a general increase of productivity in recent years from a level of about ½ sheep to the acre to more than 2 sheep to the acre in some places.

Because such spectacular results have been obtained, there is a tendency to forsake ecological principles in the application of this new knowledge. Thousands of acres of the steeper parts of this country are now being seeded and topdressed from the air and many consider that all the land use problems have been solved. However proper land use is a dynamic thing in which the solution of one set of problems produces others.

On such steep slopes it will be difficult to manage the pastures to prevent clover dominance which after a time leads to a build-up in the nitrogen status and an invasion of nitrogen-loving plants — mostly annual grasses and weeds. The stability and higher productivity could be short lived particularly if there were to be a run of drier than normal seasons. The replacement of native perennial grasses by annual species — pastures at first and weeds later — will not provide an adequate vegetative cover in times of drought. There is a need to find out how to establish a useful perennial grass species along with the clover so that use can be made of the improved nitrogen status to prevent weed invasion and also to provide a reasonable cover in dry seasons. Such problems must be recognized and solutions to them found if there is to be a stable environment at a high level of productivity. Ecological ingenuity will be required to solve them.

Soil and Ecological Surveys as a basis for the determination of land use

Because soil conservation is fundamentally an ecological problem of adjusting the system of land use to suit the environment, then it is essential to obtain all possible data concerning the various features of each environment. Complete studies of areas reveal that environmental units can be recognized, each of which by virtue of the integration of its particular soils, climate, vegetation and topography provides the basis for determination of suitable forms of land use which will provide for overcoming the problems and hazards which it presents. Such units can be recognized at different levels of generalization according to whether the determination of land use needs to be at the broad scale; that is whether the unit is suitable for agriculture or forestry; or at the narrow scale, for individual farm planning where the problem is the assessment of the suitability of areas for specific crops or pastures.

DOWNES (1949) recognized the existence of such units and their significance and called them units of land husbandry. These units were at an intermediate scale between the two extremes mentioned above and within each such unit a further subdivision of land classes was designed for the purpose of farm planning.

CHRISTIAN et al. (1952) recognised land systems as a broad unit in their study of large areas of land in northern Australia.

DOWNES et al. (1957) have now rationalised soil and ecological surveys on a reasonable basis which enables the recognition and mapping of units at different levels of intensity and at the same time provides a means of correlating the units and their relative significance at these different levels. Four units are recognised, the component, the land unit, the land system, and the geographic zone.

The component is the unit recognised in the most detailed investigation and is consequently the basic unit for considering detailed land use. A component is uniform with respect to its potential, problems and hazards. Agronomic investigations need to be based on components because they cannot be truly related to a wider range of conditions. The component is an area of land in which the climate, parent material, soil, vegetation and topography are uniform within the limits significant for a particular form of land use. In practice when studying land for potential development this is interpreted as being the most likely form of land use. Thus in areas where irrigation is likely, the component would accommodate soil variations of not greater than the type-phase level; whereas for country where grazing of the native vegetation is likely to be the land use, the range of variation within the component might be at a sub-formation level for structure and sub-association level for floristics (sensu BEADLE & COSTIN, 1952).

The land unit is an area in which there are a limited number of components occurring in a consistent sequence to form a characteristic pattern or landscape. It is the unit most commonly used for mapping.

The land system is comparable with that defined by CHRISTIAN et al. (1952) and consists of a combination of land units, land form constituting the major factor of grouping.

The geographic zone is a unit used for the initial subdivision of a large area of country and is an area in which similar land systems are included, the chief land forms of the systems being common.

To indicate the relative levels of generalisation, the scales of the final map are for components 40 chn = 1 inch, for land units 2 miles = 1 inch, for land systems 8 miles = 1 inch, and geographic zones 32 miles = 1 inch.

Surveys in which land units are mapped provide a compromise by which sufficient knowledge of the country is obtained for the determination of land use with a reasonable speed of survey. Information is obtained about all the components which occur within the land units and their relative proportions and location in relation to each other. Such a survey provides the preliminary information required for a detailed component survey if this is required at a later date for the purpose of subdivision and farm planning.

In conjunction with soil-ecological surveys at all levels, every effort is made to understand the origin and development of the land forms, soils, and the vegetation types; in fact a knowledge of the dynamics of the environment is an important aspect of the investigations if the problems and hazards under likely forms of land use are to be foreseen.

Agricultural ecology and farm planning for conservation

Agriculture abounds in ecological problems, weed control, and pasture establishment and management being excellent examples. But in addition to such facets of agriculture in which the ecological implications are obvious, it is important to understand that the agricultural use of land is an ecological problem.

Farm planning is the co-ordination of technological knowledge into a system of land husbandry in which each part of the farm is put to its best and proper use. Ideally, good farming represents a situation in which man and his animals have established a dynamically balanced ecological unit.

It is useless to consider individual technological advances in isolation because any change imposed on the environment results in other changes which can be either beneficial or detrimental from either the physical or economic aspects. Farm planning consists of a basic assessment of the natural resources of the farm and their potentiality. For this purpose the farm needs to be classified on the basis of components or the somewhat comparable units known to conservationists as land classes. The components can be readily characterised in terms of land classes.

From this basic information the alternative farming systems which can be safely and profitably used may be determined and having made the choice the detailed planning can begin.

DICKINSON & DOWNES (1953) have indicated the step by step procedure in farm planning for an area within the type of problem country discussed earlier in this paper. The major subdivision is based on a separation of land classes, since these areas will need different treatment and management because of the different problems and hazards which they present. Subsequent subdivision aims to provide paddocks of the most efficient and economical size to carry out the proposed system of farm management, to provide water supply and access to paddocks which are aimed to give convenience, but at the same time do not create hazards and subsequent erosion problems.

To indicate the complexity of such a total approach the factors influencing the final paddock size on a sheep grazing property can be elaborated. Initially the soil, slope, climate, and aspect, will determine the type of pasture and potential productivity. The type of pasture will determine the type of grazing management with respect to the grazing pressure to be applied in accordance with the season of the year, and the frequency with which it shall be left for a hay crop. To obtain the proper grazing pressure on the right occasions requires careful flock management procedure, and, according to flock management methods and breeding programme, so does the paddock size need to be designed. In addition to these criteria the water supply facilities need to be designed to enable the combined pasture and flock management programme to be carried out as planned.

So farm planning for conservation and proper use of an environment is a process of integration of knowledge of the behavior of plants and animals in relation to each other and their environment, truly a problem of applied ecology.

The whole basis of soil conservation, erosion control and reclamation is ecological. For soil conservation there is a need to assess the nature and dynamics of environments so that systems of land use can be devised to give the desired production on a permanent basis. Erosion control is basically an understanding of the reasons for the upset equilibrium and the mechanics of erosion processes to devise methods for re-establishing a balance. Reclamation is a matter of assessing the nature of the modified environment so that suitable primary and secondary colonizers can be found to occupy the available niches and so begin a succession to a new equilibrium.

REFERENCES

ANDERSON, A. J. & MOYE, D. V., 1952. Lime & Molybdenum in Clover Development on Acid Soils. *Aust. J. agric. Res.* **3,** *95.*

ANDERSON, V. G., 1941. The Origin of Dissolved Inorganic Solids in Natural Waters in Victoria. *J. Aust. chem. Inst.* **8,** *130.*

ANDERSON, V. G., 1945. Some effects of atmospheric evaporation and transpiration on the composition of Natural Waters in Aust. *J. Aust. chem. Inst.* **12,** *41.*

BEADLE, N. C. W., 1948. The Vegetation & Pastures of Western N.S.W. with special reference to Soil Erosion. N.S.W. Govt. Printer 281, pps.

BEADLE, N. C. W. & COSTIN, A. B., 1952. Ecological Classification & Nomenclature, *Proc. Linn. Soc. N.S.W.* **61,** *77.*

CHRISTIAN, C. S. & STEWART, G. A., 1953. Survey of the Katherine-Darwin Region. Land Research Series No. 1, C.S.I.R.O., Australia.

COPE, F., 1955. Catchment Salting in Victoria. Master's Thesis, Univ. of Melbourne.

COSTIN, A. B., 1954. The Ecosystems of the Monaro Region of N.S.W., N.S.W. Govt. Printer, 860 pps.

COSTIN, A. B., 1957. High Mountain Catchments in Victoria. Soil Conservation Authority of Vic.

DICKINSON, H. R. & DOWNES, R. G., 1953. The Westgate Farm Planning Project. Soil Conservation Authority of Vic.

DOWNES, R. G., 1946. Tunnelling Erosion in North Eastern Victoria. *J. C.S.I.R.O. Aust.* **19,** *283.*

DOWNES, R. G., 1949. Soil Land Use & Erosion Survey around Dookie, Vict. *C.S.I.R.O. Aust. Bull.* No. 243.

DOWNES, R. G., 1954. Cyclic Salt as a factor in the Genesis of Soils in S.E. Aust. *Aust. J. agric. Res.* **5,** *448.*

DOWNES, R. G., 1956. Conservation Problems on Solodic Soils in Victoria. *J. Soil & Water Cons.* **11,** *228.*

DOWNES, R. G., GIBBONS, F. R., ROWAN, J. M. & SIBLEY, G. T., 1957. Principles & Methods of Ecological Surveys for Land Use Purposes. Second Aust. Conf. in Soil Science, Melb., 1957.

HOLMES, L. C., LEEPER, G. W. & NICOLLS, K. D., 1939. Soil & Land Use Survey of the country around Berwick. *Proc. Roy. Soc. Vic. N.S.* **52,** *177.*

LESLIE, T. I. & HUTTON, J. T. 1958. Accession of Non Nitrogenous Ions in Rainwater in Victoria. *Aust. J. agric. Res.* **9,** *492.*

NEWMAN, R. J., 1955. Molybdenum deficiency in Central Highlands & Upper Goulburn Regions, Vict. *Vic. J. Dept. Agri.* **53,** *451.*

RATCLIFFE, F. N., 1936. Soil Drift in the Arid Pastoral Areas of Sth. Aust. *C.S.I.R.O. Aust., Pamph.* No. 64.

RATCLIFFE, F. N., 1937. Further Observations on Soil Erosion & Sand Drift with Special Reference to S.W. Queensland. *C.S.I.R.O. Aust. Pamph.* No. 70.

XXX
THE MERINO SHEEP IN AUSTRALIA

by

IAN W. MCDONALD

(C.S.I.R.O. Sheep Biology Laboratory, Prospect, N.S.W.)

The growth of the sheep population, and especially the Merino sheep, in Australia has been a remarkable phenomenon. Although a great deal of research on the Merino sheep has been and is being conducted, there has been no systematic study of the many ecological factors that have influenced this growth. This brief essay makes no pretence at presenting a detailed analysis, but rather aims to mention the salient features. Some of the statements made are no more than opinions based on accumulated knowledge of the pastoral industry, and much detailed investigation will be needed to establish the facts. Further, it is extraordinary that no detailed study of the history of the sheep in Australia has been published — the large gaps in Fig. 2 are eloquent evidence of the lack of early data. Finally, some of the comments are extrapolations from the wealth of information accruing from the active research teams throughout Australia. For these reasons it has seemed best to restrict this paper to generalized statements and not to attempt a documentation. The author is much indebted to his colleagues for their critical readings of the manuscript.

Wool Production

Wool proved to be a singularly valuable agricultural product for the Australian colonies. As a substance, it was stable, easily packed, stored and transported; it was a major raw material for the industry of the colonial power — Britain. It could be produced with little labour and little capital investment; these were important, as the aboriginal population was small and little effort was expended in adapting the aborigines to the way of life of European colonists; and land was cheap during the developmental phase. No other farm product had such unique economic advantages, and this situation encouraged the early pastoralists to attempt the improvement of of the Merino as a wool producer.

In Fig. 1 the gradual increase in wool production per sheep since 1860 is shown; the curve is derived from official figures (Commonwealth of Australia, Yearbook) for total sheep numbers and total wool production; since Merinos comprise the major fraction of the sheep population, the changes shown are essentially those for the

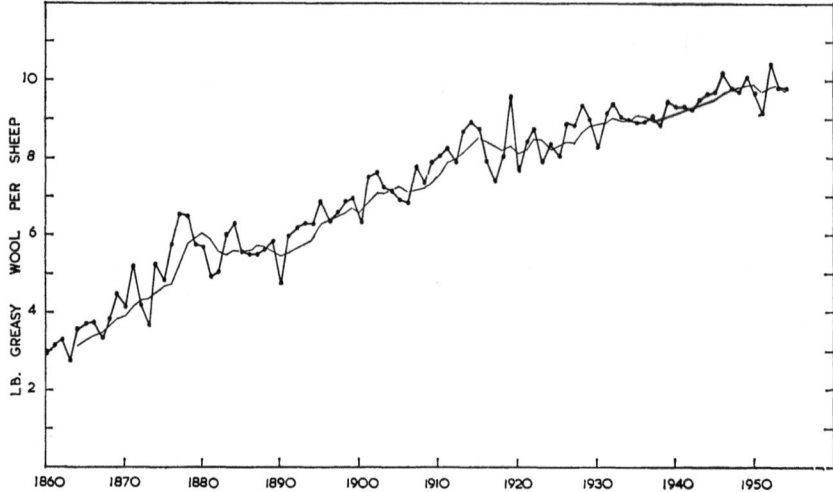

Fig. 1. Average greasy fleece weight for Australian sheep (all breeds).
The continuous line plots the progressive mean for 5-year periods. Values calculated from data for total sheep numbers and total wool produced, published in "Statistical Handbook of the Sheep and Wool Industry," Bureau of Agricultural Economics, Canberra, 1956.

Merino, at least until recent years. Many factors have contributed to the steady increase of productivity, and although early records are meagre, the following seem to have been the most important features. Perhaps the most decisive influence was the fact that the Merino was used solely for wool production, and hence the sheep were selected on this basis; conformation for mutton production, and early maturity were ignored, and it is possible that level of fecundity has also suffered from neglect. Further, very large stud flocks were established in the Riverina and in South Australia (AUSTIN, 1943); from these, new studs were developed and it became the practice for the masters of commercial flocks to purchase their rams from the studs. Genetic improvements effected in the studs were therefore soon transferred to the commercial flocks. The selection of sheep within the studs was usually entrusted to specialist sheep-classers, who exercised great influence on the large commercial flocks as well. In selecting sheep, special emphasis was placed on length of staple, fleece weight, and on body size: three factors now known to be positively correlated with each other. Recent breeding experiments have shown that the heritability of fleece weight is high, and of staple length and body weight moderately high; this, coupled with the fact that consistency of wool production is high throughout the sheep's lifetime, made visual appraisal a satisfactory basis for selecting breeding stock. In retrospect, it seems that the increase in wool production could have been much faster, but in general wool quality has taken precedence over fleece weight in selection of stud sheep (MOULE & MILLER, 1956). It is probable that during the last 2 or 3 decades there has been little genetic improvement in the Merino, and that the improved productivity is more a reflection of improvements in pasture management, disease control and sheep husbandry.

Growth of the Sheep Population

The data for numbers of sheep in Australia, shown on a logarithmic scale in Fig. 2, provide a simple growth curve which resembles, in broad outline, the growth curve of a typical bacterial culture. The existence of extensive natural pastures, suitable for sheep, made possible great increases in number giving a "logarithmic phase". Then, from the year 1891 to the present, the curve shows a general tendency toward a "stationary phase," in which there are violent fluctuations in numbers, as seen in Fig. 3, in which the numbers are plotted on an arithmetic scale. The reason for this fluctuation is easily found. Until the peak of population was achieved, it was

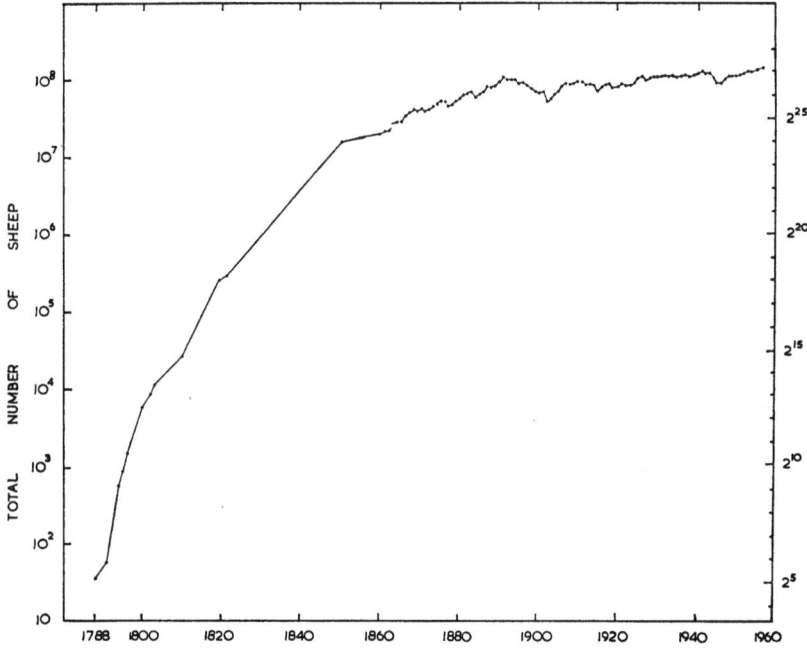

Fig. 2. Growth curve of the Australian sheep population. (Data from the Commonwealth Year Book, 1930 and 1956).

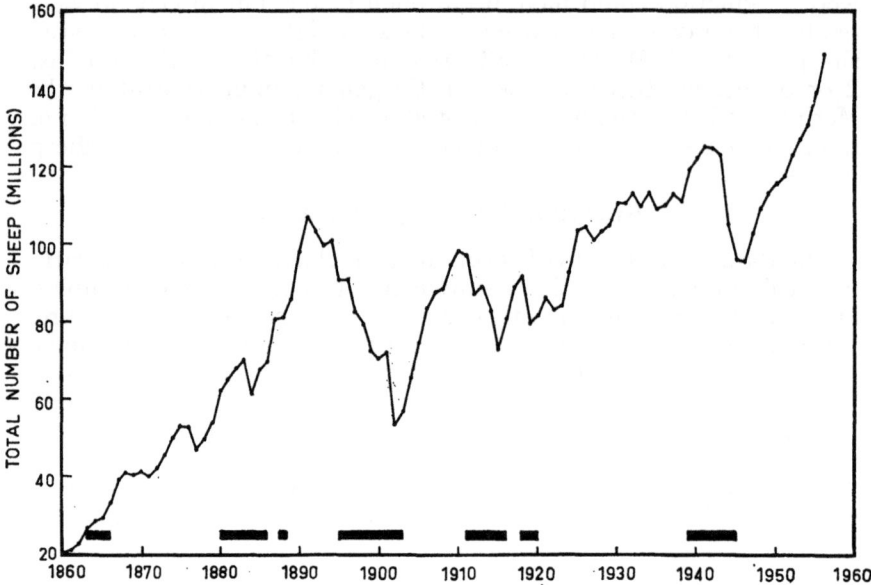

Fig. 3. Sheep numbers in Australia from 1860, showing the influence of drought periods blocks. Note that the correspondence is not exact; no precise criteria for defining a drought have yet been devised and, of course, other factors are also involved, e.g. population density, supplies of drinking water and rabbit infestation.

usually possible for flocks to be moved to new grazing lands when dry seasons provided inadequate natural pasture; when, however, the peak was reached, no grazing lands were available for agistment and heavy mortality occurred with each drought.

The first importations of sheep, in 1788, were to the area near Sydney, where a warm moist climate with low quality natural pasture combined to provide inferior conditions for sheep husbandry. When new pastures were discovered west of the Great Dividing Range, the drier climate and more nutritious pastures proved eminently suitable, and the sheep thrived and multiplied. The high quality wool produced soon attracted attention on the English markets and the stage was set for great advance in the sheep industry. The major factors, favourable and unfavourable, influencing the trend, will be briefly discussed.

Merino types

Consistent selection of Merinos for wool quality and quantity has led to the establishment of three major types, fine-, medium- and strong-wool, which have definite geographical distributions (Commonwealth of Australia, Serv. Publ.). The fine-wool sheep are small, and grow a short staple of fine (spinning counts of 70's or higher), highly crimped wool; they occur chiefly in Tasmania and the tablelands of N.S.W. and Victoria. The strong-wool sheep are large and grow a long staple of coarser wool (spinning counts of 58's and 60's) with fewer crimps; they comprise some 35% of all Merino sheep and are the chief Merinos of Western Australia, South Australia, western N.S.W. and south-western Queensland — this is especially the lower rainfall zone of the sheep lands. The largest group is the medium-wool type in which the fleece shows characters intermediate between the other two types; it is the chief type in Victoria, N.S.W. and Queensland. In the higher rainfall areas, the fine-wool sheep have proved more resistant to fleece-rot; apart from this feature, there appears to be no strong ecological ground for the regional distribution of the types, and it is therefore probable that human preferences have been chiefly responsible.

Diseases

Perhaps the most striking single factor in the ecological picture is the absence of disease transmissible from native fauna to the sheep. Not a single transmissible disease of any consequence, peculiar to the Australian continent, was encountered by the introduced European sheep. The profound influence of this fact can best be appreciated by contrasting the history of sheep in Australia with that of European sheep and cattle introduced into the southern and central regions of Africa. There is little reason to doubt that this remarkable situation was due to the complete absence from the Australian continent of native ruminants of the Order Artiodactyla. (It may be noted, in passing, that some of the Australian marsupials show striking analogies to the familiar ruminants in feeding habits, stomach structure and microbial digestion in the stomach; it is probable that the term "ruminant" could be properly applied to these species.)

All the transmissible diseases experienced by Australian sheep were in fact brought from Europe; the long sea voyage required by the sailing ships of the early 19th century no doubt helped to prevent the introduction of infectious diseases by imposing a "quarantine." However, at an early stage, diseases seriously threatened the sheep industry — especially important were those due to the bacterium *Bacillus anthracis*, and the arthropod ectoparasite, *Sarcoptes scabei*. The occurrence and severity of these diseases prompted strenuous remedial measures which not only proved effective but also laid the foundation for official disease control and research organisations which have played a decisive role in protecting the health of the flocks. Today, the major transmissible diseases are those due to infestation with internal parasites; these diseases present important ecological features as their severity is largely determined by the local climate, the quality and quantity of pasture available, and the density of the sheep population. The great economic importance of parasitic diseases has led to active research in most states of the Commonwealth.

Predators

The only important indigenous carnivorous predator of the sheep is the dingo (*Canis dingo*, MEYER); the number of these dogs was not sufficient to cause serious difficulty in the early development of the sheep flocks but is proving of economic importance in some of the sparsely settled pastoral areas of Queensland and Western Australia. The European fox (*Vulpes vulpes*, L.) was introduced into Victoria in 1868 and is now found in all States; it has occasionally been a pest, especially in preying on lambs. Although heavy losses may be experienced in individual flocks, these two carnivores have had little influence on sheep husbandry generally. The Australian crow, or raven (*Corvus coronoides*, VIGORS & HORSFIELD) is on occasion troublesome in attacking and killing newborn lambs or sick sheep; the bird possesses the unpleasant skill of destroying the resistance of its victim by pecking at the eyes. The Wedgetailed eagle (*Aquila audax*, LATHAM), has also been known to prey on the sheep. In Tasmania, a carnivorous marsupial (*Thylacinus cynocephalus*, HARRIS) attacked and killed sheep, but its predation was only of importance in the early years of the colony.

Feral sheep

It is most striking that feral sheep have never developed in Australia. All the other introduced domestic animals — horse, donkey, buffalo, cattle, camel, pig, goat, dog and cat have on occasion reverted to feral types, and some have even reached numbers which constituted a nuisance to farmers. It seems probable, though there is no proof, that the inability of the modern sheep to shed its coat is largely responsible for this situation. Sheep which have escaped muster, and hence remained unshorn, for as long as three years have been reported; it seems likely that such sheep with fleeces trailing on the ground would be ready prey to predators, would be at a disadvantage in seeking feed and water, and may have difficulty in breeding. The sheep, in Australia, seems to be uniquely dependent on man for its survival, though feral sheep of Merino origin seem to have developed in mountainous areas in New Zealand.

The Sheep Dog

The gregarious habits of sheep and their docile temperament make them very easy to handle, and man has been greatly assisted in this work by the sheep dog. Of the several breeds of sheep-dog in Europe, only the Border-Collie has established itself in Australia; this breed was probably the progenitor of the Kelpie which is now a well defined and distinct breed. In the absence of abundant cheap labour in Australia, it seems certain that the rapid growth of the sheep population could not have occurred without the extremely witting "labour" provided by these very pleasant mannered dogs.

The Climate

The distribution of sheep, shown in Fig. 4, is closely related to climate (C.S.I.R.O., 1949). The greatest density of sheep population is found in the 20"—30" rainfall belt in New South Wales and Victoria; in this region there has been a tendency for the Merino sheep to be replaced by British breeds and their crosses with Merino, but large quantities of fine and superfine Merino wool are still produced. The largest area occupied by the Merino lies in the

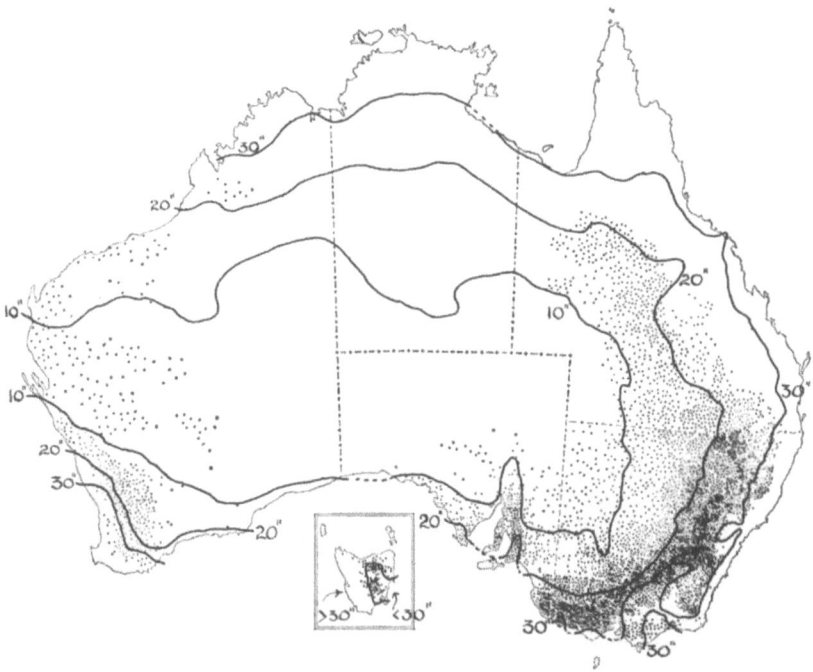

Fig. 4. Map showing the 10", 20" and 30" isohyets and the distribution of the total sheep population in Australia; each circle represents 20.000 sheep. The greatest density is found in the 20"-30" rainfall belt in south-eastern Australia. Few sheep are to be found in the arid zone with less than 10" rainfall, or in the moist tropical and subtropical areas. (Adapted from Chart II in "Statistical Handbook of the Sheep and Wool Industry," Bureau of Agricultural Economics, Canberra, 1956).

10"—20" rainfall belt extending from Central Queensland through New South Wales, Victoria and South Australia, and two isolated zones occur in Western Australia; these regions show two distinct climates: the tropical and subtropical climates in Queensland, with summer rainfall and dry warm winters; and the Mediterranean climate of the Southern states with winter rainfall and hot dry summers. Both zones are also represented in Western Australia. It is noteworthy that the majority of the fine-wool Merinos are to be found in the highest rainfall areas on the tablelands of New South Wales and Victoria, and in Tasmania.

The major factors which have led to confinement of sheep within these limits seem to be as follows. In most of the country with rainfall less than 10", pastures are too sparse and coarse, and the distances between watering places too great to enable sheep to thrive. In the high rainfall areas in the southern part of the continent, the Merino sheep tend to be replaced by the more intensive forms of agricultural production; also in these regions, the wool tends to be damaged by fleece-rot (caused by bacterial growth in the fleece) and the sheep are afflicted by internal parasites, foot-rot and fly-strike. In the warm, high rainfall zones of the eastern coastal belt the Merino is virtually absent; the agriculture comprises chiefly dairying, crop cultivation and beef cattle raising. The northern limits in the tropical zone of Queensland, Northern Territory and Western Australia seem to set by the capacity of the sheep to withstand heat, the distances between watering places, the coarseness of the summer growing grasses and the prevalence of the dingo; beef cattle dominate this region. Throughout the lower rainfall areas, the provision of drinking water for sheep has proved almost as important as the provision of feed. Indeed, during droughts, lack of water has often been the primary cause of stock losses.

Heat tolerance

The improved breeds of sheep grow wool as a continuous fibre which is not shed (F.A.O. conference). In cattle and most other mammals hair-growth occurs in cycles; at certain intervals, the growth ceases, the hair is separated from the base of the follicle and in due course is shed, while its place is taken by a newly developed fibre. The familiar summer and winter coats of most domestic animals can thus play an important role in assisting the animal to withstand extremes of heat and cold. However, in sheep on a constant diet, wool growth is in fact more rapid in summer than in winter (FERGUSON et al., 1949): the physiological basis for this seasonal effect is still not clear, and it is possible that both the higher temperatures and the longer day length of summer may contribute stimuli to increased wool growth. The thick wool covering carried by sheep during the summer reduces greatly the cooling effect of evaporation from the skin; it is questionable whether the so-called sudoriferous glands in the sheep's skin produce sufficient sweat to be of any significance in temperature regulation. It would therefore appear that the sheep relies largely on evaporation from the respiratory tract, aided by panting, for its heat loss in high ambient temperatures (MACFARLANE et al., 1956). The thick fleece does, however, serve as a very effective insulation against heat gain from solar radiation. The physiological aspects of heat regulation and tolerance of sheep are currently the subject of active research. It is however clear that the Merino sheep has a very high degree of tolerance to heat and that this fact has contributed in large measure to its capacity to thrive in the arid and semi-arid regions of Australia (MOULE, 1956).

Cold stress

The normal adult sheep is well adapted to withstand extremes of cold. However, for economic reasons, it is common for sheep to be shorn either in winter or spring, and if the newly shorn sheep are exposed to wet, cold, windy weather, very high mortality can occur. Further, the newly born lamb is susceptible to cold stress under similar conditions, and this proves to be a major cause of neo-natal mortality in many parts of the country.

Breeding season

The British breeds of sheep (except the Dorset Horn) show well defined breeding seasons, and mating occurs in the autumn. In contrast, the Merino has no well defined breeding season and this has made it possible for the sheep to be mated at the time most suitable for the locality. However, the factors governing fertility and prolificacy in the Merino are not well understood and it is by no means certain that the sheep have in fact been mated at the most favourable season of the year. There is some evidence that the high temperatures in tropical Australia have reduced fertility; of special interest are the reduced libido in rams, thought to be associated with day length, and the reduction of fertility due to rise in testicular temperature. The influences of temperature, day length, nutrition (and its association with rainfall), disease and genetic constitution on reproduction form a complex which is currently receiving much study.

Nutrition

A chapter of this book has been devoted to the ecology of sheep pastures in Australia. Here it is desired to comment on only a few features which have assisted the Merinos to be effective wool producers in environments which, on casual inspection, might appear to be unfavourable.

The type of fleece grown by a sheep is essentially determined by its genetic constitution; the variations imposed by environmental conditions are limited. The weight of clean wool produced is a function of diameter, length and number of fibres, while the quality, in terms of spinning count, is largely a reflection of the number of crimps per unit length of fibre. The various characters of the fleece are not equally affected by changes in nutrition; fibre diameter is most affected and fibre length a little less so; crimp and number of fibres are but little affected. The response in wool characters to change of nutritive level is slow compared with the response of growth in growing pigs, cattle or lambs, or of milk production in the dairy cow. The physiological basis for this delay in response is not entirely clear; however, it is known that activity of the anterior pituitary gland is essential to the maintenance of wool growth (FERGUSON, 1954), and that growth can be stimulated by administration of pituitary extract; it is also known that the adrenal corticosteroids inhibit wool growth (LINDNER & FERGUSON, 1956). It may therefore be postulated that there exists a mechanism by which changes in feed intake influence the activity of these "pacemaker" glands. Wool growth may be considered as the synthesis and "excretion" of protein by the specialised cells of the wool follicle; it may be anticipated that this synthesis will participate in the general reactions of protein metabolism in the animal's body, and probably differs fundamentally from other syntheses of protein only because it is completely irreversible — i.e. it cannot contribute to the normal catabolic reactions. The net effect then is that during periods of plethora, when feed intake of protein far exceeds the animal's need, protein can be stored in the various tissues; later, during an unfavourable season when protein intake falls, body tissues are catabolized and thus provide amino-acids for wool synthesis at a rate which is in excess of the equilibrium rate at the lower intake level. The sheep can therefore be a useful wool producer on diets which will not permit body growth and indeed can continue to grow wool while body weight is declining. Adequate data are not available to provide accurate figures for the time required to reach equilibrium after a change in feed intake, but clearly this time may be weeks or even months.

Fluctuations in diet have less effect on quality than on quantity of wool grown. Two factors are important: firstly, crimp is relatively little changed, and this is the major determinant of quality number (or count) and of the price paid by manufacturers for fleece wools. Secondly, with change of feed intake, length and cross-sectional area of fibre change in about the same proportion, thus tending to reduce the effect of diet on fleece quality.

Another feature to be stressed is the capacity of the sheep for selective grazing. Essentially, the sheep industry has developed on native pastures, which were, of course, highly adapted to the environment and, prior to the advent of European settlers, had presumably been only lightly grazed by the indigenous marsupials: these animals and the herbivorous emu (*Dromaius novae-hollandiae*, LATHAM) have never been serious competitors for the sheep, except in a few sparsely settled areas. (The Australian aborigines had no opportunity to develop herds or flocks as the continent contained no ruminants or ungulates, which provided the species for domestication in all other continents). The most important pasture plants were various species of the genus *Danthonia*, and comprised hardy, slow growing, drought resistant grasses of tuft-forming habit; they remained green for a much longer period than the rapidly growing annual species and thus provided the sheep with nutritious fodder. Provided the sheep density was kept low, their highly discriminating feeding habits enabled them to select a diet very much richer in protein than analysis of the whole pasture plants would indicate. In addition, the Merino sheep are capable of walking long distances in search of feed and water, and hence they proved able to make good body growth and wool production on apparently poor pastures. During the major part of the development of the pastoral industry, it is probable that sheep densities rarely exceeded one sheep per acre, and in many areas was very much less.

The *Danthonia* grasses and the very drought resistant salt-bushes proved unable to survive excessive stocking rates; the rabbit became a competitor with the sheep for the available fodder. By 1891 sheep numbers had risen to 107 millions and this, combined with inadequate supplies of drinking water (a perennial hazard to the sheep-industry) and a heavy rabbit population, represented overstocking which resulted in heavy losses during the following years; the stage was set for catastrophic losses in the ensuing drought of 1895—1903. By 1902, the sheep numbers had fallen to half the 1891 figures. Since that time, the sheep population has returned to new record numbers, and the modern developments in pasture improvement, rabbit control and fodder conservation have made the sheep industry less vulnerable to the vagaries of climate.

Soil erosion

In areas which have been subjected to overstocking, the sheep has contributed in large measure to soil erosion; this subject has been dealt with in another chapter.

Mineral Deficiencies in Soil

Deficiencies of phosphorus, cobalt and copper have proved critical in Australia. Deficiency of phosphorus is almost the rule. This leads to a low level of phosphate in the pasture plants, a level which is often inadequate for cattle and results in unthriftiness, low fertility, pathological changes in bone, and, indirectly, to botulism; these syndromes have been recognized and thoroughly studied in cattle in many parts of the world. It is therefore of great interest that specific phosphorus deficiency has never been recognized in sheep in Australia (MARTIN & PEIRCE, 1934). The difference in response between these closely related ruminants is probably due to several factors, of which the following are most important: the smaller size of the sheep results in a higher feed intake (hence higher total-P intake) per unit of body weight, while it has also the opposite effect of requiring a smaller skeletal weight (and hence lower need for P) per unit of body weight; also, the more selective grazing habits of the sheep enable it to consume a ration higher in P-content. The real importance of P-deficiency in the soils is the limitation placed on plant growth; in many parts of Australia, it has been possible to double and even treble the sheep numbers by the simple procedure of broadcasting phosphatic fertilizer on the pasture.

Cobalt deficiency (MARSTON, 1952), which was first discovered in Australia, is important in sheep, and, to a lesser degree, in cattle; the disease has not been found to occur naturally in any other species. In some areas, this deficiency disease is so severe that it is impossible to maintain sheep; in less severely deficient areas, unthriftiness and occasional heavy mortality is experienced. The disease is characterized by a slow, progressive wasting which has been aptly described as "starving in the midst of plenty". The disease can be easily prevented by administration of cobalt to the sheep or by the use of cobalt-containing fertilizers.

Copper deficiency (MARSTON, 1952), which often accompanies cobalt deficiency, is more widespread and is essentially a disease of sheep — cattle grazing the same pastures are much less susceptible. There are two major manifestations of the disease. Lambs born to copper deficient ewes have degenerative changes in the brain and spinal cord, and these may lead to high mortality. In the adult sheep, copper deficiency leads to a specific abnormality of the fleece; the wool fibres lose their characteristic, crimped structure and become straight, lustrous and abnormal in their physical properties. Again, the deficiency in soil or animal can easily be rectified.

Noxious plants

An interesting ecological feature is the occurrence of plants harmful to the sheep or to its wool. Most of the plants involved are indigenous, but many are introduced species. By far the most important, economically, are the various "burrs" or seed pods, which become attached to the wool and must be removed by the process of carbonizing. Other seeds, for example of the spear grasses, will even penetrate the skin and muscles of sheep. A wide variety of plants, including many genera, are toxic when consumed by sheep; the toxic principles fall into six main groups: cyanogenetic substances, oxalates, hepatotoxins, oestrogens and substances which induce photosensitization or neuropathological changes. Of particular interest in the last named category is the case of *Phalaris tuberosa;* this valuable perennial grass may, in certain localities and certain seasons, prove highly toxic, but the nature of the toxin and the circumstances leading to its occurrence are quite unknown; also, for reasons unknown, the toxic effects can be prevented by prophylactic administration of cobalt to the sheep (LEE & KUCHEL, 1953).

REFERENCES

COMMONWEALTH OF AUSTRALIA. YEARBOOK No. 42, 1956.
AUSTIN, H. B., 1943. The Merino — Past, Present and Probable. Grahame Book Co., Sydney.
MOULE, G. R. & MILLER, S. J., 1956. *Empire J. exp. Agric.* **24**, *37*.
COMMONWEALTH OF AUSTRALIA, Department of Health. Service Publications No's 5—10, Division of Veterinary Hygiene, 1950.
COMMONWEALTH SCIENTIFIC AND INDUSTRIAL RESEARCH ORGANIZATION, AUSTRALIA 1949. The Australian Environment. C.S.I.R.O., Melbourne.
F.A.O. CONFERENCE on Livestock Production under Tropical Conditions: The Livestock Industries & Agriculture in Northern Australia, 1955. Watson, Ferguson and Co., Brisbane.
FERGUSON, K. A., CARTER, H. B. & HARDY, M. H., 1949. *Aust. J. sci. Res.* B, **2,** *42*.
MACFARLANE, W. J., MORRIS, R. J. & HOWARD, B., 1956. *Nature, Lond.* **178**, *304*.
MOULE, G. R., 1956. *Aust. vet. J.* **32**, *289*.
FERGUSON, K. A., 1954. *Nature, Lond.* **174**, *411*.
LINDNER, H. R. & FERGUSON, K. A., 1956. *Nature, Lond.* **177,** *188*.
MARTIN, C. J. & PEIRCE, A. W., 1934. Coun. Sci. Ind. Res., Australia, Bull. **77**.
MARSTON, H. R., 1952. *Physiol. Rev.* **32, 66**.
LEE, H. J. & KUCHEL, R. E., 1953. *Aust. J. agric. Res.* **4,** *88*.

XXXI
ECOLOGICAL OBSERVATIONS ON PLANT COMMUNITIES GRAZED BY SHEEP IN AUSTRALIA

by

R. M. MOORE

(Division of Plant Industry, C.S.I.R.O., Canberra)

Introduction

Before European settlement plant communities in Australia were in equilibrium with the climates and soils and were adapted to the relatively light and intermittent grazing of indigenous marsupials as well as to recurrent burning. The introduction of domestic livestock and rabbits by British settlers; the felling of timber for the purpose of increasing grass growth; the consequent defoliation, trampling and manuring by confined animals profoundly altered the environments of these communities. Concurrently with the introduction of livestock, seeds of alien plants entered the country either by design — as agricultural and horticultural seeds — or accidentally as impurities in crop and pasture seeds or attached to the hides of animals. These two factors, the disturbance of the equilibrium of our native vegetation, and the introduction of alien plants pre-adapted to such disturbed areas with their modified environments, have effected marked changes in Australian plant communities. These changes occurred rapidly and over a wide area of southern Australia as a result of the sharp increase in sheep numbers following the settlement of the pastoral country. By 1880 the whole of New South Wales was settled and by 1890 sheep numbers in Australia had reached the 100 million mark.

The Major Plant Communities Grazed by Sheep

The principal plant formations grazed by sheep in Australia are temperate woodland, grassland, mulga scrub, spinifex and shrub steppe.

The Woodland Formation

This is the most important of all vegetation formations for sheep raising and most of Australia's sheep are carried on native pastures derived from woodland communities. New South Wales, which has the greatest area of savannah woodland — the sub-formation with the highest rainfall of all woodlands — had 47% of Australia's sheep in 1954.

Savannah woodlands extend from north-eastern New South Wales across the State in a south-westerly direction and cover large areas of the Tablelands, Western Slopes, and Riverina between the 25 and 15 inch rainfall isohyets. North-central and south-western Victoria, the Mount Lofty Ranges and south-eastern South Australia, eastern Tasmania and part of southern Western Australia also have large areas of savannah woodland. Hills and ridges throughout this sub-formation are covered generally with dry sclerophyll forests which until the comparatively recent discovery of a molybdenum deficiency in many of their sedimentary soils (ANDERSON & MOYE, 1952) were considered of little or no value of grazing.

Savannah woodlands are composed of several species of Eucalyptus depending on soil and climate. The principal communities are *E. blakleyi* (Red gum), *E. melliodora* (Yellow box), *E. albans* (White box), *E. pauciflora* (Snow gum) — *E. stellulata* (Black Sally), *E. camaldulensis* (River red gum), in eastern Australia; *E. leucoxylon* (Blue gum) and *E. odorata* (Peppermint) in South Australia, *E. loxophleba* (York gum) — *E. wandoo* (Wandoo) and *E. salicifolia* — *E. pauciflora* — *E. viminalis* (Ribbon gum) in Tasmania.

Individual trees of all these species are relatively short-boled and were originally so spaced as to present a park-like appearance. The ground between the trees was covered with grass and small herbaceous shrubs; the principal species throughout southern Australia being *Themeda australis* (Kangaroo grass). To increase the growth of grass the early settlers ring-barked a large proportion of the trees, often killing every one on extensive areas.

Kangaroo grass has largely disappeared under sheep grazing and is now found only in ungrazed areas, railway enclosures, cemeteries etc. This has been the most striking effect of the utilization of the woodlands and some of the other vegetation formations for sheep grazing and so complete has been the disappearance of this grass that even botanists until recently believed the Canberra grassland community, for example, to be a *Stipa-Danthonia* climax (PRYOR, 1939). Historical evidence for the predominance of Kangaroo grass in the N.S.W. savannah woodlands is given by Surveyor-General MITCHELL (1839). In volume 1, page 169 of his "Three Expeditions into Eastern Australia" MITCHELL states that after passing from the mountains to the valleys north-west of the Canoblas Mountains he found yellow oat grass (Anthistiria*) "resembling a ripe crop of grain", and in volume 2 page 168 he remarks that at Soddon 36° 36'S and 143° 30'E, "Anthistiria seen in greater abundance than anywhere before in Australia".

JAMES BACKHOUSE who visited the Adelaide plains in 1837

---

* *Anthistiria australis* was the name given by ROBERT BROWN to Kangaroo grass. See Prodromus Florae Novae Hollandiae by ROBERT BROWN, 1810.

reported "Some of the Kangaroo grass was up to our elbows and resembled two years seed meadows in England in thickness; in many places three tons of hay per acre might be mown off it". E.P.S. STURT, a Commissioner of Crown Lands for the Murray district, Victoria, writing to Lieutenant-Governor LATROBE in 1857, refers to "the long Kangaroo grass which waves to the very flaps of the saddle" (BRIDE, 1898). Observations of a number of botanists have confirmed the dominance of Kangaroo grass on ungrazed areas within savannah woodlands throughout Australia.

The rapidity with which the introduction of sheep changed the character of the herbaceous vegetation is evidenced in a letter from JOHN G. ROBERTSON of "Wandoo Vale" Wannon county, Victoria, to Lieutenant-Governor LATROBE on the 26th September, 1853. When ROBERTSON took up his run in 1852 "nothing had trodden the grass before them" — referring to his own sheep. The relatively few sheep he had made little impression on the country for 3 to 4 years but then "many of our herbaceous plants began to disappear from the pasture land, the silk grass began to show itself on the edge of the bush track and in patches here and there on the hill. The patches have grown larger every year, herbaceous plants and grasses give way to the silk grass and the little annuals beneath — which are annual peas and die in our deep clay soil with a few hot days in the spring and nothing returns to supply their place until later in the winter following" — "the long deep rooted grasses that held our strong clay hill together have died out, — the clay hills are slipping in all directions". The basaltic soils of the Wannon carried predominantly Kangaroo grass and it is almost certain that this is the species which disappeared within 3 or 4 years. The silk grass to which ROBERTSON referred was very likely Silver Grass *(Vulpia bromoides)* an introduced grass, now widespread over the whole of the Western Districts of Victoria and the annual peas were probably species of *Trifolium*. ROBERTSON, like many of the original Western District settlers, came from Tasmania and in his letter he mentions how delighted he was on taking up his run that "there was no silk grass* which had been destroying our V.D.L.** pastures where I had watched its progress with uneasiness." It is noteworthy that ROBERTSON refers to experiments with mixtures of English grasses and it is possible that the local strain of Perennial Rye grass *(Lolium perenne)* which with Subterranean clover and Yorkshire Fog predominates on the basalt soils of the Western Districts of Victoria when superphosphate is applied, originated from this or similar introductions.

The initial effect of sheep grazing on the savannah woodlands is

---

* *Vulpia bromoides* is still referred to as Silk grass in parts of Tasmania.
** V.D.L. Van Diemen's Land, former name for Tasmania.

the replacement of the tall summer-growing perennial species. Kangaroo grass *(Themeda australis)*, Plains grass *(Stipa aristiglumis)* and Tussocky Poa *(Poa caespitosa)* by other native but shorter species, notably the winter-growing Wallaby grass *(Danthonia* spp.) and Spear grass *(Stipa falcata)*. The next lower stage in degradation is marked by the disappearance of Spear grass and the dominance of the dwarf species, *Danthonia carphoides*. This stage is marked by the appearance of native summer-growing species from the interior, notably *Tripogon loliiformis, Pappophorum nigricans, Panicum effusum, Eragrostis brownii* and *Chloris truncata*.

Native perennial composites also increase in density as a result of grazing — the principal species being *Vittadinia triloba, Helichrysum apiculatum* and *Cymbonotus lawsonianus*.

Leguminous species other than trees and shrubs were rare in the original communities — the commonest being *Glycine clandestina* and *Desmodium varians* in eastern and *Oxylobium* spp., *Gastrolobium* spp. and *Kennedya* spp. in western Australia. However the application of superphosphate to the disclimax *Danthonia-Stipa* grassland promotes the growth of introduced annual legumes, principally *Medicago denticulata* on red-brown earths and *Trifolium glomeratum* and other species of *Trifolium* on podsolic soils. The increase in soil nitrogen resulting from the growth of legumes in turn leads to the invasion of other introduced species such as *Bromus* spp., Silver grass *(Vulpia bromoides)* and Barley grass *(Hordeum leporinum)*. The surface sowing of Subterranean clover so widely practised in the last two decades on the podsolic and basaltic soils of the savannah woodland has further promoted the growth of exotic annual species many of which are classed as weeds e.g. Capeweed *(Cryptostemma calendula)*, Saffron thistle *(Carthamus lanatus)*, spear thistle *(Cirsium vulgare)* and species of *Erodium*.

In the higher rainfall parts of the savannah woodland zone native pastures are being deliberately replaced by introduced pastures usually based on *Trifolium subterraneum* (Subterranean clover) but occasionally on White clover, and on perennial grasses such as *Phalaris tuberosa* or Perennial Ryegrass. For example, the area under sown pastures in New South Wales increased from 1.8 million acres in 1951 to 4.8 million acres in 1955, and in Australia from 18 m. to 27 m. acres in the same period. The majority of the areas being sown to pastures are in the savannah woodland zone. In the south-west of Western Australia grazing by sheep and the use of superphosphate and Subterranean clover has almost entirely eliminated the native perennial grasses which have been replaced by the Mediterranean and South African species, Capeweed, annual species of *Bromus, Erodium botrys* and *Vulpia bromoides*.

Disclimax native grasslands in savannah woodlands had carrying capacities of the order of $\frac{3}{4}$—1 sheep per acre. In South Australia

TRUMBLE (1935) found that annual applications of superphosphate (22% P_2O_5) increased wool production per acre from 8.2 lb. to 19.5 lb. and the carrying capacity from .75 to 2.5 sheep per acre. These increases were due largely to the increase in densities of exotic winter annuals which raised the feed level both in quantity and quality.

At approximately the 20 inch isohyet in New South Wales the red-yellow podsols of the higher rainfall areas are replaced by red-brown earths and the predominant eucalypt is Grey box *(E. woolsiana)* alone or associated with Cypress pine *(Callitris glauca)* or Poplar Box *(E. populnea)*. Kangaroo grass was prominent in this community in presettlement days when it was generally associated with *Poa caespitosa* and *Stipa aristiglumis*. The disclimax under light grazing is *Stipa falcata* and *Danthonia* sp. but both these species tend to disappear under heavy stocking when native composites, *Chloris truncata*, *Euphorbia drummondii* and *Atriplex leptocarpum* are common. Under very severe grazing the most common species are introduced annuals, *Hordeum leporinum* (Barley grass), *Medicago denticulata* (Burr medic), *Carthamus lanatus* (Saffron thistle) and *Cryptostemma calendula* (Capeweed). Grey box woodland is used extensively for wheat as well as wool growing and *Heliotropium europaeum* (Wild heliotrope), common in wheat stubbles, may cause liver damage in sheep.

Poplar box replaces Grey box as the most prominent Eucalypt in the zone between the 18 and 14 inch isohyets. Associated tree species vary with soil type. On light soils Cypress pine, on heavy soils Belah *(Casuarina lepidophloia)* and Rosewood *(Heterodendron olaefolium)* and on a laterites *Eucalyptus melanophloia* (Silver-leved ironbark) are the most common associates. Small trees and shrubs in these communities include edible species such as Kurrajong *(Sterculia diversifolia)*, Buddah *(Eremophila mitchelli)* and Yarran *(Acacia homalophylla)*. In addition there are large numbers of inedible species and some poisonous ones such as *Myoporum deserti*. Originally Kangaroo grass was common throughout Poplar box communities but at lower density levels than in the higher rainfall areas to the east. In the more northern areas the commonest herbaceous species are *Bothriochloa ewartiana* (Desert bluegrass), *Dichanthium sericeum* (Queensland bluegrass) and the less useful *Bothriochloa decipiens* (Pitted bluegrass). Further south where winter rains have a higher incidence, *Chloris acicularis* (Windmill grass), *Stipa falcata* (Spear grass) and *Danthonia* sp. predominate in moderately grazed areas. Under prolonged heavy grazing these species are replaced by *Chloris truncata*, *Stipa setacea* and a number of herbs belonging particularly to the Compositae, Cruciferae, Chenopodiaceae and Solanaceae. Associated with these species and often dominating the pasturage over extensive areas are the introduced annuals Burr Medic, Barley grass, Silver grass, Saffron thistle and Bathurst Burr (*Xanthium spinosum*). A notable feature following clearing of timber on light-textured soils has been the spread of the native Chenopod *Bassia birchii*, known as Galvanized burr because of its silvery tomentum. On very sandy soils the unpalatable Wire grass *(Aristida jerichoensis)* may form pure stands.

On heavier textured soils on which Rosewood and Belah are common trees, Plains grass *(Stipa aristiglumis)* and *Danthonia simulans* predominate under light stocking with sheep. These species tend to disappear under heavy stocking and the resultant community is dominated by *Chloris truncata, Sporobolus caroli* and introduced annuals similar to those recorded on light textured soils. Thus although the climax communities on heavy and light-textured soils are quite different, the disclimaxes under heavy utilization are similar both in native and introduced species. In all southern woodland communities the effect of heavy grazing has been to increase herbage production in the winter and spring and to decrease summer production. BIDDISCOMBE's (1954) data show that native perennial grasses make most of their growth in late summer whereas the winter annuals which largely replace the perennials under heavy grazing produce most in spring.

The Grassland Formation

The principal grasslands devoted to sheep raising are the Mitchell grasslands of northern N.S.W. and south-western Queensland, the Blue grass grasslands of Queensland, the *Stipa aristiglumis* grasslands of N.S.W. and Victoria and the *Themeda-Poa* grasslands related to the savannah woodlands of southern Australia. Mitchell grass grasslands dominated by species of *Astrebla* occupy heavy textured alkaline soils in summer rainfall zones. In higher rainfall areas (25—30 inches per annum) at the same latitude the community on heavy textured soils is a grassland dominated by Queensland Bluegrass *(Dichanthium sericeum)*.

Astrebla lappacea (Curly Mitchell grass) and *Astrebla pectinata* (Barley Mitchell grass) are the principal species on Mitchell grass grasslands but Queensland bluegrass also may be common, particularly in the eastern and higher rainfall areas of this community. Probably more important for sheep feed than the tussock forming perennials are annuals such as Flinders grass *(Iseilema* sp.) and in southern areas Burr Medic and Barley grass — the two latter especially where grazing has been heavy. Other species common where the densities of the perennials have been reduced by grazing are *Chloris truncata, Sporobolus caroli, Portulaca oleracea, Tribulus terrestris, Dactyloctenium radulans* (Button grass), *Salsola kali* and species of *Atriplex* and *Bassia*. Many of these species are prominent also in degenerate Bluegrass grassland but in places one of the serious effects of over-grazing is the spread of the unpalatable native grass, *Aristida leptopoda.* Mintweed *(Salvia reflexa)*, an introduced annual which may cause nitrate poisoning in stock may also be a serious weed problem in heavily grazed Bluegrass and Mitchell grass grasslands.

Bluegrass grasslands in the southern portion of their distribution have affinities with the *Stipa aristiglumis* grasslands of New South Wales and Victoria. Like the two grasslands described above, *S. aristiglumis* grassland is found on heavy textured soils — the principal area being the Liverpool Plains of New South Wales. The disclimax communities resulting from heavy grazing are similar on the one hand to those more southern Mitchell grass and Bluegrass communities and on the other to that of the Poplar box woodland with which *Stipa aristiglumis* grassland is often contiguous. Species common under grazing include *Chloris truncata*, Barley grass, Burr medic, *Erodium* spp., Silver grass, Bathurst Burr, Saffron thistle and in the more eastern and higher rainfall parts, Mintweed.

Themeda australis-Poa caespitosa grasslands occur in the alpine and subalpine areas of New South Wales, Victoria and Tasmania — the *Poa* being dominant; in the Tablelands of New South Wales — as scattered tracts at elevations of 1800—2000 feet where *Themeda* was dominant before the advent of domestic animals, and in the Western Districts of Victoria on basalt soils contiguous with savannah woodland. Excepting in the subalpine and alpine areas the effects of grazing have been similar to those on savannah woodlands.

The Shrub-Steppe or Saltbush Formation

Shrub-steppe is important for sheep-grazing in both New South Wales where it occupies about one third of the State and in South Australia. It is the most arid of all communities grazed by sheep and occurs in areas where the rainfall is less than 10 inches per annum and predominantly winter incident. The carrying capacity of the shrub-steppe varies but is of the order of 50 sheep per square mile.

Chenopodiaceous species predominate, the more important ones being *Atriplex vesicaria* (Bladder saltbush) and the blue bushes *Kochia sedifolia, Kochia planifolia* and *Kochia pyramidata.* The occurrence of a particular species is governed largely by soil type — *Kochia sedifolia* occurs only where limestone is within 2 feet of the surface.

The space between saltbushes and bluebushes, bare in drought, is covered with a large variety of ephemeral species following rains. The majority of these species are natives but exotic annuals such as Barley grass, Burr medic, Arabian grass *(Schismus barbatus)* and Maltese cockspur *(Centaurea melitensis)* may be common.

Bladder saltbush has been shown to be more vigorous under

intermittent grazing than in the complete absence of grazing (OSBORN, WOOD & PALTRIDGE 1932). Under heavy grazing it is replaced by the less palatable *Kochia aphylla* and with more intensive grazing by unpalatable *Bassia* spp. and annual herbs. Finally very heavily overgrazed shrub-steppe may be dominated by the completely unpalatable but drought resistant Dillon bush *(Nitraria shoberi)*.

Under extreme conditions of drought in association with heavy grazing death of saltbushes may occur over large areas and if the soil is erodable bare areas or scalds may result. RATCLIFFE (1936) investigated soil drift in South Australia and concluded that "the factor responsible for the destruction of the saltbush-bluebush cover must be recognized as the direct cause of by far the greater part of the drift occurring today. The factor in question beyond all doubt is stock". Drought and rabbits were considered of importance only in that they forced the sheep to severely graze and trample the bushes. The breaking up of the bushes by trampling was considered by RATCLIFFE to be more important than defoliation by grazing as even dead bushes protected the soil from erosion by wind. The rate of regeneration of saltbush on heavily grazed areas was studied at Koonamore, South Australia, by WOOD (1938) and found to be extremely slow — of the order of 400 yards in 4 years. According to BEADLE (1948), 90% of the bush which once occurred in the East Darling district of New South Wales has disappeared through over-grazing, particularly in times of drought.

Mulga Scrub Formation

Mulga *(Acacia aneura)* covers an extensive area in approximately the 10 inch rainfall zone, extending from western New South Wales and south central Queensland across to the coast of Western Australia. Besides Mulga the community consists of a number of other trees and shrubs of fodder value notably Needlewood *(Hakea leucoptera)*, Rosewood *(Heterodendron oleaefolium)* and *Atriplex nummularium*. The ground storey in the origin condition was dominated by *Eragrostis eriopoda*, Wire grasses (*Aristida* spp.), Spinifex (*Triodia* spp.) *Eragrostis dielsii, Eragrostis setifolia* (Never fail) and numerous other grasses and herbs. In addition to perennial species there are a large number of short-lived annuals both winter and summer growing — the ephemeral vegetation at any time depending on the time of the year in which rain falls. Under grazing the species of *Eragrostis* disappear and are replaced by Wire grasses which although unpalatable have some value in protecting the soil from erosion. Concurrently with the decrease in density of perennials there is an increase in the number of compalatable native shrubs, notably *Bassia obliquecuspsis, Calotis* spp., *Myriocephalus stuartii, Senecio gregorii* and *Helipterum* spp.

The arid conditions prevailing in mulga communities have precluded the establishment of exotics on a large scale although a number of species occur naturally, e.g. *Schismus barbatus*, *Diplotaxis tenuifolia* etc. The most obvious effects of grazing have been a reduction in perennial grasses, a consequent increase in bare areas during dry periods and a failure of mulga to regenerate particularly in the Wiluna-Meekatharia region of Western Australia.

Spinifex

A community having affinities with mulga scrub is Spinifex which covers extensive areas of northern Australia from Queensland to Western Australia. Spinifex is an arid community of low carrying capacity and is used mainly for cattle raising. The various species grouped under the name spinifex are sclerophyllous and frequently resinous grasses, belonging to the genera *Triodia* and *Plectrachne*.

The Spinifexes generally grow in mounds which may be several feet in width. As a result of age or the burning commonly practised in spinifex country the mounds may become annular through death of the centre. Scattered throughout the community are trees such as Mulga, Needlebush (*Hakea* spp.) and Eucalypts such as *E. papuana* (Ghost Gum). Associated grasses are *Chrysopogon latifolius*, *Eragrostis eriopoda*, *Eriachne* spp. and *Enneapogon nigricans*. These decrease in density under grazing and may be replaced by one of the least palatable of the spinifexes, *T. lanigera*.

Alpine Communities

The Alpine areas are of relatively little importance to the grazing industry in terms of the feeding value of their plant communities — the potential value of alpine grazing has been estimated as only 0.1 to 0.3% of value of production from grazing in New South Wales and Victoria in 1954—55 (Australian Academy of Science Report 1957) — but the importance of the 2,000 square miles of alpine country as sources of water for power and irrigation cannot be over-estimated. The important communities are the sub-alpine woodlands, dominated by Snow Gum *(Eucalyptus pauciflora)* and Snow Grass *(Poa caespitosa)*, alpine herbfields, the *Sphagnum* bog communities and the subalpine grasslands carrying snow grass. A large number of the dominant plants of these communities are unpalatable, the more palatable species having disappeared through selective grazing by livestock. The New South Wales botanist, J. A. MAIDEN (1898/1899) remarked on the abundance of *Danthonia frigida* in alpine country whereas today the species is confined to small areas difficult of access. At the present time the principal native plant available for grazing is the relatively unpalatable snow grass and graziers have adopted the practice of burning this species to remove mature material and to promote the growth of new shoots. These and accidental fires have increased the density of fire tolerant native shrubs and have burnt large areas of snow gum. The regenerating shoots of burnt snow gum are eaten by sheep and repeated defoliation leads to the death of the plant. The reduction in snow gum cover reduces the trapping and retention of snow and thus decreases water yield. The firing and grazing of bogs and fens is believed by COSTIN (1957) to result in their dessication and the same writer has emphasised the importance of such communities in maintaining a gradual and continual flow of water from the mountains. COSTIN has also indicated that sheep concentrate on areas denuded of snow grass because of the presence of introduced species such as *Rumex acetosella*. Such partially denuded areas are very liable to erosion which in turn may lead to the siltation of storage dams.

In both the arid and alpine regions the equilibrium between the vegetation and the environment is a delicate one and, moreover, the number of introduced species capable of colonizing denuded areas with such climatic extremes is limited.

General Ecological Effects of Sheep Grazing on Australian Vegetation

Surveys conducted by the writer over a number of years have shown that virgin plant communities are not invaded by introduced plants and that disturbed communities invaded by alien plants will revert to their original condition when the disturbance factor, e.g. grazing, trampling etc. is removed, and provided there has been no large-scale removal of soil by erosion. The invasion of native plant communities by alien plants occurs only where man's activities or natural disturbance factors such as landslides and floods create habitats suitable for pioneers and aliens pre-adapted to such sites by long association with man in ancient centres of civilization. In general the greater the disturbance the more complete is the replacement of native by alien. Dr. J. B. CLELAND (1928), commenting on the largely alien flora of the plain in close proximity to the city of Adelaide wrote "The original flora has disappeared for the most part over the area bounded on the one hand by the Gulf from Brighton to Outer Harbour and on the other by the foothills of the Mt. Lofty Ranges".

One of the most striking features of the original ground flora of Australian plant communities was the ubiquity of Kangaroo grass which although making its greatest growth in the summer predominated in the lower layers of many different communities even in winter rainfall zones. The rapid disappearance of this species under grazing has been equally remarkable. It has been replaced by essentially winter growing species such as *Danthonia* and *Stipa* which form stable communities under sheep grazing in most woodlands. In nearly all formations in areas receiving a component of winter rainfall, Burr medic, Barley grass and the native annual *Erodium cygnorum* are prominent. These are all early maturing species with spiny fruits which aid dissemination by sheep and possibly germination and establishment. Communities of these species provide more protein than native perennials and even though their actual growing season is short the burrs from the legume provide accessible feed during the dry period. It is significant that many of the stud merino breeding properties in Australia are located in "burr country".

It will be noted from the examples given that there is a marked similarity in the species of overgrazed areas irrespective of climatic and soil differences and of differences in the virgin plant communities. This applies both to pioneer native species and to introduced species. Of seventeen introduced plants collected by MAX KOCH on the semi-arid Mount Lyndhurst run, South Australia, in 1898, fourteen are common between Canberra and Gundagai (22—25 inches of rain).

Not all vegetation changes can be ascribed to sheep grazing — cattle grazing, crop growing and concentration of human activity generally are also responsible. However, the changes over extensive areas of southern Australia are due largely to grazing and to the grazing of sheep in particular. The overall effects of cattle and sheep grazing are hard to separate. If both types of stock are run on the one property they are frequently run together in the same paddocks. However, it is known for example that Patterson's Curse *(Echium plantagineum)* is common on areas over-grazed by cattle but is generally absent on adjacent paddocks grazed by sheep. Also in the Springsure district of Queensland there has been a marked relationship between the dominance of Spreading Spear Grass *(Aristida leptopoda)* and sheep grazing; the spear grass became prominent when sheep replaced cattle because of watering problems associated with the running of the larger animals. In general, cattle are less destructive to grass vegetation than sheep and it is noteworthy that Kangaroo grass is more persistent in cattle country.

The degree of ringbarking and clearing of trees varies in the different communities. In the woodland communities clearing has been primarily for the purpose of increasing grass growth but also for timber and firewood. In forest communities with a poorly developed herbaceous layer trees have been felled mainly because of their value as timber. In many such communities an almost continuous stratum of leguminous and proteaceous shrubs has resulted from the removal of the tree dominants. The bracken fern problem in dairying pastures is directly associated with removal of tree competition. However, even where the tree layer has not been thinned changes in the subordinate strata have been caused by the domestic stock and rabbits.

Clearing for sheep grazing purposes may have indirect effects. One of these, grasshopper swarming, has been reported by KEY (1945) and another, salting, by F. COPE (thesis University of Melbourne). The latter showed that land cleared of timber is occupied by shallower rooted shrubs and herbs with a lower water usage. Thus soil water may accumulate and seep down slopes dissolving salt in the soil en route. On reaching the lower flats further drainage being impossible, saline waters accumulate, bringing dissolved salt to the surface, where it is concentrated through evaporation. Eventually the concentration of salt is so high that vegetation is killed and the so-called salt patches develop. These patches are particularly liable to erosion because of their poor physical condition. In Western Australia and Victoria particularly, salt problems of this nature are acute.

An interesting effect of grazing was noted by SAMUEL DIXON, who in an address to the Royal Society of South Australia in 1892 said, "The only instance in which the settlement in Australia has had the effect of increasing the indigenous vegetation on a large scale occurs in the Cobar district where the Cypress Pine *(Callitris verrucosa)* has increased to such an extent that much less stock can now be carried there. This has arisen apparently from the grass being eaten off by stock so that the bush-fires no longer travel over large areas and the young plants which are easily destroyed by fire growing closely together and seeding abundantly, take complete possession of the soil to the exclusion of other plants. This is the only instance within my knowledge of any native plant largely extending its area after settlement has taken place". It is now known that there are some others, e.g. Galvanized Burr, Wire Grasses *(Aristida* sp.) and Native Lime *(Eremocitrus glauca)*.

A list of plants naturalized in New South Wales is appended to the "Handbook of the Flora of New South Wales", MOORE & BETCHE, 1893. This list shows 243 species including 24 grasses and this may be compared with ANDERSON's (1939) list which has 415 species exclusive of grasses. A. J. EWART in his Flora of Victoria (1930) states that in 1909 there were 363, in 1928, 461 and in 1930, 500 alien species in Victoria. Of the latter 102 were grasses and 50 legumes. The rate of accession of alien species being therefore roughly one every two months. Since the total number of native species listed by EWART in 1930 was 2,200, alien species comprised roughly 18% of the Victorian flora. EWART states "The native flora of Victoria, exposed as it is to the competition of imported aliens and to the presence of settlement, is in a condition of rapid flux. It is probable that less than half of the original flora will survive within a century and that many plants originally widely spread will be confined to special localities". It has been shown that there are examples of this confinement in the case of Kangaroo grass. EWART goes on to say that "the disproportionately high number (of alien species) derived from the Leguminoseae (50) and from the Gramineae (102) is an aftermath of the pastoral phase when the world was searched for fodder plants to improve our pastures". In 1930 there were 125 species of grass native to Victoria. A comparison of this figure with the one above for alien grasses; the fact that we now have a much more efficient plant introduction service and that communication between countries has been so greatly facilitated, give credence to EWART's belief that "in the near future our grass flora will be mainly foreign".

In conclusion it should be mentioned again that sheep are very efficient distributing agents for seeds and fruits, especially those with spines or awns such as the Burr medics, Barley grass, *Erodium* spp., Bathrust Burr and Noorgoora Burr. The adherence of such seeds and fruits to wool and the large movements of stock in Australia means that many introduced plants have very efficient means of reaching environments to which they are suited. The presence of seeds in wool, termed in the trade "vegetable fault", is a serious financial loss to the wool industry. As early as 1892 introduced plants lowered the value of Australian wool as evidenced in DIXON's paper to the Royal Society of South Australia in that year. Speaking of *Medicago denticulata* (Burr Medic) he said, "No introduced plant has produced greater loss to the wool industry than this . . ." DIXON also mentioned Bathurst Burr in the same context and remarked that spiny fruits in wool means "a loss of one penny per lb." The export of wool from Australia means that seeds and fruits of native and introduced plants are carried to other countries and the flora near wool scouring plants in the United Kingdom and other countries is notably influenced by wool growing in Australia.

REFERENCES

ANDERSON, A. J. & MOYE, D. V. 1952. Lime and Molybdenum in clover development on acid soils. *Aust. J. agric. Res.* **3**, *95—110*.

ANDERSON, R. H., 1939. The Naturalized Flora of N.S.W. (Excluding Gramineae). Contrib. N.S.W. Nat. Herb.

ANON., 1949. The Australian Environment. C.S.I.R.O.

ANON., 1956. Year Book of the Commonwealth of Australia. Com. Bur. Cen. & Stat. Aust.

ANON., 1957. A Report on the Condition of the High Mountain Catchments of N.S.W. and Victoria. Aust. Acad. of Science.

ANON., 1957. Pasture Statistics Australia 1950—51 to 1955—56. Bur. Agr. Econ. Aust.

ANON., 1957. The Australian Sheep Industry. *Bank N.S.W. Review* **29**, *7—10*.

AUDAS, J. W., 1929. General observations on the Australian Flora. Rept. Aust. Assoc. Adv. Sci. 1928.

BEADLE, N. C. W., 1948. The Vegetation and Pastures of Western N.S.W. with special reference to Soil Erosion. Govt. Printer, Sydney.

BIDDISCOMBE, E. F., 1953. Survey of the Natural Pastures of the Trangie District of N.S.W. with special reference to the Grazing Factor. *Aust. J. agric. Res.* **4**, *1—28*.

BIDDISCOMBE, E. F., CUTHBERTSON, E. G. & HUTCHINGS, R. J., 1954. Autecology of some Natural Pasture Species at Trangie, N.S.W. *Aust. J. Bot.* **2**, *69—98*.
BLACK, J. M., 1948. Flora of South Australia. Govt. Printer, Adelaide.
BLAKE, S. T., 1938. The Plant Communities of Western Queensland and their Relationships, with special reference to the Grazing Industry. *Proc. Roy. Soc. Qnld.* **49**, *156—204*.
BRIDE, T. F., 1898. Letters from Victorian Pioneers.
BROWN, ROBT., 1810. Prodromus Florae Novae Hollandiae.
CLELAND, J. B., BLACK, J. M. & REESE, L., 1925. The Flora of the North-East corner of South Australia, North of Cooper's Creek. *Trans. Roy. Soc. S. Aust.* **1949**, *103—120*.
CLELAND, J. B., 1928. The original Flora of the Adelaide Plains. *S. Aust. Nat.* **X**, *1—6*.
COSTIN, A. B., 1957. Some Problems Associated with Grazing in the Australian Alps. Unpublished private communication.
DIXON, SAMUEL, 1892. The effects of Settlement and Pastoral Occupation in Australia upon the Indigenous Vegetation. *Trans. & Proc. Roy. Soc. S. Aust.* **15**, *195—206*.
HAMILTON, A. G., 1892. Effect of settlement upon indigenous vegetation. Roy. Soc. N.S.W.
JESSUP, R. W., 1951. The Soils, Geology and Vegetation of North-Western South Australia. *Trans. Roy. Soc. S. Aust.* **74**, *189—273*.
KEY, K. H. L., 1945. The General Ecological Characteristics of the Outbreak Areas and Outbreak Years of the Australian Plague Locust. *C.S.I.R. Bull.* 186.
KOCH, MAX., 1898. A List of Plants Collected on the Mount Lyndhurst Run S.A. *Trans. & Proc. Roy. Soc. S. Aust.* **22**, *101—118*.
MCTAGGART, A., 1938. A Survey of the Pastures of Australia. *C.S.I.R. Bull.* **99**.
MAIDEN, J. H., 1898. A Contribution Towards a Flora of Mount Kosciusko. *Agric. Gaz. N.S.W.* **9**, *720*.
MAIDEN, J. H., 1899. A Second Contribution towards a Flora of Mount Kosciusko. *Agric. Gaz. N.S.W.* **10**, *1001*.
MITCHELL, J., 1839. Three Expeditions into Eastern Australia. Vols. 1 & 2 London.
MOORE, C. W. E., 1953. The Vegetation of the South-Eastern Riverina New South Wales.
  1. The Climax Communities. *Aust. J. Bot.* **1**, *485—547*.
  2. The Disclimax Communities. *Ibid* **1**, *548—567*.
MOORE, R. M., 1940. The Southern Pastures of Australia. *Aust. Publ. Ser. Bull.* **29**.
OSBORN, T. G. B., 1928. The biological factor in the study of vegetation with special reference to Australian conditions. *Rept. Aust. Assn. Adv. Sci.* **29**, *611—625*.
OSBORN, T. G. B. & WORD, J. G., 1931. On the Autecology of *Stipa nitida*; a study of a fodder grass in arid Australia. *Proc. Linn. Soc. N.S.W.* **56**, *299—324*.
OSBORN, T. G. B., WORD, J. G. & PALTRIDGE, T. B., 1932. On the Growth and Reaction to Grazing of the Perennial Saltbush *Altriplex vesicarium*. *Proc. Roy. Soc. N.S.W.* **57**, *377—402*.
PRYOR, L. D., 1939. The Vegetation of the Australian Capital Territory. A study of Synecology. M.Sc. Thesis. Univ. of Adelaide.
RATCLIFFE, F. N., 1936. Soil Drift in the Arid Pastoral Areas of South Australia. *C.S.I.R. Pamph.* **64**.
ROE, R., 1945. Studies on the Mitchell Grass Association in South-Western Queensland. *C.S.I.R. Bull.* 185.
TRUMBLE, H. C., 1935. The relation of pasture development to Environmental factors in South Australia. *Agric. J. S. Aust.* **38**, *1460—1487*.
WOOD, J. G., 1929. Floristics and Ecology of the Mallee. *Trans. Roy. Soc. S. Aust.* **53**, *359*.
WOOD, J. G., 1937. The Vegetation of South Australia. Govt. Printer, Adelaide.
WOOD, J. G., 1938. Regeneration of the Vegetation on the Koonamore Vegetation Reserve, 1926 to 1936. *Trans. Roy. Soc. S. Aust.* **60**, *96—111*.

XXXII
THE ECOLOGY OF SHEEP BLOWFLIES IN AUSTRALIA

by

K. R. NORRIS

(Division of Entomology, C.S.I.R.O., Canberra, A.C.T.)

The dipterous family Calliphoridae includes a number of important pest species. The adult insects, or "blowflies", may be domestic nuisances, or carry disease organisms, but in this capacity they are of far less importance than other groups of flies. On the other hand, the damage and suffering which larvae of Calliphoridae inflict upon domestic animals is of tremendous consequence to mankind. Infestation of living animals by Calliphorid larvae, in fact, occurs in all important stock-raising countries, taking various forms according to the animals attacked, and the species of flies involved.

The blowflies of Australia include no specific myiasis producers like the screw-worm fly, *Callitroga hominivorax* (COQUEREL), of north and south America, with larvae developing exclusively in lesions located by the female flies on living animals. On the other hand, as in most sheep-raising countries, there is in Australia a group of non-specific blowflies, which breed chiefly in carrion, but also at times cause cutaneous myiasis of living sheep — the infestation of the wool and skin with maggots.

In the past much discussion revolved around the reasons for the extension of the activities of blowflies from dead to living animals, but in fact it is a relatively slight extrapolation of normal blowfly activity. Wool and epidermal cells are, after all, dead tissues, admittedly resistant to decay when compared with other tissues, but nevertheless emitting odours strong enough to attract some flies (CRAGG & COLE, 1956). It has also been pointed out that the pelage of fresh carcases may be the site of a considerable proportion of blowfly oviposition (HOLDAWAY, 1932), and it is therefore not surprising that some strikes (myiases) on sheep may have no obvious predisposing causes (MACLEOD & DONNELLY, 1957), although the majority certainly originate where the normal slight attractiveness of wool is enhanced by moisture and bacterial activity, or faecal or urine staining.

The practical problem of blowfly strike

Cutaneous myiasis of sheep is of greater importance in Australia than in any other country, doubtless because of the high susceptibility of the merino. The problem appears to have assumed importance only since the turn of the century, principally, it is thought, following the introduction from abroad of the blowfly *Lucilia cuprina*.

If struck sheep are not treated, and maggot development continues, death ensues. Sheep which recover show loss of condition and reduction in yield and quality of wool. An estimate made in 1938 placed the annual cost of the sheep blowfly to Australia in terms of yield loss and expense of preventive measures at £4,000,000. Current costs would certainly not be less than this figure.

The problem of control of fly strike in Australia was approached along two main lines: the protection of sheep against strike, and the investigation of means of reducing the abundance of blowflies. Success was achieved exclusively in the former field (ROBERTS, 1952). In brief the most important procedures are as follows: The incidence of breech, or crutch, strike of ewes, which is by far the most important type of strike, can be greatly reduced by surgical removal of folds of skin from the breech, to reduce staining of the wool by excreta. This operation, known as the Mules operation, after its originator, is supplemented by regular crutching, or shearing of the wool from the breech to reduce fouling. It has also been found that medium-length tail docking, combined with the removal of the wool-bearing areas from the dorsal surface of the tail, further reduces soiling, and lowers the incidence of tail strike. Persistent insecticides such as dieldrin, jetted into the wool, also serve to reduce the incidence of crutch strike. The other important type of strike is body strike, which originates over the shoulders where scalding and fleece rot have developed, due to prolonged moistness and bacterial activity. Protection from this, and other less important forms of strike can be obtained for long periods by jetting or spraying the fleece with DDT, BHC, dieldrin, aldrin or diazinon. Highly effective dressings for struck sheep, incorporating persistent insecticides, are also available.

The problem of reducing fly abundance was approached along the following main lines: (1) Biological control, employing native and introduced hymenopterous parasites, was studied. No natural enemy was found which was capable of matching the rapid multiplication and dispersal of blowflies. (2) Carcase destruction. Burial was found to favour flies which initiate strike, unless accompanied by poisoning. Over vast areas of pastoral and bush country of Australia it is not practicable to locate all carcases, let alone destroy them by burning or burial. There has been little advocacy of carcase destruction for blowfly control since it was realised that many *L. cuprina* actually passed through their larval stages on living sheep. (3) Trapping of adult blowflies was at one time advised. However, careful experiment showed that although the incidence of strike could be significantly reduced by intensive trapping, many strikes still occurred (MACKERRAS a.o., 1936). As a means of controlling blowfly strike trapping has long been discontinued.

Investigation of the possibilities of direct attack on blowflies thus bore little fruit in Australia, and, after the development of the control measures outlined above, less intensive study was devoted to blowfly ecology. However, although a broad understanding of the subject had been obtained, there are still wide gaps in our knowledge. There is little immediate prospect of further intensive study producing practical results in Australia, in the way that biological and ecological study of *Callitroga hominivorax* in the United States made possible its eradication from the island of Curaçao, and aroused serious interest in the possibility of its eradication from Florida (BUSHLAND a.o., 1955). However, background knowledge of pests is essential in the application of scientific advances, and it is hoped that interest will be renewed in this family of flies, which, though relatively few in species, nevertheless occupies an important ecological niche with such success that throughout the world, it is almost completely unchallenged by invertebrate competitors.

Species involved in sheep myiasis in Australia

Almost all the species of Calliphoridae known to exploit carcases in sheep-raising areas of Australia have been recorded in strikes on sheep. The species of sheep myiasis flies (Table I) appear in strikes in the same succession (TILLYARD & SEDDON, 1933) as they do in carrion (FULLER, 1934), but with quite a different order of importance.

Primary flies are those which can initiate a strike on sheep: they are also the first species to lay eggs or larvae on fresh carrion.

Secondary flies only infest strikes started by other species, and also only lay on carrion after the primaries are already well established. The secondary flies rarely initiate strikes.

A group of tertiary flies is also recognised, which occur as larvae in scabs of old, partially healed strikes.

Lucilia cuprina, a green metallic blowfly, usually a little over a quarter of an inch in length, which is now widespread throughout the mainland (WATERHOUSE & PARAMONOV, 1950), and Tasmania (RYAN, 1954) and is by far the most important sheep myiasis fly.

Lucilia sericata resembles *L. cuprina* very closely in appearance and size. It is now practically cosmopolitan, though perhaps introduced into Australia and other countries through human agency.

Calliphora stygia is a robust blowfly, up to $\frac{1}{2}$ in. in length. The thorax is slate-grey, and the remainder of the body covered with golden-brown pubescence. *C. stygia* is a native species occurring in the southern half of the continent and in Tasmania.

Calliphora albifrontalis resembles the preceding species in size and appearance. It is a native species which largely replaces *C. stygia* in the south-west of the continent.

Calliphora fallax is a native species occurring throughout south-eastern Australia. It bears a superficial resemblance to *C. stygia*, but is generally slightly smaller.

Table I.
Flies involved in sheep myiasis in Australia

Stage in succession	Species	Family	Number of times recorded in 1691 strikes[1]
Primary	[2] *Lucilia cuprina* (WIEDEMANN) [2] *Lucilia sericata* (MEIGEN) *Calliphora stygia* (FABRICIUS) *Calliphora albifrontalis* MALLOCH (= *C. australis* (BOISDUVAL)) *Calliphora fallax* HARDY *Calliphora augur* (FABRICIUS) *Calliphora nociva* HARDY *Calliphora vicina* (ROBINEAU-DESVOIDY) (= *C. erythrocephala* (MEIGEN))	Calliphoridae	1376 71 } 339 1 } 550 [3]
Secondary	[4] *Chrysomya rufifacies* (MACQUART) [4] *Chrysomya micropogon* (BIGOT) *Microcalliphora varipes* (MACQUART)		128 6 4
	Sarcophaga froggatti TAYLOR *Sarcophaga misera* WALKER *Sarcophaga* sp.	Sarcophagidae	} 7
Tertiary	*Peronia rostrata* ROBINEAU-DESVOIDY *Musca hilli* JOHNSTON & BANCROFT *Musca domestica* LINNAEUS *Fannia australis* MALLOCH *Muscina stabulans* (FALLÉN)	Muscidae	32 3 23 1 1

[1] Data from MACKERRAS & FULLER (1937).
[2] Referred to by United States entomologists under the generic name *Phaenicia*.
[3] This species was not recorded from strikes at the time the statistics were compiled.
[4] Original spelling of generic name (JAMES, 1947).

Calliphora augur is a native blowfly, slightly smaller on the average than *C. stygia*. It has a slate-grey thorax, and a yellowish abdomen with a dull-blue patch down the centre of the dorsum. The last visible abdominal tergite is covered with golden tomentum. This species is most abundant in the south-eastern quadrant of the continent, and does not occur in the western half.

Calliphora nociva is a native species, differing from *C. augur* principally in having a richer blue, shining abdominal colour patch, and white, instead of golden tomentum on the last visible abdominal tergite. It occurs from south-western Australia to western New South Wales and south-western Queensland.

Calliphora vicina attains a length of ½ in. The thorax and legs are black and the abdomen blue, with silvery tesselations. This introduced species occurs throughout southern Australia and Tasmania.

Chrysomya rufifacies is a shining, green-metallic fly of about the same size as *L. cuprina*, but distinguishable to the naked eye by the dark-blue margins to the abdominal tergites. It occurs as a native species throughout the mainland of Australia, and is also found in oriental regions.

Chrysomya micropogon is a smooth, dull-blue blowfly a little less than ½ in. in length. It is exclusively Australian in distribution, and rather more restricted to tropical areas than *C. rufifacies*.

Microcalliphora varipes is smaller than the housefly. It is a green-metallic native blowfly with legs banded in dark brown and yellow. Its distribution in Australia is similar to that of *C. rufifacies*.

Sarcophaga species have the thorax striped, and the abdomen boldly chequered, with black and light-coloured patterning. As carrion flies they are most important in northern Australia.

Peronia rostrata is a blue-black native Muscid about the size of the housefly. The last visible abdominal tergite is covered with a pale, greyish-blue tomentum. This species is widespread in the southern half of Australia.

Local and seasonal variations in the abundance of sheep blowflies

From the notes given above it is evident that the species constituting the sheep blowfly fauna may differ characteristically at places a few hundred miles apart. Microgeographical variations in population density may also be great. For example there may be 500—1000 *Lucilia sericata* per acre in Canberra when there are fewer than 1 per acre a mile or so away in sheep country (WATERHOUSE & PARAMONOV, 1950). In the Canberra district *Calliphora augur*, *Calliphora stygia* and *Chrysomya rufifacies* also exhibit marked habitat preferences, while many other species are known to vary in abundance within short distances (ANDREWARTHA & BIRCH 1954; CRAGG & HOBART, 1955; GILMOUR a.o., 1946; HOLDAWAY, 1933;

MACLEOD & DONNELLY, 1957; MURRAY, 1956; RATCLIFFE, 1934.) In any district, therefore, there appear to be contours of differing population densities of blowflies, although the various species do not necessarily prefer the same localities. In some cases a species appears to seek out obvious habitat features, but often it is difficult to see why particular locations should offer greater attraction than others, particularly when all species are known to be able to fly miles in a day. Population contours also change with season (MACLEOD & DONNELLY, 1957; TILLYARD & SEDDON, 1933) and short-term weather variations (HOLDAWAY, 1933), and the results of trapping during one year may not give a typical picture of seasonal abundance (FULLER, 1934). Published information on seasonal distribution of Australian blowflies must therefore be accepted with caution. Trapping has been carried out at Canberra (FULLER, 1934a), at Moree, about 400 miles north of Canberra (GURNEY & WOODHILL, 1926), at Adelaide in South Australia (DAVIDSON, 1933), and at six localities in south-western Australia (NEWMAN, a.o., 1930).

At Canberra the seasonal variations in the abundance of the different species are fairly well established. No adults of *L. cuprina* occur during June—August: there is a population peak in December, a marked decrease in numbers in January and February, and a further moderate increase in March and April (Fig. 1). *Calliphora stygia* is present in small numbers in the winter. It may occur in very large numbers in spring and early summer, but is markedly scarcer throughout the hotter months, and (like *L. cuprina*) increases somewhat in numbers during the autumn. *Calliphora augur* is usually absent during the winter. It exhibits a population maximum later in the summer than *C. stygia*, is less depressed in numbers during January and February, and relatively more abundant during the autumn. *Chrysomya rufifacies* and *Microcalliphora varipes* are more narrowly restricted to the warmer months than *L. cuprina* and *C. augur*, and attain maximum numbers during the hottest months.

At Moree in north-western New South Wales, *C. stygia* and *C. augur* maintain their spring and autumn abundance during the relatively mild winter. In the summer their numbers are depressed even more markedly than in the Canberra area. *Chrysomya rufifacies* is most plentiful in the warmer months, but is trapped in considerable numbers throughout the year. *M. varipes* becomes a more conspicuous member of the trap catches than at Canberra, with a seasonal distribution similar to that of *C. rufifacies*.

At Adelaide, *C. nociva* replaces *C. augur*. The picture of seasonal distribution of the blowfly species is more nearly comparable with Canberra, with a briefer winter depression in fly numbers, and the *Calliphora* species scarce in autumn, as well as summer.

On the tableland of south-western Australia *Calliphora albifrontalis* is active only in the winter. It is not caught in traps there throughout the remainder of the year, though the period of activity is longer on the coastal plain. *Calliphora nociva* may be active throughout the year, although it is most abundant in autumn and spring. *C. rufifacies* and *M. varipes* are most abundant in spring and early summer: in the cooler localities they may not be represented for about two months in the winter trappings.

In the studies made at Moree, Adelaide and south-west Australia, specimens of *L. cuprina* were probably classified as *L. sericata*. However, the seasonal distribution of *L. cuprina* at Moree and Adelaide is probably similar to that recorded at Canberra, with a briefer period during which adults are absent in the winter. Available information suggests that in Western Australia *L. cuprina* occurs chiefly during the winter, spring and early summer (FULLER, 1934b).

Trapping of blowflies has not been carried out in the more northerly sheep-raising areas, but it is known that in south Queensland *Calliphora* species are a relatively inconspicuous, purely winter element in the blowfly fauna: they are absent altogether in the extreme north. *Chrysomya* species and *M. varipes* are abundant throughout the year in tropical areas.

Biology and ecology of individual species
Lucilia cuprina

More is known about the bionomics of *L. cuprina* than of any other sheep blowfly in Australia, because it has been studied intensively since its great importance in sheep strike was recognised.

In the Canberra area GILMOUR et al. (1946) liberated marked flies in sheep pasture land, and counted the numbers of marked and "wild" flies caught in 102 traps regularly placed throughout a circle of 4 miles radius surrounding the release point. Allowing for dispersal of the marked flies out of the trapped area, and their mortality within the area, it was calculated from the ratio of marked flies to "wild" flies trapped, that the natural population density of *L. cuprina* ranged from 3—4 flies to the acre in late spring, and up to 5.7 to the acre in December. Flies were fewer than 1 per 2 acres in January, and there were 1—2 flies to the acre in March. These figures agree reasonably well with the curve of seasonal distribution based on weekly trap catches (Fig. 1).

Evidence was obtained of marked local differences in population density of *L. cuprina*, which did not follow a consistent pattern in all experiments, although some areas tended to yield high catches, and others low catches with fair consistency. ANDREWARTHA & BIRCH (1954) also showed that the figures indicated non-random distribution of the population.

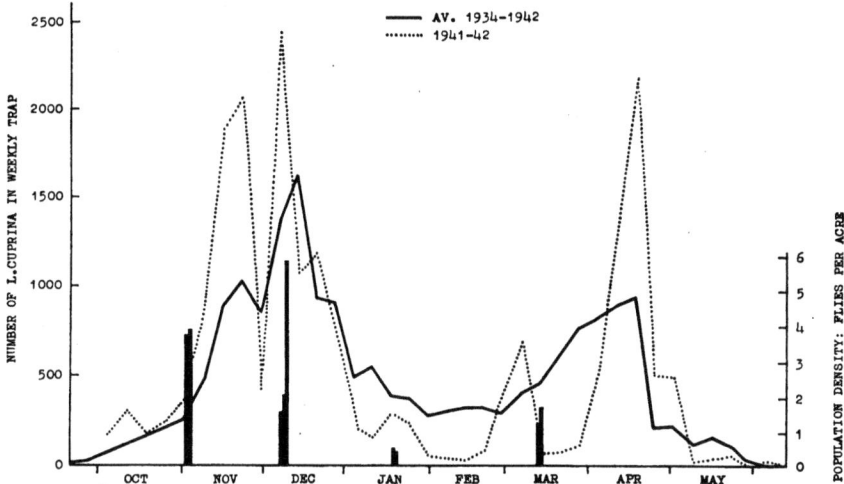

Fig. 1. Mean weekly trap catch of Lucilia cuprina over 8 years at Canberra (solid graph) and for 1941—42 (dotted graph). Population densities measured at various times during the 1941—42 season indicated by block diagrams (After GILMOUR et al. (1946)).

The habitat preferences of L. cuprina differ from those of other Calliphorids, most of which occur in greatest numbers in pastoral country carrying some timber, whereas L. cuprina may be caught in equal abundance in both timbered and treeless pastures (NORRIS).

L. cuprina released in December, immediately after emergence from the puparium and with no preliminary feeding, dispersed in the first 24 hours up to $2\frac{1}{2}$ miles from the liberation point. Some were trapped 4 miles from the centre of release within 48 hours, and may have been recoverable at greater distances had traps been placed there. Nevertheless strong concentrations of the released population remained within about a mile of the point of liberation for over a week (NORRIS).

Flies fed on cane sugar and liberated several days after emergence from the puparium were found to travel up to 4.7 miles in less than 30 hours (GILMOUR a.o., 1946). The maximum distance of flight of L. cuprina adults is not known, but with the ability to fly more than 4 miles in a day they could travel great distances if they survived for some weeks.

Doubtless adults of L. cuprina and other blowfly species are eaten by birds. Ants, predaceous wasps such as Sericophorus species (RAYMENT, 1954), Asilids and other insects also destroy some. Empusa and other fungal diseases may take a toll under some weather conditions, as they have been known to do in laboratory cultures, but on the whole it appears that the pressure for survival in this species falls most heavily on stages other than the adult.

Cultures maintained at 78° F under a regime of 9 hr light- 15 hr darkness show a well marked peak of "morning" emergence from the puparia. Similar rhythms possibly occur in nature, when temperature effects do not override them.

Approximately equal numbers of males and females emerge from most batches of puparia in cultures, and the peak of emergence of males occurs earlier than that of females. The average cage life of males is shorter than that of females (MACKERRAS, 1933).

Little information is available on the longevity of *L. cuprina* in the wild state. Large recoveries of both sexes were made in traps 9 days after a December liberation, suggesting high survival for at least that period (NORRIS). In another study one female was recaptured a month after liberation, and a male six weeks after liberation. Taking the degree of fraying of the wings of these specimens as a rough guide to longevity, it appeared that at least some of the "wild" *L. cuprina* captured at the same time had lived more than a month (NORRIS).

The sex ratio of trapped *L. cuprina* does not show a significant correlation with numbers caught. In this regard *L. cuprina* differs from the other species of carrion Calliphoridae caught in traps near Canberra (see below).

Lucilia cuprina adults require moisture and carbohydrates for survival. In the Canberra area important natural sources of carbohydrates are the honeydew of Coccids and Psyllids and the nectar of *Eucalyptus* blossoms (WEBBER, 1957).

Before eggs can be matured, the female requires a protein feed. Vertebrate carcases sometimes supply this need, but there may also be innumerable minor animal sources. For instance, in the insectary, flies were observed to suck the juices from crushed spiders and flies. Blowflies also feed on animal faeces, and the droppings of sheep, cattle, horses, poultry and rabbits may supply females of *L. cuprina* with protein adequate for egg maturation (WEBBER, 1958). Unquestionably human and carnivore faeces would also be adequate. Sheep droppings are only suitable as protein sources if the sheep have grazed on high-protein pastures (WEBBER, 1958).

Some *L. cuprina* liberated in March as unfed callows, and trapped 4 days after emergence were found to contain mature eggs. Most of the flies trapped on the 5th day contained mature eggs (NORRIS). At the time of these observations the pastures were very parched after a rainless period, and it did not appear that conditions were favourable, even to survival. A December liberation of unfed callow flies was made when there were abundant flowering plants throughout the pastures. A few with mature ovaries were taken on the 3rd and 4th days after emergence, but it was the sixth day before virtually all retrapped specimens had mature ovaries. A cold snap on the 5th day may have slowed the rate of development. Available evidence, therefore, indicates that female *L. cuprina* probably do not lay eggs before the 3rd day after emergence from the puparium, but that, on the other hand, they have little difficulty in obtaining the nutritional requirements for survival and fertility. Access to both readily assimilable carbohydrate and protein appears to result in more rapid ovarian development than protein feeding alone (MACKERRAS, 1933; WEBBER, 1958). This may explain the short pregravid period (2 days) observed by GILMOUR et al., 1946), who fed their marked specimens on cane sugar before liberation.

Males do not require a protein feed to ensure fertility (MACKERRAS, 1933), and presumably are ready to play their part in reproduction much sooner than females. However, females reject attempts at copulation until the eggs are mature, or nearly so: once fertilized they again reject attempts at copulation (BARTON BROWNE, 1958). Female specimens marked at the time of emergence in November, were captured at daily intervals, and their spermathecae examined. None was found to be fertilized until the 4th day after emergence (NORRIS). One mating is probably sufficient to fertilize all or most of the eggs developed in the course of the lifetime of the female sheep blowfly. The eggs of females which are unable to find an oviposition site may degenerate in the ovarioles and become nonviable (WEBBER, 1955).

At 78° F adequately fed, and fertilized, females lay batches of eggs at intervals of a few days. The number in a batch varies with the size of the fly, from 70 to 260 (WEBBER, 1955). At 78° F cultures of average-sized individuals of *L. cuprina* provided with abundant sugar and protein have been known to produce over 1000 eggs per female. There is no precise knowledge of the reproductive potential of "wild" flies.

In southern Australia *Lucilia cuprina* is comparatively unsuccessful in colonising carrion (FULLER, 1934a; WATERHOUSE, 1947). In experiments carried out in the Canberra area a maximum of only 654 *L. cuprina* emerged from numerous sheep carcases studied, despite a contemporary local abundance of *L. cuprina* adults to deposit eggs on the carcases. The average number of flies emerging from sheep carcases was very much lower than 654, and some carcases produced no *L. cuprina* at all. (WATERHOUSE, 1947). The species had no greater success in the carcases of birds, reptiles, hares and lambs.

On the other hand, *Lucilia cuprina* exploits living sheep for larval development more readily than other species of Calliphoridae. The live sheep has some attraction far more specific to *L. cuprina* than to other myiasis species. HOBSON (1935, 1936) demonstrated that there was a "living sheep" factor (S factor) which was essential to the attraction of *L. sericata* to fleece. A putrefactive factor (P factor) was also required, involving chemical substances known to occur in decaying animal matter. This was also found to be true for *L. cuprina* (MACKERRAS, I. M. & M. J., 1944). Moistness of the fleece is generally important. Once the female is attracted, it explores the fleece for a cavity, where the eggs are deposited in a comparatively humid, dark environment (ROGOFF & BARTON BROWNE, 1956).

The presence of ovipositing females in the fleece of a sheep is a stimulus to other females to oviposit, and many mass eggs together in one situation. Such mass oviposition may benefit the species, if humidities are unsuitable for survival of eggs and 1st instar larvae. Eggs near the centre of the mass are protected from desiccation (CRAGG, 1955), and larvae hatching there may feed on dead or dying individuals, and attain the second instar, when they are vigorous enough to attack the living tissues.

Maintenance of high humidities is necessary to ensure survival of eggs in the fleece. If this requirement is met, hatching occurs in less than 24 hours. In the establishment of the maggots, the moistening of the skin surface is important. This may be effected by rain, urine or wet faeces, or initially partly by the secretions of the maggots themselves. The young maggots feed on serous exudations from the skin, but sometimes the more advanced maggots attack the tissues so extensively that only their posterior ends protrude, keeping the spiracles in contact with the atmosphere. Further oviposition by *L. cuprina* may occur. Other species also lay their eggs or larvae, and the strike extends.

On living sheep, *L. cuprina* maggots may be fully fed as early as the 4th day after hatching. They then wander from the strike, drop from the sheep, and burrow into the soil before pupating. Later-developing *L. cuprina* maggots may be subject to attack by, or competition with *Chrysomya rufifacies* maggots (TILLYARD & SEDDON, 1933). Usually strikes are treated before this stage is reached, but many go undetected until some of the *L. cuprina* maggots have developed to a stage at which they can pupate if denied further opportunity for feeding.

WATERHOUSE (1947) took struck sheep into an insectary and collected the maggots as they matured. Table II shows the numbers of flies produced. An average of 1265 *L. cuprina* emerged, which is in marked contrast with the mean of 305 from 11 sheep carcases exposed under conditions favourable to *L. cuprina*. The transference of the sheep affected with these strikes to the insectary arrested oviposition when it was probably at a comparatively early stage. Hence the potential production of *L. cuprina* from living sheep may be even higher.

In view of its poor success in carrion, and high incidence in sheep myiasis, it has been suggested (Anon., 1940; MACKERRAS, 1936;

Table II.

Numbers and species of flies bred from natural strikes at Canberra (after WATERHOUSE (1947)).

Month strike commenced	Lucilia cuprina	Lucilia sericata	Calliphora stygia	Calliphora augur	Chrysomya rufifacies	Peronia rostrata	Musca domestica
October	980 767 499						
November	1692 37 217 433 398 956 673 1075 350		31 157 11	13 87 72 165 156			
December	*4343 398	44	168	3 156	31	301	
March	471 227 849 130						
April	*9704 2683 974 861 864 611 1432		573 658 448	2 12		4 1	203

* These sheep died.

WATERHOUSE, 1947), that the main source of *L. cuprina* population is in strikes on living sheep. This is certainly true when strike "waves" prevail, following warm, wet weather of sufficient duration to permit numerous strikes to mature one or more generations of *L. cuprina*. Such conditions are not usual, however, and although available evidence points to the living sheep as an important contributor to the population of *L. cuprina*, at times there appear to be populations of the same order as those measured by GILMOUR et al. (1946) when there has been little or no strike to account for the presence of the flies. Quantitative surveys of carrion on a specific area would be necessary to establish whether the small numbers of *Lucilia cuprina* known to be produced from carcases could account for such populations.

CRAGG (1955) found that, when a sheep was suffering from myiasis at the time of death, the numbers of *L. sericata* (the most important myiasis fly in Great Britain) produced from the carcase were as high as 20,000, far outnumbering other species. When, however, a sheep was uninfested at the time of death, *L. sericata* seldom comprised 10% of the maggot population, and at times was even completely unrepresented. The latter circumstance closely parallels Australian experience with *L. cuprina*. "Myiasis" carcases — infested with *L. cuprina* before death — have not been specifically investigated, but the large numbers of *L. cuprina* emerging in the insectary from sheep which died of fly strike (Table II) suggest a close parallel. "Myiasis" carcases would occur principally during a strike wave, when infested sheep were too numerous to permit thorough treatment of the flocks.

Presumably *L. cuprina* larvae which fall from strikes escape attack by Silphid, Staphylinid and Histerid beetles, which destroy some at least of the larvae occurring in carrion. The most common hymenopterous parasite of blowflies in Australia, *Mormoniella vitripennis* (WALKER) (Pteromalidae) attacks only puparia: the great majority of *L. cuprina* larvae wander some distance from their feeding place, and burrow underground before pupating. Hence, although a large percentage of puparia exposed in the laboratory may be attacked by *Mormoniella*, in nature they are mostly inaccessible to this parasite.

It has been presumed that *L. cuprina* passes the winter in the Canberra area in the prepupal stage (MACKERRAS, 1933). Maternal-induced diapause, such as occurs in *L. sericata* (CRAGG & COLE, 1952), was not detected in the progeny of more than 90 gravid females of *L. cuprina* trapped throughout the autumn of 1957. Diapause, if it occurs in this species, must therefore be induced in the immature stages. Less probably the development is simply cold-inhibited.

Lucilia sericata

In Australia, *Lucilia sericata* is of negligible importance as a sheep myiasis fly, in contrast with its status in Great Britain, where its incidence in strikes is as high as that of *L. cuprina* in Australia. CRAGG & COLE (1956) found that Australian *L. sericata* showed a very much weaker olfactory response to wool than either English *L. sericata* or Australian *L. cuprina*. It appears that most individuals of the Australian population of *L. sericata* lack sensory reactions enabling them to initiate attack on living sheep, although oviposition by *L. sericata* on sheep in an insectary may be stimulated by prior oviposition by *L. cuprina* (MACKERRAS, I. M. & M. J., 1944), and the larvae can thrive once established in a strike.

L. sericata survives the Canberra winter as diapausing larvae. These hatch from eggs laid in the autumn, many larvae entering diapause even if the cultures are maintained in a warm room from the time of capture of the gravid female (NORRIS).

MACKERRAS (1933) found that a pair of *L. sericata* lived for 77 days, the female laying egg batches from the 6th day onwards, until the 67th day, when the 13 th and last batch was laid. Progeny from this pair totalled 2,373.

The life cycle is slightly longer than that of *L. cuprina* at comparable temperatures (WATERHOUSE & PARAMONOV, 1950). There are doubtless other physiological differences.

The larval habitats of the heavy *L. sericata* populations found in Canberra gardens (WATERHOUSE & PARAMONOV, 1950) have not been identified. A few flies may be trapped in the sheep country, where some at least must breed, as shown by the capture of teneral flies at least 10 miles away from the city. Rural specimens of *L. sericata* show a much greater variability in body size than does the *L. cuprina* population. This may indicate that the larvae of the two species occupy different ecological niches in the rural areas.

Small numbers of "wild" flies trapped in December 1952 were stained and released in a pastoral area throughout which traps were placed up to 4 miles from the release point. Marked specimens were taken up to to $1\frac{1}{4}$ miles from the liberation site within 24 hours, and up to $3\frac{1}{2}$ miles away in 48 hours.

Calliphora stygia

Calliphora stygia is adapted to lower temperatures than other common Australian sheep blowflies. This is reflected in its geographical and seasonal distribution, its general activity curve (NICHOLSON, 1934) and the temperature tolerances of its larvae.

At Canberra, adults of *C. stygia* may appear in small numbers in the winter during any mild spell. Specimens have been found during stormy winter weather sheltering under bark and stones. Such winter adults emerge during the late autumn, larvae from winter-laid eggs not giving rise to adults until the spring (FULLER, 1934a).

In the mountains and tablelands of New South Wales *C. stygia* may attain great abundancee in spring and early summer. Tentative measurements of population density near Canberra in December 1952 showed that there were about 100 flies to the acre, assuming 100% survival of marked specimens. (MCINTYRE; NORRIS). Even allowing for a heavy mortality, the population density was obviously considerably higher than any *L. cuprina* population measured in the area. Particularly high population densities are likely to occur in the spring if severe winters destroy large numbers of sheep, because the *C. stygia* prepupae and puparia accumulate in the soil until the warm weather. For instance, samples taken over a radius of 15 ft from a sheep carcase during the winter of 1950 showed that there were about 16,000 puparia and prepupae, exclusively of *C. stygia*, hibernating in the soil. The soil surrounding several other carcases in the same paddock was similarly infested with puparia, which would all disclose flies within a few weeks of one another in spring.

C. stygia breeds in moderate numbers in carcases in spring and early summer (WATERHOUSE, 1947), and in large numbers in autumn and winter, but very few mature in November—February. This is partly caused by increased predation by *Chrysomya rufifacies* in carrion in the warmer months; but to some extent it is a direct effect of high temperatures. Cultures of *C. stygia* are destroyed if insectary temperatures approximate 100° F for brief periods. Such temperatures are frequently attained in small carcases lying in the sun or in large carcases under the combined effect of insolation and maggot activity (WATERHOUSE, 1947).

C. stygia has been recorded as initiating less than 10% of strikes in districts where it is abundant, although it is present in a higher percentage of strikes started by *L. cuprina* (MACKERRAS & FULLER, 1937). These infestations occur principally in autumn and spring. WATERHOUSE (1947) (Table II) reared up to 658 *C. stygia* from a strike. The contribution of strikes to the gross population, however, would be negligible.

"Wild" flies, stained after trapping, and released, were collected in traps up to 4 miles away within 24 hours (NORRIS). Laboratory-reared adult flies, stained and released were captured 8 miles from the point of liberation (GURNEY & WOODHILL, 1926).

Greatest numbers of this species are trapped in the Canberra district in timbered paddocks. There is a strong correlation between sex ratio and numbers of blowflies caught in a trap (Fig. 2). Possibly this reflects a preference of the species for particular habitats, which is either more strongly marked in the males, or is partly suppressed in females in search of oviposition sites and protein for egg-maturation. Males, to which protein is not a requisite, have no incentive to forsake preferred habitats.

In warm weather, males are particularly active at dawn and dusk. Individuals appear to select vantage points from which they sally at intervals to challenge passing insects. Although mating has not been observed on these occasions, it is a likely basis for such activity, possibly triggered by a certain light intensity. Out-of-doors in hot

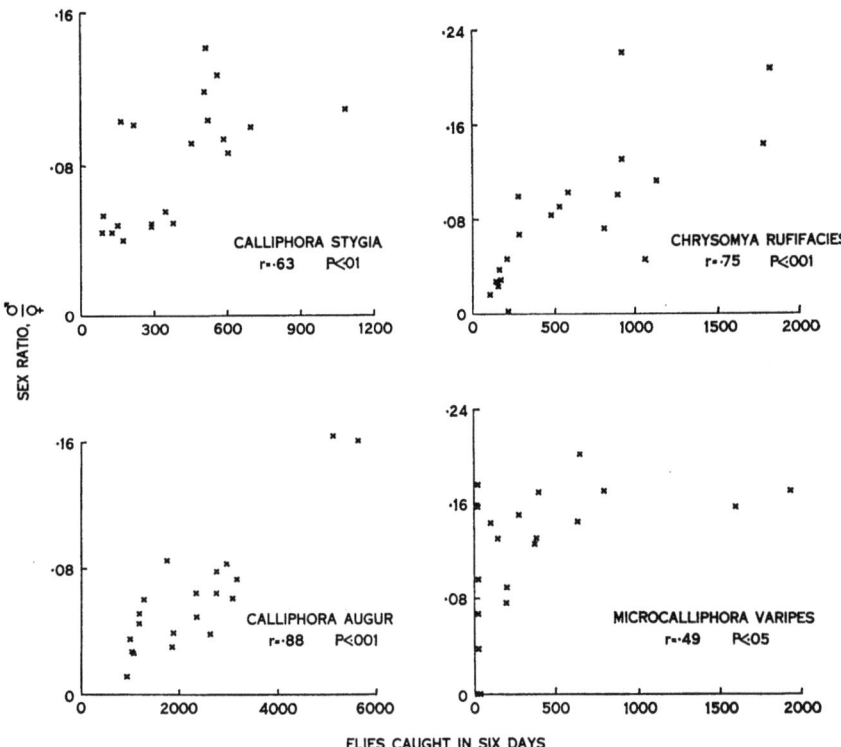

Fig. 2. Correlation of sex ratio and numbers of flies caught in six days at 20 trap sites in the A.C.T., March 1952.

weather some specimens of this species may be active well beyond the daylight hours. Indoors, females may fly at any hour during hot summer night, even in darkness.

Carbohydrates are necessary for survival, and protein for egg maturation, as indeed is true of the adults of all Calliphorids which have been maintained in culture.

C. stygia is oviparous. MILLER (1939) stated that in New Zealand *C. laemica* WHITE (currently synonymised with *C. stygia*) is oviparous in cool weather, and viviparous in hot weather, when it may larviposit on the wing. MURRAY (1956) also describes New Zealand *C. stygia* as viviparous. This is contrary to experience with Australian *C. stygia*, and re-examination of the taxonomic status of the species concerned seems necessary. The subject is relevant to the ecology of a species, as the oviparity of *C. stygia* is considered to place it at a disadvantage in comparison with ovoviviparous species in exploiting small carcases.

MACKERRAS (1933) gave the average number of eggs laid by *C. stygia* as 264. Doubtless it would vary with the size of the fly, as in *L. cuprina*.

Except during the winter, Coleopterous predators devour a proportion of the *C. stygia* larvae occurring in carrion. *Brachymeria calliphorae* (FROGGATT) (Chalcididae) may attack some larvae in late summer and autumn (FULLER, 1934a), and *Hemilexomyia abrupta* (DODD) (Diapriidae) parasitises some *C. stygia* larvae in the winter. Neither parasite is ever numerous. *C. stygia* larvae mostly migrate away from the carcase, and burrow underground before pupation. Thus the species is usually protected from the puparial parasite, *M. vitripennis*, although exposed puparia are highly susceptible to infestation.

Calliphora albifrontalis

NEWMAN et al. (1930) trapped *Calliphora albifrontalis* only in late autumn, winter and early spring in five widely-separated areas on the tableland in Western Australia. It was unrepresented in the trap catches for the remainder of the year, though other species of Calliphoridae were present. If this is a true picture of the seasonal occurrence, then *C. albifrontalis* may be adapted to survive the hot and very dry south-western Australian summer in the larval or pupal stage, unless it becomes extinct annually over wide areas, which are repopulated in the autumn from coastal regions.

NEWMAN (quoted in FULLER (1934a)) stated that *C. albifrontalis* was the most important myiasis fly in the southern parts of south-western Australia. MACKERRAS & FULLER (1937) recorded it in 45 out of 310 strikes in Western Australia.

Calliphora fallax

Calliphora fallax has a superficial resemblance to *C. stygia*, but is more closely allied to *C. augur* in important features of its reproduction. It is ovoviviparous, and the first instar larva bears a striking resemblance to that of *C. augur*.

Up to 20% of the "brown" blowflies taken in traps in the Canberra district may be *C. fallax*. For the most part these were counted as *C. stygia* by past investigators, but as *C. fallax* usually constitutes only a very small proportion of the catch, conclusions regarding *C. stygia* would not have been affected seriously.

There is a strong correlation between the numbers and sex ratio of *C. fallax* caught in traps (NORRIS).

The ecology of *C. fallax* is poorly understood. It has been bred from vertebrate carcases, dead fresh-water crayfish, and accumulations of dead bogong moths (*Agrotis infusa* (BOISDUVAL)) in mountain caves in the A.C.T. (COMMON, 1954), but has been reported from only two or three sheep myiases (FULLER, 1934a).

Calliphora augur

This species may attain great abundance during early summer, when it constitutes a large proportion of the huge population of blowflies buzzing in the bush and paddocks. It is also a household nuisance.

Marked "wild" specimens were retrapped 4 miles from the release point within 24 hours (NORRIS). Laboratory-bred specimens were recaptured 9 miles from the release point (GURNEY & WOODHILL, 1926).

It was calculated from trapping experiments that, on two successive days in December 1952, the population of *C. augur* in the A.C.T. was of the order of 350 flies to the acre (McINTYRE; NORRIS). This figure does not allow for mortality of the marked flies, but even if this had been fairly heavy, there is strong confirmation of the impression of high population density at the time of maximum seasonal abundance.

There is a high correlation between sex ratio and the numbers of flies caught in traps (Fig. 2). Like *C. stygia*, too, males of *C. augur* are intensely active at dawn and dusk in hot weather.

C. augur is ovoviviparous, laying batches of about 50 eggs, from which robust and highly active first instar larvae promptly hatch. This reproductive adaptation gives the species a great advantage over oviparous species in exploiting small, quickly perishable carcases. By the time the larvae of oviparous species have hatched, those of *C. augur* are advanced in the second instar, and the pabulum is consumed before the larvae of the oviparous species can mature.

The breeding media of *C. augur* are more diverse than those of any other Australian species. It has been known to complete its larval development in various dead invertebrates: lepidopterous larvae, masses of dead blowflies, earthworms, snails and fresh-water crayfish. It has been reared from sour milk, cheese, fermenting grain, and moistened blood and bone manure. It infests a wide variety of carcases of vertebrates of all classes, from fish to the larger domestic animals. Fifty specimens were reared from a naturally occurring carcase of *Acanthiza chrysorrhoa* QUOY & GAIMARD, a bird smaller than a sparrow. WATERHOUSE (1947) reared 8,573 *C. augur* from the carcase of a 96 lb sheep exposed in November: the numbers reared from sheep were generally much fewer, although, in carcases exposed in the summer, they were much higher than the numbers of *C. stygia*. Some hundreds of larvae were also detected in the nostrils and head cavities of a living brown hawk (*Falco berigora* VIGORS & HORSFIELD)* suffering from fowl pox. Undoubtedly *C. augur* causes traumatic myiases of other native birds and animals, and probably some at least of such human

---

* Kindly made available to the writer by Mr. A. L. DYCE, of the C.S.I.R.O. Wildlife Survey Section, Gungahlin, A.C.T.

myiases as are reported in eastern Australia. As a sheep myiasis fly *C. augur* is also more catholic in its tastes than other Calliphorids, showing higher incidence than other species in wound myiases (MACKERRAS & FULLER, 1937). Out of 603 strikes in a Canberra flock, this species was involved in 165, of which it may have initiated about a quarter. WATERHOUSE (1947) bred up to 165 *C. augur* from one strike (Table II).

Larvae of *C. augur* behave like those of *Lucilia* species and *Calliphora stygia* in wandering away from a carcase and burrowing before pupation. They therefore parallel those species in the degree to which they are vulnerable to the hymenopterous parasites. They are favoured as food by predaceous carrion beetles.

MACKERRAS (1933) advanced the opinion that *C. augur* overwintered as puparia, but this is not confirmed.

Calliphora nociva

Small numbers of this species have been reared from carrion at Canberra and Cunnamulla (south-west Queensland), but there is very little information about its bionomics.

Calliphora nociva occurred in 56% of strikes containing *Calliphora* species, and was present in nearly 50% of strikes examined from its area of distribution. Like the closely related *C. augur*, it occurs in higher incidence than other species in wound myiases (MACKERRAS & FULLER, 1937). Further investigation of its biology and ecology is highly desirable.

Calliphora vicina

There is an extensive literature on this species in the northern hemisphere. As a sheep myiasis fly in Tasmania it appears to merit further study (RYAN, 1954), but data are unavailable for Australian conditions.

Chrysomya rufifacies

The seasonal distribution, larval temperature tolerances and general activity curve (NICHOLSON, 1934) of *Chrysomya rufifacies* indicate a closer adaptation to a more tropical climate than the *Calliphora* species. *Chrysomya* belongs to the same subfamily as the American screw-worm, *Callitroga hominivorax*, which, in the United States, survives the winter only in Florida and the extreme southwestern States. Each summer adults spread north again, at an average rate of five miles a day, to repopulate vast areas from which the species died out completely during the winter (EDDY & BUSHLAND, 1956). It is possible that comparable phenomena occur in the Australian environment. Blowflies have such powers of dispersal that large areas could be reinfested annually.

At Canberra, adults of *C. rufifacies* occur for a shorter season than *Calliphora* and *Lucilia* species, and contrast with them by attaining greatest numbers in the hottest months.

Laboratory-reared flies were trapped 10 miles from the point of liberation (GURNEY & WOODHILL, 1926). In the author's own experiments some "wild" flies trapped, stained and released, were retrapped 4 miles away within 24 hours. The population density at this time, when the seasonal peak of abundance had not been attained, may have been of the order of 50 flies to the acre. There is a strong correlation between sex ratio and numbers of *G. rufifacies* caught in traps (Fig. 2).

The reproduction of this species was studied in India by ROY & SIDDONS (1939). Males have spermatozoa in their testes immediately after emergence from the puparium, and may fertilize at least 5 females. Females lay up to 386 eggs in a batch, though the mean of those studied was 210. Eggs were not deposited until fertilization had occurred, and one fertilization sufficed for the production of at least 3 egg-batches. An exceptional feature of this species was that each female produced either male offspring only, or female offspring only throughout her lifetime. This is also the case with Australian *C. rufifacies* (WATERHOUSE, unpubl.).

C. rufifacies adults may visit quite fresh carcases, but do not lay eggs upon them until considerably later. It is possible that, in keeping with the predaceous habits of the larvae, the odour of maggots of another species is usually a factor in stimulating oviposition. The larvae, which differ from primary larvae in being markedly tuberculate, are capable of completing their development in carrion in the absence of larvae of other species, but they prey upon smooth maggots if these are present. A *C. rufifacies* larva which encounters a primary larva curls itself around it, kills it by piercing it with the mouth-hooks, and consumes the body contents. In pure cultures under crowded conditions, the larvae are also cannibalistic.

Chrysomya rufifacies maggots not only devour primary maggots, but may also exert a repellent effect upon them, so that the primaries may leave a carcase before completing their feeding. This repulsive effect is particularly evident in the case of *Lucilia* larvae (FULLER, 1934a; WATERHOUSE, 1947).

The life cycle of *C. rufifacies* is shorter than that of other carrion species. At 20° C the period from egg to adult emergence is 12—14 days, compared with 14—16 days for the *Lucilia* species, 18—20 days for *C. augur*, and 20—24 days for *C. stygia* (MACKERRAS, 1933).

As many as 27,600 flies emerged from one sheep carcase (WATERHOUSE, 1947). The numbers were usually large from November to March. *C. rufifacies* does not initiate strike (MACKERRAS & FULLER, 1937). The *Chrysomya* larvae, which have been found in about 8%

of strikes, are always younger than the oldest primary larvae in the same strike. If strikes containing *C. rufifacies* are neglected, the maggots may cause extensive injury, and invade the body to a depth of several inches. According to TILLYARD & SEDDON (1933), *Chrysomya rufifacies* repels, preys upon, and competes with, primary maggots for food and space in strikes, as in carrion. Under conditions of average management, very few strikes reach the stage at which *Chrysomya* larvae could pupate, if deprived of further opportunity for feeding. Compared with carrion, the living sheep would be negligible as a source of this species.

The maggots of *C. rufifacies* pupate on or in the carcase, or close to it on the ground surface. In normally compact soil, few burrow before pupation. Most are thus extremely vulnerable to parasitization by *Mormoniella vitripennis*, though their tough puparia make them relatively immune to attack by predaceous carrion beetles.

Chrysomya micropogon

Chrysomya micropogon has smooth larvae, unlike those of *C. rufifacies*, but they are said also to prey upon *Calliphora* and *Lucilia* larvae (TILLYARD & SEDDON, 1933). *C. micropogon* is apparently an important member of the carrion fauna in northern Australia, but has been little investigated. It has been recorded as a secondary invader in only 6 strikes (MACKERRAS & FULLER, 1937).

WATERHOUSE (unpubl.) found that, unlike *C. rufifacies*, both sexes were represented in the progeny of each female *C. micropogon*.

Microcalliphora varipes

The bionomics of this small Calliphorid are poorly understood. Its seasonal distribution in the Canberra area corresponds fairly closely with that of *C. rufifacies*, extending from late October to early April, with maximum abundance in late summer.

None of the reared 12,500 specimens of this species liberated by GURNEY & WOODHILL (1926) was recaptured. In December 1952, about 3,500 "wild" flies were trapped in the Canberra district, stained, and liberated. In the 24 hours following liberation 33 specimens were retrapped $\frac{1}{2}$ mile from the release point, and a further 10 specimens up to $2\frac{1}{2}$ miles from the release point. By the end of 48 hours, some specimens had been recaptured up to $3\frac{1}{2}$ miles from the point of liberation. At this time the natural population density may have been of the order of 10—20 flies per acre, but maximum abundance for the season had not been attained. There was a positive correlation between sex ratio and numbers of flies trapped at any site (Fig. 2).

The species is unique among the blowflies under consideration in that the male postures in front of the female before mating. In this courtship display a prominent part is doubtless played by the striking pattern of the front aspect of the male, and the specialised bristles on the forelegs. Oviposition on carrion occurs only after primary maggots have undergone some development. The larvae are tuberculate, rather like those of *C. rufifacies*. FROGGATT (1918) stated that they were predaceous but FULLER (1934a) obtained no evidence of this.

Pupation, as in *C. rufifacies*, occurs on or near the carcase, and comparatively few larvae burrow underground. The puparia are therefore vulnerable to *M. vitripennis*, but their rugged cuticle gives protection from predators. WATERHOUSE (1947), raised this species from sheep carcases between October and early May. From one sheep carcase exposed in January, 78,000 *M. varipes* emerged.

M. varipes is of little consequence as a sheep myiasis fly, having been recorded in only 4 out of 1691 strikes. Its importance lies in its capacity to compete with primary blowflies in carrion.

Sarcophaga species

Members of this group are of little consequence in myiasis in Australia. No species occurs in any numbers in carrion in the Canberra area. According to FULLER (1934a), *Sarcophaga* species are secondary flies in carcases in central Queensland, and primary flies in northern Australia.

Peronia rostrata

This muscid is of problematical importance in the ecology of sheep blowflies. It is capable of completing its life cycle on carrion alone, but has been bred from a variety of other materials: pigsty debris, bone heaps, mud saturated with effluent from abattoirs (FULLER, 1934a), and fowl manure.

FROGGATT (1918) stated that it was predaceous on all species of blowfly larvae in carrion, and that it was relatively free from parasitization by *M. vitripennis*. However, it is a late-comer to carcases, compared with Calliphorids, and has a much longer developmental period. Therefore it could have little influence in the important early stages of intense competition.

According to FULLER (1934a), small numbers of adults overwinter in the Canberra area, like those of *C. stygia*. Larvae are certainly present throughout the winter in carcases and in the soil beneath them. Flies were bred throughout the year from sheep carcases in the Canberra district (WATERHOUSE, 1947). Over 22,000 emerged from a 68 lb sheep body exposed in February. The numbers from other experimental carcases were small. *P. rostrata* is classed as a tertiary fly, of no importance in sheep myiases (MACKERRAS & FULLER, 1937).

Of about 8,000 "wild" flies trapped in December in the Canberra district, stained and released, 2 were retrapped half a mile from the release point within 24 hours, and a further 8 up to 3 miles from the release point in the 24—48 hour period after release.

Inter-relations of blowflies, and other organisms

A carcase may be recognizable as a special habitat for more than a year, during which a large and diverse fauna may be associated with it. For the most part the Calliphoridae have completed their life cycle in the first week or so, and much of the fauna subsequently associated with the carcase has little relation to their success or failure. In a survey specifically of Calliphorid ecology, therefore, it would be irrelevant to consider the entire carrion complex. Attention will be concentrated particularly on the agencies which are thought to be important in the limitation of blowfly numbers.

Sources of blowfly populations

The amount of carrion available sets a maximum to the abundance of most Calliphorids. To some extent, then, the problem of regulation of blowfly numbers is secondary to population problems of vertebrate species. Furthermore, carrion-feeding mammals and birds destroy a considerable proportion of vertebrate carcases. Hence there is also an element of competition between vertebrates and blowflies in the limitation of blowfly numbers.

A regular seasonal variation in the supply of carrion in Australia has not been established, although in the southern half of the continent there may be a spring-early summer abundance of dead nestlings and fledglings, associated with the breeding season of most bird species. There appears to be no regular seasonal variation in the abundance of domestic animal carcases, nor a clear-cut seasonal fluctuation in the suitability of the entire sheep population for myiasis, as described for English conditions (MACLEOD, 1943). Cataclysmic events like droughts, or unseasonable cold snaps following shearing sometimes augment carcase numbers greatly, but these are quite irregular in occurrence.

Many factors influence the production of blowflies by those carcases not dismembered and consumed by vertebrate scavengers, or disposed of by man. Season of death determines which species are present as adult flies to deposit eggs or larvae. Mode of death influences the relative success of the various species: an animal dying of myiasis affords the best opportunity for the species involved in the myiasis: extensive wounds on a carcase may attract species earlier than they would visit a normally decomposing, intact body. The location of the body in relation to shade, or the suitability of the underlying soil or rock for the survival of the puparia may influence the chances of a blowfly species completing its life cycle. The size of the body is also a factor: only the most rapidly-developing blowflies can complete their growth in small carcases before they dry out.

There may also be more subtle influences on blowfly production. HOLDAWAY (1930) found in his experiments in France that *Chrysomya* larvae predominated if whole rabbits were used, but Sarcophagids predominated if "meat" was used. FULLER (1934a) stated that "... certain kinds of maggots are more numerous in certain dead animals. A cat always has many more *Chrysomya* larvae present than a guinea pig, whilst the guinea pig may contain only a few dozen hairy maggots and be crowded with those of *Lucilia* and *C. augur*." It is possible, therefore, that some types of naturally occurring carrion are usually neglected by *Chrysomya*, and therefore favour the production of primary flies. However, such breeding places, if they exist, have not yet been recognised. Sheep and rabbits have largely supplanted the grazing marsupials in Australia, and may reasonably be presumed to contribute most carcases for the breeding of blowflies: the bodies of neither of these animals are unfavourable to *Chrysomya* reproduction.

Competition in the limitation of blowfly numbers

There is no indication of competition between adult blowflies for requisites other than suitable sites for the deposition of their eggs or larvae. There is, in fact, strong evidence that in *Lucilia cuprina* the stress of competition for protein requirements falls on the larval, rather than on the adult stage (NICHOLSON, 1957). In all cultures at the time of their establishment from "wild" stock, *Lucilia cuprina* females have been found to require protein to enable them to mature eggs. Evidence was also advanced earlier in this chapter that, in nature, greatly augmented local populations of sheep blowflies were able to secure sufficient protein for ovarian development, even though conditions seemed unfavourable. NICHOLSON (1957) maintained cultures of *L. cuprina* in which larval food was so abundant that competition was negligible, whereas in the adult stage, the amount of protein supplied was far below that necessary for all flies to mature batches of eggs. This resulted, over a year or so, in the selection of a strain of sheep blowflies which could mature eggs with little or no protein feeding in the adult stage. Sufficient reserves had, in fact, been accumulated by the abundantly-fed larvae to enable the adults to lay some eggs. The conditions regarding protein supply in the cages were the reverse of those obtaining in nature.

Intra-specific competition amongst blowfly larvae

The secretions and excretions from one or a few maggots may not cause sufficient local liquefaction of the substratum to permit feeding and growth. Thus, up to a point, maggots of the same species derive an advantage from numbers. In nature, however, carrion is usually crowded with far more larvae than can possible reach maturity. HOLDAWAY (1930), for instance, showed that, of an initial population of 50,000 *L. sericata* in half a sheep's head, 231 flies emerged.

The action of intra-specific competition is best demonstrated under experimental conditions. FULLER (1934b) showed that, if crowding of *C. augur* maggots was progressively increased, there was a corresponding decrease in the average size of the maggots. When 25 *C. augur* maggots were placed on a small guinea pig carcase,

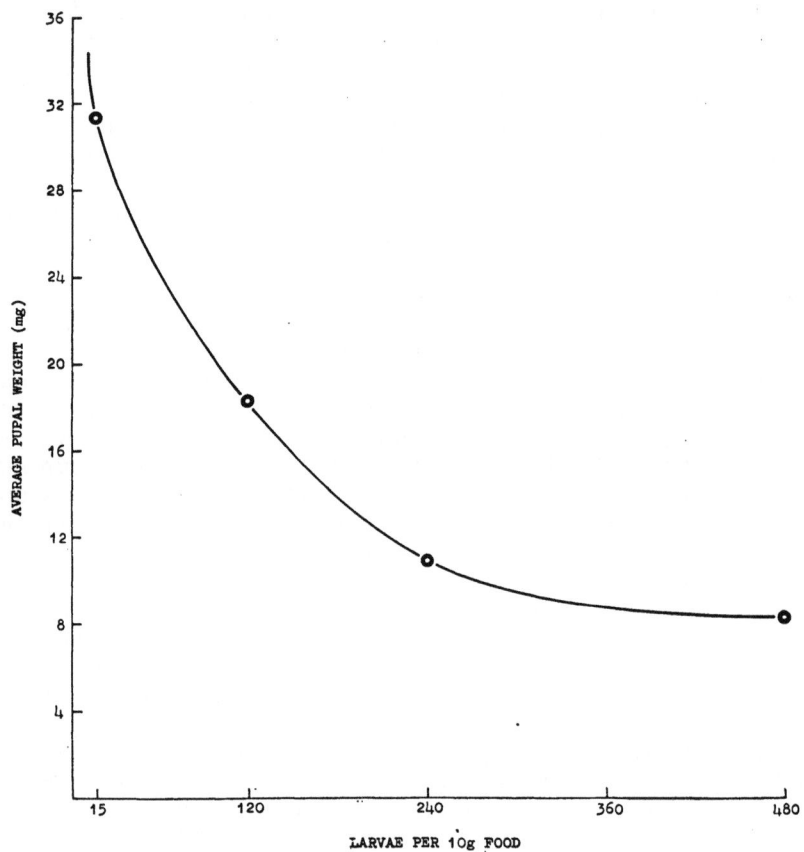

Fig. 3. Relation between puparial weight of *Lucilia cuprina* and the amount of food provided per larva (after WEBBER (1955)).

all secured as much food as they could consume, and the puparia weighed a mean of 72 mg. When 600 maggots were placed on a similar carcase, the food was exhausted before any had attained full growth, and the mean weight of the puparia was 24 mg. Comparable results are obtained through progressive increase in crowding of *Lucilia* species (FULLER, 1934a; WEBBER, 1955) (Figure 3).

Up to a certain stage of crowding, intra-specific competition leads to a reduction in mean size of mature insects rather than a reduction in numbers. It is possible, however, to overcrowd carrion with maggots of the same age and species to the extent that none at all reaches maturity.

Probably in most situations in nature intra-specific competition is secondary in importance to interspecific competition and predation, otherwise, contrary to experience, all "wild" Calliphorids would be very stunted.

Inter-specific competition amongst blowfly larvae

Inter-specific competition between primary flies (*Lucilia* and *Calliphora* species), which are all non-predaceous, likewise limits the size of the adult flies before greatly affecting larval survival (FULLER, 1934a). *Lucilia* species were the most sensitive to this factor, and likewise the most affected by the introduction of *Chrysomya rufifacies* larvae into experimental carcases, being eliminaed at a lower intensity of competition and predation than were the *Calliphora* species. *C. stygia*, in particular, suffered less than *Lucilia* species from predation by *C. rufifacies* this usually resulted in reduction in the number of individuals, but the survivors tended to approach the maximum size. Possibly some larvae of *C. stygia* from a particular age group develop a little more rapidly than their fellows, which are selectively preyed upon by *C. rufifacies*. Thus, while sufficient primary larvae are available to *Chrysomya* larvae as food, the more advanced *C. stygia* larvae complete their development unmolested. There is some evidence of a similar outcome from *C. rufifacies* predation on *C. augur* and *Lucilia* species.

It is possible that the repellent effect of *C. rufifacies* maggots on primary larvae is not always disadvantageous. TILLYARD & SEDDON (1933) described how, in strikes involving *C. rufifacies*, the remnants of the antecedent population of *L. cuprina* may be found in isolated pockets in the wool, the point of origin of the strike being invaded and completely occupied by *C. rufifacies*. In isolated pockets, *L. cuprina* larvae might remain free from attack long enough to permit them to metamorphose into viable pupae if they were ultimately so intensely repelled by the larvae of *C. rufifacies* that they were forced to desert the strike or carcase. Some such mechanism seems necessary to explain the great abundance of *L. cuprina* in the tropical regions, where it is described as swarming "on any attractive material during the spring and autumn months". Its decreasing abundance as one proceeded south-wards in Australia was ascribed to increasing competition with *Calliphora* species (MACKERRAS & FULLER, 1937). However, *C. rufifacies* is also considered to be a major factor in the precarious existence of *L. cuprina* in carrion in the south, and, to explain the great abundance of *L. cuprina* in the north in the face of heavy populations of Calliphorids with predaceous maggots, it is necessary to invoke either the increasing importance northwards of some form of carrion to which *L. cuprina* is specifically adapted, or else selective refuge of its maggots in particular portions of carcases.

Effects of predation and parasitism in the limitation of blowfly populations

Except during the winter, carcases in the Canberra area are rapidly colonised by blowflies. *Lucilia* and *Calliphora* species are first to deposit eggs or larvae. Then the early phase of intense competition among primaries is followed by the invasion of the carcase by *Chrysomya* and *Microcalliphora*. Predaceous beetles arrive at the carcase in the very early stages, attacking particularly the primary maggots. WATERHOUSE (1947) estimated that in his experiments *Saprinus* species (Histeridae) and *Creophilus erythrocephalus* (FABRICIUS) (Staphylinidae) consumed about half the primary maggots developing in a carcase when the maggots were restrained from escaping, but considered that their effect would be less in nature. On the basis of experiments in which beetles were removed from some carcases, FULLER (1934a) concluded that the predaceous beetles exerted a negligible effect on the numbers of blowflies emerging from a carcase, even under the most favourable conditions: the beetles chiefly ate maggots which were forced off the crowded carcase before completing their feeding. These maggots would mostly have died of starvation in any case.

Of the hymenopterous parasites, *Mormoniella vitripennis* is by far the most important. In the early days of blowfly research in Australia, large numbers of *M. vitripennis* were bred for distribution throughout the country, where, almost certainly, the species was already widely established, and in balance with the blowfly population. No effect on the abundance of blowflies was detected as a result of this campaign. FULLER (1934a) stated "It is obvious that there are always sufficient blowflies left after *Mormoniella* has made its attack to overcrowd the available carrion with their larvae. Therefore the parasite plays no part in reducing the (total) blowfly population, but it does influence the relative proportions of the different species, since it discriminates against *Chrysomya* and *Microcalliphora*." If the campaign of distribution of *Mormoniella* had produced any result, it would probably have increased the production of primary blowflies, which was the reverse effect to that required.

WATERHOUSE (1947) advanced evidence that 64% of both *C. rufifacies* and *M. varipes* puparia in his experiments were parasitised by *M. vitripennis*. Possibly local concentrations of *M. vitripennis* had built up in the vicinity of the experimental trays, where carrion had been exposed continuously for long periods.

JOHNSTON & TIEGS (1921) pointed out that *M. vitripennis* produced fewer eggs than most blowflies, and also lost efficiency as an agent regulating blowfly abundance through its habit of laying a number of eggs in a single puparium, and ovipositing in puparia too advanced for emergence of the adult flies to be prevented.

Climatic influences on distribution and abundance of blowflies

It is significant that there is little debate on the importance of biotic factors in regulating blowfly populations. The action of competition, in particular, is obvious, even to the casual observer. Climatic influences are nevertheless of great importance in determining the level at which competition, predation and parasitism limit the numbers of individual species.

Each blowfly species is adapted to a particular range of climatic conditions, and varies in mean abundance throughout its geographic range somewhat with the extent to which climatic conditions approach the optimum. Thus the different temperature tolerances of *Calliphora stygia* and *Chrysomya rufifacies* result in each occupying an area in Australia in which the other cannot exist. In addition, both species occur over a vast area in which there is a north to south gradient of decreasing abundance of *C. rufifacies* and increasing abundance of *C. stygia*. Probably *C. rufifacies* affects the abundance of *C. stygia* to varying degrees throughout their common range. Probably, too, the zone of permanent breeding of *C. stygia* would extend further north than it does, but for the high pressure of *Chrysomya* species upon the sparse marginal populations. However, temperature would set the ultimate boundary to the distribution of the species.

More subtle relations are suggested in the east-west distribution of the pairs of allied *Calliphora* species. Passing westward in New South Wales, there is a decreasing abundance of *C. augur*, and an increasing abundance of the very similar *C. nociva*, until, in the western half of the continent, only *C. nociva* is present. A parallel distribution exists with *C. stygia* and *C. albifrontalis*, except that, in this case, the eastern representative, *C. stygia*, is the more widespread, occurring in small numbers even in Western Australia. The reasons for these species maintaining their known distribution and relative abundance may lie in the effects of climatic factors on the intensity of inter-specific competition.

Much interest also attaches to the comparative unimportance of Australian *L. sericata* as a sheep myiasis fly. CRAGG & COLE (1956) state: "It is conceivable that under competition for breeding sites in areas with high sheep densities, strains with a high wool reaction may have been selected over a long period of time. The wool reaction is only one aspect of a fly's behaviour, and under certain conditions may not have been of major importance. Thus in . . . Australia it would seem that the strains of *L. sericata* best fitted to survive in the total environment are strains with a relatively low reaction to wool." In Australia *L. sericata* may have been circumscribed by its requirement of special environments like the gardens of Canberra, which approximate more closely than the general countryside to the English conditions in which *L. sericata* thrives. In such places, however, they would have little contact with sheep, and selection would be in the direction of enabling them to exploit available breeding media, and not towards the development or retention of a strong wool reaction.

REFERENCES

ANDREWARTHA, H. G. & BIRCH, L. C., 1954. The distribution and abundance of animals. Univ. of Chicago Press.

Anon., 1940. The prevention and treatment of blowfly strike in sheep. C.S.I.R.O. (Aust.) Pamphlet 98.

BARTON BROWNE, L. B., 1958. The relation between ovarian development and mating in *Lucilia cuprina*. *Aust. J. Sci.* **20,** 239.

BUSHLAND, R. C., LINDQUIST, A. W. & KNIPLING, E. F., 1955. Eradication of screw-worms through release of sterilized males. *Science* **122,** 287—8.

COMMON, I. F. B., 1954. A study of the ecology of the adult bogong moth, *Agrotis infusa* (Boisd.) (Lepidoptera: Noctuidae), with special reference to its behaviour during migration and aestivation. *Aust. J. Zool.* **2,** 223—63.

CRAGG, J. B., 1955. The natural history of sheep blowflies in Britain. *Ann. appl. Biol.* **42,** 197—207.

CRAGG, J. B. & COLE, P., 1952. Diapause in *Lucilia sericata* (Mg.) Diptera. *J. exp. Biol.* **29,** 600—4.

CRAGG, J. B. & COLE, P., 1956. Laboratory studies on the chemo-sensory reactions of blowflies. *Ann. appl. Biol.* **44,** 478—91.

CRAGG, J. B. & HOBART, J., 1955. A study of a field population of the blowflies *Lucilia caesar* (L.) and *L. sericata* (Mg.). *Ann. appl. Biol.* **43,** 645—63.

CRAGG, J. B. & RAMAGE, G. R., 1945. Chemotropic studies on the blowflies *Lucilia sericata* (Mg.) and *Lucilia caesar* (L.). *Parasitology* **36,** 168—75.

DAVIDSON, J., 1933. The species of blowflies in the Adelaide district of South Australia, and their seasonal occurrence. *J. Agric. S. Aust.* **36,** 1148—53.

EDDY, G. W. & BUSHLAND, R. C., 1956. Screwworms that attack livestock. Animal Diseases. U.S.D.A. Yearbook 1956.

FROGGATT, J. L., 1918. An economic study of *Nasonia brevicornis*, a Hymenopterous parasite of Muscid Diptera. *Bull. ent. Res.* **9,** 257—62.

FULLER, M. E., 1934. The insect inhabitants of carrion: A study in animal ecology. C.S.I.R.O. (Aust.) Bulletin 82.

FULLER, M. E., 1934. Observations on the flies responsible for striking sheep in Western Australia. *J. C.S.I.R.O. (Aust.)* **7,** 150—2.

GILMOUR, D., WATERHOUSE, D. F. & McINTYRE, G. A., 1946. An account of experiments undertaken to determine the natural population density of the sheep blowfly, *Lucilia cuprina* Wied. C.S.I.R.O. (Aust.) Bull. 195.
GURNEY, W. B. & WOODHILL, A. R., 1926. Investigations on sheep blowflies. Dept. Agric. N.S.W. Sci. Bull. 27.
HOBSON, R. P., 1935. Sheep blow-fly investigations. II Substances which induce *Lucilia sericata* (Mg.) to oviposit on sheep. *Ann. appl. Biol.* **22**, 294—300.
HOBSON, R. P., 1936. Sheep blow-fly investigations. III Observations on the chemotropism of *Lucilia sericata* (Mg.). *Ann. appl. Biol.*, **23**, 845—51.
HOLDAWAY, F. G., 1930. Field populations and natural control of *Lucilia sericata*. *Nature, Lond.* **126**, 648—9.
HOLDAWAY, F. G., 1932. Fly strike of sheep: a natural phenomenon. *J. C.S.I.R.O. (Aust.)* **5**, 205—11.
HOLDAWAY, F. G., 1933. Differential behaviour of *Lucilia sericata* Meig. and *Lucilia caesar* L. in natural environments. *J. anim. Ecol.* **2**, 263—5.
JAMES, M. T., 1947. The flies that cause myiasis in man. U.S.D.A. Misc. Publ. 631.
JOHNSTON, T. H. & TIEGS, O. W., 1921. On the biology and economic significance of the Chalcid parasites of Australian sheep maggot-flies. *Proc. Roy. Soc. Qld.* **33**, 99—128.
MACKERRAS, I. M., 1936. The sheep blowfly problem in Australia. Results of some recent investigations. C.S.I.R.O. Aust. Pamphlet 66.
MACKERRAS, I. M., FULLER, M. E., AUSIN, K. M. & LEFROY, E. H. B., 1936. Sheep Blowfly Investigations. The effect of trapping on the incidence of strike in sheep. *J. C.S.I.R.O. (Aust.)* **9**, 153—162.
MACKERRAS, I. M., & FULLER, M. E., 1937. A survey of the Australian sheep blowflies. *J. C.S.I.R.O. (Aust.)* **10**, 261—70.
MACKERRAS, I. M., & MACKERRAS, M. J., 1944. Sheep blowfly investigations. The attractiveness of sheep for *Lucilia cuprina*. C.S.I.R.O. (Aust.) Bull. 181.
MACKERRAS, M. J., 1933. Observations on the life-histories, nutritional requirements and fecundity of blowflies. *Bull. ent. Res.* **24**, 353—62.
MACLEOD, J., 1943. A survey of British sheep blowflies. II. Relation of strike to host and edaphic factors. *Bull. ent. Res.* **34**, 95—111.
MACLEOD, J. & DONNELLY, J., 1957. Some ecological relationships of natural populations of Calliphorine blowflies. *J. anim. Ecol.* **26**, 135—70.
McINTYRE, G. A. (Unpublished).
MILLER, D., 1939. Blow-flies (Calliphoridae) and their associates in New Zealand. Cawthron Inst. Monograph 2.
MURRAY, M. D., 1956. Observations on the biology of Calliphorids in New Zealand *N.Z. J. Sci. Technol. A.* **38**, 103—8.
NEWMAN, L. J., O'CONNER, B. A. & ANDREWARTHA, H. G., 1930. Some observations on the seasonal and regional incidence of blowflies in the south-west of Western Australia. *J. Dept. Agric. W.A.* **7**, 592—60.
NICHOLSON, A. J., 1934. The influence of temperature on the activity of sheep-blowflies. *Bull. ent. Res.* **25**, 85—99.
NICHOLSON, A. J., 1957. The self-adjustment of populations to change. Cold Spring Harbour Symposia 1957.
NORRIS, K. R. (Unpublished).
RATCLIFFE, F. N., 1934. (Communicated by JAMES RITCHIE). Sheep maggot flies in Scotland. *Scott. J. Agric.* **17**, 249—60.
RAYMENT, T., 1954. Taxonomy, morphology and biology of Sericophorine wasps. *Mem. Nat. Mus. Melbourne* 19.
ROBERTS, F. H. S., 1952. Insects affecting livestock. Angus and Robertson, Sydney.
ROGOFF, W. M. & BARTON BROWNE, L. B., 1956. The oviposition behaviour of the Australian sheep blowfly *Lucilia cuprina*. *Abst. Tenth Int. Cong. Ent.*
ROY, D. N. & SIDDONS, L. B., 1939. On the life history and bionomics of *Chrysomyia rufifacies* Macq. (Order Diptera, family Calliphoridae). *Parasitology* **31**, 442—7.

Ryan, A. F., 1954. The sheep blowfly problem in Tasmania. *Aust. vet. J.* **30,** *109—13.*
Tillyard, R. J. & Seddon, H. R., 1933. The sheep blowfly problem in Australia. C.S.I.R.O. (Aust.) Pamphlet 37.
Waterhouse, D. F., 1947. The relative importance of live sheep and of carrion as breeding grounds for the Australian sheep blowfly *Lucilia cuprina.* C.S.I.R.O. (Aust.) Bull. 217.
Waterhouse, D. F. (Unpublished).
Waterhouse, D. F. & Paramonov, S. J., 1950. The status of the two species of *Lucilia* (Diptera, Calliphoridae) attacking sheep in Australia. *Aust. J. sci. Res.* (B) **3,** *310—36.*
Webber, L. G., 1955. The relationship between larval and adult size of the Australian sheep blowfly *Lucilia cuprina* (Wied.) *Aust. J. Zool.* **3,** *346—53.*
Webber, L. G., 1957. Utilization and digestion of carbohydrates by the Australian sheep blowfly *Lucilia cuprina. Aust. J. Zool.* **5,** *164—72.*
Webber, L. G., 1958. Nutricion and reproduction in the Australian sheep blowfly *Lucilia cuprina. Aust. J. Zool.* **6,** *139—44.*

XXXIII
THE RABBIT IN AUSTRALIA

by

F. N. Ratcliffe

(Officer-in-Change, Wildlife Survey Section, Commonwealth Scientific and
Industrial Research Organization, Canberra, A.C.T.)

Although the introduction, spread, and establishment of the European rabbit (*Oryctolagus cuniculus* (L.)) in Australia has been one of the world's significant ecological events, it has not until very recently stimulated anything in the nature of planned ecological research. For nearly 100 years man has observed, discussed, and experimented with the wild rabbit in Australia, but almost always in an empirical way and in relation to his prime concern to improve methods of destruction and control. In the course of this association, as one would expect, a lot of biological information has come to light which, although mostly scrappy and anecdotal, is of considerable interest and value. This information, when pieced together, provides an outline of the ecology of the species in its Australian habitat and also indicates where more detailed and precise data are required before our ecological knowledge of the rabbit could be regarded as reasonably adequate.

In 1950 myxomatosis was successfully established in the wild rabbit population, soon spreading rapidly and with devastating effect (as will be described later in this chapter). For half-a-dozen years research was concentrated on the unique phenomenon of an introduced infection controlling a mammal pest; and it was only when the efficacy of myxomatosis started to decline, due to a host-parasite adjustment, that the need to return to a direct attack on the rabbit became obvious, and with it the need for a better ecological understanding of the animal if that attack was to be strategically sound and effective.

Early History — The Colonizing Spread*

Generally speaking, the invasion of the Australian Continent by the rabbit followed a pattern which might have been predicted in the light of the adaptability it has shown in other countries to which it has been introduced, the nature of the Australian vegetation and climate, and the fact that its original home was in the countries bordering the western Mediterranean and thus it is a species

---

* See Ratcliffe & Calaby (1958) for greater detail.

particularly well adapted to regions having a comparable climate. However, there are one or two features in this early phase of the history of the rabbit in Australia of sufficient interest to deserve special mention.

It has now been established beyond reasonable doubt that the Australian rabbit population originated from a small shipment of wild-type rabbits imported from England in 1859 and liberated on a grazing property near Geelong, Victoria. Within three years the animals had bred up sufficiently to be recognized as a potential pest. Prior to this — in fact since the arrival of the First Fleet in 1788 — domesticated rabbits must have been repeatedly brought into the country, for in the early decades of the 19th Century rabbits were being kept in every large centre of settlement and were known to have escaped or been liberated, and to have established themselves in the neighbourhood of Sydney and other towns. The fact that the 1859 importation gave rise almost at once to an aggressively spreading population, whereas the progeny of the hutch rabbits that had escaped or been set free apparently remained localized, either dying out or barely holding their own, must have been due to one of two things, or a combination of both. The domesticated varieties may have lost some of the wild type's adaptability and hardiness; and the uncleared forest that surrounded most of the coastal settlements in those early days may have acted as a deterrent to multiplication and spread — *O. cuniculus* is not by nature a forest-dwelling species. The introduction to Geelong allowed the wild-type rabbit into the continent by a back door, so to speak, giving it a foothold on the western side of the Great Dividing Range and the main forest belt, in a zone of savannah woodland merging into the plain, which the species found a very favourable habitat and which presented no serious obstacles to its spread.

From Geelong rabbits spread northward and westward as their numbers built up. They crossed Victoria and then New South Wales inland of the Dividing Range, their eastward spread over the tablelands and down the river valleys to the coast taking place, at a leisurely tempo, after the interior of the States had been invaded. One of the most remarkable features of this colonizing spread was the speed at which the vanguard advanced. By 1880, rabbits had reached and crossed the Murray River separating Victoria and New South Wales. They were first noticed on the Queensland border, approximately 500 miles to the north, in 1886, having traversed the State of New South Wales at a rate of about 70 miles a year.

The spread of the rabbit from South Australia across Western Australia was accomplished at a comparable speed. Rabbits were first seen at Fowlers Bay, S.A., in 1891 or 1892, at Eucla just over the border of Western Australia in 1894; and the advance guard reached the west coast at a point north of Geraldton in 1907. Thus the 1,100 odd miles from Fowlers Bay to the Indian Ocean was covered in 16 years.

The extraordinary speed of some of the major steps in the rabbit's colonizing spread is not easy to explain satisfactorily in biological terms. Even allowing for the fact that the dates recorded are those of the first sightings of the adventurous advance guard, progress must have been halted at times for reproduction and the rearing of litters: it is inconceivable, for example, that the individuals that were in the vanguard at the crossing of the Murray River were the same, or even of the same generation, as those that appeared on the Queensland border about seven years later.

Another interesting feature of the invasion of Australia by the rabbit is that something in the nature of a momentum seems to have been built up which carried the spreading population well beyond the regions in which the species was able to maintain itself permanently in significant numbers. Rabbits, probably in considerably diminished numbers, traversed the inland plains of Queensland, as the existence of an isolated warren colony near Normanton on the Gulf of Carpentaria bears witness. (The rabbit today is virtually confined to the south-western third of the State.) Unfortunately, this original over-optimistic extension of the northward spread has never been properly documented; and it is most unlikely now that it ever can be.

Consolidation — Ecological Consequences

The final phase — in Western Australia — of the rabbit's colonizing spread took place in the early decades of this century. Apart from the effects of control operations, minor changes in its distribution are perhaps still to be expected as a result of the development of new land use. The most likely change will be an increase in status in certain marginal regions, such as the "sandplain" in Western Australia and the areas of brigalow *(Acacia harpophylla)* scrub in Queensland.

When the rabbit had found its climatic limits, it was established as a dominant member of the fauna of the southern half of the continent, the limit of its occurrence in significant numbers extending further north in the interior than on the coasts. Certain habitat preferences, or their reverse, complicate this simple picture of a temperate and sub-tropical distribution. The rabbit is no more than a transient or casual "camper" in a closed forest association, although it will establish itself permanently along its margin and in clearings. Then there are certain types of soil which rabbits dislike and avoid, and which when they occur over substantial areas result in gaps in the distribution map of the species. The most important of these are the heavy Black Earths which occur, first patchily and then extensively, from northern central New South

Wales northward across inland Queensland. Finally, the rabbit is not well adapted to regions with a high summer rainfall, such as the coastal belt of Queensland and the extreme north-eastern corner of N.S.W. This is probably because the ground vegetation — the natural pasture — in these areas consists of a rather rank and dense grass cover.

There can be no doubt that the ecological consequences of the success and activities of the rabbit in Australia have been widespread and serious; but it is impossible to describe them in quantitative terms or with anything approaching precision. This is because no studies have been carried out to assess these effects (apart from some crude estimates of the change in stock-carrying capacity) and in any case what might be called the virgin situation, which provides the baseline for comparison, has now disappeared. Furthermore it is very hard to disentangle the effects of rabbit activity from those of the grazing of domestic stock, particularly sheep. As far as we know, the grazing habits of the rabbit, and its food-plant selection, are not importantly different from those of the sheep. The rabbit's smaller size will be reflected in the more restricted individual foraging range and in its greater ability to utilize very small plants, such as young seedlings. In addition, the rabbit displays certain habits (for example the barking of shrubs and young trees in times of adversity, presumably to get nutriment and particularly moisture from the cambium) which are not shared by the sheep and which may at certain times and places be important.

The pioneer pastoralists of Australia were in the main over-optimistic in their estimate of the carrying capacity of the natural vegetation, particularly the vegetation of the inland regions with their moderate-to-low rainfall and liability to recurrent and often severe droughts. This over-optimism led, in general, to the stocking of country at rates which we now know were excessive and unwise, resulting inevitably in a progressive deterioration of the pasture. The vegetational changes brought about by injudicious stocking, which were often aggravated by soil erosion, developed insidiously, and it took some decades before they became obvious. It so happened that the spread and build-up of the rabbit in the pastoral areas took place when the effects of over-stocking were first becoming manifest and serious; and this is what makes it so hard to distinguish between the rabbit and stocking factors. The combined effect of the rabbit and over-stocking has been particularly obvious in the western half of New South Wales. The rabbit consolidated itself and built up high numbers in this region between 1885 and 1892. The number of sheep carried in the Western Division reached a peak of over 15,000,000 in 1891. Since then the numbers have fallen by about 50%, the figures for 1911, 1931 and 1951 being 7,300,000, 6,723,000 and 6,898,000 respectively.

The effect of stocking and rabbits has not been confined to the vegetation: the native fauna has suffered as well. A review of the present-day status of marsupials (MARLOW, 1958) shows that the forms which have suffered most since white settlement, often having been reduced to rarity verging on extinction, are the medium-sized to small herbivorous ground-living species inhabiting the savannah woodland and the plains, i.e. the grazing country which has been subject to rabbit infestation. The marked reduction in status of this group of species (which has not been shared by the arboreal or ground-dwelling forest forms) must have been due, in the main, to competition from the introduced grazing animals and the changes in the ground vegetation brought about by their activities.

Before European settlement of Australia, the indigenous fauna was in balance with the native vegetation; and in the very extensive drought-risky regions of the continent (where permanent waters are few and far between) the herbivore species must have fluctuated in numbers in response to the variable conditions. Where serious droughts — presumably resulting in widespread die-offs in the ground-living animals — recurred at intervals of not much more than a decade on the average, population densities probably never remained very high for long. The white man's pattern of pastoral settlement, and infestation by the rabbit, supported as they were by the provision of permanent water on a vast scale, resulted in a form of exploitation of the vegetation that was inflexible and destructive by comparison with the light and balanced grazing of the native marsupials, and subjected these animals to competition of a peculiar intensity. It is not surprising that they have suffered an eclipse. The only species that have been successful in maintaining their status are the large and nomadic or desert-adapted forms such as the red kangaroo *(Macropus rufus)* and the euro *(M. robustus)*.

Biological Control — Myxomatosis

The control of mammal pests by pathogenic organisms has long been a cherished hope, particularly among those whose welfare is seriously affected by a species like the rabbit, the control of which calls for the expenditure of considerable time, money and intelligent effort. It is well known, however, that effective control — involving, as it must, continuing high lethality combined with a very efficient mechanism of transmission and spread — cannot normally be expected of any naturally occurring infection. Pending the time (not yet in sight) when man has learned how to breed microorganisms to his special requirements, hope for the discovery of a disease that would be of value for the control of a widely-ranging mammal rested on the occurrence of what could only be called a biological fluke — the appearance of an infection that was new to the mammal species or population concerned, and against which it had not been able to develop any mechanism of defence. The virus disease myxomatosis was a fluke of this kind.

The natural host of the myxoma virus is the South American "rabbit" *Sylvilagus brasiliensis*, in which it produces a mild non-lethal disease. At the end of last century, in Montevideo, accidental infection of laboratory rabbits — almost certainly by the agency of mosquitoes — demonstrated that the virus was highly lethal to *Oryctolagus cuniculus*. The late Dr. H. B. ARAGÃO, of Brazil, who carried out much of the pioneer research on myxomatosis, suggested nearly 30 years ago that it might be of value as a control weapon in Australia and other countries in which the European rabbit had been introduced and had become a pest.

The history of our knowledge of myxomatosis has certain features of rather special interest. The infective agent was one of the very first filterable viruses to be described as such. The disease was studied for several decades, by research workers in many parts of the world, as an infection of the European rabbit before its natural history was cleared up and the original host of the virus determined. The extreme susceptibility of *O. cuniculus* — both the wild and domesticated forms — is apparently a chance phenomenon, for other members of the Leporidae are resistant or refractory to the infection. (Populations of two species of hares, *Lepus europaeus* and *L. timidus*, were at risk in the course of widespread epizootics among wild rabbits, in Europe and Australia: about half-a-dozen cases of myxomatosis have been recorded in the former species, and one or two in the latter. Occasional individual susceptibility is the probable explanation of this.) Finally, it is worth mentioning that the research programme sponsored by the Australian Government and culminating in the establishment and spread of myxomatosis in the wild rabbit population was initiated before one aspect of the disease essential for an understanding of its epidemiology had been brought to light, i.e. that it is essentially an insect-borne infection.

Because of the cautious attitude adopted by the Commonwealth Quarantine Authorities, a comprehensive investigation of myxomatosis did not get under way in Australia until 1936, after some preliminary experiments in England had demonstrated its potential value (MARTIN, 1936). The Australian research falls naturally into two phases, the first comprising the laboratory studies, enclosure experiments, and field trials carried out by Dr. L. B. BULL* and his associates, which were terminated in 1943.

---

* Until recently Chief of the C.S.I.R.O. Division of Animal Health and Production. The Council for Scientific and Industrial Research (C.S.I.R.) became the Commonwealth Scientific and Industrial Research Organization (C.S.I.R.O.) after World War II.

The results of these early investigations have been summarized by BULL & MULES (1944). They included studies on the specificity of the infection, and the transmission of the disease from sick to healthy rabbits in the laboratory by a number of Australian insects, i.e. the stickfast flea *(Echidnophaga myrmecobii)* and four or five species of mosquitoes. Of ecological interest was the conclusion arrived at during the culminating field trials that myxomatosis was unlikely to spread effectively in natural wild rabbit populations except in the presence of adequate insect vectors (the trial liberations were made in rabbit populations carrying markedly different flea infestations) and where predation, e.g. by foxes, was not excessive (there were indications that foxes, by removing handicapped sick animals in the early stages of the disease, might tend to interfere with a developing outbreak). Because of the continuing cautious attitude of the Health Authorities, the liberations of the virus in unconfined rabbit populations had to be carried out in a low-rainfall pastoral area in inland South Australia; and the results of the trials were discouraging. Although many rabbits were killed, and individual warren colonies exterminated by the disease, the outbreaks remained strictly localized and quickly died out. It is hard to know what the history of events would have been had it been possible to make the virus liberations in a well-watered area like the Murray River Valley, where the post-war trials were carried out. The disease would presumably have established itself and spread widely in the rabbit population, creating a situation that invited intensive scientific attention; but in those anxious mid-war years the research teams needed for the task could not have been made available.

The second phase of Australian myxomatosis research covers the period since 1950 during which, after becoming established in the wild rabbit population, the infection spread until it became virtually coextensive with the distribution of its host and settled down to a pattern of endemicity punctuated by epidemic outbreaks determined by vector activity, at the same time undergoing a progressive adjustment involving changes in virulence on the part of the virus and a build-up in natural resistance on the part of the rabbit host. Those who worked on myxomatosis through this period were the privileged witnesses of an extraordinary and exciting natural phenomenon. This was the first time* that myxomatosis had been established in a natural population of the European rabbit under conditions favouring rapid transmission, and the first demon-

---

* Pre-war liberations of the virus on offshore islands in Britain and Denmark had given results comparable with those of the disappointing Australian field trials. In contrast with these, a liberation carried out privately and without publicity on an estate in Sweden, in 1938, resulted in the virtual extermination of the local rabbit population. Unfortunately this interesting outbreak was inadequately observed and documented.

stration of its ability to spread widely and effectively. (Comparable demonstrations took place shortly after in Europe, Great Britain and Tierra del Fuego.) The performance of the disease in the years 1950—1953 took place on a stage large enough and sufficiently populated to provide a spectacular biological drama. Rabbits had bred up in eastern Australia in the years immediately after the war to almost unprecedented levels; and a desperate and discouraging situation was prevailing when the trials with myxomatosis were resumed in 1950.

In the season of its "escape" from the test sites and its first wide and rapid dispersal, myxomatosis was restricted in the main to the neighbourhood of waterways and swamps. The following year saw a notable extension in the distribution of the disease. The next (1952—53) season we now know to have been one of remarkable and exceptional vector activity; and there was what can only be described as a flood of intensive disease activity, filling in the gaps left by the spread of the previous two years and carrying the infection to almost every rabbit-infested area in the eastern half of the continent. Myxomatosis, at this stage in its Australian history, was killing virtually 100% of rabbits contracting the infection; and the mortality in these three critical years must have aggregated to many hundreds of millions. Compared with its state and appearance in 1948 and 1949, the country swept by the disease was transfigured. Productivity showed a marked increase, providing a measure and demonstration of the toll that the rabbit had been taking. After due allowance had been made for other factors, REID (1953) calculated that the increase in the 1952—53 Australian wool clip attributable to the effects of myxomatosis was of the order of 70,000,000 lb.

The post-war trial-liberations of myxomatosis were carried out by the newly-established Wildlife Survey Section of the C.S.I.R.O. (RATCLIFFE et al., 1952). The sudden and surprising developments (after initial disappointment, which paralleled the discouraging results of the 1942—43 field trials) stimulated the replanning and intensification of the research programme. The cooperation of a team of virologists in the Microbiology Department of the Australian National University, led by Professor FRANK FENNER, was obtained; and this resulted in a series of highly-productive studies in immunology, on the virus and its several mutant strains, the development of host resistance, and various aspects of the epidemiology of myxomatosis under Australian conditions.

Members of the Wildlife Section found themselves involved, before long, mostly in field investigations of a predominantly entomological nature. The pattern of the 1950—51 dispersal and the subsequent activity of the disease made it quite clear that we were dealing with an infection that was being transmitted and spread by mosquitoes, with other forms of water-breeding blood-sucking Diptera sometimes playing a minor vector role. To understand precisely what was happening, what was likely to happen in future years, and whether there were any ways in which the performance of myxomatosis could be artificially stimulated, there was need for a sound knowledge of the ecology of the local mosquitoes — particularly those species which could be shown to be important as vectors. The task undertaken was onerous: the region involved was vast and varied and the mosquito fauna was virtually unknown ecologically. (We thought at the time that it had been adequately covered taxonomically; but the collections made in the course of the Section's field studies revealed many undescribed forms and have necessitated substantial revisions of certain genera.) The work was simplified to some extent when it became apparent that less than half-a-dozen species were of significant importance as vectors; but even so, intensive local studies in about ten representative areas (covering the rabbit population and its changes, disease activity, the seasonal abundance and behaviour of the prevalent mosquitoes) left substantial areas from which adequate data have not been obtained.

A lot of extremely interesting information resulted from these entomological investigations, particularly in relation to the ecology and the feeding and other behaviour of three widely-ranging species *(Anopheles annulipes, Culex annulirostris,* and *C. pipiens australicus)* which were revealed as the most important myxomatosis vectors. (Ecology and behaviour provide the key to vector efficiency, as the transmission of myxomatosis is mechanical and does not involve a cycle of multiplication in the body of the vector insect — see FENNER et al., (1952); DAY et al., (1956). Any blood-sucking insect that habitually feeds on more than one rabbit in the course of its adult life can act as a vector of myxomatosis.) As a result of these vector studies, it is possible to envisage a map of Australia on which the rabbit-infested regions are divided into (1) zones or areas which can expect an intense outbreak, with a high infection rate in the local rabbit population, annually or at any rate in most years, (2) those in which such an outbreak can be expected fairly frequently, i.e. at intervals of say five years or less, and (3) areas in which a general and intense outbreak can be expected only in an exceptional season, say once in a decade or less frequently. It became clear that vector activity was a phenomenon of a type and on a scale that make its artificial stimulation impracticable. In fact, once it had become widely dispersed and endemic, myxomatosis could be compared to a huge machine that was beyond human control.

It would be inappropriate to try and summarize here the results of the more specialized virological studies; but the ecological story would be incomplete without mention of one extremely interesting and important field experiment planned by Professor FENNER and carried out in collaboration with the Wildlife Section (FENNER et al., 1957). The myxoma virus is rather prone to spontaneous change; and one slightly attenuated strain — with a case mortality rate of about 90% in fully susceptible rabbits — was recovered from the field in the 1951—52 season. By the end of the following summer it was evident that strains of this type, which were conveniently referred to as "field strains", were becoming dominant. The prevalence of these strains of reduced virulence resulted in an appreciable number of rabbits recovering from the infection, which had a very important bearing on the progressive development, by selection, of innate resistance. The rabbits that succumbed took longer — often much longer — to die, and it was concluded that this was the explanation of the replacement in the field of the fully virulent virus by the attenuated mutant strains. Whereas the rabbit infected by what had come to be known as the standard laboratory strain died in 10—14 days and was infective to mosquitoes on the average for about five days, a typical field strain might take 25 days or more to kill a rabbit, which would remain infective to mosquitoes for some 18 days. Under conditions of random mosquito activity and feeding, such a field strain would have more than a 3-to-1 advantage over a fully virulent strain in the matter of transmission and spread.

Before this conclusion could be put to the test and confirmed in a field experiment, it was desirable to be able to work with a strain having a virulence comparable with that of the standard laboratory strain which could be distinguished in the early stage of infection from the Australian field strains. Suitable material was provided by the so-called French strain of the virus, which had been recovered from a wild rabbit in the Dordogne in 1953 and which could be distinguished from the Australian standard laboratory and field strains on the basis of the primary lesion. This permitted multiple inoculation of test rabbits in the laboratory for the identification of material recovered in the field, without which the determination of the history of the two competing strains in the experimental population would have been impracticable.

The experiment was carried out in the 1954—55 summer at Lake Urana, one of the main study areas employed in the myxomatosis investigations. The main virus liberation, involving the inoculation of over 200 juvenile non-immune rabbits, captured locally, took place between the beginning of November and mid-December. The persistence in the locality of an attenuated field strain was demonstrated by the recovery and identification of virus material from a naturally-infected rabbit early in November. Shortly after this the expected epidemic got under way, dying out by the beginning of February. (Lake Urana is an area liable to regular summer outbreaks of the disease.) In all, 45 virus samples were recovered from rabbits naturally infected during the course of the epidemic. Nine of the ten recoveries during the opening phase of the outbreak proved to be the French strain. Of the recoveries made at the peak, in mid-December, 8 out of 18 were the French strain. Between December 30 and February 2 only one sample of the French strain was recovered against 16 of the local field strain, which was the only virus demonstrably present in the final phase of the epidemic. Thus the virulent French strain, given what might be termed a flying start by mass liberation, was rapidly overtaken and finally completely supplanted by the attenuated local strain of the virus.

While only 18, or 40%, of the samples recovered during the outbreak were identified as the French strain, it was calculated that about 70% of the initial rabbit population of the experimental area had been killed by infection with this strain of the virus. This estimate is based on the difference in lethality between the two strains, and the fact that rabbits infected by the slower-killing attenuated virus are more likely to be picked up on periodic random searches.

The clear-cut results of this and a couple of similar field experiments demonstrated the selective advantage possessed by attenuated strains of the virus under natural conditions. They also indicated that the artificial dissemination of the standard virus as normally carried out by landholders (i.e. the capture, inoculation and release of a handful of rabbits — perhaps a dozen or so, but commonly fewer) was unlikely to have any significant effect on the subsequent performance of the disease while field strains were still present in the local rabbit population. Between epidemic outbreaks myxomatosis usually maintains itself at a subobservational level; and the overall picture to date leads one to believe either that the infection does not readily become extinct over an area of any size, or that the mobility of the vector population is such that when conditions become generally favourable for transmission the disease will spread rapidly from widely separated foci of persistence.

Mobility within an insect population can normally only be determined by a laborious system of marking and recapture. However, during the early dispersal phase of myxomatosis in Australia the virus itself provided a marker, isolated outbreaks indicating the terminal points of the long-range flight or drift of infected insects. Such point-outbreaks of the disease occurred in a band across western Victoria and southern South Australia (RATCLIFFE et al., loc. cit.), the aggregate distances involved pointing strongly to passive high-level drift. The most remarkable single instance of the long-range transportation of the virus was provided by the appearance of the disease in the summer of 1951—52 on a small rabbit-infested island off the Queensland coast. The rabbit population on this island is completely isolated, being separated by over

200 miles from the nearest rabbit-infested country on the mainland; and any source of infection other than by wind-borne insects had to be dismissed as to the highest degree improbable.

Not only has the myxoma virus "at large" in the Australian rabbit population become reduced in virulence, the host is acquiring a measurable degree of resistance to it, which has increased with each epidemic experienced. Thus rabbits from the Lake Urana area — the most intensively studied population in respect of myxomatosis history — are now as likely to recover as die when inoculated in the laboratory with a representative field strain of the virus that kills 9 out of 10 fully susceptible animals (MARSHALL & FENNER, 1958). Myxomatosis, in fact, no longer has the attributes of a "new" disease to the European rabbit, which is demonstrably building up a defence against it. The acquisition of innate resistance to the infection must be regarded as an irreversible process; and the position after less than a decade of host-parasite adjustment may represent the first step towards ultimate tolerance, and the complete loss by myxomatosis of any significant controlling value.

It can now (in the year 1958) be said that the effects of lowered virus virulence and increasing host resistance are becoming manifest in the general myxomatosis picture. There is an insidious and fairly general build-up in the rabbit population; and outbreaks of the disease are tending to leave more and more survivors. It is not easy to distinguish between the effects of the host-parasite adjustment and those of inadequate vector activity; for local conditions even in some of the more favoured regions have not always been conductive to mosquito breeding. "Controlling" kills are still being reported here and there, particularly when outbreaks occur during the late autumn or early winter — the severity of myxomatosis is known to be affected by temperature. There is little doubt that for many years to come myxomatosis will continue to act as an important factor in determining the levels achieved and maintained by the fluctuating rabbit population; and as seasons that favour a rapid increase in rabbits tend as a rule to favour mosquito activity, one can expect the disease to exert a buffering effect, though with increasing need for human intervention in the form of direct control measures.

General Ecology and Behaviour

If a species is a recognized and serious enemy of man and his interests, this fact inevitably affects the approach to it as an object of ecological study. Briefly, the biological information on the rabbit needed for practical purposes is that which would enable control operations to be carried out with maximum effect, particularly in low-rainfall areas where land values rarely justify intensive methods, and to develop techniques of poisoning to greater efficiency.

(Myxomatosis apart, poison provides by far the cheapest method for the mass destruction of rabbits.) Thus in addition to an understanding of the factors determining population levels and the rate of population increase in different ecological regions, and of large and small-scale movements, information was needed on individual and gregarious behaviour in relation to the location and selection of food; while a knowledge of the physiology of the wild rabbit at least insofar as it concerned nutritional requirements and water relations could be expected to throw light on many questions.

As has already been indicated, until recently the Wildlife Section's rabbit investigations were concerned almost exclusively with myxomatosis. It is therefore almost impossible to review the latest information on rabbit ecology and behaviour without reference to studies that are currently in progress, either unfinished or largely unpublished.

Field studies are seriously handicapped by the lack of a quick, convenient and reasonably accurate method of censusing a natural rabbit population — in fact of any reasonably accurate method not involving a disturbance that is liable to defeat the object of the study. This lack hinges in the main on the difficulty of live-trapping rabbits by methods used successfully for other mammal species; and it has stimulated observations on emergence behaviour in the hope that sight counts could be made reliable enough for practical purposes.

The rabbit is predominantly crepuscular and nocturnal in its habits, and the main period of activity starts in the evening or late afternoon with the emergence of the animals from their burrows or daytime cover. Apart from night transects using a spotlight (a technique that is under investigation at the present time), sight counts have to be carried out in the often very limited period between the beginning of activity and the descent of darkness. If the counts so obtained are to be acceptable as comparative indices, e.g. for the assessment of population change, one needs to be satisfied either that very nearly all of the population was active above ground at the time, or that one was sampling a reasonably constant proportion of it. The Section's observations have not been encouraging in this regard.

DUNNET (1957b), in a study of warren colonies in the neighbourhood of Canberra, found that there was considerable variation in the timetable of emergence between different and apparently comparable warrens, and also in the same warren over a period of time. Some of these variations could be explained by recognizable disturbances and trends, but some appeared to be quite fortuitous. MYKYTOWYCZ & ROWLEY (1958) in the course of continuous round-the-clock observations on a population living in semi-natural conditions in a large enclosure, found that less than 50% of the animals had become active above ground at dusk, and recorded a maximum count of 66% in the early hours of the night. Such low counts must probably be regarded as rather abnormal. MYERS (1957), working with a natural population in flat open country under conditions favouring stabilized behaviour, recorded evening counts of 90% of the subsequently determined local population, with all of the inhabitants of two medium-sized warrens being recorded on at least one occasion. In his review of recent observations by Sectional colleagues, he refers to evening observations in which between 55 and 75% of the warren populations studied by DUNNET were accounted for.

Because of the difficulty of getting certain desired biological data from natural rabbit populations, with an adequate degree of precision and without undue expense, the Wildlife Section recently initiated a series of studies on rabbit populations maintained in enclosures. These ranged in size from approximately 2 to 10 acres, the largest being used only for preliminary observations and the testing of techniques of handling and marking and examining the animals. The rabbits were allowed to breed and burrow naturally; and a system of illumination was employed (involving both flood lighting and spotlights attached to telescopes or powerful binoculars) which enabled observations to be made at night as well as during the daytime.

Several objectives were in view when the enclosure experiments were set up, including the study of the epidemiology of coccidiosis, endoparasite infections, the reproductive history and longevity of individual rabbits, and factors governing the length and incidence of the breeding season. While data on these and other aspects of rabbit biology are being steadily accumulated, observations on behaviour — especially social behaviour (MYERS & MYKYTOWYCZ, 1958) — because of their novelty and interest have tended to dominate the work. There can be no doubt that the social structure and behaviour revealed in these enclosure studies are of very real ecological significance. They help to explain the differential survival of litters of different parentage, and throw light on the incidence of intra-uterine mortality. Some of the phenomena recorded, particularly when population density had built up, must perhaps be regarded as a reflection of confinement; but even though some of their detailed manifestations may have been modified by crowding, it seems safe to assume that innate, natural behaviour patterns have been revealed and that it should be possible to generalize from the enclosure observations, if caution and common sense are employed.

The rabbits in the enclosures formed clearly-defined and stable social groups, the members of which obviously recognized the others as individuals and also as fellow members of their group. When a group increased in size, as a result of breeding, it would reform by stages into two groups, presumably because the intra-group relationships become too complicated for a rabbit's mentality when more than about half-a-dozen adults are involved. Within each group there is a rigid linear hierarchy, or peck order, among the males, and a less rigidly linear one among the females — though there are obviously high-ranking and low-ranking does. The top-ranking or dominant buck of a group is outstanding, and easily differentiated by his behaviour and freedom of movement from all the other males. Among the subordinate bucks, the second-ranking usually exhibits characteristic behaviour — that of a "pretender" — remaining as closely associated with the leader as he dares and can, obviously ready at any time to take over his role. (CLARKE (1955) found evidence of similar categories — dominants, high-ranking challengers, and subordinates — in his confined populations of voles.)

Each group occupied a particular area within the enclosure, which merits the designation of territory as trespass across its well-recognized boundaries by members of neighbouring groups was resented and dealt with. Within the group territory, individual rabbits had home ranges (i.e. areas within which most of their activities were confined) which were overlapping and not defended. The definition of the social groups and their territories weakened with the cessation of breeding in the summer, virtually breaking down when conditions became "hard".

The rabbit's social behaviour and social hierarchy are closely geared to reproductive activity. A high proportion of the contests between males by which the peck order is established and maintained spring from competition for receptive does. The dominant buck attempts and often comes close to monopolizing the females within his group. However, the as-yet-unexplained tendency for the does of a group to come into oestrus simultaneously, dropping their litters within a day or two of each other, usually gives at any rate the second-ranking male a chance of some share in paternity.

The history of the enclosure populations reveals a marked difference between high- and low-ranking females in the matter of breeding success (MYKYTOWYCZ, unpublished data). Two factors contribute to this. In the first place, subordinate does, although apparently conceiving with fair regularity, are liable to resorb their embryos. In the second place, the dominant animals have access to and monopolize the better and safer breeding sites, with the result that nestling mortality through desertion, water-logging, etc. is virtually confined to the litters of subordinate does. This latter factor may be of even greater importance under natural conditions, where nestlings and weanlings are subject to predation. It now seems likely that nesting in established warrens is the prerogative of high-ranking does. The chances of survival of litters dropped and reared in isolated "stops" (short, shallow breeding burrows) must be poor compared with those of young reared in a warren colony.

The data on reproduction that are accumulating from the enclosure studies and field observations have not yet been analysed and collated; and it would be premature to try to pick out features in the regionally variable Australian picture that differ from the pattern worked out in Britain by BRAMBELL (1944) and in New Zealand by WATSON (1957). As in those two countries, breeding generally ceases in high summer, when the females enter a period of anoestrus. Over wide areas in the rabbit-infested portion of the continent the pastures "hay off" during the summer months; and there is now a clear indication that breeding is dependent on the availability of green growing feed. It is not impossible, nor even unlikely, that the termination of the summer anoestrus is due to some specific nutritional factor in the plant growth that follows the autumn "break" (a well recognized climatic feature in the southern parts of the continent where rainfall tends to have a winter incidence). It is hoped to explore this point experimentally.

Research carried out in many parts of the world in recent years has shown that stress induced by interactions of various kinds at high densities can be a factor contributing to the regulation of mammal and bird populations through its effect on fecundity or viability. Experiments were therefore carried out on rabbits, using stressor agents suitable for laboratory study; and it was found (M. E. GRIFFITHS, unpublished data) that the wild rabbit is a very tough animal physiologically, naturally resistant to stress and capable of becoming even more resistant after being subjected to it. It seems highly unlikely, in fact, that stress is of any importance in the population dynamics of this species. A point of interest was the suggestion that subordinate does were more susceptible to stress than dominant animals; though no such difference was apparent in males.

As overseas workers have paid considerable attention to the role of the adrenals in the stress syndrome, it is worth mentioning that GRIFFITHS has been unable to demonstrate a significant increase in adrenal activity following stress in the wild rabbit. Individual variation in adrenal size and activity is exceptionally wide in this species, and apparent in a random sample of wild-caught animals.

As part of these physiological studies, the blood chemistry of representative individuals from the enclosure populations was examined from time to time to ascertain whether marked differences in population density, nutritional levels, etc. had any readily demonstrable effects. Again, the results obtained indicated physiological toughness and the maintenance of normality under all but extreme conditions. It was only when malnutrition had become obvious that blood sugar levels fell. The fall when it occurred was marked; and the animals affected were liable to collapse and die when chased for capture and examination, presenting a picture comparable with the description of "shock disease" in the snowshoe hare given by GREEN & LARSON (1938).

Insofar as they are causes of actual and significant mortality in wild rabbit populations, infestation by intestinal worms and infection by coccidia (*Eimeria* spp.) are of special interest as their incidence and intensity are likely to be density-dependent. To date, the Wildlife Section can only claim to have made preliminary observations on these, and on the equally interesting question of predation, as factors in the population ecology of the rabbit. Both the parasitic worms and the coccidia have infective stages that are sensitive to desiccation and low atmospheric humidities; and although adequate field surveys have not yet been made, on a continent with such extensive arid and semi-arid zones this fact must be reflected in a geographical restriction of the infections, at any rate at levels that are likely to be significant. (The margin of the zone in which worm infestations of sheep are troublesome and necessitate regular treatment is well marked.) There is a clear indication that when conditions favour breeding in the more arid inland portions of Australia the rate of increase of the rabbit population is greater than it is in the coastal areas; and this suggests that nematode and coccidial infections, probably in combination, may constitute a significant mortality factor. It was therefore expected that the effects of endoparasites would become manifest in our experimental enclosures when high-density populations began to lose condition and suffer from malnutrition. So far, however, this has not happened. It is possible that the final analysis of the data may reveal a slight but significant effect; but it is already clear that there was no obvious mortality attributable to worm and coccidial infections when conditions might have favoured such a development.

The significance of predation in the population dynamics of the rabbit is something that has yet to be assessed — a task which all ecologists will recognize as one of extreme difficulty. There is not very much that can be said about it at the present time. We know that when conditions favour a general increase in the rabbit population, and breeding is well under way, the activity of predators is quite inadequate to prevent the attainment and maintenance of high densities. The main question at issue, therefore, is whether predation is important in those areas where the rabbit population is relatively stable (i.e. not normally subject to violent fluctuations) and, in other areas, when the population density is moderate to low.

It is not unreasonable to believe that predation might retard to a significant degree the build-up of rabbits from low densities, extending in time the initial flat portion of the population-growth curve. The principal predators of the rabbit in Australia are the diurnal birds of prey and the introduced European fox (*Vulpes vulpes* (L.)). Both are reasonably common — sometimes abundant — and there is little doubt that both concentrate in areas where rabbits are breeding. The numbers of hawks and foxes in a region are not directly dependent on the abundance of rabbits (although they may be affected by this) as both are general predators with important alternative sources of food. Field observations indicate that the combined predation by hawks and foxes is sometimes sufficiently intense to have the suggested effect. Although the comparison is quite subjective, the rate of population increase and the densities attained within our experimental enclosures seemed to be markedly higher than would be the case in natural populations; and protection from predation provides the only obvious explanation.

Although foxes are quite capable of killing fully grown rabbits, the available evidence suggests that they are relatively unsuccessful in hunting adults. Whereas avian predation falls on the young only after they leave the nest, the fox is efficient in locating and digging out litters that have been dropped in shallow stops. (A pioneer settler, who as a young man was able to observe the arrival and establishment of the fox in south-eastern South Australia, informed the writer that there was an almost immediate and very marked reduction in the number of litters dropped and reared in stops. It is a popular belief in Australia that nesting in the warrens is a habit developed as a result of fox predation; but both warren and stop breeding are almost certainly part of the normal behaviour pattern of the rabbit.)

Avian predation on the rabbit has been incidentally studied by CALABY (1951) and DUNNET (1957a). Both showed that the hawks were successful in dealing with young rabbits only up to a certain maximum size. CALABY found that the biggest kittens that the little eagle (*Hieraaetus morphnoides* (GOULD)) was able to kill "with any consistency was about 600 to 650 g.", and concluded that rabbits were most vulnerable to attack from this species between the ages of 3 and 6 weeks. DUNNET found that the mean weight of rabbits taken by birds on his study area was around 350 to 450 g. Most of the casualties were attributable to little eagles and goshawks (*Accipiter fasciatus* (VIGORS & HORSFIELD)). His maximum of over 1,000 g. is of no great significance in relation to his colleague's observations, because the large wedge-tailed eagle (*Aquila audax* (LATHAM)) was also hunting the area.

One final point in connection with the ecology of the rabbit in Australia possibly deserves mention. This is the question of cycles (RATCLIFFE, 1955). The manager of a large sheep station in southern Queensland, who happens to be a keen naturalist and a very accurate observer, has noted the changes in the local rabbit population over a period of nearly 30 years. There is a clear suggestion in his observations of the operation of a 10-year cycle. It has not been possible to corroborate this by data obtained from other parts of Australia; but the failure is not really surprising. If there was indeed a tendency to cyclic fluctuations in numbers, one would expect it to be very largely obscured in most areas by population changes due to seasonal conditions and human interference. The demonstration of a cyclic trend behind these more obvious fluctuations would require adequate quantitative data, which are simply not available. All that can be said, therefore, is that since 1930, in an area not far from the northern limit of the rabbit in eastern Australia which enjoys fairly reliable rainfall and in which the rabbit is very little disturbed by control measures, the local population built up to peak numbers in 1931, 1941 and 1951. The first observed peak was followed by a natural "crash", the second by an exceptional drought, and the third by a devastating outbreak of myxomatosis.

REFERENCES

BRAMBELL, F. W. ROGERS, 1944. The reproduction of the wild rabbit *Oryctolagus cuniculus* (L.). *Proc. zool. Soc. Lond.* **114**, 1—45.

BULL, L. B. & MULES, M. W., 1944. An investigation of *Myxomatosis cuniculi* with special reference to the possible use of the disease to control rabbit populations in Australia. *J. Counc. Sci. Industr. Res. Aust.* **17**, 79—93.

CALABY, J. H., 1951. Notes on the little eagle; with particular reference to rabbit predation. *Emu* **51**, 33—56.

CLARKE, J. R., 1955. Influence of numbers on reproduction and survival in two experimental vole populations. *Proc. Roy. Soc. Lond.* (Ser. B) **144**, 68—85.

DAY, M. F., FENNER, FRANK & WOODROOFE, GWENDOLYN M., 1956. Further studies on the mechanism of mosquito transmission of myxomatosis in the European rabbit. *J. Hygiene* **54**, 258—283.

DUNNET, G. M., 1957a. Notes on avian predation on young rabbits, *Oryctolagus cuniculus* (L.). *C.S.I.R.O. Wildl. Res.* **2**, 66—68.

DUNNET, G. M., 1957b. Notes on emergence behaviour of the rabbit, *Oryctolagus cuniculus* (L.), and its bearing on the validity of sight counts for population estimates. *C.S.I.R.O. Wildl. Res.* **2**, 85—89.

FENNER, FRANK, DAY, M. F. & WOODROOFE, GWENDOLYN M., (1952. The mechanism of the transmission of myxomatosis in the European rabbit *(Oryctolagus cuniculus)* by the mosquito *Aedes aegypti*. *Aust. J. exp. Biol. Med. Sci.* **30**, 139—152.

FENNER, FRANK, POOLE, W. E., MARSHALL, I. D. & DYCE, A. L., 1957. Studies in the epidemiology of infectious myxomatosis of rabbits. VI. The experimental introduction of the European strain of myxoma virus into Australian wild rabbit populations. *J. Hygiene* **55**, 192—206.

GREEN, R. G. & LARSON, C. L., 1938. A description of shock disease in the snowshoe hare. *Amer. J. Hyg.* **28**, 190—212.

MARLOW, B. J., 1958. A marsupial survey of New South Wales. *C.S.I.R.O. Wildl. Res.* **3** (2):

MARSHALL, I. D. & FENNER, FRANK, 1958. Studies in the epidemiology of infectious myxomatosis of rabbits. V. Changes in the innate resistance of Australian wild rabbits exposed to myxomatosis. *J. Hygiene* **56**, 288—302.

MARTIN, C. J., 1936. Observations on Myxomatosis cuniculi (Sanarelli) made with a view to the use of the virus in the control of rabbit plagues. *C.S.I.R. Aust. Bull.* No. **96.**

MYERS, K., 1957. Some observations on the use of sight counts in estimating populations of the rabbit, *Oryctolagus cuniculus* (L.) *C.S.I.R.O. Wildl. Res.* **2,** *170—172.*

MYERS, K.& MYKYTOWYCZ, R., 1958. Social behaviour in the wild rabbit. *Nature, Lond.* **181,** *1515—1516.*

MYKYTOWYCZ, R. & ROWLEY, IAN, 1958. Continuous observations of the activity of the wild rabbit, *Oryctolagus cuniculus* (L.), during 24-hour periods. *C.S.I.R.O. Wildl. Res.* **3,** *26—31.*

RATCLIFFE, F. N., MYERS, K., FENNESSY, B. V. & CALABY, J. H., 1952. Myxomatosis in Australia. A step towards the biological control of the rabbit. *Nature, Lond.* **170,** *7—11.*

RATCLIFFE, F. N., 1955. Is there a 10-year cycle in rabbits? *Pastoral Rev. & Graziers' Rec.* **65,** *999—1001.*

RATCLIFFE, F. N. & CALABY, J. H., 1958. Rabbit. "The Australian Encyclopaedia." Vol. 7: *340—347.* (Angus & Robertson Ltd.: Sydney.)

REID, P. A., 1953. Some economic results of myxomatosis. *Quart. Rev. agric. Econ.* **6,** *93—94.*

WATSON, J. S., 1957. Reproduction of the wild rabbit, *Oryctolagus cuniculus* (L.), in Hawke's Bay, New Zealand. *N.Z. J. Sci. & Tech.* **38,** *451—482.*

XXXIV
THE BIOLOGICAL CONTROL OF PRICKLY PEAR IN AUSTRALIA

by

ALAN P. DODD

(Biological Section, Dept. of Public Lands, Brisbane)

The occupancy by prickly pear of very great areas in eastern Australia must be regarded as the most outstanding illustration of a plant invasion. Similarly, the destruction and in fact complete conquest of the weed within the space of a few years by insect enemies, chiefly through the agency of a particular insect, must be considered as the most striking achievement in the biological control of a plant pest.

Survey of the Problem

The problem of prickly pear in Australia can be stated briefly as that of a vigorous, rapidly spreading weed occupying very large tracts of potentially valuable agricultural and grazing land where mechanical and chemical methods of destruction were impracticable on account of the huge cost involved.

Prickly pears are members of the genus *Opuntia* of the Cactaceae, a plant family indigenous to North and South America. Of the various species that became acclimatised in Australia the main problem concerned the two closely related forms, *Opuntia inermis* and *O. stricta*, bushy or shrubby plants 3 to 6 feet in height, which were known as "the pest pears". *O. inermis* was introduced at some time prior to 1839; with the opening up of inland areas for grazing in the period 1850 to 1875, it was planted for hedges or was grown as a potential fodder plant for sheep and cattle at many separate points in Queensland and New South Wales. *O. stricta* was probably imported to central Queensland about 1860. Thus, these two plants obtained their footing when population was sparse and pastoral holdings of very large dimensions. Later, when the demand arose for closer settlement the more valuable land in the infested districts was overrun by the pest.

By the year 1900, 10,000,000 acres were estimated to carry prickly pear. In the next 20 years the weed advanced so rapidly that a survey in 1920 indicated that 58,000,000 acres were affected; this represented an annual increase of almost 2,500,000 acres, a figure that must have been exceeded in the later years. The peak of the invasion was reached in 1925, when the infested area must have been greater than 60,000,000 acres. Approximately one-half, or roughly 30,000,000 acres, was occupied by dense growth, completely covering the ground to the exclusion of all grass and herbage, and in bulk weighing 500 to 800 tons per acre.

About four-fifths of the infested territory was in Queensland, where the prickly pear belt was almost continuous from the southern border to a point several hundred miles northward, extending inland for 250 to 400 miles except for a relatively free coastal strip 50 to 100 miles wide; the land was mainly flat or undulating with an altitude of 600 to 1,200 feet. In some sections the dense growth was almost unbroken over many hundreds of square miles, with scarcely a habitation, a cultivated field or a head of cattle or sheep. In New South Wales the weed growth was more or less divided into separate districts situated from the Queensland border to the Hunter River, and a definite proportion of the infested land comprised poorer quality soil on hills and slopes. 90% of the main prickly pear territory had an annual rainfall of 20 to 30 inches. The vegetation was of two types throughout; on the one hand open savannah forests of well spaced trees and shrubs, mainly *Eucalyptus;* on the other, dense stands of brigalow *(Acacia)* and belar *(Casuarina)* trees, 40 to 60 feet high, such stands being known as "brigalow scrub". The treeless or semi-treeless plains and undulating grassy downs of sections of north-west New South Wales and inland Queensland were almost wholly free from prickly pear, the spread of which virtually ceased at the edge of the tree line.

The Commonwealth Prickly Pear Board and Its Work

The genesis of biological control was the appointment in 1913 by the Queensland Government of a Travelling Commission which, after visiting all countries where prickly pear was indigenous or had become established in quantity, recommended the introduction of the plant's natural insect and plant disease enemies from North and South America.

In 1920, the Governments of the Commonwealth, Queensland and New South Wales agreed to co-operate and appointed the Commonwealth Prickly Pear Board for the purpose of investigating all avenues of biological control. The Board immediately took steps to send scientific officers to America and to establish a base in Australia for the receipt of the imported enemies. Its policy comprised:
    (a) The search for and study of all insects attacking prickly pears and other Cactaceae in North and South America;
    (b) A survey of the disease organisms affecting these plants;
    (c) The investigation of the host plant restriction of prickly pear insects, including a food test programme on plants of economic importance;

(d) The selection of the insect species deemed to be of greater potential value, their collection in America and their transport to Australia;
(e) The acclimatisation of the introduced insects;
(f) The large scale rearing of certain insects, their establishment in the field, and, later, their mass distribution.

Investigation of the plant diseases showed that fungus and bacterial organisms were unlikely to be of particular value in the control campaign, and, also, that many of the pathological forms were already present in Australia. Furthermore, it would have been extremely difficult to devise satisfactory tests to indicate that disease organisms attacking prickly pear could not transfer their activities to other plants. Hence, introductions were restricted to insect enemies.

Active association was maintained with State organisations, i.e. the Department of Agriculture and the Prickly Pear Destruction Commission in New South Wales, and the Prickly Pear Land Commission in Queensland. The distribution of the established insects was a joint operation, in which the Board provided the material and the State bodies carried out the greater part of the field releases. In particular in Queensland the Board and the Prickly Pear Land Commission formed a closely-knit partnership, whereby the State organisation provided material assistance in equipping and financing field stations and incurred all expenditure concerned with the large scale distribution of *Cactoblastis* and other established insects.

The Commonwealth Prickly Pear Board ceased to function at the end of 1939, by which time it was evident that this great experiment in biological control had reached the stage of successful accomplishment where continued activities could be effectively pursued by the two States, Queensland and New South Wales, each of which took into its service the scientific officers trained by the Board.

Work in America

The Board's entomologists surveyed and investigated cactus insects in North and South America continuously from late 1920 to 1937. Greatest attention wss paid to the United States, Mexico and the Argentine; field centres for the study of prickly pear insects were maintained in the United States from 1921 to 1933, in the Argentine from 1931 to 1937, and in Mexico for intermittent periods totalling four years between 1926 and 1935. In addition to the major operations investigations were carried out in Bermuda, the larger islands of the West Indies, the Central American republics, and all countries in South America except the Guianas.

Approximately 150 different species of insects whose life cycle was restricted to cactus host plants were discovered, and many of these proved to be undescribed species. The family Cactaceae possesses a definite insect fauna, with various genera and even groups of genera confined to these plants. Most insect enemies of prickly pears do not attack other types of cacti, and this is true conversely.

The life history, habits, and host plant relations of many of the insects were studied both in the field and under cage conditions. Between 50 and 60 different species were forwarded to Australia, often in large numbers over a period of years. The total number of individual insects despatched from North and South America exceeded 500,000.

The Australian Set-Up

The headquarters of the Board was at Sherwood, an outer suburb of Brisbane; this laboratory was the receiving house and the quarantine and acclimatising centre for all material from America; here, too, host plant tests were carried out in the insectaries.

Several Field Stations were established at localities in the heart of the prickly pear territory of Queensland and New South Wales, each of which was equipped with hundreds of wire gauze or cloth-sided rearing cages set up on individual tray stands either in the open grounds or under the protection of large iron-roofed sheds. These stations acted as the breeding, liberating and mass distribution centres, and as convenient points for observing the progress in the field of experimental releases and of the general control programme.

The insect material imported from America was confined in the Sherwood quarantine insectaries and was used for attempted establishment and for host plant trials. All field releases were made with Australian-reared stocks.

The Host Restriction Question

The Board adopted a fundamental policy that all insects imported to Australia must be restricted to feeding and breeding on cactus hosts. This policy was implemented in three ways. Firstly, a study was made in the field in America of the host plant relations of each insect species. Secondly, all available information was sought from scientific sources concerning the host restriction not only of the insect itself but also of its allies. Thirdly, each kind of insect, which it was desired to introduce, was subjected to feeding and/or oviposition tests on a wide variety of economic and other plants, in order, if possible, to learn whether in the absence of its normal food plant the insect possessed the ability to breed on, or to cause appreciable injury to, any plant other than prickly pear.

The tests were carried out in America, and if these proved favourable, further experiments were undertaken under quarantine conditions in Australia. Many hundreds, even thousands, of larvae, and of adults in the case of Coleoptera and Heteroptera, of a species were used in these tests, the programme frequently being continued for two or three seasons.

With no other host available, many kinds of insects attempted to feed on the supplied plant. Thus, stem-boring larvae would penetrate or try to penetrate stems or fruit, and adult beetles would gnaw leaves and stems; such attempts were usually unsuccessful and death occurred within a few days. In some cases the insects lived longer, appeared to derive some degree of nourishment from the feeding attempts, but were unable to continue and to complete development.

In two or three instances, however, prickly pear insects were able to complete the life cycle from egg to adult on some alien host, although all evidence indicated clearly that they did not do so under natural conditions. Such insects were immediately rejected from further consideration.

Starvation tests with insects on alien host plants under conditions of close confinement must be deemed artificial or at least unnatural. But in the prickly pear control work they did serve a useful purpose, firstly in satisfying the public that adequate safeguards had, in so far as possible, been taken, and secondly in providing results that could be interpreted as evidence that the insect species was unlikely to attack plants other than prickly pear or that it possessed the ability to develop on some plant or plants other than its normal host.

General Results

The original plan of the Board called for the establishment in Australia of a prickly pear insect fauna of various species, the combined activities of which might be expected to bring about a definite measure of control. However, when it became apparent that *Cactoblastis cactorum* would bring about the destruction of the two major pest pears unaided by other insects, the policy was altered. Thereafter, emphasis was placed on the introduction of insect enemies of the less important species of *Opuntia*.

For all species of prickly pear domiciled in Australia, 15 different kinds of insects were established. This may not seem a very successful record, in view of the fact that about 50 species were introduced. Certainly it is true that repeated trials failed to acclimatise various insects. But there were other reasons why the number of established species was not greater. For example, after their introduction and study under insectary conditions, some forms were discarded because they were not considered to possess sufficient control value or because they did not pass the host plant tests. And the success of *Cactoblastis* rendered unnecessary further attempts to establish insect enemies of *O. inermis* and *O. stricta*.

Had *C. cactorum* not come into the picture, there is little doubt that the Board would have achieved its original aim. Prior to the build up of the *Cactoblastis* population, other insects, i.e. the Cochineal *Dactylopius opuntiae*, the plantsucking Coreid bugs *Chelinidea* spp., the moth borer *Olycella*, and the Red Spider *Tetranychus opuntiae*, had multiplied to very large numbers and were commencing to prove their value by thinning out and in some areas killing dense masses of *O. inermis* and *O. stricta*, by destroying fruit and hence reducing the weed's increase potential, and by generally devitalising the plant's vigour. Furthermore, it would almost certainly have been possible to establish other kinds of insects. Admittedly, without *Cactoblastis* successful control would not have been so rapid; possibly it would not have been so complete; but a large measure of control would eventually have resulted.

It is pertinent here to remark that biological control of the low-growing tiger pear, *Opuntia aurantiaca*, which was a pest in certain parts of Queensland and New South Wales and which constituted a menace possibly even more serious than *O. inermis* and *O. stricta*, was obtained per medium of a species of Cochineal, *Dactylopius*, while not only in Australia but in certain other countries *Opuntia monacantha* was brought under complete control by another species of *Dactylopius*.

The progress of the biological campaign is shown in the following chronological arrangement:

1921—25:
> The introduction, rearing in numbers and establishment in the field of several insects, i.e. the moth borer *Olycella*, the Coreid bugs *Chelinidea* spp., the Cochineal mealy bug *Dactylopius opuntiae*, and the Red Spider *Tetranychus opuntiae*.

1925—27:
> The introduction of *Cactoblastis cactorum*, large scale rearing and conducting of first liberations. *Chelinidea*, Cochineal and Red Spider increase to large numbers, spread widely, and give distinct promise of some measure of control.

1928—30:
> The mass distribution of *Cactoblastis* throughout the prickly pear infested territory and its rapid multiplication.

1930—32:
> The general collapse and destruction of the dense stands of prickly pear brought about by an enormous *Cactoblastis* population. Sudden, even total, decline of the concentrations of *Chelinidea*, Cochineal, *Olycella* and Red Spider, which were unable to thrive in competition for the available food supply.

1932—33:
> Owing to the rapid widespread destruction of prickly pear the *Cactoblastis* population falls to a low ebb on account of

starvation. Hence, heavy regrowth develops over large areas.
1933—35:
> The recovery of *Cactoblastis* and its destruction of the regrowth.

1935—40:
> The virtual complete control of the two pest pears by *Cactoblastis*. The former dense pear country is reclaimed and brought into production.

1940—57:
> Except for limited areas in New South Wales, prickly pear has been either totally destroyed or survives as a scattered plant in the community, with *Cactoblastis* maintaining effective control. Full production has brought prosperity to the former pear infested territory.

The History of Cactoblastis cactorum

Life Cycle and Habits

Cactoblastis cactorum is a moth, indigenous to the northern Argentine, southern Brazil, and adjacent Paraguay and Uruguay. It is a member of the most dominant group in the cactus insect fauna, a group of many related genera containing at least 50 species in the Pyraloid family Phycitidae, the larvae of which are internal feeders in the stems or fruit of prickly pears and other Cacti, and are either solitary or gregarious in habit. Other species of *Cactoblastis* occur in South America, while in North America the genus has its counterpart in *Melitara*, the host plants of which are prickly pears.

C. cactorum has two generations annually, a shorter summer and a longer winter life cycle; in the warmer districts in Queensland a third partial generation occurs in the autumn months. The eggs are laid in symmetrical stiff chains, known as eggsticks, the first egg being fastened to a spine or spicule on the host plant; the eggsticks contain an average of about 75 individual eggs, but the number varies greatly and may exceed 100. Among other Cactus Phycitids, the species of *Melitara* and to a lesser degree those of *Olycella* possess the same eggstick habit. The average fecundity is 100 to 120 eggs per female, but individuals may deposit as many as 300 eggs divided among several sticks. All young larvae hatching from an eggstick enter the cladode at the one point. When about one-half grown, customarily they vacate the feeding tunnels and wander over the surface for a few hours before re-entering at a new centre; this practice is associated with the division of the original colony into two or more components. The gregarious habit terminates at pupation, when the larvae make their exit from the plant singly to spin silky cocoons among debris, between destroyed fallen cladodes, under the bark of logs and in similar situations. However, larvae often select the same pupation sites; hence, cocoons may occur in clusters in favoured locations. The moths, which do not feed, are capable of surviving for 10 to 14 days; they are nocturnal in habit, and fly freely, even for distances of several miles where the host plants are scattered; females, but not males, are readily attracted to artificial lights. In the field the eggs are invariably attached to prickly pear cladodes; even where the insect has been very heavily concentrated, eggsticks have not been observed affixed to twigs or blades of grass among or adjacent to the host plant.

As prickly pears have a moisture content of approximately 90%, the growth tends to rot readily from internal injury. *Cactoblastis* larvae will eat the whole interior of the more succulent upper cladodes except the vascular bundles, leaving the thin parchment-like cuticle undamaged. But more generally their activity is accompanied by rotting resulting from various associated bacterial and fungus organisms. The larvae tunnel freely from segment to segment, even to the base of the plants; however, in older large plants they rarely burrow downward into the thick lower stems. One or two colonies of larvae are capable of destroying a small plant, but a relatively heavy population is necessary to bring about the collapse and destruction of large plants, the disintegration of which appears to be due to the concentrated attack on the upper growth causing a breakdown of the tissues of the lower stems, pathogenic organisms contributing to the decay. Although various disease agencies, usually bacterial in the younger growth and fungal in the basal stems, contribute actively to the destruction, their role is dependent on the severity of the attack by the insect.

The rapidity with which *Cactoblastis* caused the collapse of *O. inermis* and *O. stricta* varied according to the age and growth habit of the plants. Young plants and the very succulent dense growth in the brigalow and belar scrubs frequently became a rotting mass after one heavy onslaught by the larvae. Stronger growth in the more open country often required two or more successive heavy attacks before breaking down. Certain types with a low nitrogen or high starch content were less susceptible, and the process of their destruction was usually gradual over a period of a few years. As will be discussed later, the initial collapse of large plants and dense growth did not, as a rule, mean complete destruction, but was followed by the development of regrowth from the butts.

Introduction, Establishment and Distribution

One introduction only of *C. cactorum* was made. Approximately 2,750 eggs were forwarded by ship from the Argentine in early 1925 in Wardian cases containing prickly pear. Ten weeks later, in May, 1925, the consignment reached Brisbane as young larvae, which developed through the winter and produced adult moths in Sep-

PLATE I

Fig. 1. Dense prickly pear, *Opuntia inermis*, prior to insect attack; Chinchilla, Queensland; October, 1926.

ALAN P. DODD

PLATE II

Fig. 2. The same area in October, 1929, showing almost complete destruction by *Cactoblastis cactorum*.

tember—October. The females laid a total of 100,650 eggs, all of which were placed in rearing cages at the Sherwood Laboratory and at one inland Field Station. This generation completed the life cycle in February—March, 1926, and gave a return of 2,540,000 eggs. Hence, the original consignment of 2,750 eggs yielded a 900-fold increase in two generations in twelve months.

94,500 eggs were retained for cage breeding, but the majority, i.e. 2,263,000 eggs, were used for the first field liberations at 19 localities in Queensland and New South Wales. Large scale cage rearing was continued until the end of 1927, by which time 9,000,000 eggs has been liberated in batches of from 100,000 to 250,000 at various points in the infested territory. Owing to the natural increase of the insect at the initial release centres, further mass breeding had now become unnecessary.

The main distribution, in which the Queensland and New South Wales prickly pear organisations played an active role, was carried out during the three year period of 1928 to 1930. Pupae in large numbers were collected from the field into cages, where the moths emerged and oviposited, the eggsticks being collected daily. By the middle of 1929, the *Cactoblastis* population at certain liberation sites was so great that the collection of pupae was replaced by direct gathering of eggs from the field. The rapid increase of *Cactoblastis* in the early years is illustrated by the fact that at one original 1926 release point, 300,000,000 eggs were obtained in the month of February, 1930, one man collecting as many as 4,000,000 eggs or about 60,000 eggsticks in a few hours. Under these circumstances the eggs were deposited so thickly that scores of eggsticks could be taken from individual prickly pear cladodes. The total number of eggs used for distribution reached the huge figure of 3,000,000,000. In the light of later experience it is very doubtful whether so intensive an effort to secure a rapid general establishment of the insect was required; however, ample material was available and the cost of collecting and distributing the eggs was very small in view of the immensity of the prickly pear problem. In the distribution campaign, gangs of men were employed for the purposes of gathering cocoons and eggs from the field, collecting eggs from cages, lightly gumming the eggsticks on squares of paper or placing them in waxed quills, and packing the prepared eggs in suitable containers, large numbers of which were forwarded free of cost to landholders. In the very extensive dense areas of prickly pear in Queensland, motor vehicles travelled all trafficable roads and tracks, each unit having its quota of men who pinned the paper squares and waxed quills on to the host plants.

The general establishment of *Cactoblastis* was completed by 1931. Since that date, further distribution has not been necessary except on a rather limited scale for special areas.

Destruction of the Dense Infestations

The eventual success of *C. cactorum* might have been forecast within twelve months after the first trial field releases were made. But even at that stage the most optimistic scientific worker or layman could not have foreseen the rapidity and thoroughness with which the insect accomplished its vast scale destruction. The spectacle of mile upon mile of dense prickly pear collapsing en masse and disappearing within a few years was not envisaged.

Within 18 months after the first experimental liberations had been made in early 1926, the insect had increased greatly and had destroyed many plants and clumps of both *O. inermis* and *O. stricta* at widely separated localities. A year later, in several liberation areas the larvae were present in very great numbers; many hundreds of acres of dense growth had collapsed, and the insect had spread for 10 to 15 miles from the release points.

The mass distribution of 1928 to 1930 resulted in general establishment of the insect, which multiplied quickly. In consequence, the vast areas of prickly pear in all districts in Queensland and in the north-west parts of New South Wales collapsed and were reduced to decayed pulp suddenly and most spectacularly. To give an example of this striking onslaught by *Cactoblastis*, the case is quoted of one continuous belt of 100 miles of dense prickly pear in Queensland, where general distribution was undertaken in 1929; in August, 1930, colonies of larvae were lightly established but were not in sufficient numbers to have caused any destruction; yet, in August, 1932, fully 90% of the plant growth had broken down and rotted; the transformation in two years seemed truly remarkable. *Cactoblastis* needed to be very heavily concentrated to bring about destruction of the order cited. It was estimated that not less than 10,000,000 larvae were required to effect the quick breakdown of one acre of dense *O. inermis*. Where hundreds of square miles of heavy prickly pear infestation collapsed within the space of a year, the insect had to be present in enormous numbers. The last extensive tract of primary prickly pear growth in Queensland was destroyed in 1933—34.

As demolition of the host plant occurred with such celerity and on such a large scale, inevitably the very great majority of the larvae starved to death. The huge *Cactoblastis* population was suddenly decimated. As an immediate consequence there followed a big wave of regrowth which sprang up from the butts of the plants. The regrowth was common to most districts except Central Queensland where it was not an important factor. It grew quickly, much more so than the primary growth, and in many extensive areas approached within 12 to 18 months the height and density of the original infestation. However, the recovery of the insect from its setback was rapid in this succulent fresh food supply. At the end of

1934 the secondary growth had been brought under control, and subsequent recurrences were on a greatly reduced scale.

By the year 1940 not less than 95 % of the former 50,000,000 acres of prickly pear in Queensland had been wiped out, but in New South Wales destruction was not as advanced. The lesser progress in that State was due in the main to the proportion of *O. inermis* on poorer soils on timbered ridges; such growth was generally stunted and chlorotic, and was deficient in nitrogen. In this so-called "yellow pear" the rate of increase of *Cactoblastis* was comparatively slow; hence, control was more gradual. Nevertheless, destruction of these areas is now virtually complete.

In New South Wales, also, another form of "resistant" *inermis* is found on basalt soils on certain hilltops or small plateaux at elevations of 2,000 to 3,000 feet. Some of these infestations occupied hundreds of acres but their sum total represented a very small proportion of the pear-infested lands. The insect has failed signally to bring about worthwhile destruction in these basalt areas, due to a combination of food conditions and lower temperatures being unsuited to its requirements.

The Present Position

In Queensland, complete biological control of the prickly pear problem has been achieved. The position has remained static for at least ten years, and should remain so indefinitely. *O. inermis* and *O. stricta* are not now pests. The former huge areas of infestation have been eradicated or have been reduced to isolated or widely scattered plants. It would be difficult to find an area exceeding 50 acres where plants are in any way numerous. Seedlings continue to develop over a large part of the State, either as isolated individuals or in groups or small patches, to grow, to flower and to produce fruit. The plants may for a time escape the attention of *Cactoblastis*, but as the female moths fly considerable distances in search of the host, the position is remedied after a year or two. Under development the character of the former prickly pear lands has altered radically from brigalow and belar scrubs and country more or less timbered with *Eucalyptus* and other trees and shrubs affording shelter for the various stages of the insect's life cycle, more particularly the pupae which are susceptible to heat, to open grazing pastures or to cultivations. Thus, the insect's abundance from year to year fluctuates with climatic and local seasonal conditions. Not infrequently landholders may express concern at the scarcity or absence of *Cactoblastis* on prickly pear that has sprung up on their properties, and a year or so later report that colonies of larvae are present in satisfactory numbers. Re-distribution of the insect has not been found necessary. In fact the position has been so satisfactory that rearing stocks have not been maintained for this purpose.

The Government policy in Queensland is to rely entirely on continued biological control. Destruction by chemical methods of scattered plants of *inermis* and *stricta* is not enforced; rather, it is discountenanced.

The whole of the former prickly pear lands in the State, once termed Queensland's lost province, has been reclaimed and is in full production. By 1940, 22,000,000 acres of country with an average rainfall of 20 to 30 inches had been thrown open for settlement; this figure represents the compact masses of unoccupied lands heavily infested by the weed, but it does not take into account the greater acreage made up of smaller heavy infestations on grazing properties already held by settlers nor the big extent of land where prickly pear was less dense. Development proceeded rapidly, for general agricultural purposes, for dairying, for beef cattle grazing, and for wool production. Unfortunately, no separate statistics have been kept to show the use made of the reclaimed lands, their present value, and their production. The bald statement that some 50,000,000 acres have been brought from virtual uselessness to successful settlement cannot convey the real picture of the transformation scene, from stagnation under the stranglehold of the pest to thriving townships, prosperous farms growing a variety of agricultural crops, herds of dairy and beef cattle, and flocks of sheep, and the developmental improvements associated with this reclamation. Nor can a statement of this nature indicate the enormous value to the State of its re-won asset.

The position in New South Wales differs from that in Queensland. As mentioned earlier, there are certain areas where *Cactoblastis* has met with no success. There are, too, marginal lands of poorer quality where it has been deemed advisable to use chemical methods of destruction. Again, because of the lesser extent of the weed's coverage, the smaller limits of tracts of dense growth, and the poorer quality of much of the overrun land, developmental work has not been of the spectacular order as seen in the northern State. Nevertheless, the former pear areas have been brought into production for farming and for cattle and sheep grazing.

New South Wales has not followed Queensland in relying on biological agencies for the control of scattered prickly pear. Even in those sections where *C. cactorum* destroyed the dense infestations in a manner comparable to that in the adjoining State, the official attitude and practice has been to enforce destruction by poisoning of scattered plants and seedling growth, whether or not such plants were being attacked by the insect. Queensland's aim has been to control prickly pear. The expressed objective of New South Wales is to endeavour to eradicate the plant completely.

REFERENCES

DODD, A. P., 1936. The Control and Eradication of Prickly Pear in Australia; *Bull. ent. Res.* **27**, 3, *503—517*; 3 plates.

DODD, A. P., 1940. The Biological Control of Prickly Pear In Australia; Bull. 27 Herbage Publ. Series, Imperial Bur. of Pastures & Forage Crop, Great Britain, 27; *131—143*; 3 plates.

DODD, A. P., 1940. The Biological Campaign against Prickly Pear. Commonwealth Prickly Pear Board, Brisbane, Qld.; 177 pages; 37 plates, 1 map.

XXXV
AN ECOGENETIC RESEARCH PROGRAMME WITH INTRODUCED PLANTS

by

F. H. W. MORLEY and O. H. FRANKEL

(Division of Plant Industry, C.S.I.R.O., Canberra)

Aim and Purpose

The impact upon the Australian ecological scene of the introduced flora and fauna which accompanied, or followed, the European discoverers and immigrants has been the subject of many observations and discussions, but of all too few systematic studies. Even less attention has been paid to the impact which in turn the new environment has had upon the plant migrants themselves. What were they like when they came, how did they react to the new environment? Are changes discernible and, if so, how have these contributed to adaptation? Can we relate such changes to the ecological conditions of the new habitats?

Unfortunately, little is known of the precise origin, mode of introduction and establishment, and subsequent genetic adaptation even for many plants which were introduced deliberately, with the exception of the few agricultural and horticultural crops which have been intensively selected during little more than the last half century. But there is next to no information relating to the many species, whether deliberately or accidentally introduced which, genetically speaking, were left to fend for themselves and, having survived the rigours of natural selection, became part of the naturalized or the economic flora of Australia. One of us has previously collected the case histories of a few introduced species and has emphasized the interest which the process of adaptation may hold for the student of experimental evolution (FRANKEL, 1954). The present article expands this discussion and introduces an experimental programme in which adaptive processes are subjected to quantitative appraisal under a range of environmental conditions.

The experiments have the following general objectives:

(a) to study the direction and rate of changes resulting from natural selection in a range of new environments, thus relating changes in population characteristics to habitat conditions;

(b) to study the morphological and physiological elements of adaptation, i.e. to identify and evaluate characters associated with success or failure in terms of (i) survival (ii) production;

(c) to assess the contribution of recombination to adaptation by comparing the success of recombinants among a number of genotypes with that of these genotypes themselves;

(d) to examine whether natural selection in genetically heterogeneous populations is capable of extending the established habitat range of a species introduced, as is so often the case, with a relatively narrow array of genetic combinations.

The populations and the treatments to which they are exposed approximate natural conditions as closely as is compatible with efficiency of design. But both were chosen to give greater and more rapid selection responses than may be commonly occurring in nature. This of course is an experimental necessity; nor does it give cause for serious objections if it is considered that the main interest of this work lies in determining selection coefficients for particular characters rather than in the evolutionary processes per se. The experiments suffer from the inherent limitations of natural conditions in general: the environments are complex, fluid, and therefore cannot be closely defined. This restricts us to broad ecological comparisons. But once the broad pattern of plant responses is recognized, more searching studies under closely defined and controlled environmental conditions will follow.

The general purpose of this study is then to observe "evolution at work", to assay tempo and mode of adaptive processes, and to probe into the nature of adaptive responses. But there is a topical interest in these studies on introduced plants which is natural in a vast pastoral continent where every plant which is sown to pasture is an introduced plant*. It is a curious fact that "where chance or involuntary introductions established themselves over wide areas, as did so many Mediterranean plants in Australia, our own deliberate efforts on the whole seem relatively meagre". The main reason for this comparative lack of success of scientific introduction may be that "... involuntary introductions ... are often on a broad front whereas our scientific endeavours are all too frequently restricted to narrow experimental conditions", and to narrow genetic bases (FRANKEL, 1957). This study provides an opportunity for examining the potential of introduction on a broad genetic and environmental front.

The Species

The species used in the main experiments now in progress is an annual pasture legume, subterranean clover (*Trifolium subterraneum* L.). Its centre of distribution is in the Mediterranean region, but it is found in the Canary Islands and, sporadically, in localities in

---

* There may be minor exceptions – e.g. Mitchell grass (*Astrebla* spp.) has been sown on limited areas in northern New South Wales.

Western Europe. Its southern boundary is at approximately 28° latitude. It came to Australia during, or prior to, the 'seventies of the last century and during the last fifty years has become the most prominent legume of improved pastures in southern Australia. In its original habitats a plant of natural grasslands and wastelands, it probably came to Australia in forage, in packing, or as a seed impurity. It is strictly or prevailingly self-fertilized — at least under Australian conditions — hence most of its genetic variation is most likely to have come from abroad. Some sixty different types have been identified in Australia, and half a dozen have been selected and distributed commercially. It is these few highly homogeneous strains which for all practical purposes constitute the species in Australia, established now on some thirty million acres, from Western Australia to New South Wales and Tasmania, with a northern boundary roughly at 28° latitude.

Many aspects of the ecology, physiology or the mode of introduction and establishment in Australia are referred to by a number of authors, e.g., DONALD & SMITH (1937), AITKEN & DRAKE (1941), FRANKEL (1954), AITKEN (1955), MORLEY & DAVERN (1956), MORLEY, BROCK & DAVERN (1956).

The Experimental Approach

The material for the main experiments consists of two standard populations. The first is derived from twenty-two crosses involving thirteen Australian parent strains, seed from F_2 plants being mixed to give the "hybrid population". The second population is a mixture of the thirteen parent strains, the proportion of each parent being equal to its contribution to the "combined genotype" of the hybrid bulk; this is the "parent population". The crosses, originally made by our colleagues E. M. HUTTON & J. W. PEAK (HUTTON & PEAK, 1954), included combinations between types widely divergent morphologically, physiologically and also genealogically, the last indicated by a degree of hybrid sterility (MORLEY, BROCK & DAVERN, 1956). There is thus a high level of recombination in the hybrid population, in contrast with the restricted set of character combinations in the parent population. The individual components of the latter can be identified by morphological and physiological characteristics.

The experiments are sited in twelve localities along two transects (fig. 1): one in the north in a relatively high rainfall area, with predominantly summer incidence; the other in the south, with lower rainfall, and most of the effective precipitation during the winter. Both transects include sites within and beyond the established subterranean clover zone. In four of the localities duplicate sets are sited on distinctive soil types. Attempts to establish plots in the

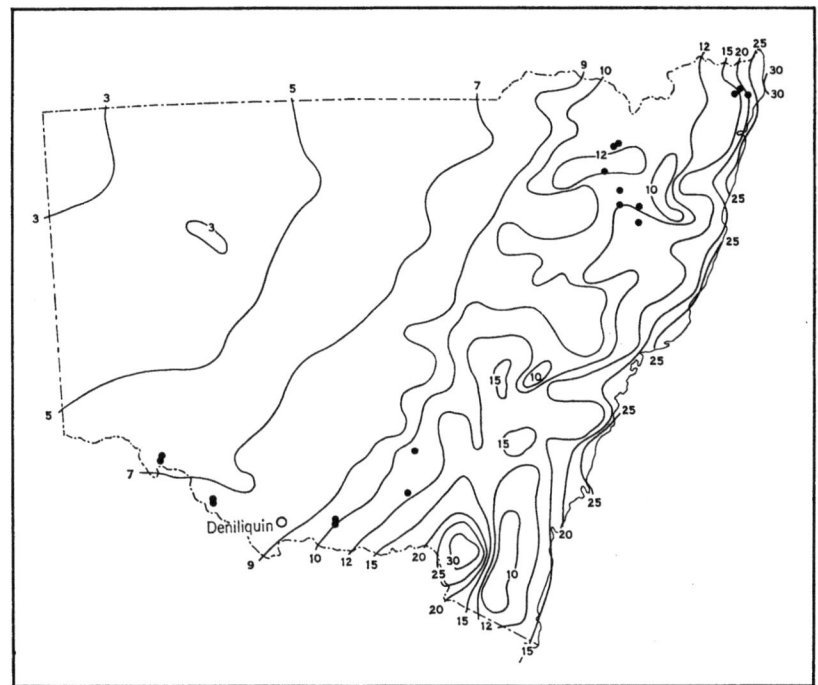

Fig. 1. Distribution of selection plots in New South Wales, showing winter rainfall isohyets.

moist subtropical zone near Lismore, N.S.W., failed owing to insect damage.

Each experiment consists of a set of four plots, 1/250 acre in size, two of the hybrid and two of the parent bulk population. Approximately 5,000 seeds were sown per plot. Seed is being collected from each plot each year and stored under refrigeration for use as required. Samples will be raised at one or more sites for comparisons within and between experimental sites.

All experimental sites were laid down in 1955.

A second project, collateral and to a degree complementary to the one so far discussed, is concerned with objective (b), viz. the study of elements of adaptation. It consists of a close study of a large range of original introductions from the Mediterranean home of the species, collected in recent years by our colleagues C. M. DONALD, J. F. MILES & C. A. NEAL SMITH in Morocco, Algeria, Portugal, Spain, Italy, Greece, Turkey, Cyprus and Israel from a diversity of habitats. In addition strains collected in Australia are included; for, although their origin is unknown, their ecological characteristics have become familiar either from farming experience, or from field experiments in various parts of Australia, or both. This study of strains in the first instance is being conducted in Canberra, but may be extended to other regions.

The first, or "natural selection", approach requires the intensity of selection to be so great that the means of characteristics will change to an extent demonstrable by our testing procedures, within a practicable experimental period. We must recognize, however, that finer adjustments to the environment may not be demonstrable because selection coefficients may be small and, in this particular material, because the lack of cross-fertilization may preclude the emergence of the best recombinants. Because of limitations of time natural mutation is practically ruled out as a mechanism of evolution (but the technique may be useful in conjunction with mutagens).

The second or "strain survey" approach is based on the assumption that genotypes existing at a particular location represent the end points of many generations of selection in a more or less stable environment. A survey of strains from contrasting habitats should thus give indications of the adaptive significance of particular characteristics. Although populations from different habitats will differ in characteristics of minor as well as of major importance, we may reasonably regard the degree of heterogeneity within and between locations as an index of the size of selection coefficients. Thus far a survey of morphological features, such as calyx colour and leaf markings, in the collections of subterranean clover, has not disclosed any clear pattern of distribution. On the other hand, a pattern is discernible for some physiological characteristics, such as vernalization requirement, time of flowering and, though less evident, seed dormancy. This distinction indicates that selection coefficients for the physiological characters are larger than those for the morphological characters. The availability of some results even at this early stage demonstrates the greatest advantage of the survey approach: it need not wait for natural selection to act.

Both approaches suffer from experimental difficulties. Different factors of the environment are frequently confounded, e.g., low temperatures are in general associated with high moisture, soil types vary with rainfall. Further, we cannot be sure that all factors of an environment have been appreciated (e.g., competitors, parasites, micro-environment). Comparisons must therefore be restricted to relatively broad differentials; more searching studies will be possible under better defined, and even controlled, environments.

Whilst it is difficult, if not impossible, to determine the effects of certain facets of the environment without the aid of expensive and complex facilities, some degree of environmental control may be achieved in the field — for example, levels of nutrients, water, parasites, and grazing pressure may all be regulated to a degree. In view of the importance of moisture limitations in many Australian environments an experiment to study the changes associated with different kinds of moisture stress has been set up at Deniliquin, in southwest New South Wales. In this area, normally too arid for subterranean clover, four irrigation regimes — zero, autumn, spring, and autumn + spring — are being applied to replicated plots of the standard hybrid and parent populations. We expect thus to eliminate confounding due to edaphic factors, and to climatic factors other than water. This experiment should yield plants adapted to water stress at different stages of development — material inviting comparative physiological studies.

Even this experiment has its limitations because of the effect of the treatments on associated and competing species and on the micro-climate.

Definition and Assessment of Adaptive Characteristics

In the past plants have been described largely on the basis of morphological features which have been found useful by taxonomists. The value of such descriptions for taxonomic purposes is not questioned, but attention is drawn to a viewpoint so forcefully expressed by LUNDEGARDH (1931):

> "Up to the present the geneticist has worked almost exclusively with morphological characters. Now it is improbable that these characters play as vital a part in survival as physiological characters; and of the selection of physiological characters we know very little." (p. 275); and "Naturally there is no attribute of the plant which is completely unimportant from the ecological standpoint; but characters which are of the greatest importance in taxonomy are often quite subordinate in ecology". (p. 280).

Subterranean clover, with its homozygous lines from a wide range of habitats, offers material eminently suitable for the study of the physiology of adaptation and of intraspecific variation in physiological characteristics.

Techniques are developed and knowledge acquired by preliminary studies on a few strains representing a wide segment of the total range of a characteristic being studied. A survey is then made of a number of strains from contrasting habitats, and the relationships between features of habitats and the degree of expression of the characteristic are examined. Eventually material from the natural selection plots will be tested and comparisons made between the products from different types of imposed habitats. In addition, if suitable material is available, or can be produced from crosses, inheritance studies are undertaken. A series of such investigations should thus present a picture of the significance of different characteristics alone and to each other.

But it is not sufficient to observe differences in behaviour in different environments, or different phenotypes emerging from the ordeal by selection. We must attempt to understand what physiological pathways have been followed to produce the observed result.

Thus, for example, in subterranean clover large genotype/environmental interactions in flowering time have been found (MORLEY & DAVERN, 1956). The real interest in such a finding lies in the interpretation of the observations in terms of the components of the physiological system — in this case vernalization, photoperiod and temperature.

Since most physiological characteristics are determined by many components, the same end point may be reached by different pathways. Thus two strains may differ in flowering-time because of differences in requirements for vernalization or photoperiod, or in the rate of post-initiation development. Strains may thus respond in different ways to different environments, and unless they have been characterized for these three components predictions of behaviour in particular environments are likely to be inexact. As an example of the practical utility of such information, a strain can be recommended for a warm coastal district only if it has a small vernalization requirement (AITKEN, 1955).

Some Results

In the selection experiments our colleague Miss V. ROGERS has noted a difference between "hybrid" and "parent" plots at Deniliquin. In those plots which are given autumn plus spring watering "other species" contributed 3% of the dry matter yield of the hybrid plots, but 17% of the parent plots. This difference may reflect a relative inefficiency for nitrogen fixation by the hybrids or, perhaps more plausibly, their higher competitive ability. The question is being investigated experimentally.

All other results to date — as yet tentative — have come from the survey of strains.

Flowering-time at Canberra shows a clear association with the rainfall of the original habitat. As might be expected, strains from moister locations are later flowering than those from more arid locations, which demonstrates the role of flowering-time in drought-escape in the original habitat. But, needless to say, in regions where proportionally more rain falls in late spring and summer than is usual in the Mediterranean area, later strains may have some advantages over early strains, since the former may derive more benefit from late spring and early summer rains.

Germination of recently harvested seeds of subterranean clover is inhibited by high temperatures, the critical temperature varying greatly between strains. MORLEY (1958) found that in general, seed of strains from warm localities germinates at higher temperatures than that of strains from cool localities. He found, however, an important exception. Seed of several strains from a cool and very moist habitat in north-west Greece will germinate readily at moderately high temperatures. Seed dormancy is valuable as it prevents germination at times unfavourable to survival of seedlings; but in north-west Greece survival is apparently reasonably assured, even after summer rains, perhaps because moisture is seldom limiting in such a habitat. Differences within a species found in this study are similar to those reported by JUHREN et al. (1956) between species from a transect of part of the Californian desert.

The effects on growth of temperature and photoperiod are being investigated in subterranean clover and in lucerne (*Medicago sativa* L.) in connexion with a project in which increased winter production of pasture species is being sought. Results obtained by MORLEY, DADAY & PEAK (1957) and those of other workers, indicate that in several species increased winter growth is more likely to come from strains adapted to warmer rather than to colder climates since the latter are winter dormant. This is an example of conflict between agronomic desirability and adaptation for survival. The winter dormant strains are likely to survive because they are more resistant to frosts, but they are unproductive at a time when feed supplies severely limit stocking rate.

Strains which are "winter dormant" are generally more frost resistant than those which lack winter dormancy. But this is not to say that frost resistance is wholly, or even partly, determined by winter dormancy. The association may be due to selection in the original habitat, and recombinants might be obtained which, while sufficiently frost resistant, give satisfactory winter growth. The regular appearance of certain associations of characters may result from a regular association of environmental features rather than from genetic linkages or common physiological pathways. Thorough genetic analyses are required to distinguish between alternatives.

Other characteristics under investigation include seed size, high temperature lesions, and the effect of temperature on relative growth rate and on net assimilation rate. Undoubtedly the emphasis and particular interest will change, with expanded knowledge and facilities, in directions at present unanticipated.

The Dialectics of Adaptation

An ecogenetic analysis should examine the relative significance of different kinds of characteristics in establishment and survival. It should distinguish criteria essential for ecological success from those which, while perhaps useful in classification, play a minor role in the performance and survival in space and in time. It should seek to understand the physiological determinants of success or failure and the genetic control to which they are subject. Knowledge such as this should give direction, understanding and strength to the synthesis of new plants for arrays of environments.

REFERENCES

AITKEN, Y., 1955. *Aust. J. agric. Sci.* **6,** *212.*
AITKEN, Y. & DRAKE, F. R., 1941. *Proc. Roy. Soc. Vict. (N.S.)* **53,** *342.*
DONALD, C. M. & SMITH, C. A. N., 1937. *J. C.S.I.R. Aust.* **10,** *277.*
FRANKEL, O. H., 1954. *Caryologia* **6** (Suppl.), *600.*
FRANKEL, O. H., 1957. *J. Aust. Inst. agric. Sci.* **23,** *302.*
HUTTON, E. M. & PEAK, J. W., 1954. *Aust. J. agric. Res.* **5,** *271.*
JUHREN, MARCELLA, WENT, F. W. & PHILLIPS, E., 1956. *Ecology* **37,** *318.*
LUNDEGARDH, H. 1931. Environment and Plant Development. London.
MORLEY, F. H. W., 1958. *Aust. J. biol. Sci.,* **11,** *261.*
MORLEY, F. H. W., BROCK, R. D. & DAVERN, C. I., 1956. *Aust. J. biol. Sci.* **9,** *1.*
MORLEY, F. H. W., DADAY, H. & PEAK, J. W., 1957. *Aust. J. agric. Sci.* **8,** *121.*
MORLEY, F. H. W. & DAVERN, C. I., 1956. *Aust. J. agric. Res.* **7,** *388.*

XXXVI
THE ECO-COMPLEX IN ITS IMPORTANCE FOR AGRICULTURAL ASSESSMENT

by

C. S. CHRISTIAN

(Chief, Division of Land Research and Regional Survey, C.S.I.R.O., Canberra)

The Australian continent has an area of about three million square miles and a range of climatic conditions from winter to summer rainfall, wet to dry and hot to cold. Until European settlement about a century and a half ago, the continent was occupied by primitive nomadic peoples who, surviving by hunting and gathering native foods, did not leave any traditional patterns of land use applicable to European development. The accumulation of scientific and geographic information about the continent and experience in land use have both been much restricted in respect to time. They have also been geographically influenced by the concentration of populations in the more easily developed areas of higher rainfall in the east, south-east and south-west which are the main areas of agricultural and industrial production but which collectively occupy only a small part of the continent. The remainder, largely arid, semi-arid or dry monsoon country, mostly has a low natural productivity, is still very sparsely populated, and in parts is still in the explorative stage of pioneering settlement.

Australians have explored widely and virtually none of these areas can be described as entirely unknown, but the amount of precise scientific data concerning them has been very limited. Broadscale studies such as those of GRIFFITH TAYLOR (1918, 1920, 1940) on climate, PRESCOTT (1931) on soils, and EDGEWORTH DAVID (1950) on geology and the C.S.I.R.O. publication "The Australian Environment" (1950) have given a general picture of the whole continent and have put it in perspective with other countries. Other investigators have provided detailed information on selected aspects of certain parts, such as that given by BEADLE (1948), BLAKE (1938), GARDNER (1944), BLACK (1943—57), TEAKLE (1938), WOOD (1937) and reports and papers from such organizations as the C.S.I.R.O. Division of Soils, the Queensland Bureau of Investigation, and in the journals of the State Departments of Agriculture and the Soil Conservation Services. It is only in recent years that an intensive effort has been made to describe and map the less developed areas of the continent in a systematic programme of regional survey and potentiality assessment. Work in this field was begun in 1946 by the C.S.I.R.O. Division of Land Research and Regional Survey, formerly the North Australian Regional Survey. As the little developed areas occupy the greater proportion of the Australian continent it is not surprising that a specific approach appropriate to this Australian scene has emerged from these investigations. This scene is essentially one in which land utilization is being planned for a modern society in hitherto little studied and little developed areas, without close environmental, sociological or economic parallels elsewhere. It has broad ecological interest apart from the principles and detailed ecological studies involved in the work itself. A discussion of some of the factors influencing the approach will help to put this in perspective.

Factors influencing the development of methods

The areas to be studied were very large by normal survey standards and had to be mapped and assessed in relatively short periods of time. The actual rate of progress by the one field survey unit over a period of ten years has been of the order of 50,000 square miles per year. Such a rate of survey and mapping was possible by the aid of aerial photography and an adequate systematic method of sampling on the ground, correlating air photo and ground characteristics. Detailed mapping of individual characteristics was not feasible. Furthermore, the mass of data obtained, and the variability of such large areas required some method of systematising of the information to permit easy interpretation and appreciation. As explained later, this led to the mapping of complexes of country, the details of which were actually described but not mapped, rather than the separate mapping of individual characteristics.

As many of the areas were only superficially known scientifically, it was necessary that specialist officers capable of adjusting their observations and standards to the new circumstances should compose the survey unit. Furthermore, for a number of reasons which emerged as the methods developed, it proved very desirable that these specialists should work together as one team with a common objective rather than making independent studies.

The fact that there was not any traditional form of agriculture or pastoral land use on which to build was of practical significance in assessing potentialities. It was not a matter of adjusting the present form of land use but one of predicting those kinds of land use which might be successful. In this connection the contrast between these parts of Australia and older countries deserves discussion. In the latter, a rural population has long occupied the land and in the course of developing the present patterns of land use many of the original variants in the environment have lost their significance through a process of adaptation of methods and species to land conditions. Those which have not been susceptible to controlled change or for which biological compensations are not easily made, have emerged as the prominent factors now considered to control adaptability, productivity, or the form of land use. Some such factors are more or less universal in their biological significance and may dictate the general categories of biological activity which might be adapted to a particular locality, as light and temperature, for example, control the season of growth and reproduction of plant species. Other variants have much more strictly local agricultural significance as the following very different examples will illustrate. In Holland the depth of the permanent water table exercises such an influence on the distribution of different kinds of production that it is artifically controlled over large areas. In many countries away from the Equator the incidence of frost is very important near the margin of distribution of summer growing crops, such as cotton and rice. By way of contrast the occurrence of short dry heatwaves may cause maize and other crops to fail in sub-tropical countries where other conditions are apparently favourable. In Australia and other dry countries, the amount of rainfall and the period and season over which it falls are dominant factors determining the distribution of arable agriculture, but these are of much less significance in the wet tropical areas of New Guinea where with altitude the significant factors may be night temperature or hours of sunshine. Each area therefore may have its own factor of particular significance locally determining success or failure of plants and animals and in the older agricultural regions these factors are now generally recognised. On the other hand in little developed areas such as those parts of Australia under discussion where land development experience is limited and adaptation of methods of land use has not proceeded very far, it is not always easy at the initial appraisal to recognise all of the factors which may limit productivity or those which are likely to remain limiting. Nor is it possible to predict how the particular assemblage of factors composing the environment will interact to influence potentialities. Thus, a soil presenting few physical problems in one climate may be a problem soil in another, so that the soil class alone means very little unless other factors of the environment are specified. The nutritive value of a pasture species can vary with climatic and soil characteristics of the environment whether it is due to physiological or genetic causes, and this may swing the balance between breeding cattle for sale as stores or breeding and fattening for direct sale as killers. The presence of one plant species poisonous to stock could degrade the pastoral value of an otherwise useful plant community. Excess fluorine in sub-surface waters, perhaps the only source of water for grazing stock in a given area, may make a pastoral region with good vegetation and good soils quite worthless. An acute soil deficiency of a major element such as phosphorus may make pasture improvement in low rainfall areas quite uneconomical in spite of other favourable characteristics or, alternatively, its amelioration may be limited by one of many possible minor chemical elements. These examples illustrate the kind of factors which may have to be taken into consideration in new areas. The assessment of land use in such new areas therefore must be deduced from a multitude of factors and the way in which they interact, and a true assessment may only be possible after field experience of land use or some form of experimentation has helped to demonstrate and more clearly define what really are the major limiting factors in the local environment.

Apart from what might be regarded as new factors or new assemblages of factors, other factors well known to be dominant controlling influences in other regions may retain their indicator significance in new areas. However, because of the combination with other factors in the environment their level of importance may change.

The existing methods of description and classification of such characteristics as climate, topography, soil or vegetation are very helpful in explaining known happenings in areas where a good deal of experience in correlation is available and to some extent may serve to place new areas in perspective. However, because the actual characteristics found useful for classification purposes are not necessarily important in all areas there can be implied or hidden characteristics of significance, not revealed or differentiated by the accepted methods of classification and description. The criteria for measuring the degree of importance under the new set of conditions may therefore need to be modified. Experience with these regional surveys in Australia has demonstrated that the routine data obtained about soils from chemical and physical laboratory examinations are inadequate to indicate the field problems which have subsequently been met by agronomists. These modifications in criteria cannot be established without intensive investigation and perhaps experimentation. It is not possible therefore to seek them for the initial reconnaissance assessments of areas. However, in the development of a long term programme of survey and assessment such as must be inevitable in Australia because of its very size, attention must be paid to the improvement of existing criteria so that as surveys progress they can be of more and more value to interpretation. Thus, accompanying the main programme of regional survey it was also found necessary to develop basic studies in certain fields related to potentiality assessment. Because of the overall significance of climate in the surveyed areas considerable attention has been paid to that field and particularly to the aspect of plant-soil-atmosphere water relationships.

In the absence of experience of land use or development, it was not desirable to approach potentiality assessments with any pre-conceived ideas of what the ultimate potential might be, even though the general form of land use might appear to be quite obvious. It was necessary to collect scientific information of a comprehensive nature which could be used as a permanent reference as more experience was gained of actual land use possibilities. The basic problem, therefore, was to first recognise areas presenting distinctive environments, to map their extent and distribution, and to record as many actual facts about them as possible, rather than merely to look for pre-conceived environments known to be suitable for particular purposes. The assessment of potentiality follows as a second stage and it has been stressed that this should not be permitted to influence the basic land or environmental classification. In this respect the approach described differs from that adopted more recently by DOWNES et al. (1957), who, working along somewhat parallel lines in a more developed part of Australia, include the most likely form of land use amongst those used for description. The word "potential" must be interpreted as including the possible problems of maintaining natural resources at their present value as well as the possibilities of developing them and increasing their productivity beyond their natural levels or the apparent limits set by the climate and the native vegetation.

With this record of basic facts it was then possible to make an initial assessment of potentialities, recognising that this assessment might very well be changed as further information or experience was gained. In order that subsequent information might be just as validly applied, it was essential that the land classification should be as fundamental as possible and based on permanent characteristics of land and not on subjective interpretations of individual characteristics. For this reason the idea of subdivision of the land surface into areas with similar or distinctive origins was introduced and from this arose the concept of land, land units and land systems which form the basis on which the present survey methods and land classifications have largely been built.

The concept of land, land units and land systems

In this concept which has been discussed by CHRISTIAN & STEWART (1947, 1952), CHRISTIAN (1952, 1957), and TAYLOR & STEWART (1956), the term land refers to all those physical and biological characteristics of the land surface which affect the possibility of land use, in other words the whole combination of factors which constitute the agro-environment. In practice, it refers more particularly to the more obvious, inherent features of the land surface, namely to topography, soils, vegetation and climate. A land unit is a particular land form, which at each of its various occurrences has associated with it the same group of soils and vegetation communities. Strict associations of characteristics are likely to occur only if the various occurrences of the land form have a common genesis. A land form is the product of particular geomorphological processes acting on particular geological material for a particular period of time. With such a physical evolution there will also have been an accompanying and interacting biological development leading up to the present biological characteristics. The combination of the physical characteristics of the land form and its biological characteristics constitute the land unit. Past and present climates must be included as major contributory factors influencing the development of the land unit as a whole.

The land system is a naturally occurring pattern of land units, geomorphologically associated and morphogenetically related. The boundary of a land system will coincide with the limits of major geological, geomorphological, climatological or biological features. The mapping of land systems, which are in effect distinctive patterns of country, and a description of the land units which compose them, permit the complexity of very large areas to be reduced in a systematic way to a small number of subdivisions permitting easier appreciation of regional variability.

Reference to one of the regional surveys made by the Division of Land Research will illustrate this. The Barkly region (C.S.I.R.O., 1954), has a total area of 120,000 sq. miles in the dry monsoon zone of Northern Australia. The annual average rainfall is 10 in. in the south and 30 in. in the north. The topography varies from flat and gently undulating plains to steeply dissected areas. Drainage is both internal and external. Thirty-two soil units corresponding to soil families were described, including various lateritic soils, grey and brown soils of heavy texture, podzolic soils, stony and highly leached sandy soils. Forty major plant communities, ranging from associations to alliances and complexes were recognised. They included grasslands, shrublands, woodlands, open forests and forests.

The region was divided into 21 major geomorphological units and 38 land systems. Each land system was composed of a number of land units varying from 2 to 6. The land systems were illustrated by diagrammatic cross-sections showing the typical arrangement of the land units. Each cross-section was accompanied by a brief digest of the relevant information concerning the location, general description, climate, geology, geomorphology and drainage of the land systems and notes on the relative proportions and nature of distribution of the component land units. Specific information was given on the topography, vegetation and soils of each land unit.

As a result of this survey a large complex region has been reduced to 145 individual units. Many of these have such obviously low potential that they do not deserve further attention at this stage. Attention can be directed to those of more immediate importance.

Some of these are small units intermixed with others within a land system. The subdivision into land systems serves not only to delineate the larger areas of useful country, but to indicate those areas within which small but important units, too small to be mapped individually occur and in what parts of the landscape they are to be found. Thus a general appreciation can be made of each land system and further investigations can be directed to where they are likely to be most profitable. The results of these investigations on any one unit can be expected to apply to the other occurrences of that unit within the land system.

Personnel and survey procedures

The survey techniques adapted by the Division of Land Research have been described by CHRISTIAN & STEWART (1947, 1952), CHRISTIAN (1952, 1957), and TAYLOR & STEWART (1956). The following brief description will be sufficient to indicate the general practices adopted and the manner in which information is obtained, as it seems that Australia offered a unique opportunity for the development of this kind of survey activity. The field survey unit comprises specialist scientific officers and a group of assistants necessary to cope with transport, messing etc. The specialist officers in the field include a geologist, a geomorphologist, a pedologist, a plant ecologist and in some instances a taxonomic botanist. The unit has the help of a climatologist who is responsible for the interpretation of available climatic data and also conducts research concerning climate-plant-soil relationships and the development of more precise methods of interpreting climate. Specialist studies in hydrology have only recently been established. Previously, the general hydrological interpretation of areas was made by the geologist and geomorphologist in combination with the other members of the unit.

There is collaboration with agronomists and pasture ecologists at headquarters and at field stations established in areas previously surveyed. The Division of National Mapping, Department of National Development, cooperates by constructing base maps and photo mosaics and is also responsible for the final preparation of maps for publication. Aerial photographs, usually of the scale of 1 : 30,000 or 1 : 50,000 are provided through this office.

Individual areas surveyed vary in size up to about 150,000 square miles. Areas of the order of 100,000 square miles are preferred as this justifies two years field work, the second year providing an opportunity for checking some of the first year's observations.

The survey procedure is briefly as follows. After reference to relevant literature and data concerning the area to be surveyed, a preliminary examination is made of the aerial photographs. This examination gives a broad subdivision of the area into its major geological and geomorphological subdivisions and, within these, the various complexes of country represented by tone patterns in the photographs. Stereoscopic examination of these patterns gives a general picture of the relief and in some instances a broad interpretation of soils and vegetation especially if they correspond to areas previously surveyed. The main objective at this stage is to recognise all distinctive patterns depicted by the photographs and to plan field traverses so as to adequately sample these on the ground. The field itinerary is drawn up to provide for the minimum amount of travel and particularly cross-country travel. The survey units on the mainland of Australia are transported by four-wheel drive motor vehicles but a similar unit operating in New Guinea finds it necessary to do most of its travelling on foot as roads are limited to restricted areas and cross-country travelling by vehicle is usually out of the question.

In the field, the survey unit navigates by the use of aerial photographs or photograph mosaics. The land units corresponding to the photo patterns and the way in which these units are associated in the field are recorded. Detailed comprehensive studies are made at type locations. The units are described in terms of their geology, geomorphology, soils and vegetation. This information not only serves to describe the elements of the landscape, but is also used on a wider basis in the interpretation of the genesis of the landscape and its various subdivisions. It is in this last aspect that collaboration between the various specialists in the field has proved so important. Each specialist is able to provide others with additional evidence on which to aggregate or separate units or explain phenomena observed. Thus, the pedologist may record a soil group transgressing a number of land units or even land systems. The fact that these may have had different histories, or different parent materials, as shown by the geomorphologist or geologist, could well imply differences in the soils not indicated by their more easily observed characteristics, used for classification purposes. Further, these differences may be correlated with different ultimate potentials, expressed for example as a difference in the mineral nutritional value of pastures, associated with a difference in stock carrying capacity or rate of growth of stock. On the other hand, the pedologist's examinations and interpretations of his soil profiles is valuable evidence for the geomorphologist's interpretation of the development of the land surface. In similar ways, the plant ecologist who in new areas does not have predetermined community classifications to guide him, may be influenced to subdivide a community because of soil or geological features of the land surface, or, alternatively, his recognition of changes in communities may guide the pedologist or the geologist to look for differences which may not otherwise have been noted. The experiences of over ten years working in this way have unquestionably proved to those participating the very considerable advantages which accrue from integrated field studies between a group of specialist officers in comparison with independent studies.

Another advantage of integration through teamwork in the field is that each specialist officer brings to the unit a knowledge of other fields than his specialist one according to his own particular background and training, so that collectively the unit may often recognise important aspects or problems of areas which could well be missed by the individual observer.

Apart from his contribution to the combined study of the landscape and its subdivision into its natural units and complexes each specialist pursues his own particular field of study as far as possible. This may have certain limitations imposed by the rate of movement of the unit as a whole, but some latitude is possible, and individual officers may spend longer periods in certain areas or visit them again at the completion of the field survey season. Such detailed specialist studies are often essential because of the new aspects found in new areas.

Each specialist officer will also consider landscape features in a broader way than his restricted specialist field dictates. Thus, the plant ecologist will concern himself with the description and evaluation of the pasture components of the vegetation and their relationship to environmental factors, as well as his general ecological studies. Likewise, the botanist or the ecologist will observe the apparent palatability of individual species if grazing stock are present and their susceptibility to grazing and degradation. In pastoral areas the plant ecologist must be particularly concerned with the stages of degradation of grazed communities as a basis for future studies of range condition standards. The soils officer must concern himself with certain hydrological interpretations of the soil profiles and may make special observations relating to permeability of soils and water penetration where there is opportunity to observe this after falls of rain. He will automatically be concerned with erodibility, stoniness, and such factors of importance to soil use. The geologist and the geomorphologist will both be concerned with major hydrological features and where bores and wells occur data relating to underground water supplies will be collected.

In the field boundaries between land units and land systems are marked on the aerial photographs covering the traverse route and symbols are given to relate them to the records of the various observers. Because of the quantity of information recorded some systematization of records is necessary, as for example, that used for describing vegetation communities (CHRISTIAN & PERRY, 1953).

Where a second field season is required it follows the general pattern of the first, but new traverses will be made, excepting where it is desired to check some previous observation, or make more detailed examinations. At the completion of the field work a more detailed and complete interpretation of the aerial photographs based on, and by extrapolation from the traverse records is made at headquarters. Decisions are made concerning what units can be mapped at the particular scale to be used. In the course of this mapping and the collation of information by the various specialists, much of the information about the region and its natural units will have been integrated. Cross-sections of each land system illustrating its land form components and their characteristics are constructed and a combined picture of the region, its distinctive environments, their characteristics, problems and limitations is developed.

Assessment of potentialities

The description, classification and mapping of the lands of the region provided by the initial survey constitute the fact-finding stage during which the major environments to be assessed are isolated and mapped and described as thoroughly as possible. This information presents a permanent framework into which more detailed information can later be fitted in a systematic way and a permanent reference for future studies and applications in a variety of fields that depend upon an understanding of the natural characteristics and variability of the region.

The assessment of potentialities represents a second stage of the investigation, although it may be done in part concurrently with the initial survey. It can be regarded as a series of steps based successively on more precise knowledge, each more closely approaching a true assessment. Assessment must be regarded as a continuing process and as new knowledge becomes available always subject to revision.

In the initial assessment the significance of observable characteristics of the environment is interpreted as well as possible by deduction or by comparison with known responses in other localities. There are definite limitations to what can be achieved at this stage. In the first instance homoclimatic studies constitute an obvious approach but the inadequacy of the usual study of this kind must be recognised (CHRISTIAN, 1958). The characteristics on which regional homoclimatic studies are made have a certain usefulness for broad zonal comparisons, but are of very much less value when land potential at a specific site is to be determined. Two aspects are involved here. First a much more precise examination of climate than is usually made for zonal studies may be necessary to expose the significant features, and second the interaction between climate and other factors such as the physical properties of the soil, or

PLATE I

Low *Eucalyptus* woodland, 35 in. rainfall country, numerous termitaria, ground flora dominated by *Themeda* and *Heteropogon*.

C. S. CHRISTIAN

PLATE II

Eucalyptus open forest between Katherine and Darwin, N.T. with remnants of Tertiary land surface in background, 50–60 in. rainfall.

drainage characteristics of the land surface. These will rarely be quite the same for the two areas being compared. Additional weaknesses are that the basis of recording climatic data in different areas is not always identical so that truly comparable data are difficult to obtain and records are often summarised as averages for long periods, a procedure which serves to camouflage the particular climatic conditions likely to influence plant or animal responses.

Ecological studies of the native vegetation may also be used for comparative purposes, even though the species present in two areas may differ. Vegetation structure alone can be of some value in this regard. Plant sociological studies of a more advanced nature scarcely have a place in this kind of work because of the very elementary knowledge of the vegetation in the new areas being examined, although indicator plants of some meaning may be identified in the course of the survey. Where assessment involves the interpretation of the suitability of an environment for the introduction of species from outside the area, a knowledge of the response and requirements of such species to specific components of climate or to other aspects of the environment may give a guide as to whether or not the species are likely to be successful. Unfortunately, autecological information of this kind even for important crop species is exceedingly rare.

The interpretation of soil characteristics plays an important part in the initial assessment but present knowledge of how to categorise soils for prediction purposes is not nearly adequate. A great deal more needs to be known about the interpretation of mineral fertility, physical responses and microbiological activity when soils are put to specified uses under particular conditions. There is almost always the need for confirmatory field tests and subsequent examination.

All of the foregoing approaches to assessment necessarily depend on the observable and recorded characteristics of the environments. It has been mentioned, however, that in new areas unsuspected factors may emerge as the limiting factors and these may only be appreciated following subsequent experimentation or land use experience. The best that can be expected from the initial assessment therefore is a broad statement on the likely scope for land use and its more obvious problems which can then be tested by experiment. It will serve to separate those areas worthy of more immediate and intensive study from those which could well be disregarded for the time being. It will be able to suggest whether an area is likely to be suitable for arable agriculture or not and the general kinds of crops which should be examined if it is, but it would be extremely hazardous to guess at the levels of production that might be expected or the particular systems of agriculture which would have to be practised in order to maintain stability of that agriculture. Likewise, in pastoral areas the estimate of stock carrying capacity is influenced by so many factors, including the standards of management of both herds and vegetation, and the stage of improvement of areas in terms of amount of fencing to control stock, and the number and placement of permanent watering points for stock, that a true estimate of this cannot be made until precise studies of the effect of grazing and climatic crises on the vegetation have been made. The initial assessment therefore indicates where further investigations should be made and the general nature of such investigations. It can indicate the more obvious problems which will have to be studied, but cannot hope to define all the problems which may be met in the subsequent period of field experimentation. It can rarely hope to indicate actual levels of production.

It follows that in the absence of established principles concerning either arable agriculture or range management, there is nearly always the need for actual field tests to confirm or deny the initial assessments and to expose or define more precisely specific problems associated with particular land units. Land classification by the regional survey will indicate where these tests can most appropriately be made and the areas to which results from any test location are likely to be applicable. In arable areas these tests in the first place take the form of simple crop, variety, fertilizer and time of planting experiments, but may become more specialised according to the responses and interactions observed. In pastoral areas where stock are dependent upon native vegetation there is usually much less established information available for application. In consequence there is always the need to study many aspects of plant and animal ecology before the desirable relationships between land units and animal populations can be established and defined. These later studies will in most circumstances lead to a clarification of problems and a specialised research programme, the results of which may require a substantial modification of the initial estimates of potential.

Examples of the eco-complex in potentiality assessment

Investigations have been started in several regions in Northern and Central Australia in areas mapped and assessed by these regional survey methods. Examples from these will indicate some of the ways in which the field testing of initial assessments has revealed the importance of other factors or the interaction between various features of the environment. The locations quoted represent type areas for a large part of little developed Australia.

The Katherine Research Station in the Northern Territory was established within a land system selected for its possibilities for dryland agriculture in the dry monsoon zone of Northern Australia. The soils are mostly light to medium textured and large areas are topographically suitable for cultivation. On the basis of a broad assessment of climate there was every justification to believe that crops such as peanuts, sorghum and cotton could be grown successfully in that area. Once a very obvious phosphorus deficiency was overcome it was soon demonstrated that the first two crops were well adapted to the region. Cotton, however, has given most variable results over a number of years and from experiment to experiment, and there are still unexplained differences within the agronomic environment contributing to this variation. The Katherine climate has long been compared with that of Kano in Nigeria. The assumption has been that the two areas are homoclimatic. More detailed studies by R. O. SLATYER (unpublished) have indicated some significant differences, not obvious in mean values, particularly in short term variations within the season. These variations in combination with the physical responses of the soils cultivated at Katherine produce environmental changes which are apparently much more severe than those experienced by crops at Kano. These changes are particularly significant during the establishment period of crops, for the soil, which has a narrow range of available water, fluctuates rapidly in moisture content. In the case of cotton, it is believed that these changes resulting from the interaction of climate and soil physical properties are partly responsible for the variation in cotton yields. The effect is in part through the proportion of flowers which persist and produce bolls, but this also appears to be influenced by the fertility status of the soil, particularly in respect to nitrogen. With this particular crop in mind, precise analyses of environmental factors and their interactions will be necessary before consistent production can be achieved.

At the Kimberley Research Station on the Ord River in the north-west of the continent, the investigations are concerned with the development of an agriculture on heavy-textured soils under irrigation. Similar situations occur in other parts of Northern Australia. Here, because of the availability of irrigation water shortages of soil moisture can be eliminated. Another factor in this environment has complicated investigations with cotton. In this area native host plants support a variety of insect pests of cotton, which until recently damaged experimental plantings so seriously and so variably that the true effects of other experimental treatments could not be determined. Over the last three years new insecticides have been used, insect damage has decreased and good yields have been obtained. If these insecticides remain effective in seasons with prolonged wet periods such as were experienced in the earlier stages of the investigations, a more systematic investigation of the agronomy of the crop will be possible.

This last point introduces the factor of variability between seasons. On both these research stations, experience has shown the absolute necessity for conducting experiments for a sufficient number of years to adequately sample the variations in seasons likely to be experienced before conclusions are drawn. Predictions of production based on climate must take this aspect of variability into account, especially in those areas where a crop is near the margin of its potential area of distribution.

Work with various crops at the Kimberley Research Station illustrates some interesting interactions between the soil, plant nutrition, species and cultivation methods. Under irrigation the most successful crop is sugar cane, which gives yields comparable to the best sugar cane areas of Queensland. Providing moderate amounts of phosphatic and nitrogenous fertilizers are applied there is no plant nutrition problem. On the other hand, rice grown under paddy conditions develops a necrotic tipping and yellowing which varies in severity from season to season and with variety to variety. The symptoms disappear following heavy applications of nitrogenous fertilizer but their effect is not prolonged. Likewise, irrigated pastures show a rapid reduction in yield after the first good period of growth. It is suspected that these responses may be related to anaerobic conditions or an interaction between anaerobic and wet conditions in the soil induced by prolonged rainfall periods or paddy irrigation. It is interesting to note that the native vegetation in this area is mainly tussock grassland with a sparse scattering of low trees and shrubs but even with applied fertilizer and irrigation the production of sown pasture has been very disappointing.

With irrigation water available the possibilities of production of annual crops in both the summer and winter seasons or, as referred to locally, the wet and the dry seasons, arises. As the winter is almost totally dry it offers scope for far greater control of moisture relationships than does the summer during which rainfall averaging about 30 inches per annum falls in a very irregular manner throughout the season. Certain crops grown in the winter rainfall zone of southern Australia such as safflower do reasonably well, but crops such as cotton, peanuts and sorghum normally grown in the summer are retarded during the June—July—August period of moderately low temperatures and harvesting may thereby be delayed until damage from early storms becomes a hazard. The achievement of dry season production potentials may involve an extensive programme of variety selection and adaptation. For example, with rice the Japonica group of varieties grow much better during the winter than in the summer and the reverse is generally true of the Indica varieties. However, such well differentiated groups of varieties do not occur in other crops under investigation and more detailed studies of variety-environment interaction will be necessary.

The arid pastoral areas present quite a different group of problems. At Alice Springs in the centre of the continent, C.S.I.R.O. has begun investigations in the fields of plant ecology, climatology and hydrology, all related to the maintenance, utilization and improvement of the grazed native vegetation, and the more efficient utilization of water, which is the main limiting factor in this arid area. Here, the first objective must be effective utilization of the vegetation without degradation and any assessment of productivity in terms of stock products per unit area must be in keeping with this objective. Even without stocking certain fluctuations occur in the vegetation communities as a result of climatic crises. The work of the ecologist involves a segregation of these effects from those induced by the grazing animal and a recognition of the trend taking place in the vegetation community at any particular time. There is need to study the successive stages of degradation and of management practices which may retard or reverse the trend.

The two most extensive plant communities in the arid centre are the desert woodland communities dominated by *Acacia* spp. and particularly *Acacia aneura*, and the sclerophyllous hummock grasslands dominated by a group of grasses known collectively as spinifex and porcupine grass, belonging to the genera *Triodia* and *Plectrachne*. These communities vary in floristics and structure from place to place and the concept of land units and the particular community associated with each, provides a more systematic basis for the examination of trends.

Consideration of ultimate potential must speculate beyond the satisfactory use of the present vegetation. It should not necessarily be assumed that this native vegetation is utilising water and other plant nutrients or the environment as a whole in the most efficient manner possible. For example, the components of the spinifex communities are widely spaced with often up to 90% bare ground between the grass hummocks. As a large proportion of the total rainfall in these areas occurs in light falls which cannot penetrate the soil very deeply, it is inevitable that much of it is lost by direct evaporation from these bare spaces. It is conceivable that a community such as the spinifex grasslands induces a more arid environment than might be expected from the amount of rainfall received. Should this be so, there may be scope for greater change and increase in productivity than would appear possible at first sight. Such speculations can only be decided by more precise information and for this reason detailed studies of the fate of water received at a typical spinifex site and of the microclimatic profile of the community have been initiated. Concurrently with this, certain empirical practices are being tested to determine if other species will survive in this environment and if they can be maintained in competition with the native components in the spinifex community. From these combined investigations it is hoped to be able to state more precisely what is the true potential of the areas occupied by spinifex and to what extent this can be economically achieved.

A somewhat similar question arises in regard to the desert woodland community. *Acacia aneura,* commonly called mulga, is renowned as a valuable fodder plant, particularly in drought periods. Associated with these woodlands are ground storey species which in many areas probably contribute a greater proportion of stock feed in normal years than does the mulga. *Acacia aneura* is a very valuable species, some types being much more palatable and useful than others. Furthermore, individual trees are pruned by grazing and communities may reach the stage where most of the growth is out of reach of the stock, although the fall of leaves and pods undoubtedly helps to keep stock alive during droughts. The variation within the species and in the density of the tree and ground storey components and their respective levels of production of useful fodder in relation to their utilization of water and nutrients raises a number of questions concerning what is the most effective community for any specific land units. Detailed examinations of the species of the communities, their water regimes and microclimates above and below ground could provide information necessary for more precise thinking along these lines, and are necessary before the real potential of the environment can be assessed.

Another important aspect influencing the true potential of these arid areas is the extent to which control can be exercised of the water they receive to obtain more effective production than that which is given by the natural vegetation. Wholesale destruction and change of the vegetation is not practicable because of the cost involved and the low return per unit area. Certain parts of the topography, however, are more favoured than others. They receive run-off from the surrounding country and their total water resources may be much greater than the average rainfall of the area. Under uncontrolled conditions much of this water may be lost to underground aquifers; some of it may be recoverable for small scale irrigation purposes but much of it may be contaminated with saline water. A programme of hydrological investigations is necessary and has been commenced to examine the degree to which the potential and production may be increased through exercising some control of run-off to direct it to the most favourable habitats, or by recovery of water from underground aquifers.

A different kind of problem in pasture assessment has emerged at the Katherine Research Station. Here stock gain weight on native pastures for almost half the year and lose weight for the other half. When mature these pastures are very low in protein. They are also deficient at most stages in phosphorus and stock fed supplementary phosphate show substantial increases in liveweight over stock not supplemented (NORMAN, 1957). Also in this northern part of Australia the ectoparasite *Boophilus microplus* is a common pest. NORMAN (1957) has shown that the control of this pest leads to still greater increases.

Any assumption concerning the true livestock productivity from pastures in these areas would be considerably in error if the implications of these two features of the eco-complex, one biochemical, the other zoological, were not fully considered.

The various illustrations which have been given demonstrate the wide range of factors and interactions which may determine the potential of different environments or limit the achievement of the theoretical potential. In a parallel way, it could be expected that each land unit will present its own particular sets of problems. The examples also serve to show that it will be a rare instance when the true potential of a new area can be predicted from a reconnaissance examination, unless there is very close association between that area and similar adjacent ones in which experience has already been gained. The stage of autecological knowledge of plant and animal species is such that it is quite futile to attempt to define all the factors which may be important at a particular location. It is for this reason that assessment of potential must be considered as an ever-progressing procedure. It is also for this reason that the need arises to subdivide regions into areas representing distinctive environments, using whatever means are possible including description of the more obvious characteristics of the environment and other approaches such as the genesis of the landscape and its components as a basis for aggregating similar parts of the landscape, or segregating those which might be expected to be different in spite of their superficial similarity.

Economic aspects

It finally remains to draw attention to the possible difference between purely theoretical and practical potentialities. The ecologist and the agronomist assessing land use potential are primarily concerned with the biological possibilities of the environment. Before these possibilities can be put to practical use, it must be demonstrated that they represent acceptable practices from the economic, sociological and political points of view. Agronomically, it has been shown that probably the best form of plant production at the Kimberley Research Station is sugar cane. The existence of international agreements and quotas on the world markets and the fact that the industry is concentrated along the east coastal strip of Queensland may mean that use will not be made of this potential in the Kimberley area at least for some time to come. Thus, the potential of this area to grow sugar cane could remain largely a theoretical one.

Certain parts of Northern Australia at present devoted to cattle raising on an extensive range system could produce fodder if favoured parts of the topography were cultivated. Under present conditions the beef cattle raised under these extensive conditions are moved to fattening areas on foot over distances of several hundred miles. Unless improved transport systems can be developed in these very extensive and relatively low producing areas there seems little possibility of utilising the more intensive possibilities of production which would be possible by cultivation of selected areas.

Agronomically many parts of Northern Australia are suitable for growth of rice. The production of this crop under Australian conditions will necessitate mechanization at every stage. This requirement introduces economic and skilled labour aspects which do not normally enter into tropical rice production. The biological possibility of growing rice, characteristic of such large parts of the tropics, in this particular area presents practical aspects which virtually make it a new industry with a host of new technical problems.

These three examples illustrate ways in which aspects other than purely biological ones will govern the final determination of regional assessments. The ecologist looking for realistic assessments therefore must extend the scope of his science to include practical ecology of man as well as the study of plants and animals, on which man subsists.

REFERENCES

BEADLE, N. C. W., 1948. The vegetation and pastures of New South Wales with special reference to soil erosion. N.S.W. Government Printer, Sydney.
BLACK, J. M., 1943—57. Flora of South Australia. 4 parts. 2nd edition. Government Printer, Adelaide.
BLAKE, S. T., 1938. The plant communities of western Queensland and their relationships, with special reference to the grazing industry. *Proc. Roy. Soc. Qld.* **49,** 156—204.
CHRISTIAN, C. S., 1952. Regional land surveys. *J. Aust. Inst. agric. Sci.* **18,** 140—6.
CHRISTIAN, C. S., 1957. The concept of land units and land systems. 9th Pacific Science Congress, Bangkok.
CHRISTIAN, C. S., 1958. The nature of climatological problems encountered by the Land Research Section, CSIRO. Proc. Canberra Symposium on Arid Zone Climatology and Microclimatology, UNESCO, Paris.
CHRISTIAN, C. S. & PERRY, R. A., 1953. The systematic description of plant communities by the use of symbols. *J. Ecol.* **41,** 100—5.
CHRISTIAN, C. S. & STEWART, G. A., 1947. North Australia Regional Survey 1946, Katherine-Darwin Region, General Report on land classification and development of local industries. Mimeo. Melbourne.
CHRISTIAN, C. S. & STEWART, G. A., 1952. General report on survey of Katherine-Darwin region, 1946. CSIRO Aust. Land Res. Ser. No. 1.
C.S.I.R.O., 1950. The Australian environment, CSIRO, Melb. 2nd ed. 1950.
C.S.I.R.O., 1954. Survey of Barkly region 1947—48. CSIRO Aust. Land Res. Ser. No. 3.
DAVID, T. EDGEWORTH, 1950. Geology of the Commonwealth of Australia, edited and much supplemented by W. R. Browne. 3 vols. Arnold, London.
DOWNES, R. G., GIBBONS, F. R., ROWAN, J. F. & SIBLEY, G. T., 1957. Principles and methods of ecological surveys for land use purposes. Second Aust. Conf. Soil Sci., Melbourne, 1957.

GARDNER, C. A., 1944. The vegetation of Western Australia. *J. Roy. Soc. W. Aust.* **28**, *11—87*.
NORMAN, M. J. T., 1957. Weight responses to tick control and phosphate supplementation in beef cattle at Katherine, N. T. *J. Aust. Inst. agric. Sci.* **23**, *344—5*.
PRESCOTT, J. A., 1931. The soils of Australia in relation to vegetation and climate. C.S.I.R. Aust. Bull. No. 52.
TAYLOR, B. W. & STEWART, G. A., 1956. Vegetation mapping in the Territories of Papua and New Guinea conducted by C.S.I.R.O. Paper presented at UNESCO Symposium on methods of study of tropical vegetation, Kandy 1956.
TAYLOR, GRIFFITH, 1918. The Australian environment (especially as controlled by rainfall). Adv. Counc. Sci. and Industr. Memoir No. 1. Government Printer, Melbourne.
TAYLOR, GRIFFITH, 1920. Australian Meteorology. Clarendon Press, Oxford.
TAYLOR, GRIFFITH, 1940. Australia. Methuen & Co., London.
TEAKLE, L. J. H., 1938. A regional classification of the soils of Western Australia. *J. Roy. Soc. W. Aust.* **24**, *123—195*.
WOOD, J. G., 1937. Vegetation of South Australia. Government Printer, Adelaide.

XXXVII
NATURE CONSERVATION IN AUSTRALIA

by

ROBERT CARRICK and ALEC B. COSTIN
(Wildlife Section and Division of Plant Industry, C.S.I.R.O., Canberra, A.C.T.)

Introduction

Australia and its Territories possess natural assets of plant, and especially of animal, life that are rich in variety and numbers and often unique in scientific value and aesthetic interest. These vast regions extending from the tropics of New Guinea through the wide range of climates and habitats of Australia proper to ice-bound Antarctica, and long separated by ocean barriers from the rest of the world, have evolved their own living froms, with little interference until recently from civilised man. The aborigines were few in number and largely nomadic; they possessed no grazing animals and did not cultivate the land; they harvested rather than exploited the fauna, except possibly by use of the fire-stick, which destroys habitat and kills more animals than are used; and it is doubtful if their primitive hunting methods exerted enough pressure to jeopardise the survival of any species. The endemic warm-blooded fauna is dominated by marsupial herbivores and phytophagous and insectivorous birds; apart from the dingo, there were no alien introductions to upset the balance of what was, only 150 years ago, one large nature reserve.

Compared with the densely-populated agrarian and industrialised countries of the Old World, and even with more recently developed North America and South Africa, Australia has had the advantage of a small human population which has left untouched large tracts of the more difficult habitats, especially desert and forest. The widespread and fundamental changes in grass lands due to sheep, rabbits and cattle, the practice of burning and other measures for pasture improvement, and the depredations of white man and alien predators such as the fox, feral cat and pig on defenceless marsupials and birds, have all taken their toll of species and numbers. But enough remains to make this country the envy of other, more heavily exploited parts of the world. The tempo of change is increasing, however, and this is an appropriate time to examine what there is to conserve, which elements merit most consideration, and how best to ensure that the economic claims of agriculture, forestry, hydro-electric development, industry and expanding settlement do not unnecessarily deplete or completely deprive Australia and the world of an irreplaceable heritage.

Nature conservation is a broad term often used to encompass not only the biological aspects with which this chapter is primarily concerned, but also geological, physiographic, scenic and even anthropological values which are not included here. However, preservation of flora and fauna, and of rocks, landscape and scenery, so often requires the same approach and methods, that measures for the protection of one automatically ensure the others. Conservation of the natural environment, and the proper balance of its components — water, soil, plants, animals — are basic to the preservation of every species, and nature management entails an understanding of complex ecological processes. Also, it is the real effects and safeguards of legislation on land and habitats, rather than the terms used to denote reserved areas, that decide how effective is the protection of plant and animal communities or individual species. In this essay the conservation value of a wide nominal range of such areas throughout Australia will be considered. The questions in mind are:

Why preserve plants and animals in their natural state?

Which elements of the Australian flora and fauna most merit and require protection?

What is being done at present?

How can nature conservation in Australia be improved?

The case for conservation

Whether a plant or animal community, or a single species, merits conservation depends on its value from several points of view. Those which occupy a unique scientific position, as primitive types or as examples of highly specialised or unusual adaptations, rank high. Even though relatively few people may study or observe them in their wild state, their contribution to scientific knowledge evokes a world-wide interest which the country possessing them should recognise by effective conservation. Mere rarity, unaccompanied by some outstanding scientific or aesthetic attribute, does not seem a very cogent argument for preservation of a species; changes must occur in the course of evolution, and the conservation effort has to aim first at the more important targets, not merely the preservation of the entire status quo.

The range of natural habitats, and their inhabitants, steadily recede in the face of economic progress; every acre of usable land tends to become highly modified, and too late the necessity for reference areas in their primitive condition becomes apparent. Study of these enables natural potentialities to be assessed, gives guidance on methods of maintaining long-term soil fertility, suggests remedial measures for depleted land, and assists in the evaluation of different forms of land use in terms of change from the original state.

The aesthetic appeal of most living things is the wider value, and is part of our standard of living. Provision of scenic, floral and faunal reserves in their unspoilt state is recognised as an important contribution to the pleasure and well-being of the human community, especially the urban one. The comments of an Australian psychiatrist (Sydney Morning Herald, October 24, 1957) are pertinent: "A city child probably needs three months a year in the country to balance the effects of the "concrete jungles" on his mental stability." "It is hard to imagine that mature minds can develop in city children who have been denied these influences of nature." The educational and recreative properties of peaceful and entirely natural surroundings are inestimable, especially in a country which is becoming increasingly urban and industrial, and in an age when physical progress and economic values tend to make us lose perspective on our biological past, and future.

Thus there is a case — scientific, economic, aesthetic and social — for preserving in their natural state an adequate and accessible number of reserves which are representative of the range of biological communities and which contain species of outstanding scientific value and popular interest. These areas, in which nature management takes precedence over all other forms of land use, are the hard core of the nation's conservation plan; they help to develop an interest in country, flora and fauna which becomes reflected everywhere in the public's appreciation of these assets.

The Australian flora and fauna

Australian plant and animal life is renowned for its distinctive eucalypt, marsupial, and avian elements, of high scientific interest as well as widespread popular appeal. These are rich in variety, numbers and distribution, and are supplemented by many forms of life more akin to those found elsewhere. Conservation covers all flora and fauna, and the tendency to refer principally to some groups, which stems from their scientific uniqueness and greater popular appeal, should not obscure the claims of more humble forms. The emphasis on preservation of habitat, rather than individuals or even species, ensures protection for all.

The liberation, intentional or otherwise, of alien plants and animals, wild and domestic, has had serious adverse effects not only on crops and stock but also on the endemic flora and fauna, rivalling the more direct results of agricultural and industrial development and often presenting extremely difficult conservation problems. Also, some native species, however important and attractive, weaken their claims for protection by their activities as economic pests.

The wide range of regional climatic types throughout Australia ensures equally varied vegetation and animal life, which have often become highly adapted to make the most of a variable and unpredictable rainfall. Rapid growth and fruiting of annuals after rains, associated with flushes of invertebrate foods, lead to wide nomadism and opportunist breeding by herbivorous marsupials and seed-eating, insectivorous and water-frequenting birds. This poses questions of conservation management not present under more equable conditions. The generally mild winters, with some evergreen foliage and insect food available throughout the year, support a high proportion of vegetarian and insectivorous mammals and birds, without recourse to hibernation or extensive migration. However, seasonal movements, especially by some birds, are sufficiently extensive to necessitate an inter-State and international outlook on conservation practices.

Flora

A description of Australian plant communities is also an outline of the habitats of land animals. The vegetation shows a broad semi-concentric zonation according to rainfall; desert and semi-desert occupy the dry interior, with progressively moister communities towards the Great Dividing Range and the coasts. Most of the main forms of world vegetation occur in Australia, and the genus *Eucalyptus*, with about 600 species in Australia and 7 in New Guinea, is native to the region. In the mainland and Tasmania, there are some 12,000 species of vascular plants, 85% of which are endemic. The yellow blossoms of about 600 species, half the world total, of wattles, *Acacia*, brighten many Australian bush scenes, and provide a seed harvest for many animals. The 2000 or more species of xerophytic shrubs and flowers of Western Australian sand-plains offer an unsurpassed floral pageant in spring, with many unusual flower forms and brilliant colours; the desert fringes of South Australia and Western Australia after rains, and the high mountain moors of the Great Dividing Range in summer, have rich carpets of wild-flowers, with Compositae outstanding. To name only one plant with special claims to conservation, the unique pitcher-plant, *Cephalotus follicularis* is restricted to swamps in a small area of the south-west.

The desert communities comprise hummock grassland, in which spinifex, *Triodia*, and cane-grass, *Spinifex*, are conspicuous; scrubs, dominated by shrubs or small trees such as mulga, *Acacia aneura*, myall, *A. pendula*, and gidgee, *A. cambagei*, and in more heavily textured soils the dwarf chenopodiaceous shrubs saltbush, *Atriplex*, and bluebush, *Kochia*, are developed; lighter soils along the southern semi-desert fringe support mallee, dominated by multi-stemmed eucalypts up to about 25 feet high.

The sub-humid areas contain a variety of woodland, savannah and grassland associations, depending on climatic and soil type. Single-stemmed eucalypts rising to 50 or 100 feet form parklands with a well-developed grass or shrub layer, and grade through savannahs with scattered trees and shrubs into drier tussock grasslands. The characteristic species of northern grasslands are Mitchell grass, *Astrebla*, blue grass, *Dichanthium*, and Flinders grass, *Iseilema;* in the south, spear grass, *Stipa*, wallaby grass, *Danthonia*, and kangaroo grass, *Themeda*.

As rainfall availability increases, taller and denser sclerophyll forest and rainforest vegetation develop. The former, dominated almost entirely by eucalypts, is widespread on the coast and tablelands in most of the southern States. Space precludes a just description of the variety and beauty of the gums, but mention must be made of the mountain ash, *Eucalyptus regnans*, the tallest hardwood in the world, which not infrequently tops 300 ft. in Victoria and has an established record of 375 ft. Rainforests, restricted to the highest rainfall areas and the better soils of Queensland and northern New South Wales, are tropical and subtropical, with a large number of Indo-Malaysian species; temperate rainforest, dominated mainly by Beech, *Nothofagus*, occurs in Tasmania. Within the sclerophyll forest areas, mainly in the south, heaths and scrubs assume dominance under conditions of soil or local climate unfavourable for trees; species of she-oak, *Casuarina*, needle-wood, *Hakea*, grass tree, *Xanthorrhoea*, honeysuckle, *Banksia*, and tea-tree, *Leptospermum*, are typical.

At the higher levels in New South Wales, Victoria and Tasmania, mainly above 4,000—5,000 feet, the forests grade into subalpine woodland and alpine herbfield associated with a variety of grassland, heath, bog, fen, and fjaeldmark (fellfield) communities. Coastal areas include a wide range of lagoon, estuarine, sand dune and headland habitats, with a corresponding variety of marsh, mangrove, heath, scrub, and herbaceous vegetation.

Some of the original plant communities, notably grasslands, have been much altered by grazing of stock and rabbits, by pasture and soil improvement, and by cropping. The spread of introduced weeds often further alienates the natural habitat, as do burning, logging and ring-barking of trees and removal of scrub. Stands of exotic softwoods are a further necessary, but foreign, addition to the endemic vegetation, of little interest to native fauna. However, considerable tracts of country, especially forest, desert, mountains and coast, remain basically unaltered, and it is essential that an adequate range of habitats should be made secure from other claims of land use and retained as reference areas in their original condition.

Fauna

The primitive egg-laying monotremes — platypus, *Ornithorhynchus anatinus*, and spiny ant-eaters or echidnas, *Tachyglossus*, — are among the most scientifically-interesting animals in existence.

This order has no counterpart, even in the fossil record, outside Australia and New Guinea. Their unique position as the only living links between reptile and mammal, their retiring and harmless nature, their unusual appearance and habits, have won for them complete protection by law and the force of public opinion, so they are still by no means uncommon. The semi-aquatic platypus requires a waterway stream or lake, clear or muddy, cold or warm — with a bank that can be burrowed; if not directly disturbed, it lives around settlements. The spiny ant-eaters inhabit open forest and scrub, feed on ants, etc., can withstand considerable starvation, and use their highly protective coat of dorsal spines by curling up or digging in, effective defence against any animal predator. While adequate habitats remain, the monotremes seem safe.

Marsupials comprise 119 of the 229 species of native mammals (excluding whales). Their relatively primitive position and classic adaptive radiation of feeding habits and movement; their unusual breeding behaviour and wide range of attractive forms designed for jumping, climbing, gliding, burrowing and running; their abundance and (for mammals) comparative visibility and sometimes tameness, have earned a good measure of protection from direct destruction except where economic considerations prevail. Rare and vanishing marsupials of distinctive merit include the carnivorous and dog-like Tasmanian wolf or tiger, *Thylacinus cynocephalus;* the banded ant-eater or numbat, *Myrmecobius fasciatus*, and the marsupial mole, *Notoryctes typhlops*. The highly popular koala "bear" *Phascolarctos cinereus*, once widely distributed in eastern Australia, became reduced by disease and exploitation for skins to a remnant of its former strength, but vigorous protection of this vulnerable species, including re-introduction to earlier haunts in Victoria, has turned the tide. A recent survey of the status of marsupials in New South Wales by MARLOW (1958) emphasises that pastoral development and the introduction of placental carnivores are the main causes of a serious decline, especially in marsupials of open grassland and scrub; of 52 species, 22 are now extinct (not recorded since 1910) or rare (not recorded since 1954). Species dwelling in forests, especially possums (Phalangeridae) have better withstood white occupation, for large tracts of this habitat are as yet undisturbed.

The placental mammals endemic to Australia include 67 rodents and 41 bats, as well as the dingo which is considered to have been brought by the first aboriginal settlers from the north. These have been unfortunately augmented by introduced and feral aliens, including fox, cat, pig, rabbit, hare and rats, etc. which deplete pastures and destroy the more defenceless native animals.

This region is particularly rich in birds, from New Guinea birds-of-paradise to Antarctic penguins and petrels. Australia proper has about 650 native species, some 570 of which breed there. Although alienation of habitat, and some excessive killing, have reduced the numbers and range of some, none have suffered extinction and few have become very rare through white occupation.

Several species of unusual scientific merit are confined to, or occur mainly in, Australia, and their future largely depends on this country. The flightless emu, *Dromaius novae-hollandiae*, the second-largest living bird, and the cassowary, *Casuarius casuarius*, are primitive; the emu can be an agricultural nuisance, which indicates its healthy numerical status. Specialised adaptations, mainly related to breeding, include the mound-building of the mallee-fowl, *Leipoa ocellata*, and other megapodes, which use fermenting vegetation and solar heat to hatch their eggs in an "incubator"; the display and vocal mimicry of the superb lyre-bird, *Menura novae-hollandiae;* and the remarkable culture of decorated bowers by the bower-birds (Ptilonorhynchidae).

In numbers, variety, colour, form and voice, the Australian avifauna is the equal of any. The dominant groups reflect the vegetation and prolific insect life. Honey-eaters, (Meliphagidae) number 70 species and depend on flowering eucalypts, etc.; 60 parrots and cockatoos (Psittacidae), often highly colourful, eat seeds and fruits, and mainly nest in the plentiful holes in gum-trees; 36 birds of prey take insects, reptiles and the smaller birds and mammals; 21 weaver-finches (Ploceidae) require grass and other seeds; and 19 flycatchers and 18 robins (Muscicapidae), 14 cuckoos (Cuculidae), 10 kingfishers (Alcedinidae) and many other families are mainly insectivorous. Water-birds and sea-birds are strongly represented, and the little or fairy penguin, *Eudyptula minor*, breeds plentifully around the southern half of Australia and Tasmania. The sub-Antarctic islands, Heard and Macquarie, and Australian Antarctic Territory contain many petrels and penguins, including several of the finest colonies of the emperor penguin, *Aptenodytes forsteri*.

Reptiles number about 240 lizards, including goannas 8 feet long, 140 snakes and 2 crocodiles. There are over 100 frogs and toads; 2,200 fishes, only 180 of which are freshwater; and 10,000 molluscs.

Neoceratodus forsteri is Australia's representative of the lung-fishes (Dipnoi); and Queensland also possesses, in the corals, molluscs and bizarre fishes of the Great Barrier Reef, a wealth of marine life equally accessible and attractive to the biologist and the layman. At least 40,000 Australian insects are known, so immensely abundant that they are a staple food of many birds, reptiles and mammals, including even aboriginal man.

Legislation

In Australia, each State has sovereign powers over its own land, flora and fauna; the Australian Capital Territory, the Northern

Territory, Papua and New Guinea, Australian Antarctic Territory, and several small islands come under the jurisdiction of two Commonwealth Government Departments. This diversity of control is the basic reason for the present wide differences in legislation and achievements in the broad field of nature conservation throughout Australia. In each State, a considerable body of naturalists and national parks enthusiasts has sought to have land reserved and habitats retained in their natural state, as well as laws passed to protect plants and animals from indirect destruction. Even in a single State, each of these two main aspects of nature conservation may be affected by several enactments and their subsequent amendments and proclamations. There is considerable inconsistency between States and Territories in regard to the Acts which cover the same subject, and the government department responsible for the organisation and resources for conservation varies accordingly. Differences in biological coverage, in terminology, and in the actual conservation value of reserved areas which bear similar titles, further combine to make it very difficult indeed to obtain a clear, comprehensive, and up-to-date picture, and Australia-wide generalisation is impossible.

National parks, although primarily set aside for their scenic attractions, are an important contribution to flora and fauna conservation, for, properly managed, they ensure preservation of large tracts of natural habitats, especially those of lesser economic value. The present situation in Australia is summarised in Table I, from which the diversity of development and control by different governments is obvious. Standards differ enormously, according to the grasp of national park principles by the particular department or trustee board concerned, and also in relation to available finance. Queensland, where the National Parks Association movement in Australia was pioneered in 1932, has an effective Act which has been translated into well-managed parks (GROOM 1949). Victoria is following suit, and a National Parks Act and Authority are also under consideration in New South Wales. Practices which detract from national park standards and reduce the nature-conservation value of parks, such as settlement, playground developments, grazing (with repeated burning), tree-felling, etc. are not uncommon; they stem partly from the inadequacy of the controlling body, which too often lacks biological, or even any scientific, advice, and which has to raise somehow the funds which it cannot obtain from government.

Very briefly, these parks include a fair representation of subtropical, temperate and eucalypt rainforest, some alpine and other high mountain vegetation, and some coastal eucalypt and heath associations, from Queensland to Tasmania. Sub-tropical islands of the Great Barrier Reef, a stretch of Victorian mallee, true desert in

TABLE I.
ACTS, AUTHORITIES, AREAS AND COSTS OF NATIONAL PARKS* IN AUSTRALIA

STATE or TERRITORY Acts	Authorities (Department)	Number of Major* National Parks	Approx. Area of Nat. Parks / Total Area (square miles)	Park Area as % of Area of State	Population (1957)	Park Area per capita (acres)	Annual expenditure per capita (pence)
New South Wales Various, and Kosciusko State Park Act, 1944	Trustee Boards (Department of Lands)	14	2,494 / 309,433	0.81	3,624,308	0.4	3.8
Victoria National Parks Act, 1956	National Parks Authority (Premier)	13	489 / 87,884	0.56	2,673,639	0.1	3.2
Queensland State Forests and National Parks Act, 1906—1957	National Parks Branch (Department of Forestry)	14*	1,231 / 670,500	0.18	1,404,016	0.6	8.0
South Australia National Park Act, 1891. National Pleasure Resorts Act, 1914—55. Flora and Fauna Board Act.	Trustee Boards (Tourist Bureau; Flora and Fauna Board)	5	250 / 380,070	0.07	873,863	0.2	
Western Australia Parks and Reserves Act, 1895	Trustee Boards (National Parks Board; Department of Lands)	6	69 / 975,920	0.007	698,553	0.7	
Tasmania Scenery Preservation Act, 1915—54	Trustee Boards under Scenery Preservation Board (Department of Lands)	8	842 / 26,215	3.2	327,896	1.7	11.0
Northern Territory National Parks and Gardens Ordinance, 1955	Reserve Board (Administrator)	2	488 / 523,620	0.09	29,301	11.0	
Australian Capital Territory		0	nil / 939		37,866		
Commonwealth		62	5,863 / 2,974,581	0.2	9,669,442	0.33	

* Wide differences of size, quality, legislation and control make comparison between "national parks" of States and Territories very difficult; Queensland has 250 parks, Tasmania has 2 scenic roads and 50 reserves, and so on. Current developments in some States and Territories will materially improve the figures quoted here. In Papua and New Guinea, the establishment of national parks and reserves is under consideration.

central Australia, and karri, *Eucalyptus diversicolor*, forest in Western Australia, are also represented. There is a notable lack of savannah parklands and grasslands, heathland and semi-desert, e.g. mulga, and some types of eucalypt forest, e.g. mountain ash and Murray red gum, *E. camaldulensis*. While progress toward more adequate national parks, with greater security from conflicting forms of land use and with better management along natural lines, will undoubtedly contribute much to nature conservation in general, it is clear that a substantial range of equally secure nature reserves, selected to meet the particular needs of flora and fauna, is essential.

Flora reserves, as such, are seldom the subject of special legislation in Australia. In New South Wales, the Forestry Act 1916—49 empowers the Governor to dedicate any area of Crown lands or State forest as a native flora reserve, which can be revoked only by Act of Parliament; but logging can still be included in the working plan, and few such reserves have been dedicated. Flowers, or plants generally, are frequently covered by laws for other types of reserve such as fauna, scenery preservation, or national parks. All States and Territories, except Tasmania, Northern Territory and Papua—New Guinea have Acts, administered by a variety of Departments, prohibiting the picking of wild flowers and native plants, or certain species of them, throughout the whole or part of the State or Territory. Licences to collect, and even to sell, may be granted.

Fauna reserves, sanctuaries, or districts can be declared under the appropriate Act for each State (Table II). The number and size, and the systematic and regional coverage, of these reserves vary greatly between States, as does the effective degree of security from interference with habitat and even destruction of fauna. The dedicated faunal reserve of New South Wales is outstanding in that it is Crown land under the control of the Fauna Protection Panel, is specifically safeguarded from all forms of economic development of the land, and can be revoked only by Act of Parliament. However, only six faunal reserves have been dedicated, all on or near the coast, and, for example, there is no provision yet for such a notable bird species as the mallee-fowl. Fauna sanctuaries and districts throughout Australia, while they serve an immensely useful purpose, are seldom legally covered against economic pressures, though public opinion can often prevent violation. Even so unique and world-famous a reserve as the lyre-bird sanctuary at Sherbrooke Forest near Melbourne was recently threatened by proposed forestry activities, but public agitation resulted in an influential Committee of Management being appointed by the Minister of Forests.

Legislation for protection of the individual animal is complicated (Table II). The State Acts and Territory Ordinances vary in their definitions of fauna, sometimes omit to define terms, and may include "aquatic" mammals, birds, reptiles. frogs and apparently

TABLE
FAUNA LEGISLATION AND

State or Territory Acts Authorities (Department) Staff	Fauna Covered by Acts	Degree of Protection*
New South Wales Fauna Protection Act, 1948. Chief Guardian of Fauna (Chief Secretary's Dept.); Fauna Protection Panel of 14. Field biologist and honorary rangers (also police, teachers, public servants).	Mammals (native, introduced, imported, aquatic; **not** domestic or any rats except water rats, or mice) Birds, (native, introduced, imported; not domestic).	Complete, except 8 mammals and 32 birds unprotected throughout State or part. Above list amended, and open or close seasons declared, by proclamation.
Victoria. Game Act, 1928–57 (new act in preparation). Game (Koala) Act, 1938. Director, Fisheries and Game Dept. Several biologists and inspectors.	Mammals (**not** domestic). Birds (**not** domestic).	(See F. and G. Dept. Circular No. 7, January, 1958) Complete, except 15 mammals plus all native rodents except water rat and 21 birds wholly unprotected. Open season for game by declaration.
Queensland Fauna Conservation Act, 1952. Fauna Officer (Department of Agric. and Stock). Biologist, fauna officers (police, staff of Agriculture and Stock, Lands) and honorary protectors.	Mammals (wild, native, migratory, introduced; **not** marine, mice or rats except water rats). Birds (wild, native, migratory, introduced).	Permanently protected — koala, platypus, echidna. Protected — all except 11 mammals and 33 + birds with open seasons for commerce and sport; alterable by proclamation. Pest Fauna — 5 + mammals and 17 + birds.
South Australia Animals and Birds Protection Act, 1919–38. Chief Inspector of Fisheries and Game (Department of Agriculture). Inspectors (include all police).	Mammals (native, introduced; includes seals). Birds.	Complete, except: Partly protected — 4 + mammals and 11 + birds with fixed open seasons in whole or part of State; Unprotected — 8 + mammals and 34 birds.
Western Australia Fauna Protection Act 1950–54. Chief Warden of Fauna (Fisheries Department), Fauna Protection Advisory Committee of 6. Biologists; honorary wardens (include all police, Fisheries and Forests staff).	Vertebrates (wild, indigenous or introduced). Mammals, birds, reptiles, frogs, (but whales, seals and fish covered by Fisheries Act and Whaling Act).	Complete, except 22 mammals, 33 birds, 10 + reptiles and 1 toad not protected in whole or part of State. Above list amended, and close or open seasons declared, by proclamation).
Tasmania Animals and Birds Protection Act, 1928–55. Commissioner of Police, Animals and Birds Protection Board of 11. Secretary; officers (include all police).	Animals, i.e. mammals, except whales and seals (indigenous or exotic). Birds.	Wholly protected — 13 mammals and all birds except in next two groups. Partly protected — 6 + mammals and 9 + birds, with open and close seasons. Unprotected — 4 mammals and 26 birds.
Northern Territory Birds Protection Ordinance 1928–40 (A mammal ordinance is under consideration). (Animal Industry Branch). Inspector of Birds (include all police).	Birds (wild, indigenous or imported).	Protected during Whole Year — all birds except in next two groups. Protected during Part of Year — 6 + geese, ducks, pigeons. Not Protected — 13 + birds. Above lists amended by notice in Gazette.
Papua and New Guinea Animals and Birds Protection Ordinance 1952–53. Director of Agriculture Stock and Fisheries. Rangers.	Animals and Birds (other than livestock capable of being mustered into a yard, and domestic pets).	Prohibits capture, destruction, traffic in fauna, without licence. Natives may take for domestic use. Administrator can declare unprotected fauna and open seasons.
Australian Capital Territory Animals and Birds Protection Ordinance 1918–1937. (Department of Interior) Inspectors.	Animals and Birds (except domestic pets).	Prohibits capture, killing, sale, etc. except 7 + mammal pests, venomous snakes, domestic pests and parasites, and 6 + bird pests.

* Acts and Ordinances may list species individually or use terms which cover several species; in the latter case a plus sign follows the number.

RESERVES IN AUSTRALIA

Title	Legal Status	Fauna Reserves		Number of Reserves
		Type of Land of Security	Fauna Protected	
Faunal Reserve	Dedicated by Governor	Crown land. Panel controls all activities; no mining, felling, selling or lease, grazing, or soil experiments.	All fauna covered by Act. Panel can authorize exceptions.	6 (more pending)
District	Proclaimed by Governor	Any land Not safe from other use.	All fauna covered by Act. Exceptions by proclamation.	1, and one mile around all schools.
Sanctuary	Proclaimed by Governor	Any land Not safe from other use.	Protected mammals (including deer) and birds. Exceptions by proclamation.	Many including many waterfowl refuges.
Sanctuary	Declared by Governor	Any land Not safe from other uses except National Parks.	All fauna covered by Act. Except pest fauna on holdings and by proclamation.	Many, including all State Forests and National Parks.
Sanctuary (= closed area)	Declared by Governor.	Crown land. Private land with consent of owner or occupier. Not safe from other use.	All fauna covered by Act.	48
Sanctuary	Reserved by Governor	Crown land. Private land by agreement with owner. Not safe from other use.	All fauna covered by Act.	73
Sanctuary	Declared by Governor	Crown land. Private land with owner's consent. Not safe from other use.	All animals and birds covered by Act. Board can authorize exceptions.	c. 50
District	Declared by Governor.			
Bird Protection District.	Declared by Administrator.	Crown land, sea shore, or private land on request of owner. Not safe from other use.	All birds except those not protected under Ordinance.	

Conservation strips are being purchased in populated areas.

No provision for reserves.

(Australian Capital Territory) invertebrates, or may exclude "marine" mammals, whales and/or seals, mice and/or rats (except water rats), and so on. The koala is the only animal which has a Protection Act all to itself, in Victoria; this accords with its very high popularity, and with the need to give individual protection to a vulnerable species with a valuable fur. Other animals are variously scheduled as permanently protected (most species); partly protected (game species and some economic pests; dates and districts of open seasons vary); and unprotected (pest species). Opinions regarding economic status, and the seasons and areas in which it should be legal to take partly-protected fauna, often differ, both within and between States. Fauna Acts and Ordinances make provision for licenses for scientific collecting, and some States permit commercial collecting and export of birds such as parrots, cockatoos and finches, a contentious point with other States mainly because the practice undermines conservation propaganda and the holders of permits operate wastefully and do not respect State boundaries.

The advisory panels constituted under the Fauna Acts of three States (New South Wales, Western Australia, Tasmania) bring a wide range of interest, scientific knowledge and impartial opinion to bear on the problems which administration of such an Act can raise. Several Acts stress the necessity for research and education, but only Victoria provides the necessary finance to appoint an adequate staff of qualified biologists and to undertake a comprehensive programme of field work. The same State employs full-time inspectors, a service which is performed by other public servants acting ex officio and by honorary rangers elsewhere. However, the impossibility of policing Acts of this nature, especially in a vast country with a small population, is evident, and it is widely recognised that education of public goodwill toward fauna is the best policy.

Special problems

Most problems of nature conservation are universal, common to all countries and soluble by the application of basic conservation principles. Australia's historical, political, economic and ecological background frequently alters the emphasis of the factor involved, and necessitates a different approach or degree of compromise. Consideration of several current and recent problems, involving both individual species and reservation areas, will point the way toward more effective measures. These selected examples by no means exhaust the available list.

Species

Mountain Ash, *Eucalyptus regnans*. The finest stand of Australia's largest gum in the Mt. Field National Park, Tasmania, has been felled for paper pulp. An enabling bill was carried through Parliament despite public protests. Thus, even the protection of an Act of Parliament, in a State with an advanced national parks system, proved inadequate in the face of economic pressure. There is an equally good stand in a water catchment area of Victoria, but this species requires the protection of a nature reserve and influential public vigilance.

Anemone Buttercup, *Ranunculus anemoneus* and **Giant Wallaby Grass,** *Danthonia frigida*. This large alpine buttercup was reduced by the selective grazing of cattle to the verge of extinction on the Kosciusko Plateau, and it appears to have been eliminated from the Victorian Alps. In 1944, the fortunate cessation of grazing in the summit area of Kosciusko was followed by the reappearance of this and many other alpine flowers. Giant wallaby grass is also recuperating where grazing no longer occurs; fifty years ago, sheep sheltered in its tussocks, but around 1935 botanists could not find it on Kosciusko (COSTIN, 1958).

Waratah, *Telopea speciosissima*. The wildflowers of the Hawkesbury sandstone vegetation near Sydney present a spring pageant which has steadily deteriorated owing to illegal picking and marketing, despite protective legislation. Massed displays of waratah, the floral emblem of New South Wales, Christmas bells, *Blandfordia*, flannel flower, *Actinotus*, native rose, *Boronia*, and many others were once a familiar sight, but commercial florists exploit public appreciation of these wildflowers. During the Waratah Spring Festival in Sydney it is necessary to have special guards in Kuringai Chase National Park to prevent despoliation. Education of the public, stricter control of trading in wildflowers, and more effective management of reserves are necessary if this attractive feature of Sydney's environs is to be retained.

Platypus, *Ornithorhynchus anatinus*. It was once permissible to take platypus skins for rug-making, but rigid legislation, with heavy penalties, has given complete protection everywhere during the present century. It is the consistency of all State and Territory laws which makes illegal trapping impossible to conceal.

Koala, *Phascolarctos cinereus*. The history of the koala — specialised, slow-breeding and defenceless against man's methods of trapping — can be paralleled in many countries, but the ending is happier than some. Once plentiful throughout eastern Australia, it was reduced some thirty years ago to near-extinction in Victoria and rarity in New South Wales by disease and exploitation for fur. In 1924, two million skins were exported, and in Queensland alone in 1927 ten thousand licensed trappers took half-a-million koalas, half the 1921 total (CHISHOLM, 1958). Again, complete and universal protection by law, including prohibition of export, backed by strong public sentiment and measures of rehabilitation, are steadily restoring the situation. In Victoria, which had an estimated koala population of only 500 in 1920, a policy by the Fisheries and Game Department of concentration and protection on Phillip Island and of regular "seeding" to suitable habitats throughout the State, has put the most popular marsupial back on the map.

Kangaroos, Wallaroos, Wallabies, Macropodidae. The larger herbivorous and browsing marsupials are economic pests in so far as they compete with sheep and other stock for pasture, or prevent regeneration of trees. Settlement and pastoralisation have accounted for some species, but several of those surviving flourish to a degree which regularly elicits demands for extensive control, at present somewhat unsatisfactorily achieved by declaration of local open seasons. There is a large export skin trade, but the common species can withstand considerable harvesting without risk to their survival in numbers. These marsupials are important in the general conservation plan; small reserved areas would not be adequate, especially for far-ranging species and in view of the unreliable distribution of rainfall over the country they occupy; and there seems little possibility of regaining large, or indeed any, tracts of pastoral country as reserves. The present situation may be regarded as reasonably satisfactory from the conservation viewpoint, but it could deteriorate as the result of increased efficiency of control methods, or closer settlement.

Humpback Whale, *Megaptera nodosa*. Commercial whaling is practised off both eastern and western coasts of Australia, during spring and early summer when the whales are on their northern migration to breed. The aim is to harvest without depleting the stock, eminently a subject for research, which is in progress at the C.S.I.R.O. Division of Fisheries and Oceanography and the Zoology Department, University of Sydney.

Emu, *Dromaius novae-hollandiae* and **Cassowary,** *Casuarius casuarius*. These two flightless and primitive species illustrate the difference in long-term conservation problem between a bird (emu) of the open plains, where it does sufficient damage to be unprotected in Western Australia (where a bounty is paid on birds and eggs) and South Australia (open season May to August), and a bird of the scrub and forest where reserves are more easily obtained. The emu poses the same problem as the marsupials of the plains, and reservation of valleys at higher altitudes where it also flourishes may be one means of ensuring its continued existence under natural conditions.

Australian Bustard, *Eupodotis australis*. Despite complete legal protection and propaganda, illegal shooting of the rather tame "plain turkey" continues, and it is difficult to be optimistic about its future outside of reserves. It looks like a game bird, is good eating, and lacks the outstanding scientific or aesthetic attributes

PLATE 1

Plate I. Ayers Rock near Alice Springs in the Northern Territory is part of a large national park recently declared. This giant pre-Cambrian monolith has a perimeter of five miles and rises 1100 feet from the semi-desert plain. It is sacred to aborigines and has caves decorated with their paintings.

(Photo by R. MUNYARD.)

R. CARRICK & A. B. COSTIN

PLATE II

Plate II. A mob of grey kangaroos, *Macropus major*, one of the largest species of marsupials. Its abundance and nomadic habits bring this herbivore into conflict with domestic stock in pastoral areas. (Photo by Australian News and Information Bureau.)

PLATE III

Plate III. Koala or native bear, *Phascolarctos cinereus*, with young one. Once seriously threatened by disease and trapping, this popular marsupial is now rigidly protected throughout Australia.
(Photo by Australian News and Information Bureau.)

Plate IV. Male superb lyre-bird, *Menura novae-hollandiae*, searching for food. The large foot is used to scrape the litter of the forest floor so that crustacea, worms and other invertebrates can be picked up. (Photo by R. Carrick.)

which elicit public support for protection. Because of the advance of settlement it is now plentiful only in the grasslands of the north, and its hope would seem to lie in adequate reservation of that habitat before that part of Australia becomes more densely populated.

Australian Gannet, *Sula serrator*. The largest of the five Australian colonies, at Cat Island, Tasmania, is unfortunately also the most accessible; cray fishermen, seeking to augment the supplies of penguins, mutton-birds etc. which they use for bait, raided this gannet colony so remorselessly that it was reduced from about 2,500 pairs to less than 50 pairs by 1952. Since then, the Fauna Board of Tasmania has employed a warden to guard the gannets during the breeding season, probably the only action of its kind in Australia at present. The solution to the general problem of depredations by cray fishermen on wild birds and marsupials (scrub and pasture are deliberately burned, and macropods shot at a later date when they are attracted to the green feed) is surely to provide them with alternative bait; cheap, tinned meats are reported as promising.

Wedge-tailed Eagle, *Aquila audax*. The largest Australian bird of prey is unprotected, except in Tasmania (where it is not common), the Australian Capital Territory, and Papua and New Guinea. Bounties are paid in Western Australia and Queensland, and large numbers are killed annually throughout Australia without making any appreciable difference to its abundance. The main argument for control is alleged depredations among lambs, but rabbits and dingo pups are also credited to it. The "eagle-hawk" inhabits a wide range of country and is in no danger, but a sound scientific assessment of its real economic status in different situations is desirable. Various other birds of prey are unprotected in each State and Territory, and the Northern Territory inclusion of "hawks" in this category leaves a lot to be desired.

Parrots and **Cockatoos,** Psittacidae, and **Weaver Finches,** Ploceidae. Commercial trading in these aviary birds (and in some marsupials), made possible by the issue of collecting permits by Queensland, South Australia and Western Australia, and export licenses by the Commonwealth Customs, occasions some controversy. While trade is confined to the common species there is no serious conservation problem, although depletion of more accessible places is unfair to those who enjoy watching free-living birds. Wasteful collecting, involving unnecessary mortality and destruction of nest-holes of parrots, as well as transgression of State boundaries, certainly occurs. However, without condoning an outlook or practice inconsistent with good nature conservation, the eventual answer to problems of this kind will come from better and wider education of the public and from increased co-operation between States on all conservation fronts.

Reserves

Royal National Park, New South Wales. This area of 36,880 acres of coastal eucalypt forest and rainforest gullies was established in 1878, only six years after Yellowstone National Park, U.S.A., the first in the world. But the Royal National Park Trust has lacked the resources and standards to manage the park as a conservation area, and uncontrolled settlement, playground development, railway and roads, military camp and operations, clearing and removal of timber, coupled with vandalism, shooting and bushfires, have detracted from the natural values of the area. Sydney requires extensive nature reserves near the city, where appreciation of flora, fauna and scenery are paramount, and Royal National Park could still supply much of this need, given the necessary finance, staff, legal security, controlling authority and management plan according to national park principles.

Kosciusko State Park, New South Wales. In 1944 the Kosciusko State Park Act set aside nearly one-and-a-half million acres, the highest mountain area in Australia, with magnificent scenery and glacial features, and the best alpine, sub-alpine and high altitude eucalypt vegetation in the country. A Trust under the Minister for Lands was set up, consisting of ten part-time members who included tourist and grazing interests but no trained biologist. The Act provides for a primitive area not exceeding one-tenth of the park, excludes forestry and mining except as approved by the Trust, empowers the Trust to construct roads, tracks, ski trails, hostels, etc. and grant snow leases (for summer grazing) and permissive occupancies. There is free public access for riding, camping, snow sports and fishing. The Trust inherited a century of summer grazing and burning of vegetation; snow leases were its only source of regular income, and quite inadequate. The adverse effects of this practice were stressed in a report by the Australian Academy of Science (TURNER et al. 1957), and in 1958 grazing above 4,500 feet and burning at all levels ceased. This mountain area is also the nation's most valuable water resource, and since 1949 the Snowy Mountains Authority has carried out vast damming, river diversion and water-catchment works for irrigation and hydro-electric purposes. These operations have, of necessity, impaired natural values in part of the park area, and scientists and others have questioned whether all the hydro-electric (as distinct from irrigation) works are entirely justified, and whether the extensive engineering activities are being carried out with minimum loss to vegetation and soil. The Academy of Science Report states: "We recommend in the interests of erosion control and of national park values, that at these elevations and especially above 6,000 feet, any constructional work should only be undertaken if it can be proved that it is essential to an efficient irrigation scheme." A submission to the Trust in 1958 to have a primitive area declared, including the Kosciusko Plateau, was unsuccessful, but the appointment of a resident manager qualified to maintain park interests is proceeding. The history of this national park repeats the experience of many other countries; tourist and appropriate recreational facilities are desirable, grazing is not, essential economic developments must proceed, but unless the controlling authority has a clear conception of park management for its primary scenic and nature conservation purposes, is given adequate finance and staff, and has a well-informed and appreciative community behind it, unnecessary deterioration from economic pressures is likely to occur.

Macquarie Marshes, New South Wales. An anabranch system of the Macquarie River on the central plain of New South Wales floods extensively after rain on the eastern range, and this marsh, which has been declared a sanctuary, is a noted breeding-place for waterfowl, which include ibises, spoonbills, herons, egrets, black swan and ducks. The Burrendong Dam near Wellington, now under construction, will hold water for an irrigation area upstream from the marshes and so prevent the flooding which stimulates nesting. The construction of levees in the marshes to flood selected areas from the limited river flow would be costly, and money is not available for this purely as a fauna reserve project. The Macquarie Marshes are in the most important locust outbreak centre in eastern Australia, and since ibises are well known to feed upon locust swarms it has been suggested that the expense is justified on economic grounds. However, these predator-prey relationships are not simple, and recent investigation does not suggest that either locust swarms or hatching hoppers are significantly reduced by bird predation. The marshes are also used to raise beef cattle, and this may be a case where the conservation authority could relax its normal anti-grazing principle and seek the co-operation of all the agricultural interests involved in order to secure some flooding and a measure of water-bird breeding which would be little affected by the presence of stock.

Sherbrooke Forest Reserve, Victoria. Few nations possess an ornithological show-piece with the sheer attraction and public accessibility of the superb lyre-birds of Sherbrooke Forest, near Melbourne. For half-a-century they have been accustomed to the close presence of people, and the male will perform his display-dance and pour out his medley of mimicry to a circle of admiring onlookers and photographers a few feet away. It has taken years of careful conditioning to achive this, and elsewhere along the forests of the eastern ranges the lyre-bird is normally shy and unapproachable. Yet in 1957, there was a move to clear part of Sherbrooke Forest Park, which covers only 1983 acres, and substitute more "useful" conifers. Public outcry prevented it and led to the appointment under the Forests Act, 1957, of a Committee of Management (which, however, has no qualified wildlife scientist or ornithologist, probably because this sanctuary has been maintained by amateur effort and interest to date). Retention of this unique reserve is assured, but this is a classic example of the risks run when nature reserves are not covered by unequivocal legislation which excludes all competitive forms of land management.

Lamington National Park, Queensland. Since 1915, this 48,800 acre area of subtropical mountain scenery on the McPherson Range has been developed along natural lines by graded access tracks which enable visitors to enjoy the spectacular forests and bird life at close quarters. In 1947 a government tourist committee recommended that a road be put through the park and a hostel sited within it; but, despite the proximity to Brisbane, conservationists and the responsible branch of the Forestry Department were able to retain the nature character of the park.

Macquarie Island, Tasmania. Politically Australian, but lying halfway between New Zealand and the Antarctic Continent, Macquarie Island is a striking illustration of several principles of fauna conservation (CARRICK, 1957). It was discovered in 1810 by sealers searching for the New Zealand Fur Seal, *Arctocephalus forsteri;* by 1813 180,000 skins had been taken, in 1820 none were left, nor was the species seen there again until 1948. The numbers are increasing annually, and breeding has occurred since 1954; even after complete destruction, recolonisation to the former abundance is possible because the habitat remains intact. The elephant seal, *Mirounga leonina*, and the royal penguin, *Eudyptes chrysolophus schlegeli*, were harvested in more economical fashion by sealers, but, thanks to the efforts of Sir DOUGLAS MAWSON (1922) the Island was declared a sanctuary in 1933. However, alien introductions are still doing irreparable damage; the rabbit, *Oryctolagus cuniculus*, denudes vegetation and causes soil erosion, while feral cats and the weka, *Gallirallus australis scotti*, from New Zealand have reduced the former extensive colonies of nine species of burrowing petrels to three much less abundant species.

King's Park, Western Australia. A capital city which possesses a 1,000 acre reservation of light sclerophyll forest with a spring display of Western Australian wildflowers is fortunate indeed. Yet there is a constant battle to prevent firewood cutting, rubbish dumping, planting of exotics, special clearing and lately the construction of a swimming pool. The vigilance of local naturalists has to make good the scientific biological representation and clear conception of management for preservation of natural values which are lacking on the trustee board.

Montebello Islands, Western Australia. The problem raised in

1952 when the British and Australian Governments arranged to conduct atom bomb tests at these islands, close to the important faunal reserve of Barrow Island, brings conservation of nature into the atomic age. The unexpected and unilateral nature of the decision disturbed fauna conservationists and authorities. Official secrecy on some military matters, or simply poor inter-departmental liaison, is always liable to present conservation with the fait accompli. It is up to the conservation authorities to take measures to ensure that the location, importance and aims of nature reserves are well known to all potential competitors, especially military and public service departments.

Research

The basis of all good conservation is knowledge, which falls broadly into two categories — surveys of the resources and more detailed ecological research into plant and animal associations and species. In general, botany is ahead of zoology in this field, as several chapters in this volume reveal; much soil and water conservation study has been done, the Australian floral map is reasonably well known, and the requirements of individual plant species are often well understood. However, selection of the best areas for nature reserves will necessitate extensive and detailed botanical surveys, especially in the drier regions which have received less attention from agriculturist and forester, and scientific management of the varied habitats in them will provide research problems for many years to come.

The manpower available for animal ecological research in Australia, especially in the wildlife field, has been small indeed, but the last decade has seen some notable beginnings which are likely to expand. This is not to decry earlier contributions, especially by ornithologists, but it was possible for McGILL (1948) to reveal extensive gaps in bird distribution data, and accurate knowledge of bird movements based on banding has been accumulating for only a few years. The Wildlife Survey Section, C.S.I.R.O., was founded in 1949, and although heavily preoccupied with economic work it is effectively advancing research in conservation, mostly published in its periodical "C.S.I.R.O. Wildlife Research" which commenced in 1956. Other chapters of this book by CARRICK (bird-banding) and FRITH (duck ecology), review some of the results. The fauna authorities of Victoria, Western Australia, Queensland and Northern Territory are engaged in field research on wild ducks and other subjects and also surveys of sanctuaries; since 1948 the Australian National Antarctic Research Expeditions have undertaken biological research, including studies of seals and birds; and the chapter by WARING outlines the research on marsupial ecology in progress in Western Australia and elsewhere. Reference collections of mammals and birds are seriously deficient, and an extensive programme of large-scale collecting, with up-to-date documentation, is an essential part of wildlife surveys and a pre-requisite to the adequately-illustrated field guides and handbooks required to catalyse further study. This very brief statement does bare justice to much effort, by museums and amateur naturalists in particular, but it has to be said that first principles, rather than detailed knowledge, will be the guide for years to come in the drafting of wildlife legislation and in the selection and management of faunal reserves.

Conclusion

The merits and needs of nature conservation in Australia are evident enough from the foregoing account of its potential, its current legislation, and some of its problems. The encouraging fact is that so much of this distinctive flora and fauna still remains and can be conserved, provided the real threats are appreciated in the right quarters, and the necessary measures are taken in time. The rising tempo of population increase, of agricultural extension into marginal land and new areas, of irrigation, hydro-electric and mining developments, of greater leisure and mobility, all require space and reduce the opportunity to obtain nature reserves in both quantity and quality. Yet these very changes increase the necessity for conservation of both spacious and special natural environments for public recreation and scientific reference. The immediate future is critical; if unnecessary impoverishment is to be prevented, and if centres of population are to have these amenities where they can derive most benefit from them, an adequate system of national parks and nature reserves will have to be secured before contending claims for the same land prevent it.

Although the sixth-commandment outlook on fauna protection is being seen in perspective and the concept of habitat preservation is gaining ground, the need for reserved areas to have full legal security, as well as a management plan backed by finance and trained staff, requires to be more widely understood. Nowhere is this more necessary than in government departments dealing with land and finance, where too often nature conservation is the Cinderella of land uses, not regarded as quite legitimate. The right note is struck by ISAAC (1957) in a review of the natural resources of Victoria, in which the chapter "Flora, Fauna and Natural Scenery" is on a par with others which deal with agriculture, water supplies, minerals and forests. Much public demand, education and scientific advice will be required before governments concede that man's need for nature is their responsibility just as are his economic welfare and more sophisticated pleasures.

A special difficulty arises from the subdivision of the Australian mainland, a coherent biological unit, into politically separate States and Territories whose present approach and standards of nature conservation, and whose resources to support it, vary so widely. This is part of a much wider question, and immediate progress has to be sought within the present political framework. There is so much room for improvement in national parks, flora and fauna legislation and practice in Commonwealth Territories, and the future importance of these areas is so great, that an office in the Federal Government seems justified. This could become a national reference centre, to facilitate close liaison and co-operation between all States and Territories and to integrate education and programmes of survey and research. Such an office might also administer the Commonwealth funds which will be necessary to ensure that the nation's best assets of natural beauty do not go by default but are maintained to the highest standard. Through State-Commonwealth co-operation and mutual responsibility for a series of key parks and reserves, there could materialise a truly national system of nature conservation which accords with the world's conception of Australia as a biological entity richly endowed with its own unique flora and fauna.

Acknowledgements

We are much indebted to the officials in charge of national parks, flora and fauna in all States and Territories for information; equally to many societies, associations and individuals for their assistance. Several colleagues in C.S.I.R.O. have given helpful discussion and comment, and the mimeographed submissions (1957) by Mr. F. N. RATCLIFFE on "The Conservation of Australian Fauna and Flora" to the Australian Academy of Science, and reports (1953 and 1957) by Dr. D. L. SERVENTY to the Standing Committee on Pacific Conservation of the Pacific Science Association, have been most helpful.

REFERENCES

CARRICK, R., 1957. The wildlife of Macquarie Island. *Aust. Mus. Mag.* 12, 255—60.
CHISHOLM, A. J., 1958. The Australian Encyclopaedia. Articles by various authors on national parks, plants, protection of flora, fauna, marsupials and protection of fauna. Angus and Robertson, Sydney.
COSTIN, A. B., 1958. The grazing factor and the maintenance of catchment values in the Australian Alps. C.S.I.R.O. Div. Plant Industry Tech. Paper No. 10.
GROOM, A., 1949. One Mountain after Another. Angus and Robertson, Sydney.
ISAAC, C. E., 1957. Natural Resources of Victoria. Natural Resources Conservation League of Victoria, Melbourne.
MARLOW, B. J., 1958. A marsupial survey of New South Wales. *C.S.I.R.O. Wildl. Res.* 3, 71—114.
MAWSON, SIR D., 1922. Macquarie Island and its future. *Papers and Proc. Roy. Soc. Tasmania*: 40—54.
MCGILL, A. R., 1948. The need for more definite distribution data. *Emu* 48, 127—40.
TURNER, J. S., et al., 1957. A Report on the Condition of the High Mountain Catchments of New South Wales and Victoria. Aust. Acad. Sci. Canberra.

INDEX

Aborigines, age in Australia, 35, 38, 44
  culture, 37, 47
  cultural sequence, 37
  ecology, 39
  hunting, 42
  implements, 43, 47
  origin, 35
    native beliefs, 46
  physical types, 39
  spread, 45
Acacia, 17, 30, 276, 281, 285, 291, 452, 504, 601, 609
  genus, fossil record, 297
  Grassland, 61
Acridoidea, 206, 207
Adaptation, saltatory, 78
Adelaide, 53, 64
Africa, rodent Jerboas, 87
  Springhare (Pedetes), 87
Agamidae, (dragon lizard), 115
  distribution, 122
Agricultural assessment, eco-complex, 587, 598
  economic aspects, 603
  potentialities, 596
  regional survey, 588
  survey procedure, 593
  techniques, 593
Agriculture, tropical 66
Ahamitermes (Termites), 218
Aire méditerranéenne, 266
Alice Springs, 65
Alpine woodland, 26
Amitermes (Termites), 218
  building of meridional nests, reasons, 219
America, rodent Jerboas, 87
Ancestral Beings, 40
Animals, giant, 46
Anomalopterigidae (moas), 93
Antarctic origin, 10, 28, 151, 155, 168, 254, 275, 430
Antarctica, 162
  marsupial migration, 70
Apterygidae(kiwis), 93
Archaean Shield 15
Arid areas, control of the water, 602
  environment, factors, retarding colonisation of bare areas, 456
Arid areas, nitrogen sources, non-symbiotic organisms, 458
    symbiotic organisms, 458
  plant communities, 454
  soil erosion, 455
    nitrogen, 457
    organic matter, 457
  vegetation, effect of aborigines, 454
    effect of white man, 454
    effect of grazing animals, 454
Aridity, 18, 52, 61, 97, 138, 197, 240, 277, 281, 298, 303, 383, 397, 452
Arnhem Land, 42, 47
Artesian water, 56
Aru Islands, 137
Arunta Desert, 61
Asian continental shelf, 26
Atherton Plateau, 43
  Tableland, 41, 71
Atrichornithidae (scrub-birds), 91
Australia bororientalis, 184
  deserta, 179, 183
  Felix, 63
  latitude, 57
  merorientalis, 185
  physiography, 15
  sylvatica 179, 181, 184, 185
  uniqueness in biology, 9
Australian aboriginal, ecological agent, 43
  environment, development, 26
  shelf, 26
  vegetation change, 70
Australo-Papuan continental fauna, western limits, 27
Australoid man, introduction of the dog, 45
  southward migration, 45
  peoples of the deserts, 46
Avifauna, Australo-Papuan, distinctive elements, 90
  relationships, 92
    with Africa, 92
    with Asia, 92
    with Bali, 90
    with Borneo, 92
    with Celebes-Molluccas, 90, 92

Avifauna, relationships with New Caledonia, 93
　　with New Guinea, 92, 93
　　with New Zealand, 93, 94
　　with Nicobars, 90
　　with Philippines, 90
　　with Polynesia, 91
　　with Solomons Islands, 90
　　with South America, 93
　　with Timor, 93
Axe, edge-ground stone, 43
　　kodja shipped edge, 43

Babinda, wet lands, 39
Bali, 137
Ballarat region, gold, 64
Balranald, 63
Bandicoots, 80
　　recognition, 81
Banksia, 91, 277, 284, 293, 610
Barrinean negrito tribes, 41
Bass Strait, 137
Bassian province, 181, 182
Bathurst, gold, 64
　　Plains, 62
Bees, native 43
Beings, ancestral, 40
Belt of high pressure, 57
Billabongs, oxbow lakes, 17
Bioclimat Méditerranéen, 263
Bioclimatologie, 260
Bird-banding scheme, 369
Bird species, habitat specialisation, 94, 95
Birds, 89
　　annual behaviour circle, 97
　　banded and recovered numbers, 370, 371
　　breeding potential, 373
　　　　seasons, 100, 101
　　　　zones, 101, 102
　　build-up during good seasons, 111
　　clutch size increased, 112
　　colonisation, 91
　　concentration in areas, favourable to rearing of young, 111
　　daylength, 96
　　desert mulga, 95
　　desert spinifex, 95
　　distribution in Australia, 94, 95
　　failure in breed during droughts, 109
　　gonad development, 96
　　good seasons, 111
　　increase in number of broods, 111
　　losses during droughts, 108
　　losses of life during droughts, 109
　　mallee, 95

Birds, mangroves, 95
　　marshes, 95
　　migration, 96
　　moult, 107
　　movements, environmental factor, 100
　　　　within Australia, 377
　　nomads, 99
　　number of species, 90
　　old world families, 92
　　psittacine, enzootic infection, 413
　　　　isolation of virus, 413
　　rainfall and breeding, 102
　　rainforest, 95
　　reduced clutch size during droughts, 109
　　reproduction, 108
　　Savannah grassland, 95
　　　　woodland, 95
　　sclerophyll forest, 95
　　seasonal cycle, 96
　　　　movements, 98
　　sedentary species, 99
　　social behaviour, 373
　　south-north migrants, 99
　　Subantarctic Sea-, migrations and breeding, 374
　　swamps, 95
　　territorial behaviour, 371
　　trans-equatorial migration, 374
　　Trans-Tasman migration, 377
Birth right, 40
Blepharoceridae (Diptera), 158
Blowflies, (Calliphoridae), adult insects, 514
　　biology and ecology of individual species, 520
　　breeding media, 531
　　climatic influences, 541
　　control of strike, 515
　　larvae, 514
　　Mules operation, 515
　　numbers and sex ratio, 529
　　and species bred from natural strikes, 525
　　reproduction, 533
　　sheep myiasis, 516
Blowfly larvae, interspecific competition, 539
　　puparial weight and amount of food, 538
Blow fly numbers, competition in the limitation, 537
　　populations limitations, effects of predation and parasitism, 540
　　sources 536
　　reduction, main lines, 515
　　strike, problem, 514

Blue Lake, 46
Boidae (Phytons and Boas), 116
Borneo, 137
Breeding, environmental factor, birds, 106
Brisbane, 63, 64, 65, 66
Broken Hill, 67
Browne Crater Lake, 46
Bukumari (bookoomuri), 46
Bunya Mountains, 43

Cactoblastis cactorum (Phyticidae), distribution, 573
  establishment, 573
  introduction, 572
  life cycle and habits, 571
Cainozoic Era, 70
Cairns 66, 107
Callaeidae (wattled crows), 93
Canning Basin, 42
  desert, 42
Cape Leeuwin, 53
  York, 42
    Peninsula, 53
    point of entry from New Guinea, 122
Capricorn, Tropic of, 57
Casuariidae (cassowaries), 90
Casuarina, 31, 299, 610
Cat, 46
Cataclysmic episode, 48
Cats, dasyurid, 78
Cattle, 66
  Lands, 53
Celebes, 71, 137
Central Basin, 15
Ceram, 137
Cercopoidea (Frog-hoppers), 152
Chelyidae (chelyid fresh-water tortoises), 115
  distribution, 123
Chernozyems, 67
Chile, 159
Cicadas, 152
Cicadelloidea (Leafhoppers), 153
  geographical distribution, 150
Cinnamomum, 30, 284
Cladocera (water fleas), 247
Climat aride de l'intérieur, 260
  méditerranéen aride, 267
    humide, 268
    limites, 267
    perhumide, 268
    semi-aride, 268
    subhumide, 268
Climate, 22
  tertiary, 30
Climates, classification, 23

Climatic change, quarternary, 287
  controls, 56
  fluctuations, vegetation, 283
  zones, 22
Climatology of Australia, 259
  recent, 32
Cloncurry, 71
Coal, 55, 62, 63, 67
Colubridae (common snakes), 116
  distribution, 120
  invasion pathways, 121
Conchostraca (Entomostraca), 246
Conopophagidae, 93
Controls, climatic, 56
  geologic, 54
  topographic, 54
Convict settlements, 53
Convicts, 64
Cooktown, 71
Copepoda (Entomostraca), 247
Copper, 60, 64, 67
  Tasmania, 66
Coptotermes (Termites), 215
  Eucalyptus trees, 221
  mound building, 216
  nesting behaviour, 216
  pests to forest trees, 221
Cotton, 66, 599, 600
Crabs (Decapoda), 248
Cracticidae (magpies, butcher-birds), 91
Crayfish (Decapoda), 248
  groups, 255
  terrestrial existance, 249
Creation Myths, 40
Crocodilidae (Crocodiles), 116
  distribution, 123
Crow man, 46
Crustacea, causing damage, 250
  distribution, 250
  dry conditions, 248
  freshwater, 246
  hot water, 249
  origin of the freshwater fauna, 253
  utilisation, 250
  zoogeographical regions, 252
Culture, Kartan, 37
  Mudukian, 37, 48
  Murundian, 37, 48
  Pirrian, 37
  sequence, 37
  Tartangan, 37
Cuscus, 71
Cyprinidae (carp tribe), 136
Cyprinodontidae (killifishes), 136

Dactylopsila (striped-possum), 71
Dampier Peninsula, 47

Darling river, 63, 65
Daru, 72
Darwin, 65
Dasyuridae, 75
  carnivorous, 77
  insectivorous, 77
Dendrolagus (tree-kangaroo), 71
Dentition, dasyurid, 78
Desert, animal forms and adaptation, 40, 80, 96, 133, 138, 183, 195, 215, 237, 249, 327, 383, 398, 495
  Gibber, 25
  Grassland, 25
  mulga, Birds, 95
  pebbles, layer, 60
  plant forms and adaptation, 276, 287, 299, 452, 463
  spinifex, Birds, 95
  vegetation, 282
Digging stick, alteration flora, 42
  womans, 43
Dingo, 44
Dinornithidae, 93
Diprotodon, 72
  elimination, 43
Diptera, altitudinal zonation in the distribution, 167
  Chilean forms, 167
  distribution, 169
  faunal division, 179
  influence of recent immigrants, 167
  Malayan elements, 166
  paleoantarcts, 188
  Papuan elements, 166
  Patagonian forms, 167
  peculiarities of families, 169
  relationship to Chile, 187
    to New Zealand, 188
    to Patagonia, 187
    to South America, 168
    to Tasmania, 188
  systematic position, 169
  typical Tasmanian fauna, 168
  zoogeographical aspects, 164
Divide, 62
  river, 63
Dog, ecological agent, 44
  introduction, 45
Drainage, 16, 17, 141, 233, 384
  map, 16
Dromaeidae (Emus), 90
Droughts, 109
Ducks, nomadism, 380
  wild-, breeding, effect of water level, 393
    breeding seasons, 391
    food habits, 389

Ducks, wild, in inland, 383
  movements, 388
  sexual cycle, 394
Dunes, fixed, 60

Eagle men, 46
Ecological agent, aboriginal, 43
  dog, 44
Ecology, Human, 52
  primitive aboriginal man, 36
Elapidae (front-fanged snakes), 116
  distribution, 120
Emu, 45
Entomostraca (Crustacea), resting eggs, 246
Environmentalism, 52
  Taylor, 12
Ephemeroptera (Mayflies), 158
Equator, 56
Eremean province, floristic character, 276
  vegetation, 280
Erosion, Soil, 472
Eucalyptus, 17, 70, 71, 91
  association, 471
  distribution, 295, 464
  forests, 61
  genus, history, 296
    origin, 297
  highest elevations, 462
  hybridization, 466
  interbreeding groups and hybridism, 469
  mixed stands, 469
  morphology, 461
  number of hybrids, 467
  variability in populations, 465
  Wallace's line, 461
  woody plant (Mallee), 463
Euro (Macropus robustus), population control, 328
European settlement, 33
Europeans, arrival with sheep, cattle and rabbit, 43
Eutheria, 73
Evaporation, 21, 59
Expansion, centres, 53, 63
  pastoral, 64
Exploration, end of, 66
  first steps, 62
Eyre, Lake, 17

Fauna, Freshwater-, comparison with other continents, 136
  indigenous, balance with native vegetation, 549
  pleistocene, 31

Fauna, present, 32
   tertiary, 30
Faunal lists, differences, 45
   province, definition, 129
Faunula, Autochthonian, 235
   Bassian, 237
   Caurine, 238
   Eremian, 236
   Euronotian 237
   Torresian 237
Fiji 118
Firestick, alteration flora 42
Fishes, adaptive radiation 144
   artesian water 139
   drought adapted 138
   fossil, 136
   freshwater-, 136
      alliances with overseas genera and species, 137
      principal genera, 145, 146, 147, 148
   hot water, 140
   native freshwater, conservation, 137
   pop-eye disease, 139
   spawning, 148
   Wallace's Line, 136
Flinders Range, 60
Flora see also Plants and Vegetation
   alteration by digging stick, 42
      by firestick, 42
   course of evolution, 433
   early recent, 31
   life form composition, 432
   lower tertiary, 283
   Monaro region, 431
      biological spectra, 432
   pleistocene, 31
   present, 32
   quarternary, 285
      environmental changes, 286
   tertiary, 30
   upper tertiary, 285
Fluvifaunula, definition, 141
   Jardinean, 144, 234
   Krefftian, 144, 234
   Leichhardtian, 142, 234
   Lessonian, 143, 234
   Mitchellian, 143, 234
   Sturtian, 143, 234
   Tobinian, 143, 234
   Vlaminghian, 142, 235
Fly river, 72
Forests of Eucalyptus, 61
   of Jarrah, 61
   of Karri, 61
   rain, 17, 26
      birds, 95
   sclerophyll, 17, 26

Forest, sclerophill, birds, 95
Fossils, cretaceous, 16
   jurassic, 16
   pleistocene, 31
Free settlers, 63
Frogs, adult behaviour, 404
   size, 408
   breeding season, 399
   desert adaptations, 398
   ecology, 396
   feeding, 407
   habitat selection, 407
   larval life, 401
   morphological adaptations, 397
   physiology, 405
   site of oviposition, 400

Gaimardian fluvifaunula, 144
Gambier, Mount, 47
Gastropods, freshwater operculate, 228
   pulmonate, 225
Gecko, distinct endemic genera, 121
Geelong, 63
Gekkonidae (Geckos), 115
   distribution, 121
Geographical Zoology, 182
Geological controls, 54
Geo-syncline, 60
Giant animals, 46
Gibber Desert, 25
Glaciation, pleistocene, 30
Goannas, arboreal life, 122
   pleistocene, 122
   terrestrial life, 122
   water life, 122
Gold, Ballarat region, 64
   Bathurst, 64
Goldfields, Kalgoorlie, 66
   Perth, 66
Gold-Rushes, 64
Gonad development, birds, 96
Goondiwindi 47
Goulburn, 63
   Plains, 62
Gourinae (giant ornamental pigeons), 90
Grafton, 66
Grallinidae (mud-nest builders), 91
Grasshoppers, biotic community, 206
   climate, 197
   drought-enduring, 194
   drought-evading, 194
   ecological characteristics, 193
      cliffs, 206
   effect of other animals, 201
   environment, 196
   faunal regions, 206
   fen, 205

Grasshoppers grassland, 205
  heath, 204
  human influence, 202
  intraregional biogeography, 207
  invasion, 207
  major habitat categories, 203
  procryptic mechanism, 196
  prolonged geographical isolation, 192
  rainforest, 203
  saltbush, 205
  sclerophyll forest, 204
  soil, 198
  structure of vegetation, 196
  term, 192
  topography, 199
  typical life-cycle, 195
  vegetation, 200
Great Arid hypothesis, 47
  Divide, elevation, 30
Grevillea, 91
Greyian fluvifaunula, 142
Guano, 60
Hall Sound, 72
Halmahera, 137
Hamada, 60
Hammer-stones, Kartan man, 49
  Tartangan phase, 49
Heaths, sand-, 279
High Moor, 26
Highlands, Eastern, 15
High mountains, alpine tract, 429
  adaptation, 430
  flora, 443
  geographical affinities, 430
  hydrological research, 448
  land use, 442
  management requirements for water production, 450
  native fauna, 443
  pasture establishment, costs, 444
  plant communities, 434
  recreational values, 447
  soil groups, 429
  soils, 434, 443
  subalpine tract, 429
  Tasmania, 430
  temperatures, 429
  types of vegetation for water needs, 447
  use of water, 445
  vegetation, 427
  water, hydro-electric power, 446
  irrigation in dry areas, 446
  winds, 429
Hobart, 53, 62
Homoptera, indicators of past geography, 150

Homoptera, permian, 150
  triassic, 150
Honeycomb, 43
Human Ecology, 52
Hydro-electric power, 67
Hypsiprymnodon (Musky Rat-Kangaroo), 85, 86

Ice Age, 70
Implements, Kartan, 44
  Tartangan, 44
Insect fauna of oceanic islands, 160
Insects, paleogeographical conclusions, 151
  sources of the present Australian fauna, 189
  of present New Zealand fauna, 189
  with aquatic larvae, 158
  with a southern distribution, 158
  zoogeography, 150
Invertebrates and vertebrates, distribution, identical causes, 159
Irrigation, 59, 67

Jardinean fluvifaunula, 144, 234
Jarrah, forests, 61, 279
Jerboa, dasyurid pouched- (Antechinomys), 87
  rodent, Africa, 87
    America, 87

Kakatoeinae (large seed-eating cockatoos), 90
Kalgoorlie, 67
  goldfields, 66
Kangaroo, 45
  family (Macropodidae), 85
  Island, 44, 45
  tree, (Dendrolagus), 86
  giant (Macropus), 87
Karri, forests, 61, 279
Kartan culture, 37
  implements, 44
  man, hammer-stones, 49
Kimberleyan culture, 47
Kiwis (Apterygidae), 93
Koala, 75
  classification, 84
  myology, 84
  survey of a population, 317
Kodja axe-hammer, 49
Komodo Dragon (Varanus komodoensis) 122
Kosciusko highlands, 39
Koskiusko, Mount, 16
Krefftian fluvifaunula, 144, 234
Kujani, 41

Lachlan river, 62, 63
Lake, Blue, 46
   Browne Crater, 46
   Colongulac, 44
   Daviumbo, 72
   Eyre, 17, 40, 48, 49, 60, 64, 65
   George, 63
   Menindee, 44, 48
   Urana, 556
Land Companies, 63, 64
   connections, postulated, 28
   conservation, 484
     agricultural ecology, 484
     farm planning, 484
   continuity Australia-Antarctica-South America, 162
   forms, geographic zone, 483
   husbandry units, 483
   masses, other, relationship to, 26
   system, major factor of grouping, 483
   unit for mapping, 483
   use, the component, basic unit, 483
     determination, 482
Leafhopper Fauna, 154
   post-triassic chronology, 154
Leafhoppers, Australian-Indian association, 155
   Australian-South African-New Zealand-South American association, 155
   endemic groups, 155
   late Indo-Malayan invasion, 156
Leichhardtian fluvifaunula, 142, 234
Leipoa ocellata, 90
Lessonian fluvifaunula, 143, 234
Lightning Rocks, 48
Livistona, Palm, 15
Lombok, 137
Lord Howe Island, 29
Loriinae (honey lories), 90

Macassar Strait, 137
Macdonnell Ranges, 65
Macropodidae (Kangaroo family), 85
Madagascar, 118
Magpies banded, 372
Main Divide, watershed, 17
Maize, 62
Maize-swine economy, 67
Mallee, birds, 95
   Scrub (desert steppe), 26
Malurinae (warbler-like birds), 91
Mammal assemblage, pleistocene, 45
   fauna, disappearence, 46
   pests, biological control, 549
Mammals, egg-laying, 72
   general, 29, 31, 33, 44, 70, 315 331, 500, 545, 611

Mammals, introduced, 34, 487, 500, 545
   Native Land-, Australia, 77
   New Guinea, 77
   placental, 33, 44, 487, 500, 545
Man, primitive aboriginal, ecology, 36
Mangroves, birds, 95
   palaeotropic, 278
Marshes, birds, 95
Marsupial anteater (Myrmecobius), 75, 79
   life, zoogeographical dispersal, 69
   Lion (Thylacoleo carnifex), 85
   mice, 79
   migration, Antarctica, 70
   mole, 80
   Order, origin, 69
   wolf (Thylacinus), 75, 79
Marsupials, adolescent, hormone induced organ transformation, 361
   allantoic placentation, 347
   American ancestry, 69
     continent, 75
   anatomy, female reproductive system, 334
     male reproductive system, 337
   changes following oestrus, 343
   classification, 73, 75
   conservation, 317
   development of young in the pouch 356
   diversity in early Tertiary, 161
   ecology, 314
   economic aspects, 316
   embryonic development, 346
   environmental changes, 320
   evolutionary deployment, 75
   female reproduction system, effects of hormones, 362
   hormon control of sex differentiation, 360
   individual physiology, 319
   Koala, 75
   marking studies, 317
   migration during the Upper Cretataceous, 162
   migratory route, 69
   milk, 357
   oestrous cycle, 339
   oestrus, 343
   origin, 73
   Papuan region, 71
   parturition, 351
   past distribution, 319
   phylogeny, 69
   pouch young, growth, 358
   pregnancy, anatomical and histological changes, 352

Marsupials, pregnancy, gestation periode, 345
  pro-oestrus, 342
  radiation, 71
  reproduction, 332
  reproductive cycles, 339
    periodicity in the female, 337
    periodicity in the male, 339
  South America, 79
  trapping studies, 317
  yolk-sac, 346
Mastotermes (Termites), 222
  attacking potatoes, 222
  damage to crops, 222
    to fruit trees, 222
  most primitive living termite, 211
Mecoptera (Scorpion Flies), 159
Megaloptera (Dobson Flies), 158
Megapodiidae, 90
Melbourne, 53, 63
Meliphagidae (honeyeaters), 94
  pollinators of Eucalyptus, Banksia, Grevillea, 91
Melville Island, 64
Menuridae (lyrebirds), 90
Merauke, 72
Merino, effective wool producers, 495
  fine wool sheep, 490
  medium wool sheep, 491
  strong wool sheep, 491
Merinosheep, 13, 487
  diseases, 491
  growth of population, 489
  predators, 492
  wool producer, 487
Merino types, 490
Micropsittinae (pigmy parrots), 90
Migration, Birds, 96
  initial trend, 38
  southward, Australoid man, 45
Mildura, 67
Millstones, 49
Mindanao, 137
Mining, 55
Mitchell Grass, 61
Mitchellian fluviafaunula, 143
Moas (Anomalopterigidae), 93
Mole, Cape Golden- (Chrysochloris), 80
  marsupial (Notoryctes), 80
  pouched, 79, 80
  rodent (Sphalax), 80
  true (Talpa), 80
Mollusca, African origin, 243
  climatic factors, 241
  freshwater, 225, 226
    desiccation, 239
    faunal regions, 233

Mollusca, Land, 229, 230
  zoogeography, 236
  Melanesian origin, 241
  native fauna, 238
  New Zealand origin, 242
  newer Asian origin, 242
  older Asian origin, 243
  South American origin, 243
  terrestrial faunal regions, 235
Molybdenum, 60
Monotremes (egg- laying mammals), 72, 73, 610, 619
  origin, 72
Moult, geographic variation, 107
Mound springs, 65
Mt. Gambier, 47
  Isa, 72
  Kosciusko, 67
  Mann, 48
Mudukian culture, 37, 48
Mulga, 61
  Bush, vegetation, 281
  Scrub (desert steppe), 26
Murray-Darling system, streams, 17
Murray river, 17, 41, 47, 63, 67
Murrayians, Southern Australoids, 41
Murrumbidgee river, 63
Murundian culture, 37, 48
Muscicapidae (Passeres), 12, 91
Musgrave Ranges, 48
Musky rat Kangaroo (Hypsiprymnodon) 85, 86
Mussels, freshwater, 225
Myxoma virus, degree of resistance to, 556
  reduced virulence, 556
  strains of reduced virulence (field strains), 554
Myxomatosis, 13, 549
  entomological investigations, 553
  long-range transportation, 555
  mortality, 552
  point outbreaks, 555
  research, 551
  spread by mosquitoes, 553

Narangga, 41
Nasutitermes (Termites), 220
  change in vegetation, 220
  light intensity, 220
Nature conservation, Anemone Buttercup (Ranunculus anemoneus), 619
  Australian Bustard (Eupodotis australis) 620
  Australian Gannet (Sula serrator), 621
  Cassowary (Casuarius casuarius), 620
  Cockatoos (Psittacidae), 621

Nature conservation, Emu (Dromaius novae-hollandiae), 620
 fauna, 610
  reserves, 615, 616, 617
 Flora, 609
 Giant Wallaby grass (Danthonia frigida), 619
 Humpback whale (Megaptera nodosa) 620
 Kangaroos (Macropodidae), 620
 Koala (Phascolarctos cinereus), 619
 legislation, 612
 Mountain Ash (Eucalyptus regnans), 618
 National parks, 613, 614
 Parrots (Psittacidae), 621
 Platypus (Ornithorhynchus anatinus) 619
 research, 625
 reserves, Kings Park, 624
  Kosciusko State Park, 622
  Lamington National Park, 624
  Macquarie Island, 624
  Macquarie Marshes, 623
  Montebello Islands, 624
  Royal National Park, 622
  Sherbrooke forest, 623
 Wallabies (Macropodidae), 620
 Wallaroos (Macropodidae), 620
 Waratah (Telopea speciosissima), 619
 Weaver Finches (Ploceidae), 621
 Wedge-tailed eagle, (Aquila audax), 621
Nestorinae (parrots), 93
New Guinea, 29, 71, 85, 118
 fauna, 72
 tropical forests, 70
New South Wales, 65, 67
New Zealand, 28, 69, 70, 118, 159
Newcastle, 62, 67
Ngadadjara, 46
Ngadjuri, 41
Ngarkat, 41
Northam, 58
Northern Territory, 65, 67
Notomys (hopping rodents), 87
Notothenium, 72
Nukunu, 41
Nullarbor Plain, 42, 64

Oceanic islands, Insect fauna, 160
Odonata (Dragonflies), 158
Oppossum, 48
 Cretaceous period, 161
Opuntia inermis (Yellow pear), 575
 resistant form, 575
Oranges, 62

Oriomo river, 72
Ornithosis, 412
 cross infection, 422
 enzootic, South Australia, 416
 epidemiology, 423
 epizootic 1938-1939, 415
 fatal and non-fatal epizootics, 424
 infection, common species, 418
  domestic birds, 421
 mutton bird (Puffinus tenuirostris), 422
 pigeons, 421
 population levels, 425
Orographic map, 16
Ostracoda (Entomostraca), 247
 parasitic, 257

Pademelon (scrub-wallabie), 87
Palaeoclimatology, 29
Paleobotanic evidence, vegetation, 283
Palm Valley, 71
Palorchestes (giant kangaroo), 72
Papua, 137
Papualand, 72
Papuan Region, North Queensland, 71
Paradisaeidae (birds of paradise), 91
Parramatta, 53
Passeres, Muscicapidae, 12
Pastoral occupation, end of new, 66
Peaches, 62
Peanuts, 599, 600
Pedionominae (collared hemipodes), 90
Penguins, special bands, 371
Peramelidae, 80, 81
Perth, 47, 53, 64
 goldfields, 66
Phalanger genus, 71
 possum, 71
Phalangeridae (Australian Possums), 75, 82
 radiation, 82
Phalangers, classification, 84
 Myology, 84
Phosphorus, 60
Phreatoicoidea (Isopoda), 248
Physiography, 15
Phytogeography 275, 291, 430, 452, 464
Phytonidae, distribution, 121
 ecological types, 121
Pintubi, 46
Pirrian culture, 37, 47
 rapid spread, 46
 users, 45
Pitjandjara, 46
Plant Formations, Characteristics, 25
Plants see also Flora and Vegetation,
 adaptation to drought, 276

Plants, distribution during late Tertiary-
    Recent, 298
  fire resistance, 277
  introduced, adaptation, 581
    adaptive characteristics, 583
    ecogenetic research programme, 578
    effects of temperature on growth, 585
    flowering time, 584
    germination of seeds, 584
    natural selection, 582
    photoperiod, 585
    species, 579
    strain survey, 582
    winter dormant, 585
Plecoptera (Stoneflies), 158
Pleistocene, 70
  chronology, 31
  climatic succession, 31
  fauna, 31
  flora, 31
  fossils, 31
  glaciation, 30
  mammal assemblage, 45
  sea level, 30
  vulcanism, 30
Pluviosité annuelle, 260
Population density, lines of equal, 12
  distribution, 53
Port Essington, 64
  Keats, 107
  Moresby, 72
  Phillip 62, 63
Portland, 63
Possibilism, 52
Possums, Australian (Phalangeridae), 82
Prawns (Decapoda), 248
Prickly pear (Opuntia), advanced weed, 565
  biological campaign, 570
  control, 565
  Board's work, 566
  destruction of dense infestations, 574
  present position, 575
Proto-man type, 39
Psittacosis, 412
Ptilonorhynchidae (bower-birds), 91
Pygopodidae (legless lizards), 115
  distribution, 122

Quarternary climatic change, 287
Queensland, 63, 64, 65, 85
Quokka (Setonix brachyurus), 86
  annual activity, 323
  control of population size, 324
  damage to pine plantations, 316, 317
  daily activity, 323
  ecological projects, 320

Quokka, reproduction, 323
  Rottnest Island, 320
  starvation, 326
  summer dehydration, 327
  trace element deficiency, 325

Rabbit, 46
  behaviour, 556
  climatic limits, 547
  data on reproduction, 560
  dominant member of the fauna, 547
  habits, 557
  high- and low-ranking, 559
  mortality by coccidia, 561
    by intestinal worms, 561
  natural host of Myxoma virus, 550
  period of activity, 557
  population, origin, 546
  predation, significance in population dynamics, 561
  predators, 561
  reproductive activity, 559
  social behaviour, 558
Rain Forests, 17, 26
  birds, 95
  burning of, 43
Rainfall, Antarctic system, 20
  Australia, 67
  Breeding, Birds, 102
  Bingara, Bourke, Coonabarrabran, Sydney, 105
  distribution, 18
  effectiveness, 21
  factor, 58
  humidity, 127
  initiating breeding factor, birds, 97
  tropical system, 20
Reptiles, Bassian (south easterncoastal) zone, 129
  continent wide distributions, 126
  distribution, 118, 119
    compared to that of birds, 128
  patterns, 124
  distributional barriers, 128
  Eyrean (interior) zone, 129
  factors governing distribution, 126
  families, 115
  influence by physiography, 128
  influence by vegetation associations 127
  inland distribution patterns, 125
  linked with soils, 128
  living relics, 118
  peripheral fauna, 131
  rainfall, 127
  river systems, 127
  specialized and restricted ranges, 126

Reptiles, surface water, 127
  temperature, 127
  Torresian (northern coastal) zone, 129
  Wallace's line, 117
  westward movement, 117
Rice, 64, 67, 600, 604
Rigo district, 72
Rock carvings, 46
  wallabie, 48
Rockhampton, 66
Roeburne, 58
Rottnest Island, Quokka, 320

Salt pans, vegetation, 282
Saltbush area, 60
  fodder plant, 61
Saltatory adaptation, 78
Sangjir Islands, 137
Savannah Grassland, 26
  birds, 95
Savannah Woodland, 26
  birds, 95
Scincidae (smooth-skinned Lizards), 116
  distribution, 122
Sclerophyll forest, 26
  birds, 95
Sclerophyllous grass steppe, 26
  hummock grasslands, 601
  Shrub genera, origin, 293
Sea, invasions, 16
  level, pleistocene, 30
Sécheresse, intensité, 265
Settlements, convict, 53
  Future, 68
Sheep, 66, 67
  alpine communities, 508
  blowflies, 514
    local and seasonal abundance, 518
  bluegrass grasslands, 506
  breeding season, 495
  changing the herbaceous vegetation, 502
  characters of fleece, 496
  cold stress, 495
  cutaneous myiasis, 514
  death of saltbushes, 507
  distributing seeds and fruits, 512
  distribution, map of isohyets, 493
    related to climate, 493
  dog, 492
  feral, 492
  fleece grown, genetic constitution, 496
  fluctuation in diet, 496
  grassland formation, 505
  grazing, general effects on vegetation, 509
  heat tolerance, 494

Sheep, Kangaroo grass, 501
  leguminous species, 503
  Merino, 13
  Mulga Scrub formation, 507
  noxious plants, 498
  nutrition, 495
  plant communities grazed, 500
  regeneration of saltbush, 507
  saltbush formation, 506
  Savannah woodlands, 501
  seeds in wool "vegetable fault", 512
  selective grazing, 496
  shrub-steppe, 506
  soil deficiency of Cobalt, 498
    of copper, 498
    of phosphorus, 497
  soil drift, 507
  erosion 497
  spinifex community, 508
  woodland formation 500
  wool characters, 496
Shells, freshwater, 225
Silver-lead, Tasmania, 66
Simuliidae, 158
Snails, calcareous epiphragm, 239
  effects of soil-type, 240
  endodontid, 229
  Land, 229
  land-, operculate groups, 229
    pulmonate groups, 229
  zonitoid, 232
Snake fauna, development, 116, 117
  radiation, 117
Snakes, potency of the venom, 117
Snowy Mountains, 67
Soils, 22
Soil, black earths, 309
  deltaic formations, 313
  desert loams, 310
    sandhills, 311
    sandplains, 311
  Central Eastern Lowlands, 304
  Characteristics, 24
  Chernozems, 309
  conditions, 59
  conservation problems, 475
  ecology, 303
  Eastern Highlands, 304
  erosion, ecological catastrophes, 472
    effect of settlement, 474
  hydrology, 478
  introduced species, Opuntia (prickly pear) 476
  introduced species, domestic sheep, 476
  introduced species, Pinus radiata

(Monterey Pine), 476
Soil, erosion, introduced species, rabbit, 476
    introduced species, Trifolium subterraneum (subterranean clover), 476
    man's activity, 477
    mechanism of tunnel erosion, 480
    need of soil conservation, 473
    new people, 472
    result of physical forces, 472
    salinization, 478
    soils deficient in molybdenum, 481
    tunnel erosion, 480
    tunnel erosion, reasons, 480
  fertility, 303
  finer textured red earths 312
  Great Western Plateau 303
  grey-brown desert 311
  grumusols, 310
  halomorphic (solonchak, solonetz, solod forms), 313
  krasnozems (red loams), 312
  lateritic, 279
    podzolic, 312
  map, 308
  minor elements, 307
  phosphate levels, 306
  podzolic, 309
  potash deficiency, 307
  red-brown earths, 310
  red calcareous desert, 311
  skeletal, 313
  solonised brown, 311
  subsoil texture, 307
  Terra Rossa, 309
  tidal marshes,
  yellow, 312
Sorghum, 599, 600
South Africa, 69, 70
  America, 69, 70
  Australia, 65, 66
Southern Australoids (Murrayians), 41
Sparselands 61
Sphenodon (Sphenodontinae), 118
Springhare (Pedetes), Africa, 87
Strigopinae (flightless parrots), 93
Sturt's Stony Desert, 64
Sturtian fluvifaunula, 143
Subregions (postulated), ecological significance, 132
Sugarcane, 66, 600
Sundaland, 72
Superphosphate, 60, 67
Swamps, birds, 95
Sydney, 62, 67
  first settlement, 52
Syncarida (Crustacea), 248

Tamworth, 63
Tartanga Island, 45
Tartangan culture, 37
  implements, 44
  phase, hammer-stones, 49
Tasmania, 29, 39, 62, 65, 137, 159
  copper mines, 66
  fauna, 44
  implements, 44
  silver-lead, 66
  tin mines, 66
Tasmanians, negritic, 41
Temperature, 17
Tennant Creek, 58, 61, 65
Termites, 211
  Ahamitermes, 218
  Amitermes, 218
  Coptotermes, 215
  distribution, ecological factors, 213
  fossil species, 211
  living, Australian 212 region,
    Ethiopian region, 212
    Indo-Malayan region, 212
    nearctic region, 212
    Malagasy region, 212
    neotropical region, 212
    palaearctic region, 212
    Papuan region, 212
  Mastotermes, 222
  Nasutitermes, 220
  soil fertility, 221
  Tumulitermes, 221
Territories, tribal, 40
Tertiary, climate, 30
  fauna, 30
  flora, 30
  vulcanism, 30
Thylogale stigmatica (Wallaby), 72
Tiger cat, 78
Timor, 137
Tin, Tasmania, 66
Tirari, 49
Tobinian fluvifaunula, 143
Topographic Controls, 54
Torres Strait, 137
Torresian barrier, 72
  province, 181
  shallows, 71
Tortoises, horned Land-(Meliolania), 118
Trace elements, 67
Trade Winds, 57
Tribal territories, 40
Trichoptera (Caddis Flies), 158
Tropical agriculture, 66
Tumulitermes (Termites), 221
Turtle, pitted shelled (Carettochelys), 118

Typhlopidae (worm snakes), 116
  distribution, 121
  colonization from New Guinea, 121

Van Diemens Land, 63
Varanidae (Goannas), 116
  distribution, 122
Vertebrates and invertebrates, distribution, identical causes, 159
Vegetation see also Flora and Plants,
  bioclimate, 269
  bogs, 440
    overgrazed by cattle, 441
  change, 70
  climatic fluctuations, 283
  community interrelationships, 442
  Desert Grassland, 275
  deserts, 282
  Eremean province, 280
  fens, overgrazing damage, 441
    Carex alliance, 441
  fjaeldmark, 438
  Gibber Desert, 275
  grass steppe, 275
  heath, 440
  high moor, 275
  mountains, 427
  Mallee Scrubs, 275
  Mulga Bush, 281
    scrub, 275
  natural, 60
  paleobotanic evidence, 283
  rain forest, 275
  répartition des étages, 271
  salt pans, 282
  Savannah Grassland, 275
    Woodland, 275
  Sclerophyll Forest, 275
  short alpine herbfield, 438
  Sod tussock grassland, 439
  subalpine woodland, 435
  trees, defoliation by sheep, 436
  tall alpine herbfield, 436
  typically Australian genera, 292
  Western Australia, 274
Vegetation Zones, 25
Victoria, 63, 65, 66, 67
Vlaminghian fluvifaunula, 142
Vulcanism, pleistocene, 30
  tertiary

Wailpi, 41
Wallabia agilis, 72
Wallabie, 45
  brush- (Wallabia), 87
  Hare- (Lagorchestes), 86
  nail-tailed (Onychogalea), 86
  Rock- (Petrogale), 48, 86
  scrub- (Thylogale), 87
  short tailed (Setonix), 86
Wallace's line, 27, 92, 137
Wassi Kussa river, 72
Water, artesian, 56
Waterfowl habitat, 384
  utilization, 386
Wattles (Acacia), 70
Weber's line, 27, 137
West Australia, 65, 66
Westralian province, 181
Western Australia, 67
Wiradjuri belief, 46
Wollongong, 67
Wombat, classification, 84
  myology, 84
  alpine, 26
  communities, 601
Worms, histriobdellid, 257
Wyndham, 56

Xenicidae (wrens), 93

Yorke Peninsula, 41

Zink, 60, 67
Zoogeographic regions, 92
  definition, 129
Zoogeograhic sub-region (Faunal province), definition, 129
Zoological geography, 182

If you have any concerns about our products,
you can contact us on
ProductSafety@springernature.com

In case Publisher is established outside the EU,
the EU authorized representative is:
**Springer Nature Customer Service Center GmbH
Europaplatz 3, 69115 Heidelberg, Germany**

Printed by Libri Plureos GmbH
in Hamburg, Germany